Monoclonal Antibodies: Principles and Practice

Third Edition

To Emanuela

Monoclonal Antibodies: Principles and Practice

Production and Application of Monoclonal Antibodies in Cell Biology, Biochemistry and Immunology

Third Edition

JAMES W. GODING

Department of Pathology and Immunology,
Monash University, Alfred Hospital,
Prahran, Victoria, Australia

ACADEMIC PRESS

Harcourt Brace and Company, Publishers

London San Diego New York Boston Sydney Tokyo Toronto

ACADEMIC PRESS LIMITED
24–28 Oval Road
LONDON NW1 7DX

U.S. Edition Published by
ACADEMIC PRESS INC.
San Diego, CA 92101

This book is printed on acid free paper

First edition published 1984
Second edition 1986
Reprinted 1988, 1993

A catalogue record for this book is available from the British Library

ISBN 0-12-287023-9

Typeset by Servis Filmsetting, Manchester
Printed in Great Britain by WBC Book Manufacturers, Bridgend, Mid Glamorgan

Contents

Preface to First Edition

Ever since the beginnings of experimental immunology at the end of the nineteenth century, scientists have exploited the specificity of antibodies to detect, isolate and analyze biological material. The power of antibodies as probes for biological structure underwent a quantum increase in 1975, when Köhler and Milstein published their classic paper on the production of monoclonal antibodies of predefined specificity. Paradoxically, the very success of monoclonal antibodies has generated a literature that is now so vast and scattered that it has become difficult for the non-specialist to obtain a perspective.

I believe that the need has arisen for an integrated text which treats the application of monoclonal antibodies as a subject in its own right. In any field, the best results cannot be achieved until the basic principles are understood. I have therefore tried to integrate theory and practice, and to write in such a way that the reader will be able to adapt and change procedures to suit individual needs and resources. One of the recurring themes in the history of scientific discovery is the improvement in precision and scope of analytical systems. I have therefore emphasized the factors that set the limits to the performance of each system, and indicated ways in which these limits might be extended or bypassed.

I am deeply indebted to many people for help in various aspects of this book. In particular, I thank Leonard and Leonore Herzenberg, in whose laboratory I was first introduced to monoclonal antibodies, and to Vernon Oi, Jeff Ledbetter and Israel Pecht, from whom I learned so much. Many people helped by discussions, critical comments and reading portions of the manuscript. I am grateful to Gustav Nossal, Donald Metcalf, Alan Harris, Grant Morahan, Nick Gough, Suzanne Cory, Roland Scollay and Richard Haugland for many helpful suggestions, and to Jenny Taylor for patient, accurate and cheerful help with typing and word processing. The illustrations were drawn by Richard Mahoney and Peter Maltezos, and were photographed by Sonya Belan.

I would welcome suggestions, criticism or correction from readers of this book, so that any future editions might be as accurate and useful as possible.

James W. Goding *Melbourne, 1983*

Preface to Second Edition

It has been very pleasing to see the acceptance of the first edition of this book. However, almost as soon as the manuscript had been sent to the publishers, I found myself compiling a list of ways in which it could be improved. These involved clarification of detail, the discovery of exceptions to generalizations, the rapid advance of knowledge, and correction of errors.

It is said that the best way to neutralize a scientific opponent is to make sure he or she writes a book. I have therefore chosen to update the book selectively, concentrating on areas of practical importance, rather than attempting to give the last word on each of the many topics covered.

Once again, I ask the reader of this book to offer suggestions, criticism or correction.

James W. Goding *Melbourne, January 1985*

Preface to Third Edition

It has been nearly 10 years since the publication of the second edition of this book. So much has changed! The elucidation of the three-dimensional structure of the histocompatibility antigens has given us a much clearer understanding of the mechanisms by which antigen is presented to T lymphocytes. The mechanisms of intracellular processing of antigen for presentation to T cells are now reasonably well understood; 10 years ago they were almost totally unknown. There have been major advances in the understanding of the mechanisms of activation and tolerance in B and T lymphocytes. Other areas of major progress include the purification, cloning and sequencing of numerous cytokines or interleukins that are secreted by or act on lymphocytes, and the molecular characterization of their receptors and associated intracellular signalling mechanisms. We are now beginning to understand in detail the molecular and cellular basis of the somatic hypermutation processes that result in affinity maturation in secondary antibody responses.

There have been remarkable advances in the knowledge of the biosynthesis and transport of membrane and secretory proteins. Many processes, such as folding and disulfide bond formation, which were previously regarded as occurring spontaneously, are now known to be assisted by a variety of other proteins, including chaperonins and protein disulfide isomerase. A new class of membrane proteins, attached to the membrane via phosphatidyl inositol, has been identified and characterized.

The techniques of recombinant DNA have become far more widespread in their application. They are no longer the exclusive domain of the specialist, but an integral part of most immunological laboratories. The use of these techniques has helped to elucidate many important areas relating to antibody synthesis. Some scientists feel that the newer techniques of recombinant DNA will soon replace the older techniques of monoclonal antibody production based on cell fusion. However, there are still major problems to be solved in these approaches, including solubility, affinity and stability of antibodies produced in bacteria. Notwithstanding the widespread use of recombinant DNA techniques, the production of monoclonal antibodies by gene manipulation in bacteria still requires a high degree of specialist expertise, and it is currently much simpler to use the 'classical' technique based on cell fusion. Only time will tell whether production of monoclonal antibodies by expression in bacterial systems will replace the 'classical' technique of lymphocyte fusion for production of monoclonal antibodies.

At the practical level, the basic procedures for production of monoclonal antibodies have not changed greatly in the last 10 years, although there have been numerous technical refinements that have made the process easier. In addition, there is a better understanding of some previously mysterious areas, such as why so many investigators have had difficulties making rat monoclonal antibodies. The use of non-radioactive techniques for detection and characterization of antigens is reaching maturity, and in many cases is now even more sensitive and convenient than those procedures involving radioactivity. Reliable production of human monoclonal antibodies remains elusive, despite numerous claims to the contrary. Attempts to 'humanize' mouse monoclonal antibodies have been only partially successful, and still require a great deal of specialized expertise.

I have attempted to respond to all these changes by appropriate revisions of the book. Every section of the book has been scrutinized and revised; most of it has been completely rewritten. I have included a great deal more background on basic immunology, and revised the text to reflect major changes in the techniques of molecular genetics. An extra chapter on the use of antibodies in immuno-histology has been added.

Most of the procedures described in this book can be performed in any reasonably well equipped laboratory without the need for equipment costing many tens of thousands of dollars. I have resisted the temptation to add numerous highly specialized and expensive techniques, such as the purification of monoclonal antibodies by high-pressure liquid chromatography. Some procedures described in earlier editions are now obsolescent or have not fulfilled their early promise, and have been deleted.

My role over the last 10 years as Chairman of a large department has, of necessity, made me more removed from the laboratory bench and my ability to speak from 'hands on' experience is less than it used to be. The advantages of a single-author book in terms of uniformity and cohesion may soon be outweighed by the logistical impossibility of one individual covering all of the necessary areas.

I express my gratitude to those who have written to me to point out errors and make suggestions for future editions, and I would like to encourage readers of this edition to do likewise.

I also express my gratitude to Anna Abramovici and Danielle Haliczer for help with retyping the original manuscript, Fay Bower and Tom Girke for numerous computer searches of the literature, Mark Malin and Leon Martin for help with computing and in numerous other ways, and Richard Boyd, Wayne Hancock, Helen Ramm and Denis Nyguay for advice regarding flow cytometry and immunohistology.

I am grateful to Peter Hudson and Bob Irving for their help in writing Chapter 14, which is devoted to bacteriophage display libraries for production of monoclonal antibodies. Their help in an area in which I have had little personal experience is greatly appreciated.

Finally, I express my immense gratitude to my wife, Emanuela Handman, for her help and support in countless ways.

James W. Goding *Melbourne, 1995*

Georges Köhler (1946–1995)

Georges Köhler died on 1 March 1995, at the age of 48. Together with Cesar Milstein and Niels Jerne, he was awarded the Nobel Prize in Physiology or Medicine in 1984. I first met him when he was a teacher at an EMBO course on cell fusion, held in Naples in 1977. He was a quiet and modest man and it was a privilege to have been taught by him. The invention of the technique for making monoclonal antibodies has had an enormous impact on medicine and basic biological science, and has been the catalyst for numerous other discoveries. It will undoubtedly be the key to numerous discoveries in the future. In reflecting upon his death at such an early age, we should take solace in the knowledge that his place in history is assured.

List of Abbreviations

Å	angstrom unit (10^{-1} nm)
AMCA	7-amino-4-methylcoumarin-3-acetic acid
AP	ammonium persulfate
Bis	*N*,*N*′-methylenebisacrylamide
BLOTTO	5% nonfat powdered milk in PBS
BSA	bovine serum albumin
c.p.m.	counts per minute
DATD	*N*,*N*′ diallyltartardiamide
DEAE	diethylaminoethyl
DME	Dulbecco's modified Eagle's medium
DMSO	dimethyl sulfoxide
DTT	dithiothreitol
ECL	enhanced chemiluminescence
EDTA	ethylenediamine tetra-acetic acid
ELISA	enzyme-linked immunosorbent assay
FACS	fluorescence-activated cell sorter
FCS	fetal calf serum
FITC	fluorescein isothiocyanate
h	hours
HAT	hypoxanthine, aminopterin, thymidine
HEPES	*N*-2-hydroxyethylpiperazine-*N*′-2 ethane sulfonic acid
HGPRT	hypoxanthine guanosine phosphoribosyl transferase
HPLC	high performance (pressure) liquid chromatography
HT	hypoxanthine, thymidine
IEF	isoelectric focusing
Ig	immunoglobulin

IPTG	isopropyl-β-D-thiogalactoside
M	moles per litre
2ME	2-mercaptoethanol
MEM	minimal Eagle's medium
MHC	major histocompatability complex
min	minute
mM	millimoles per litre
M_r	relative molecular mass; molecular weight
NEM	*N*-ethyl maleimide
NEPHGE	non-equilibrium pH gradient electrophoresis
nm	nanometre
PAGE	polyacrylamide gel electrophoresis
PBS	phosphate-buffered saline
PEG	polyethylene glycol
PLP	periodiate-lysine-paraformaldehyde
PMSF	phenylmethylsulfonyl fluoride
RIA	radioimmunoassay
SCID	severe combined immunodeficiency
SDS	sodium dodecyl sulfate
TBS	tris-buffered saline
TCA	trichloracetic acid
TEMED	*N*,*N*,*N*′,*N*′-tetramethylene-ethylenediamine
TLCK	tosyl-L-phenylalanine-chloromethyl ketone
Tris	Tris (hydroxymethyl) aminoethane
TRITC	tetramethylrhodamine isothiocyanate
XRITC	a derivative of rhodamine isothiocyanate

1 Introduction

Genetic engineering is now a mature field. It is possible to study, control and manipulate genes in ways that continue to amaze even those working in the field. The power of the tools that have become available has resulted in an explosion of discoveries. The development of a technique for the production of monoclonal antibodies has been an integral part of this revolution. By combining the nuclei of normal antibody-forming cells with those of their malignant counterparts, Köhler and Milstein (1975) developed an unprecedentedly powerful way of analysing and purifying individual molecules within the enormously complex mixtures encountered in biological material. They were awarded the Nobel Prize in Physiology and Medicine for this work in 1984.

Monoclonal antibodies have had revolutionary consequences in biological research and clinical medicine, and have resulted in the generation of an industry with an economic impact measured in billions of dollars (Chien and Silverstein, 1993). It should be emphasized, for the benefit of those who believe that science should be planned to produce economically useful outcomes, that monoclonal antibodies were the result of basic research into the mechanism of generation of antibody diversity. It had no practical application in mind and, indeed, it took some time before the full scientific and economic impact was appreciated, even by the inventors of the technique (see Milstein, 1980, 1986; Melchers, 1995).

Köhler and Milstein's monumental achievement was the culmination of many seemingly unrelated discoveries by other workers (Tables 1.1 and 1.2; Silverstein, 1989; Gallagher *et al.*, 1995). Of particular importance were the proof of the clonal selection theory (Nossal and Lederberg, 1958), the development of cell fusion techniques (Okada, 1962; Littlefield, 1964), the artificial induction of plasmacytomas (Potter and Boyce, 1962) and their

Table 1.1 Landmarks in the history of antibody research

1847	Urinary protein in myeloma	Bence-Jones
1890	Discovery of antibodies	von Behring and Kitasato
1900	'Side-chain' theory formulated	Ehrlich
	Discovery of ABO blood groups	Landsteiner
1955–56	Allotypes	Grubb, Oudin
1956	Classification of Bence-Jones proteins into two groups (now called κ and λ in honour of their discoverers)	Korngold and Lipari
1957	Clonal selection theory	Burnet (an earlier precursor was proposed by Talmage)
1958	One cell; one antibody	Nossal and Lederberg
	Cell fusion by Sendai virus	Okada
1959	Elucidation of disulphide-bonded chain structure of antibodies	Edelman
1960	Discovery of spontaneous cell fusion	Barski
1962	Demonstration that Bence-Jones proteins are antibody light chains	Edelman and Gally
	Induction of plasmacytomas by mineral oil	Potter and Boyce
1962–3	Controlled proteolytic cleavage of IgG, identification of Fab and Fc; topographic relationship between light and heavy chains	Porter, Fleischman, Pain and Press
1964	Use of mutant cells and selective media to isolate hybrids	Littlefield
1965	Amino acid sequencing reveals that *N*-terminal half of light chains is variable; *C*-terminal constant	Hilschmann and Craig
	Postulate of two genes; one polypeptide	Dreyer and Bennett
1968	Demonstration that B cells are the precursors of antibody-secreting cells, but that they need 'help' from T cells	Mitchell and Miller
1969	First complete amino acid sequence of an immunoglobulin; concept of domains	Edelman and colleagues
1970	Hypervariable regions	Wu and Kabat
	Growth of plasmacytomas in continuous culture	Horibata and Harris
1973	Fusion of mouse and rat myeloma cells with preservation of secretion of both immunoglobulins sets the stage for production of monoclonal antibodies	Cotton and Milstein
1973–74	Discovery that helper and cytotoxic T cells recognize antigen in intimate association with class II and class I (respectively) products of the major histocompatibility complex	Rosenthal and Shevach; Katz, Hamaoka and Benacerraf; Zinkernagel and Doherty
1975	Construction of hybridomas secreting antibody of predefined specificity	Köhler and Milstein
1976	Demonstration of DNA rearrangements in antibody-forming cells	Tonegawa and colleagues
	Use of polyethylene glycol for cell fusion	Pontecorvo
1977	Cloning and sequencing λ genes; J segments	Tonegawa and colleagues
1977–80	Multiple germ-line genes for V regions	Many authors
	Discovery of D segments	Group of L. Hood
1980	Mechanism of insertion of membrane Ig	Groups led by L. Hood and R. Wall

Table 1.1 cont.

1982	Discovery that a chromosome translocation in plasma cell tumors involves the oncogene *myc*	Many authors
1983	Discovery of 5′ 'enhancer' elements in Ig genes	Many authors
	Identification of the T cell receptor protein	Groups of Marrack and Kappler and groups of Reinherz and Schlossman
1984	Cloning of genes for T cell receptor	Groups of M. Davis and T. Mak
1980s	Elucidation of processing pathways for presentation of antigen to T cells; antigens synthesized within the cell are transported into the endoplasmic reticulum and loaded into grooves of class I MHC molecules; antigens arising outside the cell are endocytosed into a discrete vesicle and loaded into grooves of MHC class II molecules	Many authors
1980s, 1990s	Purification, cloning and sequencing of numerous cytokines (interleukins)	Many authors
1987	Elucidation of three-dimensional structure of class I major histocompatibility antigens, and demonstration of peptide-binding groove	Bjorkman, Wiley and Strominger
1990s	Elucidation of role of CD40 and its ligand in B cell activation	Groups of Banchereau and Ledbetter
1990s	Elucidation of molecular and cellular mechanisms of affinity maturation and role of germinal centres	Many groups, notably those of Milstein, Neuberger, Rajewsky, Kelsoe, MacLennan and Gray
1993	Direct demonstration of somatic mutation in individual B cells and localization of these cells in lymphoid tissues	Group of Rajewsky
1994	Elucidation of the three-dimensional structure of class II major histocompatibility antigens	Group of Wiley and Strominger
1994	Engineering of mice containing large blocks of human heavy and light chain immunoglobulin genes	Many authors
1990s	Antibody expression libraries in *Escherichia coli*	Many authors
1990s	Epitope libraries, phage display libraries of antibodies	Many authors
1995	Elucidation of the three-dimensional structure of the β chain of the T cell receptor	Bentley, Boulot, Karjalainen and Mariuzza

For further information, see Silverstein (1989) and Gallagher *et al.* (1995).

adaptation to tissue culture (Horibata and Harris, 1970). Finally, the demonstration that it was possible to fuse two different plasma cell tumour lines with retention of both antibody products (Cotton and Milstein, 1973) paved the way for subsequent developments.

In order to appreciate the revolutionary impact of monoclonal antibodies, it is necessary to understand the problems and limitations of conventional serology. Suffice it to say that prior to 1975, the production of antibodies was considered by some to be a black art practised by immunologists. While the specificity of antibodies provided a way of overcoming the enormous complexity of biological material, the production of highly specific antisera was difficult and unreliable. It required highly purified antigen. The uncertainties about the specificity of individual antisera led to many prolonged and acrimonious debates. All that has now changed. It is now possible to produce unlimited quantities of exquisitely specific antibodies against virtually any molecule, regardless of the purity of the immunizing antigen. The fine specificity, degree of cross-reaction, affinity and physical properties of antibodies may be selected to suit individual needs.

It would be wrong to think that monoclonal antibodies will completely replace conventional serology. The production of monoclonal antibodies involves a great deal of work and a high level of commitment. There will often be occasions when the effort may not be justified. Fortunately, a wide range of monoclonal antibodies is becoming available commercially. Polyclonal antibodies may be preferred when the antigen to be recognized is denatured or altered in some other way. It would be unwise to use monoclonal antibodies for the detection of molecules in genetically diverse species without thorough testing to ensure that some molecules do not escape detection owing to genetic polymorphism. For all these reasons, the book concludes with a chapter on the production of conventional antibodies.

I have written this book because I believe that previous accounts of the production, and particularly the usage, of monoclonal antibodies have been too dogmatic and inflexible. 'Recipes' have been given which work if followed to the letter, but little attention has been given to the underlying principles. In the real world, one has to adapt each procedure to an individual biological problem.

I have therefore tried to emphasize the important variables that make for success or failure in the use of antibodies, and those points of refinement that allow the capabilities of a system to be pushed to the limit. I have also tried to point out areas in which the literature gives misleading impressions, and a few situations in which published procedures are unreliable.

This book thus represents the distillation and critical evaluation of many hundreds of publications relating to the production and use of antibodies, together with some of my own experience. Immunochemistry also has an oral tradition, and a surprising number of key elements of knowledge are not easily accessible from the literature. I have incorporated some of these elements where appropriate; in many cases no citation is possible.

It is not possible to cover all possible applications of antibodies in one book, nor would it be wise to attempt to do so. I have therefore restricted the book to the 'core' techniques of production and handling of antibodies, and their use in studies of antigen analysis, purification and localization. I have tried to avoid techniques that require expensive or specialized equipment. All the techniques described in this book can be carried out in a modestly equipped laboratory. I have deliberately minimized discussions and protocols concerning genetically engineered antibodies, because this field requires

Table 1.2 Selected Nobel prizes in immunology

Year	Recipient	Discovery
1901	E. von Behring	Serum therapy (antibodies)
1908	P. Ehrlich and E. Metchnikoff	Work on immunity
1919	J. Bordet	Discoveries relating to immunity
1930	K. Landsteiner	Human blood groups
1960	F. M. Burnet and P. B. Medawar	Acquired immunological tolerance
1972	G. M. Edelman and R. R. Porter	Chemical structure of antibodies
1977	R. Yalow, R. Guillemin and R. Schally	Radioimmunoassay (Yalow) and discovery of peptide hormones in the brain
1980	B. Benacerraf, J. Dausset and G. Snell	Histocompatibility antigens
1984	C. Milstein, G. Köhler and N. Jerne	Monoclonal antibodies (Milstein and Köhler); theoretical contributions (Jerne)
1987	S. Tonegawa	Mechanism of action of antibody genes

highly specialized knowledge and skills, and is changing very rapidly. It is neither possible nor desirable to cite all the literature. I have chosen references which explain basic principles, or illustrate the use of these principles in practical situations.

Finally, I have tried to write a book that would be useful to a wide range of biologists. The use of antibodies is becoming increasingly important in botany, cell biology, embryology, endocrinology, enzymology, forensic studies, genetics, haematology, medicine, microbiology, molecular biology, neurobiology and parasitology. It is not realistic to expect workers in these fields to be intimately acquainted with the minutiae of immunochemistry and yet it is very likely that they will need to use antibodies at some point in their research.

This book is not intended to be read sequentially from cover to cover. It is expected that most readers will dip into individual chapters or sections to retrieve specific pieces of information. Accordingly, each chapter is more or less self-contained, and there is a certain amount of repetition of some material, slanted to the needs of the particular context, with cross-referencing to other sections where appropriate.

The biological sciences have now reached a point in their development where the edges of subdisciplines are virtually nonexistent, and it is often necessary to draw upon knowledge and techniques that are well outside traditional boundaries. This book is therefore very broad in scope. I hope that readers will find it useful.

References

Chien, S. and Silverstein, S. C. (1993) Economic impact of applications of monoclonal antibodies to medicine and biology. *FASEB J.* **7**, 1426–1430.

Cotton, R. G. H. and Milstein, C. (1973) Fusion of two immunoglobulin-producing myeloma cells. *Nature* **244**, 42–43.

Gallagher, R. B., Gilder, J., Nossal, G. J. V. and Salvatore, G. (1995) 'Immunology: The Making of a Modern Science'. Academic Press, London.

Horibata, K. and Harris, A. W. (1970) Mouse myelomas and lymphomas in culture. *Exp. Cell Res.* **60**, 61–77.

Köhler, G. and Milstein, C. (1975) Continuous cultures of fused cells secreting antibody of predefined specificity. *Nature* **256**, 495–497.

Littlefield, J. W. (1964) Selection of hybrids from matings of fibroblasts *in vitro* and their presumed recombinants. *Science* **145**, 709–710.

Melchers, F. (1995) Georges Köhler (1946–1995). *Nature* **374**, 498.

Milstein, C. (1980) Monoclonal antibodies. *Scientific American* **243**, 66–74.

Milstein, C. (1986) From antibody structure to immunological diversification of the immune response. (Nobel Lecture) *Science* **231**, 1261–1268.

Nossal, G. J. V. and Lederberg, J. (1958) Antibody production by single cells. *Nature* **181**, 1419–1420.

Okada, Y. (1962) Analysis of giant polynuclear cell formation by HVJ virus from Ehrlich's ascites tumor cells. I. Microscopic observation of giant polynuclear cell formation. *Exp. Cell Res.* **26**, 98–107.

Potter, M. and Boyce, C. R. (1962) Induction of plasma cell neoplasms in strain BALB/c mice with mineral oil adjuvants. *Nature* **193**, 1086–1087.

Silverstein, A. M. (1989) 'A History of Immunology'. Academic Press, San Diego.

2 The Antibody Response

In the first year of life, a child will develop many infections, particularly of the respiratory tract, middle ear or gastrointestinal system. A baby unfortunate enough to have been born without a functioning immune system is unlikely to survive the first year because of the inability to fight these infections. The importance of the immune system is also vividly illustrated by the almost invariably fatal acquired immune deficiency syndrome (AIDS), in which the human immunodeficiency virus (HIV) destroys the immune system.

The purpose of this chapter is to provide a brief outline of the basics of the immune response, and to set the stage for the more detailed chapters to follow. One of the easiest ways to understand the concepts and nomenclature of a complex system is to look at how the relevant concepts developed. This chapter therefore begins with a brief account of the history of antibody research, and then moves on to look at a typical antibody response.

2.1 Early History

It has been known since antiquity that recovery from certain diseases results in specific immunity. The survivors are resistant to contracting the same disease. In other words, the immune system has *memory*. The survivors are, of course, still equally prone to contracting many other diseases, which defines the second key concept of immunological *specificity*. The exquisite specificity of antibodies is their most important feature, and forms the basis of most of this book. How did the current vision of the immune system begin?

2.1.1 Phagocytosis

Perhaps the earliest attempts to understand the bodily defences against bacteria came from Metchnikoff, who began his studies of the ingestion of micro-organisms and particulate matter by cells (phagocytosis) in the 1870s. His first studies used marine metazoans such as starfish, but were soon extended to human disease, where he studied phagocytosis of the causative organisms of anthrax, typhus and tuberculosis (see Metchnikoff, 1891; Bulloch, 1938; Silverstein, 1989). Phagocytosis is now recognized as the cellular arm of what can be termed 'innate' defenses, which are available immediately and require no immunization, but have very limited flexibility or specificity. Later in this chapter, we will see that phagocytosis is greatly facilitated by antibodies and complement.

2.1.2 Discovery of Antibodies and Complement

There is a long line of evidence that blood has the ability to kill micro-organisms. John Hunter (1794) observed that blood did not putrefy as rapidly as other tissues. In 1888, Nuttall observed that the serum from normal animals could sometimes kill micro-organisms. He named this activity 'alexin', and showed that it was destroyed by heating to 56°C. The activity was later renamed 'complement' by Ehrlich and this is the term by which it is currently known.

Bordet, working in Metchnikoff's laboratory in 1895, made important extensions to Nuttall's work by showing that killing of bacteria by immune serum involved two separable activities. One activity (complement) did not require immunization, but was destroyed by treatment at 56°C for 30 minutes (Bordet, 1895). The other required immunization but was stable to heating at 56°C for 30 minutes. The latter activity was subsequently identified as antibody. Bordet went on to show that lysis of red blood cells by immune serum was functionally equivalent to lysis of bacteria by immune serum and complement (Bordet, 1898).

The discovery of antibodies is generally accredited to von Behring and Kitasato (see Bulloch, 1938; Cohen and Porter, 1964; Silverstein, 1989). In 1890, they showed that immunity to diphtheria and tetanus is due to antibodies to the toxins secreted by these organisms. In 1894, Fraenkel and Sobernheim found that anti-cholera serum, even when heated to 70°C for 1 h, could protect normal guinea-pigs against a lethal dose of cholera vibrios.

The discovery of antibodies was put into clinical usage almost immediately. In 1891, a child dying of diphtheria was given immune serum and made a dramatic recovery (see Bulloch, 1938). Although 'serum therapy' was sometimes dramatically effective, it proved to be somewhat risky. Side-effects included a syndrome that became known as 'serum sickness', comprising skin rashes and arthritis, which was caused by the development of massive amounts of antigen-antibody complexes, which were composed of the foreign serum and antibodies to it. Occasionally, sudden death occurred from bronchospasm and circulatory collapse, a syndrome that subsequently became known as 'anaphylaxis'. Anaphylaxis is now known to be caused by degranulation of mast cells by IgE antibodies.

2.1.3 Emergence of the Concept of Complementarity and Binding Between Antigen and Antibody: Contributions of Ehrlich and Landsteiner

The discovery of antibodies by von Behring and Kitasato led to a stream of new discoveries (Browning, 1955). In 1891, Ehrlich demonstrated that there was a clear quantitative relationship between the amount of the plant toxins ricin or abrin and the amount of antibody needed to neutralize them, suggesting a chemical reaction between antigen and antibody. The concept of binding of antibodies to antigen via three-dimensional complementarity of surfaces is explicitly shown in the diagrams that accompany Ehrlich's Croonian Lecture to the Royal Society in 1900. The notion that antibodies must bind to antigen was also implied by the experiments of Gruber and Durham in 1896, who discovered that immune serum could agglutinate bacteria, and those of Kraus in 1897, who showed that mixing of immune serum with soluble immunizing antigen could cause a precipitate to form.

Perhaps more than any other individual, Ehrlich contributed to the notion that antigen–antibody interactions were chemical reactions that involved specific binding, and could be studied using the chemical techniques (Browning, 1955). His dictum of *'corpora non agunt nisi fixata'* (substances do not act unless bound; Ehrlich, 1913) was given in the context of chemotherapy, but also epitomizes the concept of binding and stereospecific complementarity.

Although we might now regard the binding of antibodies to antigen as obvious, there was an opposing view championed by no lesser figures than Bordet and Landsteiner. It was pointed out that antigen–antibody reactions did not follow simple chemical kinetics, and were sometimes reversible. It was argued that the antibody–antigen reaction was not a true chemical process but was more akin to adsorption or a poorly defined colloidal reaction (see Bulloch 1938). Both parties were partly right, and the debate reflected inadequate understanding of chemical bonding. The evidence, such as it was, suggested that the antigen–antibody bond was not what we now term covalent. The contemporary concepts of adsorption did not involve significant specificity. The confusion can now be seen to reflect an incomplete understanding of the fact that macromolecules could bind to each other in a very strong and specific but reversible (noncovalent) manner owing to the summation of many weak interactions, which is our current concept of how antibodies bind to antigen.

The first indication that enzymes had defined structures came with Sumner's demonstration in 1926 that urease could be crystallized, although this finding was highly controversial and was not awarded the Nobel Prize until 20 years later. The first demonstration that proteins had defined and unique sequences was not made until 1953 (Sanger and Thompson, 1953) and the first three-dimensional structure of a protein was published by Kendrew and colleagues in 1958. The debate concerning the mechanism of action of antibodies thus reflected the inadequate state of knowledge of protein chemistry at a time when it was not yet proven that proteins were well-defined entities with unique sequences of amino acids and well-defined three-dimensional structures.

Landsteiner discovered the ABO blood groups in 1900 which, together with the existing knowledge that citrate could prevent blood clotting, allowed blood transfusion to become a clinical reality. In a large body of work, Landsteiner set the foundations for rigorous and quantitative study of the specificity of antibodies (Landsteiner, 1945; reviewed by Silverstein, 1989).

Towards the end of an extraordinarily long and distinguished career, Landsteiner also discovered the Rh blood groups (Landsteiner and Wiener, 1940) and made the important discovery that delayed-type hypersensitivity could be transferred between individuals by cells but not by serum (Landsteiner and Chase, 1942). The latter discovery was one of the first indications of the existence of a previously undefined arm of the immune system, thymus-derived lymphocytes (T cells).

2.1.4 The Controversy Between Cellular and Humoral Immunity

Not all observations concerning immunity could be explained by Metchnikoff's ideas on phagocytosis. There soon arose a somewhat acrimonious debate concerning whether cellular (phagocytosis) or humoral (carried in serum) defenses were more important (see Silverstein, 1989). The humoral theory was championed by Ehrlich. As pointed out by Silverstein (1989) it is somewhat ironic that Metchnikoff, in what appeared to be a somewhat *ad hoc* defence of his cellular theories, proposed the idea that immunization led to the formation of substances called 'stimulins' that enhanced phagocytosis, and that Bordet, working in Metchnikoff's laboratory, made important contributions to the knowledge of complement.

Part of the problem stemmed from the different experimental systems. *Vibrio cholerae* is easily lysed by antibody and complement, while many other bacteria are not. Experiments using *V. cholerae* tended to support the humoral theory. Conversely, many of Metchnikoff's experiments were carried out with the anthrax bacillus, which is highly resistant to immune lysis (see Silverstein, 1989).

The subsequent work of Sir Almroth Wright, published in 1903, showing that antibodies and phagocytosis interact to provide enhanced phagocytosis ('opsoninization') helped integrate the seemingly disparate views concerning cellular and humoral immunity. The story has been told countless times how Wright attempted, with only partial success, to translate these findings into clinical practice, earning him the nickname 'Sir Almost Right' and immortalization as Sir Colenso Ridgeon in Shaw's play 'The Doctor's Dilemma'.

It is now easy to see that Metchnikoff and Ehrlich were both right, and that they were studying different aspects of the immune system, in which antibodies, complement and phagocytosis all helped contribute to immunity. In 1908, Metchnikoff and Ehrlich were jointly awarded the Nobel Prize for their work. The current realization that all components of the body must be synthesized by cells and that the cellular and humoral arms of the immune response interact synergistically makes the debate between cellular and humoral immunity seem somewhat irrelevant. A third 'cellular' arm involving thymus-derived lymphocytes (T cells), was discovered a half-century later.

2.1.5 Innate Defenses

Innate defenses are those that do not require immunization. They consist of phagocytic cells and a series of proteins including lysozyme and complement, which recognize bacterial products which are not part of the vertebrate body, such as teichoic acid, peptidoglycan, exopolysaccharide, lipopolysaccharide (LPS) and unmethylated DNA (see Marrack and Kappler, 1994; Krieg *et al.*, 1995), and *N*-formyl methionine, which is only found in prokaryotes and in mitochondria. The release of *N*-formyl methionine signals cellular damage or bacterial infection, and is recognized by specific receptors on the surface of phagocytic cells, initiating an inflammatory response. Lysozyme is an enzyme present in bodily secretions, which digests components of bacterial cell walls that are unique to bacteria (Fig. 2.1).

2.1.6 Cellular Basis of the Adaptive Immune Response: Need for Both Bone-marrow-Derived Lymphocytes (B cells) and Thymus-derived Lymphocytes (T cells)

The *adaptive* part of the immune response is that part which requires immunization, and is carried out by lymphocytes. Immunological research in the first half of the twentieth century was dominated by the study of antibodies. The study of the cellular basis of immunity was largely neglected until the 1950s, when Gowans showed that immunity was carried by small round cells called lymphocytes (reviewed by Gowans and McGregor, 1965; Gowans, 1995).

The 1950s also marked the beginnings of attempts to explain the cellular basis of immunity in terms of the emerging concepts of gene action. The demonstration that proteins had defined sequences (Sanger and Thompson, 1953), and the 'Central Dogma' that information flowed in one way from DNA to RNA to protein were particularly influential, because they highlighted the serious difficulties in the prevailing 'instructionist' theories, in which the structure of antigen was proposed to direct the amino acid sequence or folding of antibodies (reviewed by Silverstein, 1989). These theories were replaced by a series of 'selectionist' theories by Jerne, Talmage, Lederberg, and most successfully by Burnet, whose 'Clonal Selection Theory' (1957, 1959, 1968) will be explored in Chapter 3.

In the late 1950s and early 1960s, it gradually became apparent that immunity involved a third arm to the immune response (now known as T cells), which could be transferred between individuals by cells but not by serum (Gowans *et al*., 1962). As mentioned earlier, the first glimpse of T cells was probably seen by Landsteiner and Chase in 1942, who showed that delayed-type hypersensitivity could be transferred between animals by cells but not by serum. However, it was nearly 20 years when Miller (1961) showed that this activity could be abrogated by neonatal thymectomy. The stunted and unhealthy nature of neonatally thymectomized mice and their inability to mount most immune responses reflected their deficiency of T cells. Neonatal thymectomy may thus be regarded as the first animal model of AIDS. Around the same time, Szenberg and Warner (1962) demonstrated that the immune responses in birds could be dissociated into two types, one of which was abrogated by bursectomy and the other by thymectomy.

Subsequent work by Claman *et al.* (1966) and Mitchell and Miller (1968) showed that antibody-secreting cells are derived from precursors from the bone marrow, which require 'help' from thymus-derived lymphocytes. By this time, the notion that there were two types of lymphocytes had been generally accepted. They were named B and T cells by Roitt *et al.* (1969).

The immune system could then be thought of as having three arms: an 'innate' arm made up of phagocytic cells and two 'adaptive' arms (Fig. 2.1), one comprising B cells, which are the precursors of antibody-forming cells, and the other comprising T cells, which do not make conventional antibodies but regulate the action of other cells.

Most antibody responses require T cell help, and thus require activation of both T and B cells (reviewed by Parker, 1993). T cell help is now known to involve a multi-step process that requires both the secretion by T cells of polypeptide hormones known as lymphokines or cytokines, which act by binding to specific receptors on B cells, and direct cell–cell contact between T and B cells, by which specific signals are exchanged in a two-way dialogue (Clark and Ledbetter, 1994). As we will see, T and B lymphocytes are activated by quite different mechanisms.

Antigenic stimulation has two major consequences. These may be divided

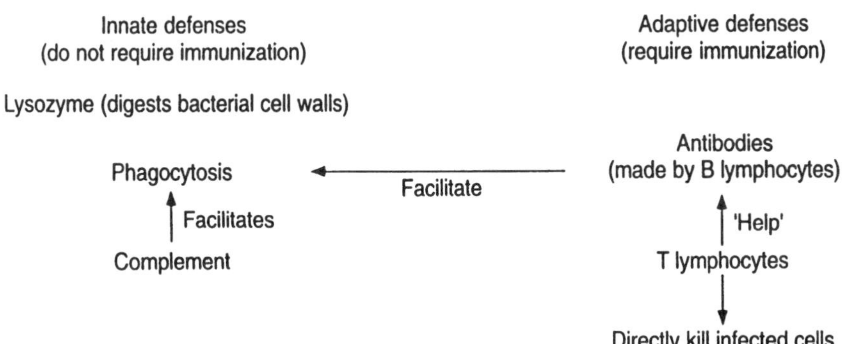

Fig. 2.1. Innate and adaptive defenses. Innate defenses are available immediately but lack flexibility because they can only attack structures that are never present in the body. In contrast, adaptive defenses provided by the immune system are slower to act but are far more flexible, as they can respond to virtually any structure, even if it is extremely similar to self. The innate and adaptive defenses interact in many ways, such as the facilitation of phagocytosis by antibodies.

into 'effector' and 'memory' responses. Effector responses include the production of antibodies by B cells and subsequent phagocytosis, activation of complement and mast cell degranulation. On the T cell side, effector responses include secretion of small polypeptides known as cytokines (also known as lymphokines) and the production of cytotoxic T cells, which are capable of killing virally infected or foreign cells by direct contact.

'Memory' refers to the more rapid and vigorous response seen after a second encounter with antigen. There is currently some controversy over its cellular basis. The traditional view has been that memory reflects the generation of long-lived T and B 'memory' cells. This view has recently been challenged. There is now considerable evidence that memory fades rapidly in the absence of continuing presence of antigen (reviewed by Gray, 1993, Sprent, 1994), suggesting that long-lived 'memory' lymphocytes do not exist. If this view is correct, the persistence of memory could be due to either persistence of antigen or constant re-encounter with the same or closely related antigens from the environment (Matzinger, 1994).

2.1.7 T-independent Antigens

Some B cell antigens are said to be 'T independent', in that they appear to be able to activate B cells directly and without obvious T cell help. These antigens are generally polymeric bacterial products, typically polysaccharides or lipopolysaccharides, or synthetic polysaccharides such as Ficoll or dextran. Responses to T-independent antigens tend to be confined to the IgM class of antibody and provoke little immunological memory (Colle *et al.*, 1988; Zhang *et al.*, 1988; Kolb *et al.*, 1993).

2.2 Structure of Antibodies

Work carried out in the 1950s and early 1960s, notably by Porter and Edelman, led to the current picture of the structure of antibodies (Cohen and Porter, 1964; Edelman *et al.*, 1969). The prototype antibody class is IgG, which is the most abundant form in serum. It consists of a symmetrical structure of two light chains and two heavy chains (H_2L_2), held together by strong noncovalent forces as well as disulfide bonds (Fig. 2.2).

Portions of the amino-terminal 110 or so amino acids of the heavy and light chains contribute to the structure and specificity of the *antigen-binding site,* which comprises six short loops (three from each heavy chain and three from each light chain) that protrude from the end of the molecule, comprising an area of about 1500 Å^2. Binding of antigen is tight, specific and reversible, with dissociation constants of the order of 10^{-6} to 10^{-10} M or less. Antibodies with dissociation constants of greater than 10^{-6} M are difficult to detect, and probably of little biological significance. The antigen combining site is discussed in more detail in Chapter 5.

Antibody light chains can be divided into two broad types, κ and λ. Each antibody molecule has one or other but rarely both (i.e. κ_2H_2 or λ_2H_2; but see Pauza *et al.*, 1993 and Giachino *et al.*, 1995). Each chain consists of a series of 'domains' or 'homology units' of about 110 amino acids, which are closely related in amino acid sequence and probably arose by repeated duplication of an ancestral gene. Comparison of the amino acid sequences from individual antibody molecules showed that all κ light chains have a common 'constant'

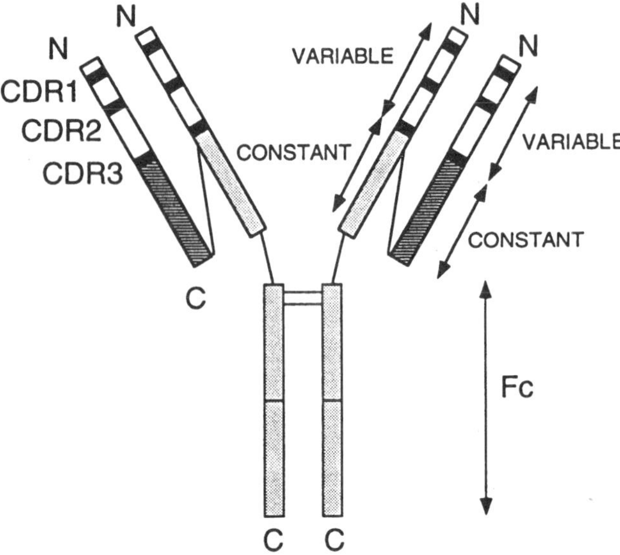

Fig. 2.2. Schematic structure of the IgG molecule. Hypervariable regions are shown in black.

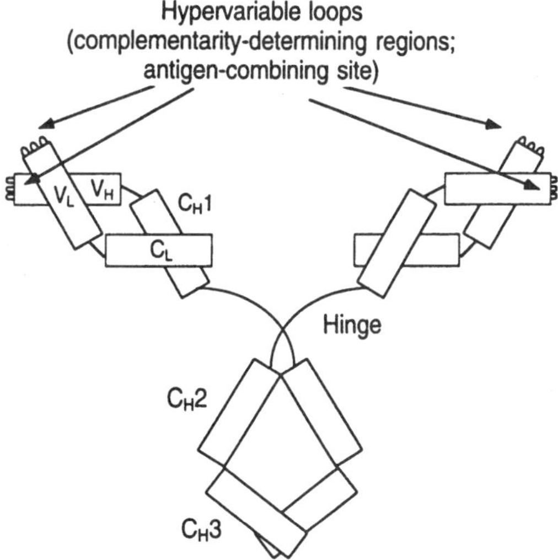

Fig. 2.3. Schematic representation of the folded structure of an IgG molecule to illustrate the relationships between individual domains and the way in which the hypervariable regions protrude from one end of the molecule to form the antigen-combining site.

domain (C_κ) comprising the carboxy-terminal 110 or so amino acids, while the amino-terminal 110 or so amino acids of the κ chains show multiple differences between individual molecules. The amino-terminal region is therefore known as the 'variable' domain (V_κ). Within the variable regions of the light and heavy chains, there are three *hypervariable* regions (Fig. 2.2). Lambda light chains can also be divided into V_λ and C_λ domains.

There are five *classes* of antibody, which are defined by the their type of heavy chain: γ chains for IgG, μ chains for IgM, δ chains for IgD, α chains for IgA and ε chains for IgE. In any one molecule, the two heavy chains will always come from the same class. As for the light chains, the first 110 or so amino acids of the heavy chains constitute the variable domain (V_H). The constant regions of the heavy chains are longer than those of light chains, and consist of a series of three or four constant domains, depending on the class of antibody (see Chapter 5).

Each class of antibody can exist in *secretory* or *membrane-bound* forms. The membrane-bound form is attached via the *C* terminus of the heavy chain, which is slightly longer than the corresponding region of the secretory forms, and is hydrophobic in character, allowing it to remain embedded in the lipid bilayer of the membranes of B cells. These forms of antibody act as receptors for antigen.

X-ray crystallography has shown that each 110-amino-acid homology unit of antibodies forms a compact folded domain, predominantly in the form of β-pleated sheets (Fig. 2.3). The domains have a similar three-dimensional structures, but in each variable domain the hypervariable regions form three loops that protrude from one end of the molecule, forming the *antigen-combining site* (Fig. 2.3). Because of their involvement in binding antigen, the hypervariable regions are also known as the *complementarity determining regions* (CDRs). Each combining site is therefore made up of six CDRs, three from the light chain and three from the heavy chain.

The angle between the two variable portions of IgG is itself quite variable, owing to a flexible 'hinge' region comprising a short additional sequence that lies between the first and second constant domains. Digestion of IgG with proteolytic enzymes can result in specific cleavages in the hinge region, generating two large fragments which define and delineate important structural and functional portions of the molecule. The portion of the constant region between the hinge and the carboxy terminus is known as the Fc (crystallizable) region, while the remainder is known as the Fab (antibody combining site) region (Fig. 2.2; see also Fig. 5.1).

2.3 A Typical Antibody Response

The most fundamental property of the immune system is the ability to distinguish self from nonself. An incoming substance will be regarded as antigenic if it differs from self. We will explore the question of what is self in Section 2.4, but for now, let us consider the case of the first encounter of a bacterial infection, where the bacterium may be safely assumed to be nonself.

Upon entry into the body, at least a few of the bacteria will die, or be killed by lysozyme or pre-existing antibody and complement, or killed intracellularly following phagocytosis. Whatever the means of death of the bacteria, the fragments will ultimately be phagocytosed. Some phagocytic cells are specialized to present foreign proteins to T and B lymphocytes.

Within 2–3 days, the first specific IgM antibody from the *primary response* will be detectable. IgM is a large pentameric or hexameric polymer of the basic H_2L_2 structure referred to above. Many bacterial antigens are polymeric, with multiple repeating motifs. The polymeric nature of IgM allows it to bind tightly to polymeric antigens, because even if one antigen-combining site lets go, the others will stay attached (Fig. 2.4). Thus, even if the affinity of the individual sites is low (as is often the case for primary responses), the overall strength of binding, which is known as *avidity*, will be very high (see Metzger, 1970 for further discussion). IgM is also particularly efficient at activating complement, and thus acts as an excellent primary defence.

Three or four days after the primary antigenic challenge, the immune response starts to switch to IgG, a process which requires T cells (Fig. 2.5). Like IgM, IgG antibodies interact with complement and kill bacteria. In addition, IgG greatly facilitates phagocytosis of any remaining bacteria by binding

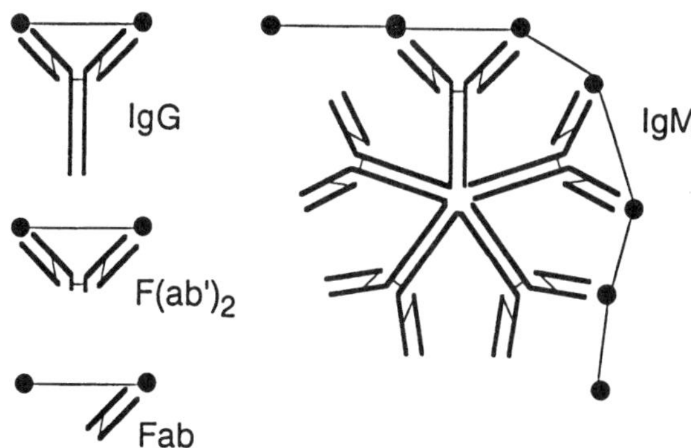

Fig. 2.4. Effect of valency on strength of binding of antibody to antigen. For a multivalent antigen, the pentameric structure of IgM will allow strong binding, even if the affinity of each individual binding site is low. The divalent binding of antigen to IgG also strengthens binding, because if one site releases antigen, the other may still retain it. Divalent F(ab')$_2$ fragments of IgG will also bind strongly for the same reason, while binding to monovalent Fab fragments is much weaker. The fragmentation of antibodies and its effect on strength of binding of antigen will be discussed in more detail in Chapters 5 and 9 (see especially Section 9.5.1).

to specific receptors for IgG (Fc receptors) that are present on the surface of many phagocytic cells. Specific IgG antibodies also neutralize bacterial toxins by binding to them. The antibody response will, if successful, neutralize toxins and kill bacteria, and thus defend the body.

As early as 3–4 days after the injection of antigen in a primary response, the affinity of the IgG starts to increase (reviewed by Nossal, 1992a, Gray, 1993; Küppers *et al.*, 1993; MacLennan, 1994). Affinity maturation is a kind of Darwinian selection at the somatic cell level. Hypermutation of the variable portions of antibody genes and selective survival of those B cells with membrane immunoglobulin of sufficient affinity to bind the steadily diminishing amount of antigen results in a progressively improved 'fit' between antigen and antibody (Fig. 2.6; see Küppers *et al.*, 1993).

If there is no additional input of antigen, the antibody response usually declines gradually over a period of weeks. However, antigenic stimulation is often sustained over very long periods by the persistence of very small amounts of antigen that remain attached to the surface of follicular dendritic cells in the lymphoid follicles (B cell areas) of spleen and lymph nodes, via specific antibody attached to Fc receptors (reviewed by Nossal, 1992a, Gray, 1993 and MacLennan, 1994). The decline in antibody may be greatly retarded if the antigen is given in a 'depot' form where it is very gradually released, as is the case when Freund's adjuvant is used. If, some time after the initial dose of antigen, a further dose is given, the 'secondary response' is usually much greater than the primary response and occurs more rapidly (Fig. 2.5).

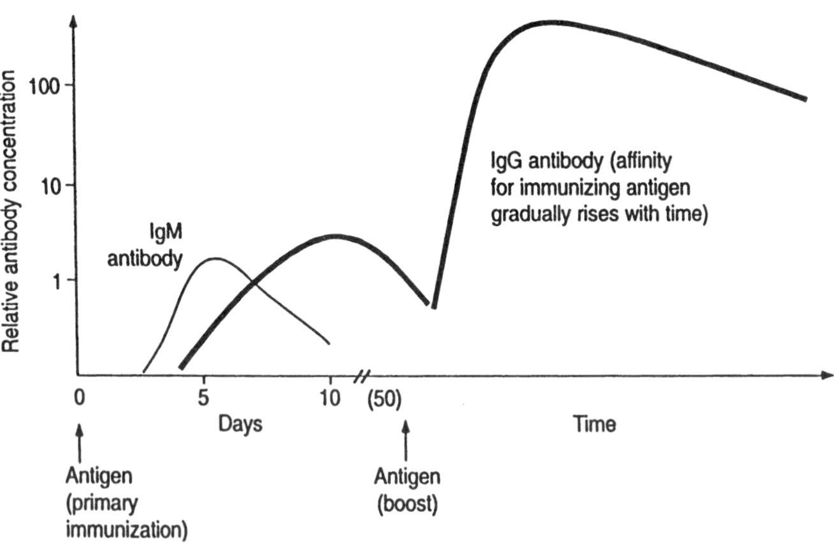

Fig. 2.5. A typical antibody response.

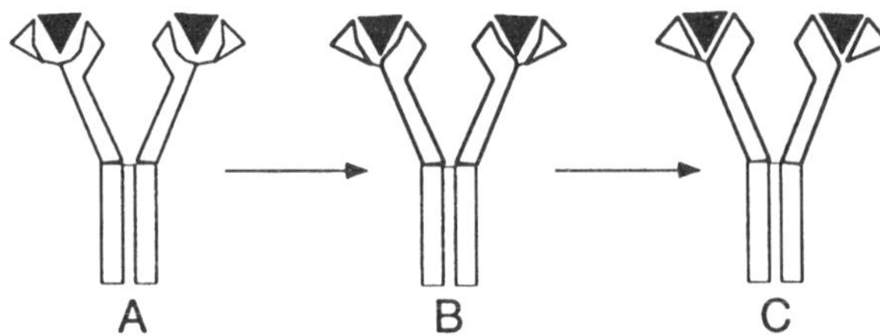

Fig. 2.6. As the immune response proceeds, somatic mutations in the lymphoid follicles combined with antigen-induced selection result in an increasingly good fit between antigen and antibody. This process is known as affinity maturation.

2.4 What Distinguishes Self from Nonself?

As a general rule, the immune system does not respond to self, a concept that Ehrlich called *'horror autotoxicus'* (reviewed by Silverstein, 1989). What, then, is self? Chemical differences between self and nonself are often extremely subtle. Even though the DNA sequences of different humans are typically 99% identical, grafts between unrelated humans are virtually always rejected by the immune system, because they are recognized as nonself. A single amino acid substitution can often be detected by the immune system. How can the immune system make such fine discrimination? *Perhaps the most useful concept is that self is what is present all the time.* Nonself appears sporadically. The most common example of nonself would be a protein that has an amino acid sequence that is different to that encountered in self.

A moment's reflection can tell us that the ability to distinguish between self and nonself is unlikely to be encoded in the germline, but must be learned by the immune system during development. A child inherits half its genes (haplotype) from the mother and half from the father, and is tolerant to the products of both sets of received genes. However, the genes that the child inherits from the father cannot have been preprogrammed to be tolerant to the mother, because the father's genes could not have anticipated who the father would marry. Providing that the union was not incestuous (as in inbred mice), a father will not be tolerant to the mother's antigens, as shown by the fact that skin grafted between the mother and father would always be rejected. Each haplotype of the child must therefore learn what is in the other haplotype, and indeed what is in its own haplotype. In other words, self-tolerance is learned by the immune system as it develops in each individual.

An experiment of nature illustrates this concept. 'Freemartin' cattle are genetic females that have been masculinized by intra-uterine exposure to Müllerian inhibiting substance from a twin male fetus that shares the same placenta (Hunter, 1995). While the intersex nature of these calves has no immunological significance in itself, it allows the identification of calves that, as a result of the shared intra-uterine circulation, are stable haemopoietic chimeras (genetic mosaics). These calves are tolerant to each other's cells, indicating that they have acquired tolerance *in utero* (Owen, 1945).

The immune system thus defines self as the sum of all the protein and other macromolecular structures that are present in the body and available to the immune system in sufficient concentration to learn of its presence. However, there are many 'holes' in self tolerance. In the case of T cells, the major site of tolerance induction is in the thymus. The immune system will not be tolerant to substances that are present only at minute concentrations in the body or which are sequestered away from the immune system ('immune ignorance'; reviewed by Miller and Heath, 1993). If a self antigen is not present in sufficient amounts in the thymus, T cells may emerge that are not tolerant to that antigen.

The immune system sometimes appears to have a second chance to 'tolerize' such T cells. This form of tolerance is known as 'peripheral tolerance' and may occur in both T and B cells (reviewed by Miller and Morahan, 1992; Nossal, 1992b). The cellular basis for self tolerance will be examined in Chapter 3.

The phenomenon of self tolerance means that it is usually difficult to generate antibodies to self antigens, although sometimes self-tolerance can be destroyed or bypassed (see Chapters 3 and 4).

2.5 Effector Functions of Antibodies

In addition to the specific binding of antigen, antibodies have a number of *effector functions*, which generally involve binding to cells or to complement. All effector functions of antibodies are mediated by the constant region of the heavy chain, generally via the Fc portion (see Fig. 5.1), and are highly dependent on antibody class (see Chapter 5). This result was anticipated by Ehrlich, who postulated distinct sites on antibodies for antigen binding and effector functions (Ehrlich, 1900).

Binding and activation of complement is almost exclusively confined to IgM and IgG antibodies and results in lysis of cells, induction of inflammation (via complement fragments C3a and C5a), and facilitation of phagocytosis. Of these, the facilitation of phagocytosis is probably the most important, and occurs via receptors for complement on phagocytic cells.

A second very important antibody effector mechanism involves the specific binding of their Fc portions to specific receptors (Fc receptors) on the surface of cells (reviewed by Metzger, 1990; Hulett and Hogarth, 1994). The biological consequences of binding of antibody to Fc receptors depends on the class of antibody, the type of Fc receptor and the type of cell bearing it, and includes facilitation of phagocytosis (IgG), activation of mast cells (IgE) and transport across epithelia (IgA and IgG) (reviewed by Metzger, 1990; Hulett and Hogarth, 1994).

2.6 Tumour Immunity

Cancer cells contain mutated self genes and hence might be perceived as foreign. This concept is termed 'immunological surveillance' (see Burnet, 1970). Until recently, there was little in the way of convincing evidence, but there is now new data to suggest that it may have some validity (Boon, 1993; Boon *et al.*, 1994; Kawakami *et al.*, 1994; Tsomides and Eisen, 1994). At least some of the mutations found in cancer cells can be recognized by the immune system (Chen *et al.*, 1992). As pointed out by Houghton (1994), 'The boundary between self and nonself is not well-defined, and the concept of cancer antigens fits more with a self/altered self paradigm than with the nonself paradigm for antigens recognized by infectious diseases'.

In some instances, the antitumour response is directed at mutated self, while in others, it appears to be directed to nonmutated self proteins which are selectively expressed in the neoplastic cells and their normal counterparts (Houghton, 1994). Many mutations may escape detection because they fail to result in peptides that can be appropriately presented to the immune system.

The biological value of any such recognition of cancer cells would almost certainly be limited by the numerous possible 'escape routes' available to the tumour, including mutation or down-regulation of the gene encoding the mutant peptide, or down-regulation of the machinery that presents it to the immune system. This would be especially likely if the mutation were not essential for maintaining the malignant or metastatic phenotype (Houghton, 1994).

If tumour immunity were highly effective, we might expect that cancer would usually be nipped in the bud by the immune system. There is little evidence that this happens.

2.7 The Cost of the Immune System: Autoimmunity and Transplant Rejection

The complexity of the immune system undoubtedly reflects its great survival value. Microbial invaders and their toxins are rapidly neutralized and destroyed, using flexible combinations of antibody, phagocytosis, and T cell-mediated killing of cells supporting the growth of viruses or other intracellular pathogens. Although the immune system uses thousands of genes and many different cell types, this is a small price for immediate, short-term and long-term defences that can deal with any external threat, as long as it can be recognized as nonself. In nature, there is generally little pressure for economy of genes or cells.

There are two much more important prices for having a good immune system. First, the protection against infectious disease given by the immune system makes transplantation between unrelated individuals impossible without the use of immunosuppressive drugs. Second, in a few individuals, the immune system turns its attack to the host.

Paroxysmal cold haemoglobinuria is a disease in which episodes of haemolysis are triggered by exposure to cold. In 1904, Donath and Landsteiner provided evidence from three patients that this disease was caused by haemolytic auto-antibodies to self red cells. However, most workers found the concept of Ehrlich's *horror autotoxicus* so persuasive that there was great reluctance to accept the possibility that autoimmunity might occur. Perhaps the thinking was oriented excessively along the lines of what would make sense for normal physiology, and insufficient consideration was given to the possibility that the immune system might make a mistake or be misled. Whatever the reason, the concept of auto-immunity met very great resistance, and it took nearly half a century before the concept of auto-immunity became generally accepted (Silverstein, 1989; Mackay, 1995).

There is now good evidence that juvenile onset diabetes, rheumatoid arthritis, systemic lupus erythematosus, myasthenia gravis, thyrotoxicosis and numerous other diseases are auto-immune in nature. The reasons why immunological tolerance breaks down in these diseases are still poorly understood (Miller and Morahan, 1992; Nossal, 1992b; Sercarz and Datta, 1994; Theofilopoulos, 1995a,b). There is some evidence that auto-immunity is driven by antigen rather than being caused by an autonomous 'forbidden clone' of lymphocytes, as envisaged by Burnet (1959), because many auto-antibodies have somatic mutations characteristic of antigen-driven secondary responses (Behar *et al.*, 1991; van Es *et al.*, 1991; Diamond *et al.*, 1992).

In a few instances, notably thyrotoxicosis, myasthenia gravis, bullous pemphigoid and pemphigus vulgaris, the antibodies are directly pathogenic (Naparstek and Plotz, 1993). However, in the great majority of instances it is not clear whether the auto-antibodies are directly involved in the pathogenesis of the disease or are perhaps an indirect reflection of the inciting antigen or some other epiphenomenon. The understanding of auto-immune disease represents a major challenge for contemporary immunologists, but there is a widespread contemporary belief that we are now close to acquiring adequate knowledge and tools to be able to make real progress.

Whatever the answers to these questions, the evolution of the immune system has given biologists a set of specific and versatile tools. The exquisite specificity of antibodies allows the molecular dissection of mixtures of molecules that would be far too complex to analyse by conventional chemistry. The remainder of this book will focus on the generation of antibodies, their structure, function and application to biological problems.

References

Behar, S. M., Lustgarten, D. L., Corbet, S. and Scharff, M. D. (1991) Characterization of somatically mutated S107-VH_{11}-encoded anti-DNA antibodies derived from autoimmune (NZB × NZW)F_1 mice. *J. Exp. Med.* **173**, 731–741.

Boon, T. (1993) Teaching the immune system to fight cancer. *Scientific American*, March 1993, 32–39.

Boon, T., Cerottini, J-C., van den Eynde, B., van der Bruggen, P. and van Pel, A. (1994) Tumor antigens recognized by T lymphocytes. *Annu. Rev. Immunol.* **12**, 337–366.

Bordet, J. (1895) Les Leucocytes et les propriétés actives du sérum ches les vaccinés. *Ann. de l'Inst. Pasteur* ix, 462–506.

Bordet, J. (1898) Sur l'agglutination et la dissolution des globules rouges par le sérum d'animaux injectés de sang défibriné. *Ann. de l'Inst. Pasteur* **xii**, 688–695.

Browning, C. H. (1955) Emil Behring and Paul Ehrlich: Their contributions to science. *Nature* **175**, 570–575; 616–619.

Bulloch, W. (1938) 'The History of Bacteriology' (University of London Heath Clark Lectures, 1936). Dover Publications, New York, reprinted 1979.

Burnet, F. M. (1957). A modification of Jerne's theory of antibody production using the concept of clonal selection. *Austral. J. Sci.* **20**, 67–69.

Burnet, F. M. (1959) 'The Clonal Selection Theory of Acquired Immunity'. Cambridge University Press, Cambridge.

Burnet, F. M. (1968). 'Changing Patterns: An Atypical Autobiography'. William Heinemann, London.

Burnet, F. M. (1970) 'Immunological Surveillance'. Pergamon Press, New York.

Chen, W., Peace, D. J., Ravira, D. K., You, S.-G. and Cheever, M. A. (1992) T-cell immunity to the joining region of p $210^{BCR\text{-}ABL}$ protein. *Proc. Nat. Acad. Sci. USA* **89**, 1468–1474.

Clark, E. A. and Ledbetter, J. A. (1994) How B and T cells talk to each other. *Nature* **367**, 425–428.

Claman, H. N., Chaperon, E. A. and Triplett, R. F. (1966) Thymus-marrow cell combinations. Synergism in antibody production. *Proc. Soc. Exp. Biol. Med.* **122**, 1167–1171.

Cohen, S. and Porter, R. R. (1964) Structure and biological activity of immunoglobulins. *Adv. Immunol.* **3**, 287–349.

Colle, J.-H., Truffa-Bachi, P. and Freitas, A. A. (1988) Secondary antibody responses to thymus-independent antigens. Decline and life-span of memory. *Eur. J. Immunol.* **18**, 1307–1314.

Diamond, B., Katz, J. B., Paul, E., Aranow, C., Lustgarten, D. and Scharff., M. D. (1992) The role of somatic mutation in the pathogenic anti-DNA response. *Annu. Rev. Immunol.* **10**, 731–757.

Donath, J. and Landsteiner, K. (1904) Ueber paroxysmale Hämaglobinurie. *Munch. med. Wchnschr.* **51**, 1590–1593.

Edelman, G. M., Cunningham, B. A., Gall, W. E., Gottlieb, P. D. Rutishauser, U. and Waxdal, M. J. (1969) The covalent structure of an entire γG immunoglobulin molecule. *Proc. Natl Acad. Sci. USA* **63**, 78–85.

Ehrlich, P. (1900) Croonian Lecture: On immunity with special reference to cell life. *Proc. Roy. Soc.* **66**, 424–448.

Ehrlich, P. (1913) Address in Pathology on Chemotherapeutics: Scientific Principles, Methods and Results. *Lancet* **ii**, 445–451.

van Es, J. H., Gmelig-Meyling, F. H., van de Akker, W. R., Aanstoot, H., Derksen, R. H. and Logtenberg, T. (1991) Somatic mutations in the variable regions of a human IgG anti-double-stranded DNA autoantibody suggests a role for antigen in the induction of systemic lupus erythematosus. *J. Exp. Med.* **173**, 461–470.

Fraenkel, C. and Sobernheim, P. (1984) Versuche über das Zustandekommen der Künstlichen Immunitaät. *Hyg. Rundschau, Berl.* iv, 97–145.

Giachino, C., Padovan, E. and Lanzavecchia, A. (1995) $\kappa^+\lambda^+$ dual receptor B cells are present in the human peripheral repertoire. *J. Exp. Med.* **181**, 1245–1250.

Gowans, J. L. and McGregor, D. D. (1965) The immunological activities of lymphocytes. *Progr. Allergy* **9**, 1–78.

Gowans, J. L., McGregor, D. D., Cowen, D. M. and Ford, C. E. (1962) Initiation of immune responses by small lymphocytes. *Nature* **196**, 651–655.

Gowans, J. L. (1995) The mysterious lymphocyte. In: 'Immunology: The Making of a Modern Science'. Gallagher, R. B., Gilder, J., Nossal, G. J. V. and Salvatore, G. (eds), pp. 65–74. Academic Press, London.

Gray, D. (1993) Immunological memory. *Annu. Rev. Immunol.* **11**, 49–78.

Gruber, M. and Durham, H. E. (1896) Ein neue Methode zur raschen Erkennung des Choleravibrio und des Typhusbacillus. *München med. Wehnschr*. xliii, 285–286.

Houghton, A. N. (1994) Cancer antigens: Immune recognition of self and altered self. *J. Exp. Med.* **180**, 1–4.

Hulett, M. D. and Hogarth, P. M. (1994) Molecular basis of Fc receptor function. *Adv. Immunol.* **57**, 1–127.

Hunter, J. (1794) 'Treatise on the Blood, Inflammation and Gunshot wounds'. Webster, Philadelphia, Pennsylvania, 1823.

Hunter, R. H. F. (1995) 'Sex Determination, Differentiation and Intersexuality in Placental Mammals'. Cambridge University Press, Cambridge.

Kawakami, Y., Eliyahu, S., Delgado, C. H., Robbins, P. F., Rivoltini, L., Topalian, S. L., Miki, T. and Rosenberg, S. A. (1994) Cloning of the gene for a shared human melanoma antigen recognised by autologous T cells infiltrating into the tumor. *Proc. Natl Acad. Sci. USA* **91**, 3515–3519.

Kendrew, J. C., Bodo, G., Dintzis, H. M., Parrish, H., Wyckoff, H. and Phillips, D. C. (1958) A three-dimensional model of the myoglobin molecule obtained by X-ray analysis. *Nature* **181**, 662–666.

Kolb, C., Fuchs, B. and Weiler, E. (1993) The thymus-independent antigen α-(1,3) dextran elicits proliferation of precursors for specific IgM antibody-producing cells (memory cells), which are revealed by LPS stimulation in soft agar cultures and detected by immunoblot. *Eur. J. Immunol.* **23**, 2959–2966.

Kraus, R. (1897) Veber spezifische Reaktionen in keimfreien Filtration aus Cholera-Typhus-Pestbouillonculture erzeugt durch homologes Serum. *Weiner klin. Wehnschr*. x, 736–739.

Krieg, A. M., Yi, A-K., Maston, S., Waldschimidt, T. J., Bishop, G. A., Teasdale, R., Koretzky, G. A. and Klinman, D. M. (1995) CpG motifs in bacterial DNA trigger direct B-cell activation. *Nature* **374**, 546–549.

Küppers, R., Zhao, M., Hansmann, M-L. and Rajewsky, K. (1993) Tracing B cell development in human germinal centres by molecular analysis of single cells picked from histological sections. *EMBO J.* **12**, 4955–4967.

Landsteiner, K. (1945) The Specificity of Serological Reactions. Harvard University Press, Boston. Reprinted by Dover Publications, New York, 1962.

Landsteiner, K. and Chase, M. W. (1942) Experiments on transfer of cutaneous sensitivity to simple compounds. *Proc. Soc. Exp. Biol. Med.* **49**, 688–690.

Landsteiner, K. and Wiener, A. S. (1940) Agglutinable factor in human blood recognised by immune sera for Rhesus blood. *Proc. Soc. Exp. Biol. Med.* **43**, 223.

Mackay, I. R. (1995) Roots of and routes to autoimmunity. In: 'Immunology. The Making of a Modern Science', R. B. Gallagher, G. J. V. Nossal and G. Salvatore, (eds), pp. 49–62. Academic Press, London.

MacLennan, I. C. M. (1994) Germinal centres. *Annu. Rev. Immunol.* **12**, 117–139.

Marrack, P. and Kappler, J. W. (1994) Subversion of the immune system by pathogens. *Cell* **76**, 323–332.

Matzinger, P. (1994) Memories are made of this? *Nature* **369**, 605–606.

Metchnikoff, E. (1891) 'Lectures on the Comparative Pathology of Inflammation'. Delivered at the Pasteur Institute. Reprinted by Dover Publications, New York, 1968.

Metzger, H. (1970) Structure and function of γ-M macroglobulins. *Adv. Immunol.* **12**, 57–116.

Metzger, H. (ed.) (1990) 'Fc Receptors and the Action of Antibodies'. American Society for Microbiology, Washington, D.C.

Miller, J. F. A. P. (1961) Immunological function of the thymus. *Lancet* **ii**, 748–749.

Miller, J. F. A. P. and Morahan, G. (1992) Peripheral T cell tolerance. *Annu. Rev. Immunol.* **10**, 51–69.

Miller, J. F. A. P. and Heath, W. R. (1993) Self-ignorance in the peripheral T cell pool. *Immunol. Rev.* **133**, 131–150.

Mitchell, G. F. and Miller, J. F. A. P. (1968) Cell to cell interaction in the immune response. II. The source of hemolysin-forming cells in irradiated mice given bone marrow and thymus or thoracic duct lymphocytes. *J. Exp. Med.* **128**, 821–837.

Naparstek, Y. and Plotz, P. H. (1993) The role of autoantibodies in autoimmune disease. *Annu. Rev. Immunol.* **11**, 79–104.

Nossal, G. J. V. (1992a) The molecular and cellular basis of affinity maturation in the antibody response. *Cell* **68**, 1–2.

Nossal, G. J. V. (1992b) Cellular and molecular mechanisms of B cell tolerance. *Adv. Immunol.* **52**, 283–332.

Nuttall, G. (1888) Experimente über die bacterien feindlichen Einflüsse des thierischen Körpers. *Ztschr. f. Hyg.* iv, 353–394.

Owen, R. D. (1945) Immunogenetic consequences of vascular anastomosis between bovine twins. *Science* **102**, 400–401.

Parker, D. C. (1993) T cell-dependent B-cell activation. *Annu. Rev. Immunol.* **11**, 331–360.

Pauza, J. E., Rehmann, J. A. and LeBien, T. W., (1993) Unusual patterns of immunoglobulin gene rearrangement and expression during human B cell ontogeny: human B cells can simultaneously express cell surface κ and λ light chains. *J. Exp. Med.* **178**, 139–149.

Roitt, I. M., Greaves, M. F., Torrigiani, G., Brostoff, J. and Playfair, J. H. L. (1969) The cellular basis of immunological responses. *Lancet* **ii**, 367–371.

Sanger, F. and Thompson, E. O. P. (1953) The amino acid sequence in the glycyl chain of insulin. *Biochem. J.* **53**, 353–374.

Sercarz, E. E. and Datta, S. K. (1994) Mechanisms of autoimmunization: perspective from the mid-90s. *Curr. Op. Immunol.* **6**, 875–881.

Silverstein, A. M. (1989) 'A History of Immunology'. Academic Press, New York.

Sprent, J. (1994) T and B memory cells. *Cell* **76**, 315–322.

Sumner, J. B. (1926) The isolation and crystallization of the enzyme urease. *J. Biol. Chem.* **69**, 435–441.

Szenberg, A. and Warner, N. L. (1962) Dissociation of immunological responsiveness in fowls with a hormonally arrested development of lymphoid tissue. *Nature* **194**, 146–147.

Theofilopoulos, A. N. (1995a) The basis of autoimmunity: Part I. Mechanisms of aberrant self-recognition. *Immunol. Today* **16**, 90–98.

Theofilopoulos, A. N. (1995b) The basis of autoimmunity: Part II. Genetic predisposition. *Immunol. Today* **16**, 150–159.

Tsomides, T. J. and Eisen, H. N. (1994) T-cell antigens in cancer. *Proc. Natl. Acad. Sci. USA* **91**, 3487–3491.

Zhang, J., Liu, Y.-J., MacLennan, I. C. M., Gray, D. and Lane, P. J. L. (1988) B cell memory to thymus-independent antigens type 1 and type 2: The role of lipopolysaccharide in B memory induction. *Eur. J. Immunol.* **18**, 1417–1424.

3 Cellular Basis of the Immune System

3.1 The Clonal Selection Theory

The immune system is amazingly versatile. It can respond to almost any substance by making specific antibodies, provided that the substance is seen as nonself. How is this achieved? The mechanisms by which this specificity and diversity is produced has fascinated generations of immunologists.

In 1900, Ehrlich presented a paper to the Royal Society in London, in which he proposed a prescient model of the cellular origin of antibodies. This model was remarkably similar to the currently accepted picture of the clonal selection theory (see Kindt and Capra, 1984).

However, the time was not ripe to ask the right questions concerning the origin of antibody diversity until there was a clear understanding of the fact that proteins had defined sequences (Sanger and Thompson, 1953). Prior to the 1950s, it was widely believed that antigen must somehow 'instruct' the specificity of antibodies by providing some sort of template (Silverstein, 1989; Nossal, 1995). The gradual realization that the primary amino acid sequence could be sufficient to specify all the biological activities of a protein (reviewed by Anfinsen, 1973) created great difficulties for instructionist theories. Even more importantly, the elucidation of the structure of DNA, the genetic code, the mechanism of protein synthesis and the realization that the flow of sequence information from DNA to protein was essentially one-way, provided the final nails in the coffin of instructionist theories.

Although there were important precursor theories, notably by Jerne and by Talmage (Talmage, 1995), the clonal selection theory as proposed by Burnet (1957, 1959, 1968) provided the best basis for understanding immunological diversity in the framework of the emerging clarity of concepts of gene action,

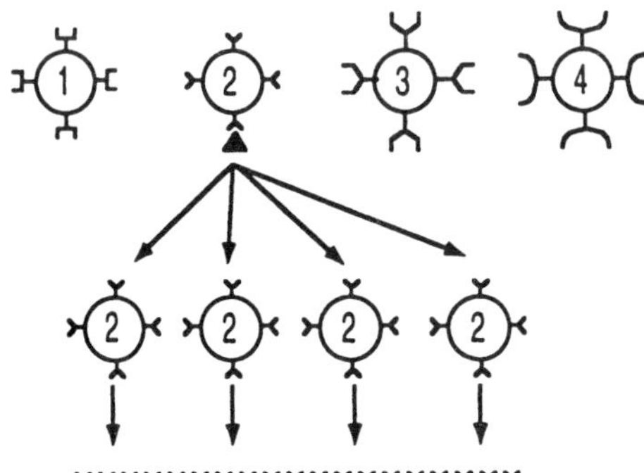

Fig. 3.1. The clonal selection theory. Interaction of antigen (solid triangle) with specific receptor immunoglobulins on the surface of clone 2 leads to proliferation and differentiation into cells which secrete antibody of the same specificity as the receptor.

and has been confirmed in all essentials (Fig. 3.1). The clonal selection theory avoided the difficulties of the instructionist theories by postulating that each lymphocyte had a unique receptor specificity, and was thus precommitted to making only one antibody after appropriate stimulation. Precommitment in specificity would be passed on to the progeny of the cell after division. The main detail that Ehrlich did not foresee was that each cell was precommitted to a single specificity, which was hardly surprising, given the state of knowledge of genetics and protein chemistry at the time.

The clonal selection theory raised several new issues. By what mechanism could individual clones be committed to a particular specificity? The postulated transmission of clonal specificity to daughter cells seemed to imply a unique genetic constitution for each clone. By the 1950s, the concept of 'one gene, one enzyme' of Beadle and Tatum was well accepted. Did this mean that there had to be a separate gene for each antibody? If so, how would clonal commitment to a particular antibody gene be made? Given the enormous diversity of antibodies, could there be sufficient germline genes to encode for all of them, or was diversity generated by some sort of somatic process from a small number of genes, or even single genes? In what way were membrane forms of antibodies different from the circulating forms? The answers to these questions had to wait a further 20 years (reviewed by Kindt and Capra, 1984). It is now apparent that there are hundreds of genes for antibody light and heavy chain variable regions, but single copies of the constant region genes. Clonal precommitment to a particular specificity is now known to involve an unprecedented genetic mechanism, in which clone-specific somatic recombination during differentiation of lymphocytes from stem cells brings a particular variable region gene closer to a constant region gene. This mechanism was predicted by Dreyer and Bennett in 1965, and directly demonstrated by Tonegawa and colleagues in 1976 (Tonegawa, 1983). A more detailed account of antibody genes will be given in Chapter 6.

Evidence in support of the clonal selection theory was first provided by Nossal and Lederberg (1958), who showed that single antibody-forming cells from rats immunized with two different antigens made antibody to one or other antigen, but rarely both. However, the clonal selection theory remained controversial until the mid-1960s, and it was not until the Cold Spring Harbor Conference of 1967 that it was generally accepted (Burnet, 1968).

By the late 1960s, it became apparent that lymphocytes could be divided into B and T cells. Clonal selection applies to both types of lymphocyte.

As described in Chapter 2, B lymphocytes are the precursors of antibody-forming cells, and are formed in the bone marrow. They bear 'receptor' forms of antibody on their surfaces, and the unique rearrangement of antibody genes in each cell ensures that the cell and its progeny are committed to that specificity during subsequent divisions, although a certain amount of 'fine tuning' of specificity occurs by somatic mutation and further selection during secondary responses, which results in improved affinity of antibody for antigen.

T lymphocytes arise in the thymus, and possess antigen-specific membrane receptors which are similar to antibodies but are encoded by a distinct set of genes. Like B cells, each T cell has a unique rearrangement of its antigen-receptor genes. Unlike B cells, T cells have no somatic mutation after commitment to their initial specificity, most likely because the harm that could be caused by the possible generation of autoreactive T cells as a result of random mutations would be much more serious than for B cells.

The clonal selection theory provides the conceptual framework for all that follows in this book. We will return to clonal selection towards the end of this chapter.

3.2 Lymphocyte Development—General Aspects

Prior to the pioneering work of Gowans, it had been felt that lymphocytes had something to do with the immune system. Gowans showed that lymphocytes *were* the immune system (Gowans *et al.*, 1962; Gowans and McGregor, 1965).

Like all haemopoietic cells, lymphocytes are produced by the body throughout life, and go through a complex set of differentiation steps before reaching maturity. Lymphocyte differentiation can be thought of as occurring in two stages. The first stage involves differentiation from stem cells, and occurs in the *primary lymphoid organs*, which are the bone marrow for B cells and the thymus for T cells, and occurs relatively normally in germ-free animals in which there is no external antigenic stimulation. The second phase of lymphocyte differentiation takes place in the *secondary lymphoid organs* (spleen, lymph nodes and Peyer's patches) and requires antigenic stimulation, resulting in further differentiation and activation into *effector lymphocytes*. In the case of B cells, this involves differentiation into antibody-secreting cells, while activation of T cells results in secretion of cytokines or differentiation into cytotoxic 'killer' cells.

3.2.1 Lymphocyte Surface Antigens Recognized by Monoclonal Antibodies ('CD' Antigens)

In addition to receptors for antigen, lymphocytes have membrane receptors for many different cytokines, as well as receptors mediating adhesion, migration, and a variety of other functions, usually involving cell–cell communication. As lymphocytes develop and differentiate into more specialized subsets, they acquire or lose individual membrane receptor types. The CD (cluster determinant) antigens recognized by monoclonal antibodies to lymphocyte membrane receptors have been extremely useful in defining lymphocyte differentiation pathways. There are now about 100 different CD antigens, of which perhaps 50% currently have a known biological function. Even when their function is not yet known, CD antigens are very useful as signposts to define particular stages of differentiation.

3.2.2 Lymphocyte Differentiation in the Primary Lymphoid Organs

Lymphocytes develop by differentiation from multipotential stem cells that arise in the yolk sac, the fetal liver, and the adult bone marrow (reviewed by von Boehmer and Rajewsky, 1993). During development in the primary lymphoid organs, lymphocytes undergo a series of maturational stages in which they rearrange their receptor genes and thereby become committed to a particular specificity (see Chapter 6). It is here that they receive their first 'education' concerning what is self and nonself. They then migrate to the secondary lymphoid organs as mature lymphocytes, ready for action.

Although there are important differences, the pivotal steps in B and T cell differentiation are similar, and involve rearrangements of antibody or T-cell receptor genes. These rearrangements are stochastic in nature. Many of them will be nonproductive, leading either to no product or defective products. The newly emerging lymphocyte undergoes positive selection to detect productive rearrangement, which is usually followed by the shutting down of further rearrangements in that clone and progression to the next stage of differentiation (see Melchers *et al.*, 1994). Failure to produce a functional product generally results in death of the cell by apoptosis.

A further consequence of the random nature of receptor gene rearrangements is that it is inevitable that some clones will have receptors with reactivity against self. These clones are generally eliminated (negative selection) or functionally silenced (anergy; Nossal, 1992b, 1994), or may be excluded from lymphoid follicles (Cyster *et al.*, 1994).

There is now a large body of evidence that during their development both T and B lymphocytes undergo both positive and negative selection by antigen (von Boehmer, 1994; Nossal, 1994). Negative selection is concerned with the deletion of auto-reactive cells or cells which have receptors that are unable to function. Positive selection selects for functional receptors early in development, and for activation and survival during immune responses.

3.3 B Cells

3.3.1 B Cell Receptors for Antigen

B cell receptors for antigen are almost identical in structure to secreted antibodies. The only structural difference is that the *C*-terminal region of the heavy chains contain a short hydrophobic stretch which spans the lipid bilayer of the membrane. In addition to the antibody module that recognizes antigen, B cell receptors have short transmembrane chains (Ig-α and Ig-β) that are involved in signal transduction (Cambier *et al.*, 1994; Pleiman *et al.*, 1994; Reth, 1994). Membrane IgM may also be associated with prohibitin and a prohibitin-related protein (Terashima *et al.*, 1994); membrane IgD may also be associated with two other as yet unidentified proteins (Kim *et al.*, 1994). The precise pathways of signalling from the receptors to the interior of the cell are not yet fully understood but involve associated protein kinases (Pleiman *et al.*, 1994), and lead to rapid cleavage of phosphatidyl inositol and mobilization of calcium (reviewed by Weiss and Littman, 1994; Cambier *et al.*, 1993, 1994; Peaker, 1994).

3.3.2 B Cells Recognize Shapes of Proteins and Other Molecules

Antibodies bind to antigen via precise complementarity between the opposing surfaces and are thus dependent on the conformation of antigen. B cell receptors are essentially identical to secreted antibodies in terms of specificity and binding mechanisms, and B cells will therefore only recognize antigen that is in an appropriately folded conformation. This has important implications for both activation and tolerance in B cells (Benjamin *et al.*, 1984; Goodnow *et al.*, 1989).

T cells recognize protein antigens in a fundamentally different way to B cells. In order to be ‘seen’ by T cells, protein antigens must be proteolytically cleaved into short peptides that bind in an extended conformation in the peptide-binding groove of the major histocompatibility antigens (Madden *et al.*, 1991; see Section 3.4.3). The conformation in which antigen is seen by T cells is therefore quite different from that of the native protein, and the initial conformation of the protein is probably only important to the extent that it influences proteolytic processing.

3.3.3 Early B Cell Development: Acquisition of Receptor Specificity

In birds, B cell development occurs in the bursa of Fabricius, a discrete lymphoid organ situated near the cloaca. In mammals, the bone marrow is the bursal equivalent. Stem cells differentiate into B lymphocytes in a series of discrete and recognizable stages, which are generally identified by changes in the arrangement and expression of immunoglobulin genes (Melchers *et al.*, 1994).

The heavy chain genes are the first to rearrange. Production of cytoplasmic μ chains is followed by their pairing with a surrogate light chain ($V_{preB}/\lambda5$) made up of products of the $\lambda5$ locus, which is a λ-like gene closely linked to the λ locus but which does not rearrange (Melchers *et al.*, 1993, 1994). The μ and surrogate light chain complex then moves to the cell surface. It has been postulated that the $\mu/V_{preB}/\lambda5$ complex interacts with a complementary structure on marrow stromal cells which results in further differentiation (Melchers *et al.*, 1993, 1994; Karasuyama *et al.*, 1994). Deletion of the gene for the surrogate light chain greatly impairs the efficiency of B cell production, but does not abolish it completely (Kitamura *et al.*, 1992).

The next step in B cell differentiation involves rearrangement and expression of the antibody light chain genes. Early work suggested that the light chain rearrangement is hierarchical (κ first, then λ if the κ rearrangement is nonproductive) but more recent data has questioned this concept (Chen and Alt, 1993). Productive rearrangement of one light chain type usually prevents further rearrangement of the other type (but see Rolink *et al.*, 1993). The great majority of B cells express either κ or λ light chains, although a few B cells may express both (Pauza *et al.*, 1993). Shortly after productive rearrangement of a κ or λ light chain gene, the immature B cell expresses IgM on the membrane.

In recent years, a distinct but numerically minor B cell subset has been identified. All B cells but no T cells express the CD19 antigen. All T cells express the CD5 antigen. However, a subset of B cells express CD5. This subset arises early in ontogeny, is highly represented in peritoneal fluid, and tends to synthesize IgM autoantibodies. It has been suggested that $CD5^+$ B cells may represent a distinct parallel lineage to the more conventional B cells, although this point is controversial (Haughton *et al.*, 1993). Kantor and Herzenberg (1993) have proposed that $CD5^+$ B cells be named B-1, and the 'conventional' B cells be named B-2 (Herzenberg and Kantor, 1993; Kantor, 1993a, b). Interestingly, virtually all chronic lymphoid leukaemia cells belong to the $CD5^+/CD19^+$ subset (Schroeder and Dighiero, 1994).

3.3.4 Induction of Tolerance in B Cells

In order for an immune response to occur, the immunizing antigen must be recognized as nonself. As far as the immune system is concerned, nonself may be defined as everything that has not been recognized as self. In other words, recognition of a given antigen as nonself is the default situation, which may be expected to pertain if the lymphocyte population has not previously seen the antigen, either because the antigen has never been present at sufficient concentrations or because it has not been in a form that can be recognized by lymphocytes. However, B cells will often (but not always) have been rendered nonresponsive by an early encounter with self antigen, or a later encounter with antigen in the absence of additional activating signals (Nossal, 1992b; Goodnow, 1992). Antigens that have been 'seen' by developing B cell populations are generally regarded as self (see section 2.4 for further discussion).

Self tolerance, like induction of an immune response, probably requires cross-linking of antigen receptors on lymphocytes. Any substance that cannot cross-link lymphocyte surface receptors (e.g. steroid hormones) will probably not be capable of inducing natural tolerance or immunity unless presented in an artificially multivalent form. In these cases, the immune system could be said to be 'ignorant' of the antigen (see Miller and Heath, 1993).

Because antibodies recognize shapes of molecules, and B cells use antibodies as receptors for antigen, it follows that B cells will not be tolerant to denatured or unfolded self unless they have been exposed to it during their development.

It has sometimes been argued that because most antigens require T cell help in order to generate an antibody response, self tolerance might be confined to T cells and there may be no such thing as B cell tolerance. There is a compelling argument against this point of view (Goodnow, 1989; Goodnow *et al.*, 1989). If a mouse is immunized with a protein from another strain of mouse in which there is a very slight difference in amino acid sequence, it is generally possible to raise antisera that are highly specific for the immunizing allele and which do not react with the self allele. *As the T cell help signal delivered by T cells to B cells is not antigen-specific, a lack of tolerance in B cells with specificity for the self allele would allow the generation of equal quantities of antibody to the self allele; this is not observed. It therefore follows that B cells can be tolerant to at least some self antigens.* B cell tolerance to membrane-associated proteins is generally robust because of the strong propensity to cross-link the receptors of any B cell encountering it (Hartley *et al.*, 1991). In contrast, B cell tolerance to soluble monomeric antigens may be weaker and more easily overcome (Goodnow 1992).

The practical importance of B cell tolerance lies in the possibility that it may be difficult or impossible to generate antibodies to certain antigens. This is particularly likely for antigens that are present in fairly high concentrations in the body, and where the immunizing antigen is very similar or identical to self. For example, self tolerance often makes it hard to generate antibodies to highly conserved proteins that differ little between species. This point will be explored further in Chapter 4.

3.3.5 Intermediate B Cell Differentiation; Development of Mature B Cells

The next major step in the development of the B cell involves the acquisition of membrane IgD, which signals maturity. The great majority of mature peripheral B cells possess both IgM and IgD (Goding and Layton, 1976, Klein *et al.*, 1993), and the typical mature peripheral B cell has about 10 times as much IgD as IgM (Havran *et al.*, 1984). Quantitatively speaking, IgD is therefore the major B receptor on most B cells. The specificity of membrane IgM and IgD on individual B cells is identical (Goding and Layton, 1976) because of the use of the same light chains and the same heavy chain variable-region genes.

Shortly after acquiring IgD, the mature B cells leave the bone marrow and migrate to the secondary lymphoid organs.

Upon activation by antigen and T cells, IgD is rapidly lost (Bourgois *et al.*,1977). Its orderly acquisition and loss of expression would suggest an important biological role for this IgD. It is therefore somewhat surprising that deletion of the δ chain locus results in only minimal changes to B cell function (Nitschke *et al.*, 1993; Roes and Rajewsky, 1993).

3.3.6 Late B Cell Development: Antigen-driven B cell Responses

In the mouse, about 5×10^7 mature B cells leave the bone marrow each day. Since there are only about $2–3 \times 10^8$ mature peripheral B cells, this implies that the peripheral B cell pool is renewed every 4–5 days. However, average lifespan of the typical mature peripheral B cell is of the order of months, so it must be concluded that most of the newly produced B cells from the marrow die soon after leaving the marrow (see Forster and Rajewsky, 1990; Gray, 1993; Matzinger, 1994). It has been suggested that the long-lived peripheral population must have been selected in some way, because the variable gene usage of the peripheral pool is different from that in the bone marrow (Gu *et al.*, 1991). The nature of this (hypothesized) selection step is currently unknown.

3.3.7 Activation of B Cells

Activation of B cells leads to antibody production and generation of B cell memory. The signals for B cell activation are now being elucidated (Fig. 3.2). The most important activation signals are antigen-induced cross-linking of membrane immunoglobulin and the binding of the CD40 ligand to the CD40 molecule on the B cell surface. Cytokines secreted by T cells are also important, particularly for switching of antibody class.

Cross-linking of B cell receptors with antigen leads to rapid mobilization of calcium, cleavage of phosphatidyl inositol and activation of tyrosine kinases,

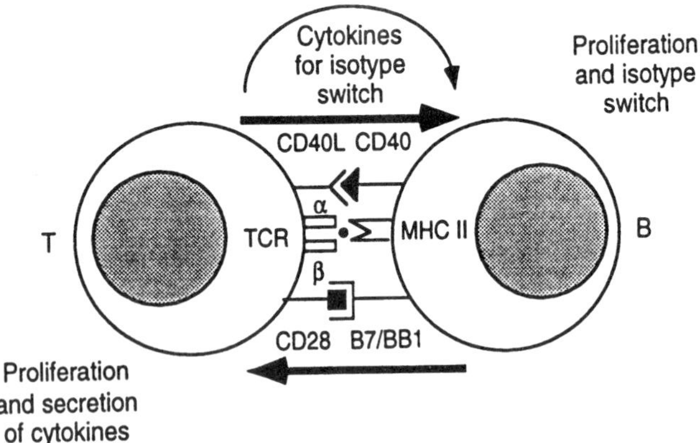

Fig. 3.2. Dialogue between T and B cells The antigenic peptide is the black dot in the centre of the picture, and is bound in the groove of the class II major histocompatibility complex (MHC II) molecule on the surface of the B cell. The peptide–MHC complex is seen by the T cell receptor (TCR). Recognition of the MHC–peptide complex by the T cell causes T expression of CD40 ligand (CD40L) on the T cell. Cross-linking of MHC II molecules leads to expression of B7/BB1 (CD80) on the B cell. Binding of CD40L to CD40 on the B cell, and binding of BB1/B7 to CD28 on the T cell amplifies the initial activation and leads to proliferation of both cell types. Cross-linking of CD28 molecules on the activated T cell induces secretion of cytokines needed for isotype switching in B cells. The bold arrows indicate signals mediated by direct cell–cell contact, while the light arrow indicates signals mediated by secreted soluble cytokines. After Clark and Ledbetter, 1994.

but the intracellular pathways to the nucleus in B cells are still inadequately understood (Pleiman *et al.*, 1994). While the importance of the CD40 system in B cell activation is now well documented (Gray *et al.*, 1994; Foy *et al.*, 1994; Banchereau *et al.*, 1994), little is known about the intracellular signalling pathways that follow from it (Parker, 1993).

Upon activation by antigen and T cells, B cells lose IgD, begin to secrete IgM, and undergo division. After 3–4 days, under the influence of cytokines secreted by T cells, many B cells will switch to the production of other classes of antibody, notably IgG, IgA and IgE (Section 3.3.10).

B cell activation is followed by several rounds of cell division. The daughter cells of these divisions may be either antibody-secreting cells or memory B cells, a somewhat poorly defined entity (see below; reviewed by Gray, 1993, 1994). The signals that determine the balance between numbers of memory cells and antibody-secreting cells are not yet known, although there is some evidence that the presence of large amounts of free antigen tends to favour antibody secretion rather than memory. It appears likely that memory cells and antibody-forming cells arise by asymmetrical division from a common parent cell. The possibility that memory B cells and antibody-secreting cells are distinct lineages has not been completely ruled out, but it appears unlikely.

3.3.8 Antibody-secreting Cells (Plasma Cells)

Antibody-forming cells are generally considered to be terminally differentiated 'end' cells incapable of further division, with a lifespan of a few days at the most (Nossal and Ada, 1971). The antibody-secreting B cell is often called a 'plasma cell', which is, historically speaking, a morphological entity consisting of a B cell with an abundant basophilic cytoplasm containing large amounts of secretory immunoglobulin, a prominent rough endoplasmic reticulum and an eccentric nucleus with a perinuclear halo (the Golgi apparatus). However, the term plasma cell is often used much more loosely to mean any antibody-secreting cell. The great majority of antibody-secreting cells are small to medium-sized lymphocytes, and do not have the morphology of the typical plasma cell (Nossal and Ada, 1971). As the classical plasma cell morphology probably has little functional significance, the broader use of the term to encompass any antibody-secreting cell appears more appropriate.

3.3.9 Affinity Maturation and B cell Memory

The generation of antibody-forming cells and memory cells is intimately associated with a process in which somatic hypermutation of antibody variable region genes is coupled to antigen-mediated B cell selection, and which leads to the emergence of clones secreting higher and higher affinity antibody.

Affinity maturation begins very early in immune responses, perhaps as early as 3–4 days after the initiation of the primary response (Gray, 1993), and coincides with the switch from IgM to other classes, particularly IgG. The enzymic mechanisms by which the variable regions of the genes of the expressed antibody are hyper-mutated are still not well defined (Sharpe *et al.*, 1991; Betz *et al.*, 1994; see Chapter 6), but the anatomical site where somatic hypermutation in B cells occurs is now known to be the germinal centres of secondary lymphoid organs (Küppers *et al.*, 1993; Gray, 1993; MacLennan, 1994).

Somatic hypermutation in the follicles is essentially random with regard to antigen specificity. The mutated receptors could have either lower or higher affinity for antigen than the original unmutated receptors, or could even acquire reactivity to self, with the potential for autoimmunity (see Berek, 1993). The latter possibility suggests that there may be a need for a further round of tolerance induction at this stage (Nossal, 1995). Mutated B cells compete for the small amounts of antigen that are attached to the follicular dendritic cells, with the highest affinity cells competing most successfully, and the less successful cells dying of apoptosis. Activation of high-affinity B cells by the binding of antigen to membrane immunoglobulin, together with a signal from the CD40 molecule (Gray *et al.*, 1994, Foy *et al.*, 1994), appears to

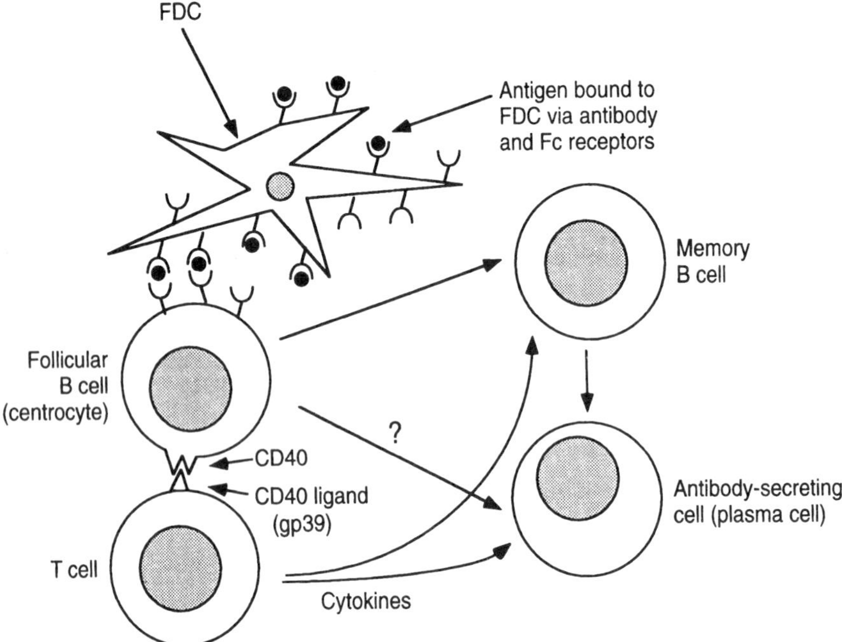

Fig. 3.3. Cellular mechanism of antibody affinity maturation. B cells in the germinal centres compete for binding to small amounts of antigen attached to follicular dendritic cells (FDC) via antibody and Fc receptors. T cells and the binding of the CD40 ligand to the CD40 molecule on B cells are also required. B cells with the highest affinity receptors survive, while lower-affinity B cells die of apoptosis.

rescue them from apoptosis (Nunez *et al.*, 1991; Gray, 1993; Küppers *et al.*, 1993; MacLennan, 1994).

The choice between life and death of individual B cells is correlated with the transient turning off of the *bcl-2* gene during this critical selection window. However, the role of the *bcl-2* gene in this process is still unclear. While it has been shown that lower-affinity B cells that would ordinarily die can be 'rescued' by fusion with plasmacytomas transfected with the *bcl-2* gene (Ray and Diamond, 1994), affinity maturation proceeds fairly normally in mice in which the *bcl-2* gene is constitutively expressed in the lymphoid follicles, suggesting that it is not the down-regulation of the *bcl-2* gene itself that mediates negative selection of lower-affinity B cells (Smith *et al.*, 1994). The so-called 'tingible body macrophages' which are a prominent feature of germinal centres (Nossal and Ada, 1971) contain the remains of apoptotic B cells that have not bound antigen with sufficient affinity.

The occurrence of a much more rapid, vigorous and high-affinity antibody response after a second encounter with antigen is termed 'immunological memory' (reviewed by Gray, 1993). As mentioned earlier, it was widely believed that memory involved the production of long-lived memory B and T cells which could survive for months or years without antigen. This notion was difficult to prove rigorously, however, because it was hard to be sure that all antigen had been eliminated. More recently, the conventional view of long-lived memory B lymphocytes has come under serious challenge. There is now a considerable body of evidence that memory is only sustained by the presence of antigen, whether it is the original antigen or incidental environmental antigens that are closely related to the original antigen and cross-react with it (Gray, 1993; Matzinger, 1994 and references therein; Sprent, 1994). When rigorous measures are taken to eliminate residual antigen, immunological memory tends to fade over a period of weeks. The immune system may be rather like a library with limited space for books; old books must be discarded in order to make way for the new (Matzinger, 1994).

Although the term 'virgin B cell' is often used, there is currently no way to tell that a B cell has not encountered antigen. There is no unambiguous way of identifying virginity. What, then, is a memory B cell? A memory B cell has been defined by Kocks and Rajewsky (1989) as 'a surface immunoglobulin expressing B lymphocyte which has been selected in a pathway of antigen-driven proliferation and somatic hypermutation and is ready to produce a secondary response upon antigenic challenge'. The relationships between memory cells, activated B cells and antibody-secreting cells are not yet completely clear. B cells can differentiate into antibody-secreting cells, but these are thought to be end cells incapable of further division. Memory B cells are most unambiguously identifiable by the presence of mutated antibody variable genes (Nicholson *et al.*, 1995). At present, there is no universally accepted surface marker for memory B cells, although the memory-containing population seems to be contained mainly or exclusively in the IgD^- population (Nicholson *et al.*, 1995).

3.3.10 Antibody Class Switching

About 3–4 days after the initial encounter with antigen, the B cell response begins to switch away from IgM to IgG and other antibody classes. Class switching occurs in the lymphoid follicles, and is closely associated with somatic hypermutation and affinity maturation of antibody classes other than IgM.

In addition to antigen, class switching requires the binding of the CD40 ligand to CD40 (Villa *et al.*, 1994; Geha and Rosen, 1994; Renshaw *et al.*, 1994) and the presence of cytokines from T cells (Finkelman *et al.*, 1990; Coffman *et al.*, 1993). The subclass to which the B cell switches is heavily influenced by the cytokine type, and also depends on the species of animal. In the mouse, switching to IgG1 and IgE is favoured by interleukin-4 (IL-4), which tends to suppress IgG2b and IgG3. Switching to IgG2a is favoured by interferon-γ (IFN-γ), while switching to IgA appears to be favoured by transforming growth factor β (TGF-β) (Finkelman *et al.*, 1990; Coffman *et al.*, 1993).

Although the distinction is not completely clear-cut, T cells can be divided into Th1, which tend to secrete IL-2 and IFN-γ, and Th2, which tend to secrete IL-4, 5, 10 and 13 (see Mosmann and Coffman, 1989, 1992). There is some evidence that Th1 cells tend to favour help for IgG2a, while Th2 cells tend to help the production of IgG1 and IgE. Freund's complete adjuvant is said to be a strong activator of Th1 cells, while alum tends to activate Th2 cells (Grun and Maurer, 1989; Audibert and Lise, 1993).

The molecular mechanism of antibody class switching involves site-specific genetic recombination events at the switch regions that lie just 5′ of each constant region gene, and appears to require a class switch control region at the 3′ end of the immunoglobulin heavy chain locus (Esser and Radbruch, 1990; Cogne *et al.*, 1994; see Chapter 6).

3.4 T Cells

3.4.1 Function of T Cells

The main function of T cells is regulation of other cell types. T cells can be subdivided on the basis of the surface proteins CD4 and CD8. Mature T cells possess CD4 or CD8 but not both. $CD4^+$ T cells act by secreting cytokines that act on many other cell types. $CD8^+$ T cells are capable of differentiating into cytotoxic killer cells, which may be important in the elimination of virally infected cells and foreign grafts.

Most antigens cannot activate B cells to produce antibodies without the help of T cells and are therefore said to be *T dependent*. As mentioned earlier, the way that T cells recognize antigen is fundamentally different to that of B cells. In order to understand the production of antibodies, we must explore their development and the mechanisms by which T cells are activated.

3.4.2 The T Cell Receptor for Antigen

The great majority of T cells possess antigen receptors with variable 'recognition subunits' that consist of disulfide-linked α and β chains, each of molecular weight about 40 000 daltons. A small minority of T cells possess similar antigen receptors made up of disulfide-bonded γ and δ chains. In each case, the recognition subunits of the T cell receptor are noncovalently associated with a group of proteins known as the CD3 complex and the ζ (zeta) chain, which is involved in signal transduction via protein kinases in a manner similar to membrane immunoglobulin on B cells (Weiss, 1993; Weiss and Littman, 1994).

Binding of antigen leads to activation of protein tyrosine kinases that are associated with the receptor, and thence to a series of downstream events, particularly the phosphorylation and activation of phospholipase C-γ, which leads in turn to the cleavage of phosphatidyl inositol into a phosphorylated sugar (inositol tris phosphate) which binds to a receptor on the endoplasmic reticulum, causing calcium to be released, and a lipid (diacyl glycerol) which binds to and activates protein kinase C.

T cells that emerge from the thymus-bearing αβ T cell receptors and either CD4 or CD8 are specialized to be helper or cytotoxic precursor cells, respectively. However, a minor proportion of T cells (1–10% in mice and humans; up to 30% in ruminants and chickens) emerge with γδ receptors (Strominger, 1989; Raulet, 1989; Hein and Mackay, 1991; Havran and Boismenu, 1994). Many of these γδ T cells lack both CD4 and CD8. They are particularly prominent in skin, uterus, gut and lungs, suggesting a possible defensive role in dealing with microbial invaders at these interfaces. The specificity of γδ T cells is not yet completely understood and there is some evidence that they may not behave in a major histocompatibility complex (MHC)-restricted manner (see below). Their biological role is almost completely unknown (Strominger, 1989; Raulet, 1989; Allison and Havran, 1991; Haas *et al.*, 1993). An interesting possibility is that they recognize carbohydrate or other nonpeptide antigens (see Chapter 4).

3.4.3 Antigen Recognition by T Cells Differs Fundamentally from B Cells: T Cells Recognize Short Linear Peptide Sequences

In the early 1970s, it was discovered that T cells do not see antigen alone, but in intimate association with products of the major histocompatibility complex (MHC; Katz *et al.*, 1973; Rosenthal and Shevach, 1973; Zinkernagel and Doherty, 1974, 1977). This phenomenon has been termed 'MHC restriction'. Its structural basis was solved in 1987 when the three-dimensional structure of class I histocompatibility antigens was elucidated by Bjorkman *et al.* Histocompatibility antigens comprise a 'platform' with a groove bounded by two α-helices and a floor made up of β-pleated sheet. The groove is capable of binding short peptides of about eight amino acids in length. Subsequently, the structure of the class II MHC proteins was found to be very similar, but the bound peptides are slightly longer (Brown *et al.*, 1993). In each case, the peptide is in an extended conformation within the groove (Madden *et al.*, 1991), with some residues bound in hydrophobic pockets in the floor of the groove (anchor residues) and others protruding upwards, and available for recognition by the T cell receptor.

In both class I restricted (cytotoxic) and class II-restricted (helper) T cells, the T cell receptor recognizes a complex surface bounded by the α-helices of the MHC, and containing the peptide in the central groove, like the sausage in a hot-dog roll. MHC molecules can thus be seen as 'peptide receptors' (Rammensee *et al.*, 1993; Engelhard, 1994).

These findings have major implications for the nature of antigenicity. First, the native conformation of protein antigens is probably only relevant to T cell recognition to the extent that it influences proteolytic processing of the antigen. Second, the question must be asked whether T cells are capable of recognizing nonpeptide antigens (Chapter 4). Third, a physical explanation for the MHC-linked immune response genes is apparent (see Section 3.4.6). Only a subset of peptides from any protein will bind to the groove, and if a protein is not capable of generating a peptide that binds to the groove in a particular MHC type, individuals with that MHC type will be genetic nonresponders to that antigen.

MHC class I molecules are present on the surface of virtually all nucleated cells. Class II proteins are present on the surface of B lymphocytes, some macrophages, Langerhans cells in skin, and a few other cell types, and can often be induced in cells that normally do not express class II by the cytokine IFN-γ. The mechanism by which peptides enter the groove is different for MHC class I and class II MHC proteins (reviewed by Germain and Margulies, 1993; Neefjes and Momburg, 1993).

In the case of class I molecules, large cytoplasmic multi-subunit structures known as proteasomes break up cytoplasmic proteins into small peptides, which are transported via specific transporter molecules into the lumen of the endoplasmic reticulum, where they bind to the MHC class I molecules. Interestingly, the binding of peptide is required for the stable assembly of MHC molecules (Townsend *et al.*, 1990; Bijlmakers and Pleogh, 1993). From this mechanism of loading, it is apparent that class I MHC molecules will present to the T cell any peptides that are synthesized within the cell, including intracellular pathogens such as viruses. *In other words, class I MHC molecules allow T cells to 'see' what is going in inside cells, and to respond to intracellular pathogens.* Peptide:MHC class I complexes are mainly seen by $CD8^+$ T cells, which upon activation differentiate into cytotoxic T cells capable of lysing the cells that they recognize, and hence destroying the source of viruses by killing the cell that makes them.

The peptide transporters that carry peptides from the cytoplasm into the endoplasmic reticulum seem to have some specificity for individual peptide sequences and polymorphism of these transporters may produce differences in the availability of individual antigens for recognition by T cells. This is a further source of gaps in the T cell repertoire, where lack of response may be due to failure of transport of individual peptides (reviewed by Monaco, 1992).

The loading mechanism for MHC class II molecules is quite different to that of class I molecules. Proteins from *outside* the cell are internalized by fluid-phase endocytosis, endocytosis facilitated by binding to membrane immunoglobulin on B cells, or endocytosis of antigen–antibody complexes bound to Fc receptors. They are then processed into short peptides in specialized acidic endosomal structures in which the class II MHC molecules are loaded. The loaded MHC class II molecules are then transported to the cell surface, where they are seen by $CD4^+$ helper T cells, which upon activation secrete cytokines that help activate B cells and induce antibody class switching.

3.4.4 T Cell Development and Tolerance

The development of T cells in the thymus is intimately associated with the acquisition of self-tolerance. However, as pointed out by Sercarz and Datta (1994), a significant proportion of T cells directed against self determinants emerge from the thymus without becoming tolerant (see below). It is presumed that there are additional peripheral fail safe mechanisms of tolerance induction to prevent these cells from producing autoimmunity. Unlike B cells, T cells do not undergo somatic hypermutation, which presumably would be much too dangerous because of the resultant risk of autoimmunity. The classical neonatal thymectomy experiment of Miller (1961) showed that the thymus is the source of T cells and that there is no significant movement of T cells out of the thymus before birth in the mouse. Neonatal thymectomy results in severe depletion of peripheral T cells, with profound impairment in skin graft rejection and greatly reduced antibody responses to T cell-dependent antigens. Let us now consider what happens in the thymus.

Cells destined to become T cells migrate from the bone marrow to the thymus, where they undergo a stepwise series of maturation events. The earliest immigrants are found in the subcapsular region and are negative for most T cell markers, although they may have low levels of CD4, which is then lost, generating 'triple negative' cells ($CD3^-$, $CD4^-$, $CD8^-$). They then migrate into the cortex and asynchronously acquire both CD4 and CD8 to become 'double positives'. T cell receptor expression, as assessed by expression of CD3 or $\alpha\beta$ or $\gamma\delta$ chains, occurs soon afterwards. From here, T cells begin their education by antigen, and are subject to *positive selection* for lower affinity for self MHC, which results in cell survival by a *bcl-2*-independent mechanism, *negative selection* for excessively high affinity for self MHC, resulting in cell death by apoptosis, and movement to the medulla with concomitant loss of either CD4 or CD8. If the T cell receptor 'sees' antigen in conjunction with MHC class I, the cell will retain CD8 and lose CD4; if it 'sees' antigen in conjunction with MHC class II, it will retain CD4 and lose CD8.

The details of the selective process in the thymus are still under debate (see von Boehmer and Rajewsky, 1993), but the net outcome is that mature $CD4^+$ or $CD8^+$ T cells emerge into the peripheral lymphoid organs, largely depleted of cells with strong autoreactivity. It appears unlikely that all possible self antigens would be present in the thymus at sufficient concentrations to guarantee complete tolerance to self and there is evidence that a further round of self–nonself discrimination may occur in the periphery (see Miller and Morahan, 1992).

There is evidence that the acquisition of self-tolerance to peptides in the thymus profoundly affects the available T cell repertoire in the periphery (Pullen *et al.*, 1988), resulting in further nonresponsiveness to individual antigens. Thus, gaps of nonresponsiveness in T cells can arise from at least three different sources: clonal deletion caused by self tolerance, the inability of certain MHC alleles to bind certain peptides (see Section 3.4.6), and the failure of peptide transporters to transport certain peptides.

3.4.5 Mechanism of Activation of T Cells: Antigen-Presenting Cells

Activation of T cells requires several signals (Fig. 3.2). First, there is a requirement for an antigen-specific signal to be delivered to the T cell receptor. This signal requires a suitably processed peptide, bound to an MHC molecule, and a T cell with a receptor with appropriate specificity to receive such a signal. While the potential repertoire for T cells is random, the massive amount of selection undergone in the thymus and possibly also in the periphery effectively edits this repertoire to a much smaller size in the interests of preventing autoimmunity.

A second obligatory signal for T cell activation is via the CD28 molecule, which must bind to a complementary ligand (B7 or CTLA-4) on an antigen-presenting cell, such as a B cell, macrophage, dendritic cell or other cell type expressing B7 or CTLA-4 (B7-2) (Linsley and Ledbetter, 1993; Allison, 1994; Caux *et al.*, 1994). If CD28 ligation does not occur, an encounter between the T cell and antigen may cause inactivation rather than activation.

Many different cell types may present antigen to lymphocytes but some cells may be regarded as 'professional antigen-presenting cells'. B cells, macrophages and Langerhans cells in the skin are capable of presenting antigen to T cells. 'Non-professional' cells may be able to present antigen to T cells under certain circumstances, particularly if expression of the relevant accessory molecules are induced by inflammatory cytokines such as IFN-γ. Some antigen-presenting cells may be highly specialized, such as follicular dendritic cells, which are specifically dedicated to presenting antigen in the form of antigen–antibody complexes to B cells in lymphoid follicles.

In summary, T cell activation is a complex process requiring several signals. T cells do not respond to naked antigen, but require highly processed antigen to be presented to them bound to MHC molecules on a cell surface. The antigen-presenting cell must also possess accessory molecules, such as a ligand for the CD28 molecule on the T cell. Additional signals from soluble cytokines may also be required.

3.4.6 Genetic Control of the Immune Response: the Immune Response Genes

When inbred mice are immunized with certain antigens, particularly synthetic peptides or proteins that closely resemble self, it is found that some strains respond and others do not (Benacerraf and McDevitt, 1972). The genes responsible were found to be encoded within the MHC, and have been termed Immune Response (Ir) Genes. They act in a dominant manner, in which heterozygotes between responders and nonresponders are responders.

For many years, it was thought that the MHC-linked Ir genes may encode the T cell receptor. However, it is now known that the T cell receptor is encoded by a group of genes that are distinct from and unlinked to the MHC. The discovery of the phenomenon of MHC restriction (Section 3.4.3) and the elucidation of the structure and peptide-binding properties of the MHC molecules has made it clear that the Ir gene products are the histocompatibility antigens themselves, and that the MHC molecules act as receptors for peptides (Rammensee *et al.*, 1993; Engelhard, 1994). Responsiveness or nonresponsiveness reflects the binding or lack of binding of individual peptides to individual MHC molecules.

From a practical point of view, the importance of the MHC-linked immune response genes is that occasionally it may be found that a particular inbred strain of mouse may not respond to a particular antigen. This problem may usually be solved by simply immunizing a different strain of mouse with a different MHC type.

Genetic factors are also important in controlling the immune response in a nonspecific way. While hard data are scarce, there is a general feeling among immunologists that SJL, 129/J and A/J mice are particularly good responders, while C57BL mice are poorer responders. BALB/c mice are probably somewhere in between.

All mouse myeloma cells commonly used for hybridoma production are of BALB/c origin and it is generally easiest to use BALB/c mice as the spleen donor. If an adequate response cannot be obtained, consideration should be given to varying the strain of mice for immunization. However, strain variation is one of the least likely causes of failure of production of monoclonal antibodies.

3.5 Summary—Cellular Mechanisms in Antibody Production

In this Chapter, we have reviewed the complex web of cellular interactions that lead to the immune response. We have seen that the immune response represents the outcome of interactions between antigen-presenting cells, T cells and B cells. For most antigens, there is an obligatory requirement for T cell help, which in turn requires processing of antigen and presentation of peptides in the groove of MHC molecules. Activation of lymphocytes also requires signalling via accessory molecules, CD40 in the case of B cells and CD28 in the case of T cells. T cells possess gaps of nonresponsiveness in their repertoire, owing to deletion of some clones by self-tolerance, failure of some MHC molecules to bind certain peptides, and failure to transport some peptides from the cytoplasm to the lumen of the endoplasmic reticulum.

In the following chapter, we will explore the features of antigen that determine the outcome of encounters with the immune system.

References

Allison, J. P. (1994) CD28-B7 interactions in T-cell activation. *Curr. Op. Immunol.* **6**, 414–419.

Allison, J. P. and Havran, W. L. (1991) The immunobiology of T cells with invariant γδ antigen receptors. *Annu. Rev. Immunol.* **9**, 679–705.

Anfinsen, C. B. (1973) Principles that govern the folding of protein chains. *Science* **181**, 223–230.

Audibert, F. M. and Lise, L. D. (1993) Adjuvants: current status, clinical perspectives and future prospects. *Immunol. Today* **14**, 281–284.

Banchereau, J., Bazan, F., Blanchard, D., Briere, F., Galizzi, J. P., van Kooten, C., Liu, Y. J., Rousset, F. and Saeland, S. (1994) The CD40 antigen and its ligand. *Annu. Rev. Immunol.* **12**, 881–922.

Beadle, G. W. and Tatum, E. L. (1941) Genetic control of biochemical reactions in *Neurospora*. *Proc. Natl Acad. Sci. USA* **27**, 499–506.

Benacerraf, B. and McDevitt, H. O. (1972) Histocompatibility-linked immune response genes. *Science* **175**, 273–279.

Berek, C. (1993) Somatic mutation and memory. *Curr. Op. in Immunol.* **5**, 218–222.

Benjamin, D. C., Berzofsky, J. A., East, I. J., Gurd, F. R. N., Hannum, C., Leach, S. J., Margoliash, E., Michael, J. G., Miller, A., Prager, E. M., Reichlin, M., Sercarz, E. E., Smith-Gill, S. J., Todd, P. E. and Wilson, A. C. (1984). The antigenic structure of proteins: a reappraisal. *Annu. Rev. Immunol.* **2**, 67–101.

Betz, A. G., Milstein, C., Gonzalez-Fernandez, A., Pannell, R., Larson, T. and Neuberger, M. S. (1994) Elements regulating somatic hypermutation of an immunoglobulin κ gene: critical role for the intron enhancer/matrix attachment region. *Cell* **77**, 239–248.

Bijlmakers, M-J. and Pleogh, H. L. (1993) Putting together an MHC class I molecule. *Curr. Op. Immunol.* **5**, 21–26.

Bjorkman, P. J., Saper, M. A., Samraoui, B., Bennett, W. S., Strominger, J. L. and Wiley, D. C. (1987) Structure of the human class I histocompatibility antigen, HLA-A2. *Nature* **329**, 506–512.

Bourgois, A., Kitajima, K., Hunter, I. R. and Askonas, B. A. (1977). Surface immunoglobulins of lipopolysaccharide stimulated spleen cells. The behaviour of IgM, IgD and IgG. *Eur. J. Immunol.* **7**, 151–153.

Brown, J. H., Jardetzky, T. S., Gorga, J. C., Stern, L. J., Urban, R. G., Strominger, J. L. and Wiley, D. C. (1993) Three-dimensional structure of the human class II histocompatibility antigen HLA-DR1. *Nature* **364**, 33–39.

Burnet, F. M. (1957). A modification of Jerne's theory of antibody production using the concept of clonal selection. *Austral. J. Sci.* **20**, 67–69.

Burnet, F. M. (1959) 'The Clonal Selection Theory of Acquired Immunity'. Cambridge University Press, Cambridge.

Burnet, F. M. (1968). 'Changing Patterns: An Atypical Autobiography.' William Heinemann, London.

Cambier, J. C., Pleiman, C. M. and Clark, M. R. (1994) Signal transduction by the B cell receptor and its coreceptors. *Annu. Rev. Immunol.* **12**, 457–486.

Caux, C., Vanbervliet, B., Massacrier, C., Azuma, M., Okumura, K., Lanier, L. L. and Banchereau, J. (1994) B70/B7-2 is identical to CD86 and is the major functional ligand for CD28 expressed on human dendritic cells. *J. Exp. Med.* **180**, 1841–1848.

Chen, J. and Alt, F. W. (1993) Gene rearrangement and B-cell development. *Curr. Op. Immunol.* **5**, 194–200.

Clark, E. A. and Ledbetter, J. A. (1994) How B and T cells talk to each other. *Nature* **367**, 425–428.

Coffman, R. L., Lebman, D. A. and Rothman, P. (1993) The mechanism and regulation of immunoglobulin isotype switching. *Adv. Immunol.* **54**, 229–270.

Cogne, M., Lansford, R., Bottaro, A., Zhang, J., Gorman, J., Young, F., Cheng, H-L. and Alt, F. W. (1994) A class switch control region at the 3′ end of the immunoglobulin heavy chain locus. *Cell* **77**, 737–747.

Cyster, J. G., Hartley, S. B. and Goodnow, C. C. (1994) Competition for follicular niches excludes self-reactive cells from the recirculating B-cell repertoire. *Nature* **371**, 389–395.

Dreyer, W. J. and Bennett, J. C. (1965) The molecular basis of antibody formation, a paradox. *Proc. Natl Acad. Sci. USA* **54**, 864–869.

Ehrlich, P. (1900) Croonian Lecture: On immunity with special reference to cell life. *Proc. Roy. Soc.* **66**, 424–448.

Engelhard, V. H. (1994) How cells present antigens. *Scientific American* **271**, 44–51.

Esser, C. and Radbruch, A. (1990) Immunoglobulin class switching: molecular and cellular analysis. *Annu. Rev. Immunol.* **8**, 717–785.

Finkelman, F. D., Holmes, J., Katona, I. M., Urban, F. F., Beckmann, M. P., Park, L. S., Schooley, K. A., Coffman, R. L., Mosmann, T. R. and Paul, W. E. (1990) Lymphokine control of *in vivo* immunoglobulin isotype selection. *Annu. Rev. Immunol.* **8**, 303–333.

Forster, I. and Rajewsky, K. (1990) The bulk of the peripheral B cell pool in mice is stable and not rapidly renewed from the bone marrow. *Proc. Natl Acad. Sci. USA* **87**, 4781–1784.

Foy, T. M., Laman, J. D., Ledbetter, J. A., Aruffo, A., Claassen, E. and Noelle, R. J. (1994) gp39-CD40 interactions are essential for germinal center formation and the development of B cell memory. *J. Exp. Med.* **180**, 157–164.

Geha, R. S. and Rosen, F. S. (1994) The genetic basis of immunoglobulin-class switching. *New Eng. J. Med.* **330**, 1008–1009.

Germain, R. N. and Margulies, D. H. (1993) The biochemistry and cell biology of antigen processing and presentation. *Annu. Rev. Immunol.* **11**, 403–450.

Goding, J. W. and Layton, J. E. (1976) Antigen-induced co-capping of IgM and IgD-like receptors on murine B cells. *J. Exp. Med.* **144**, 852–857.

Goodnow, C. C. (1989) Cellular mechanisms of self-tolerance. *Curr. Op. Immunol.* **2**, 226–236.

Goodnow, C. C. (1992) Transgenic mice and analysis of B-cell tolerance. *Annu. Rev. Immunol.* **10**, 489–518.

Goodnow, C. C., Crosbie, J., Adelstein, S., Lavoie, T. B., Smith-Gill, S. J., Mason, D. Y., Jorgensen, H., Brink, R. A., Pritchard-Briscoe, H., Loughnoin, M., Loblay, R. H., Trent, R. J. and Basten, A. (1989) Clonal silencing of self-reactive B lymphocytes in a transgenic mouse model. *Cold Spring Harbor Symp. Quant Biol.* LIV, 907–920.

Gowans, J. L. and McGregor, D. D. (1965) The immunological activities of lymphocytes. *Progr. Allergy* **9**, 1–78.

Gowans, J. L., McGregor, D. D., Cowen, D. M. and Ford, C. E. (1962) Initiation of immune responses by small lymphocytes. *Nature* **196**, 651–655.

Gray, D. (1993) Immunological memory. *Annu. Rev. Immunol.* **11**, 49–77.

Gray, D. (1994) Regulation of immunological memory. *Curr. Op. Immunol.* **6**, 425–430.

Gray, D., Dullforce, P. and Jainandunsing, S. (1994) Memory B cell development but not germinal center formation is impaired by in vivo blockade of CD40-CD40 ligand interaction. *J. Exp. Med.* **180**, 141–156.

Grun, J. L. and Maurer, P. H. (1989) Different T helper subsets elicited in mice utilizing two different adjuvant vehicles: the role of endogenous interleukin 1 in proliferative responses. *Cell. Immunol.* **121**, 134–145.

Gu, H., Tarlinton, D., Muller, W., Rajewsky, K. and Forster, I. (1991) Most peripheral B cells in mice are ligand-selected. *J. Exp. Med.* **173**, 1357–1371.

Haas, W., Pereira, P. and Tonegawa, S. (1993) Gamma/Delta Cells. *Annu. Rev. Immunol.* **11**, 637–685.

Hartley, S. B., Crosbie, J., Brink, R., Kantor, A. B., Basten, A. and Goodnow, C. C. (1991) Elimination from peripheral lymphoid tissues of self-reactive B lymphocytes recognizing membrane-bound antigens. *Nature* **353**, 765–769.

Haughton, G., Arnold, L. W., Whitmore, A. C. and Clarke, S. H. (1993) B-1 cells are made, not born. *Immunol. Today* **14**, 84–91.

Havran, W. L. and Boismenu, R. (1994) Activation and function of γδ T cells. *Curr. Op. Immunol.* **6**, 442–446.

Havran, W. L., Di Giusto, D. L. and Cambier, J. C. (1984). mlgM:mlgD ratios on B cells: mean mIgD expression exceeds mIgM by 10-fold on most splenic B cells. *J. Immunol.* **132**, 1712–1716.

Hein, W. R. and Mackay, C. R. (1991) Prominence of the γδ T cells in the ruminant immune system. *Immunol. Today* **12**, 30–34.

Herzenberg, L. A. and Kantor, A. B. (1993) B-cell lineages exist in the mouse. *Immunol. Today* **14**, 79–83.

Hozumi, N. and Tonegawa, S. (1976) Evidence for somatic rearrangement of immunoglobulin genes coding for variable and constant regions. *Proc. Natl Acad. Sci. USA* **73**, 3628–3632.

Kantor, A. (1993a) A new nomenclature for B cells. *Immunol. Today* **12**, 388.

Kantor, A. B. (1993b) The development and repertoire of B-1 cells (CD5 B cells). *Immunol. Today* **12**, 389–391.

Kantor, A. B. and Herzenberg, L. A. (1993) Origin of B cell lineages. *Annu. Rev. Immunol.* **11**, 501–538.

Karasuyama, H., Rolink, A., Shinkai, Y., Young, F., Alt, F. W. and Melchers, F. (1994) The expression of $V_{pre\text{-}B}/\lambda 5$ surrogate light chain in early bone marrow precursor B cells of normal and B cell-deficient mutant mice. *Cell* **77**, 133–143.

Katz, D. H., Hamaoka, T. and Benacerraf, B. (1973) Cell interactions between histocompatible T and B lymphocytes. Failure of physiologic cooperative interactions between T and B lymphocytes from allogeneic donor strains in humoral responses to hapten-protein conjugates. *J. Exp. Med.* **137** 1405, 1418.

Kim, K-M., Adachi, T.k, Nielsen, P. J., Terashima, M., Lamers, M. C., Köhler, G. and Reth, M. (1994) Two new proteins preferentially associated with membrane immunoglobulin D. *EMBO J.* **13**, 3793–3800.

Kindt, T. J. and Capra, J. D. (1984) 'The Antibody Enigma'. Plenum Press, New York.

Kitamura, D., Kudo, A, Schaal, S., *et al.* (1992) A critical role of λ5 protein in B cell development. *Cell* **69**, 823–831.

Klein, U., Küppers, R. and Rajewsky, K. (1993) Human IgM^+IgD^+ B cells, the major B cell subset in the peripheral blood, express Vκ genes with no or little somatic mutation throughout life. *Eur. J. Immunol.* **23**, 3272–3277.

Kocks, C. and Rajewsky, K. (1989) Stable expression and somatic hypermutation of antibody V genes in B-cell developmental pathways. *Annu. Rev. Immunol.* **7**, 537–559.

Küppers, R., Zhao, M., Hansmann, M-L. and Rajewsky, K. (1993) Tracing B cell development in human germinal centres by molecular analysis of single cells picked from histological sections. *EMBO J.* **12**, 4955–4967.

Linsley, P. S. and Ledbetter, J. A. (1993) The role of the CD28 receptor during T cell responses to antigen. *Annu. Rev. Immunol.* **11**, 191–212.

Machennan, I. C. M. (1994) Germinal centres. *Ann. Rev. Immunol.* **12**, 117–139.

Madden, D. R., Gorga, J. C., Strominger, J. L. and Wiley, D. C. (1991) The structure of HLA-B27 reveals nonamer self-peptides bound in an extended conformation. *Nature* **353**, 321–325.

Matzinger, P. (1994) Memories are made of this? *Nature* **369**, 605–606.

Melchers, F., Karasuyama, H., Haasner, D., Bauer, S., Kudo, A., Sakaguchi, N., Jameson, B. and Rolink, A. (1993) The surrogate light chain in B-cell development. *Immunol. Today* **14**, 60–68.

Melchers, F., Haasnaer, D., Grawunder, U., Kalberer, C., Karasuyama, H, Winkler, T. and Rolink, A. G. (1994) Roles of IgH and L chains and of surrogate H and L chains in the development of cells of the B lymphocyte lineage. *Annu. Rev. Immunol.* **12**, 209–225.

Miller, J. F. A. P. (1961) Immunological function of the thymus. *Lancet* **ii**, 748–749.

Miller, J. F. A. P. and Heath, W. R. (1993) Self-ignorance in the peripheral T cell pool. *Immunol. Rev*. **133**, 131–150.

Miller, J. F. A. P. and Morahan, G. (1992) Peripheral T cell tolerance. *Ann. Rev. Immunol*. **10**, 51–69.

Monaco, J. J. (1992) Antigen presentation. Not so groovy after all? *Curr. Op. Immunol.* **2**, 433–435.

Mosmann, T. R. and Coffman, R. L. (1989) Th1 and Th2 cells: Different patterns of lymphokine secretion lead to different functional properties. *Annu. Rev. Immunol.* **7**, 145–173.

Neefjes, J. J. and Momburg, F. (1993) Cell biology of antigen presentation. *Curr. Op. Immunol.* **4**, 27–34.

Nicholson, I. C., Brisco, M. J. and Zola, H. (1995) Memory B lymphocytes in human tonsil do not express immunoglobulin D. *J. Immunol.* **154**, 1105–1113.

Nitschke, L., Kosco, M. H., Köhler, G. and Lamers, M. C. (1993) Immunoglobulin D-deficient mice can mount normal immune responses to thymus-independent and -dependent antigens. *Proc. Natl Acad. Sci. USA* **90**, 1887–1891.

Nossal, G. J. V. (1992a) The molecular and cellular basis of affinity maturation in the antibody response. *Cell* **68**, 1–2.

Nossal, G. J. V. (1992b) Cellular and molecular mechanisms of B cell tolerance. *Adv. Immunol.* **52**, 283–332.

Nossal, G. J. V. (1994) Negative selection of lymphocytes. *Cell* **76**, 229–239.

Nossal, G. J. V. (1995) Choices following antigen entry: Antibody formation or immunological tolerance? *Ann. Rev. Immunol.* **13**, 1–27.

Nossal, G. J. V. and Ada, G. L. (1971) 'Antigens, Lymphoid Cells, and the Immune Response'. Academic Press, New York.

Nossal, G. J. V. and Lederberg, J. (1958) Antibody production by single cells. *Nature* **181**, 1419–1420.

Nunez, G., Hockenberry, D., McDonnell, T. J., Sorenson, C. M. and Korsmeyer, S. J. (1991) Bcl-2 maintains B cell memory. *Nature* **353**, 71–73.

Parker, D. C. (1993) T cell-dependent B-cell activation. *Annu. Rev. Immunol.* **11**, 331–360.

Pauza, J. E., Rehmann, J. A. and LeBien, T. W. (1993) Unusual patterns of immunoglobulin gene rearrangement and expression during human B cell ontogeny: human B cells can simultaneously express cell surface κ and λ light chains. *J. Exp. Med.* **178**, 139–149.

Peaker, C. J. G. (1994) Transmembrane signalling by the B-cell antigen receptor. *Curr. Op. Immunol.* **6**, 359–363.

Pleiman, C. M., Abrams, C., Timson Gauen, L., Bedzyk, W., Jongstra, J., Shaw, A. S. and Cambier, J. C. (1994) Distinct p53/56^{lyn} and p59^{fyn} domains associate with non-phosphorylated and phosphorylated Ig-α. *Proc. Natl Acad. Sci. USA* **91**, 4268–4272.

Pullen, A. M., Marrack, P. and Kappler, J. W. (1988) The T-cell repertoire is heavily influenced by tolerance to polymorphic self-antigens. *Nature* **335**, 796–801.

Rammensee, H-G., Falk, K. and Rotzschke, O. (1993) MHC molecules as peptide receptors. *Curr. Op. Immunol.* **5**, 35–44.

Raulet, D. H. (1989) The structure, function and molecular genetics of the γδ T cell receptor. *Annu. Rev. Immunol.* **7**, 175–207.

Ray, S. and Diamond, B. (1994) Generation of a fusion partner to sample the repertoire of splenic B cells destined for apoptosis. *Proc. Natl Acad. Sci. USA* **91**, 5548–5551.

Renshaw, B. R., Fanslow, W. C. III, Armitage, R. J., Campbell, K. A., Liggitt, D., Wright, B., Davison, B. L. and Maliszewski, C. R. (1994) Humoral immune responses in CD40 ligand-deficient mice. *J. Exp. Med.* **180**, 1889–1900.

Reth, M. (1994) B cell antigen receptors. *Curr. Op. Immunol.* **6**, 3–8.

Roes, J. and Rajewsky, K. (1993) Immunoglobulin D (IgD)-deficient mice reveal an auxiliary receptor function for IgD in antigen-mediated recruitment of B cells. *J. Exp. Med.* **177**, 45–55.

Rolink, A., Grawunder, U., Haasner, D., Strasser, A. and Melchers, F. (1993) Immature surface Ig+ B cells can continue to rearrange κ and λ chain gene loci. *J. Exp. Med.* **178**, 1263–1270.

Rosenthal, A. S. and Shevach, E. M. (1973) Function of macrophages in antigen recognition by guinea pig T lymphocytes. I. Requirement for histocompatible macrophages and lymphocytes. *J. Exp. Med.* **138**, 1194–1212.

Sanger, F. and Thompson, E. O. P. (1953) The amino acid sequence in the glycyl chain of insulin. *Biochem. J.* **53**, 353–374.

Schroeder, H. W. and Dighiero, G. (1994) The pathogenesis of chronic lymphocytic leukemia: analysis of the antibody repertoire. *Immunol. Today* **15**, 288–294.

Sercarz, E. E. and Datta, S. K. (1994) Mechanisms of autoimmunization: perspective from the mid–90s. *Curr. Op. in Immunol.* **6**, 875–881.

Sharpe, J. J., Milstein, C., Jarvis, J. M. and Neuberger, M. S. (1991) Somatic mutation of immunoglobulin κ may depend on sequences 3′ of $C_κ$ and occurs on passenger transgenes. *EMBO J.* **10**, 2139–2145.

Silverstein, A. M. (1989) 'A History of Immunology'. Academic Press, San Diego.

Smith, K. G. C., Weiss, U., Rajewsky, K., Nossal, G. J. V. and Tarlinton, D. M. (1994) Bcl–2 increases memory B cell recruitment but does not perturb selection in germinal centers. *Immunity* **1**, 803–813.

Sprent, J. (1994) T and B memory cells. *Cell* **76**, 315–322.

Strominger, J. L. (1989) The γδ T cell receptor and class Ib MHC-related proteins: enigmatic molecules of immune recognition. *Cell* **57**, 895–898.

Talmage, D. W. (1995) Origins of cell selection theories of antibody formation. *In* 'Immunology. The Making of a Modern Science', Gallagher, R. B., Gilder, J., Nossal, G. J. V. and Salvatore, G. (eds), pp. 23–38. Academic Press, London.

Terashima, M., Kim, K-M., Adachi, T., Nielsen, P. J., Reth, M., Köhler, G. and

Lamers, M. C. (1994) The IgM antigen receptor of B lymphocytes is associated with prohibitin and a prohibitin-related protein. *EMBO J.* **13**, 3782–3792.

Tonegawa, S. (1983) Somatic generation of antibody diversity. *Nature* **302**, 575–581.

Townsend, A., Elliott, T., Cerundolo, V., Foster, L., Barber, B. and Tse, A. (1990) Assembly of MHC class I molecules analyzed *in vitro*. *Cell* **62**, 285–295.

Villa, A., Notarangelo, L. D., Di Santo, J. P., Macchi, P. P., Strina, D., Frattini, A., Lucchini, F., Patrosse, C. M., Giliani, S., Mantuano, E., Agosti, S., Nocera, G., Kroczek, R. A., Fischer, A., Ugazio, A. G., Basile, G. de S. and Vezzoni, P. (1994) Organization of the human CD40L gene: Implications for molecular defects in X-chromosome-linked hyper-IgM syndrome and prenatal diagnosis. *Proc. Natl Acad. Sci. USA* **91**, 2110–2114.

von Boehmer, H. (1994) Positive selection of lymphocytes. *Cell* **76**, 219–228.

von Boehmer, H. and Rajewsky, K. (1993) Lymphocyte development. *Curr. Op. Immunol.* **5**, 175–176.

Weiss, A. (1993) T cell antigen receptor signal transduction: a tale of tails and cytoplasmic protein–tyrosine kinases. *Cell* **73**, 209–212.

Weiss, A, and Littman, D. R. (1994) Signal transduction by lymphocyte antigen receptors. *Cell* **76**, 263–274.

Zinkernagel. R. M. and Doherty, P. C. (1974) Restriction of *in vitro* T cell-mediated cytotoxicity in lymphocytic choriomeningitis within a syngeneic or semiallogeneic system. *Nature* **248**, 701–702.

Zinkerngel, R. M. and Doherty, P. C. (1979) MHC-restricted cytotoxic T cells: Studies on the biological role of polymorphic major transplantation antigens determining T-cell restriction-specificity, function, and responsiveness. *Adv. Immunol.* **27**, 51–177.

4 Nature of Antigens

An antigen may be defined as a substance that is seen by the immune system. However, the initiation and the effector phases of the immune response do not always coincide. It is possible for an antigen to show a strong immune reaction with antibodies or T cells, and yet the same substance, on its own, may not be capable of initiating the response. It is therefore necessary to introduce the term *immunogenicity*, which refers to the extent to which an antigen is capable of *initiating* an immune response. All immunogens are antigenic, but not all antigens are immunogenic.

In order to be immunogenic, some antigens must be coupled to macromolecular 'carriers', usually proteins. This is especially the case for small molecules such as drugs or other compounds of molecular weight less than a few thousand. Such small molecules are termed *haptens*. To be immunogenic, an antigen must also be capable of being seen by the immune system. There must be lymphocytes with appropriate receptors, and these lymphocytes must be capable of being activated by the antigen. In many situations, lymphocytes will fail to respond because of self tolerance.

In the case of B cells, it appears likely (although there is still no firm proof) that cross-linking of membrane immunoglobulin is essential for activation. Polymeric structures are particularly strong immunogens for B cells, and immunogenicity of an antigen is often enhanced by rendering it polymeric by self- or antibody-induced aggregation. In addition, at least for secondary responses of classes other than IgM, B cells require a signal through the CD40 molecule (see Chapter 3).

T cell activation requires a complex series of steps, including proteolytic processing of antigen followed by the binding of short antigenic peptides in the grooves of class I or class II MHC molecules. Gaps in the T cell repertoire may result from the failure of some peptides to be transported from the cytoplasm into the endoplasmic reticulum where loading of class I molecules occurs (Monaco, 1992). Other holes of nonresponsiveness are caused by the fact that only certain peptides will bind to certain alleles of certain MHC products (Engelhard, 1994a, b). The structural requirements for a given peptide to bind to MHC products, which is essential for that peptide to be seen by T cells, have been termed *agretopes* (*an*tigen *re*striction *to*pes). In addition to the requirement for appropriate agretopes on the antigen, the T cell response may have a third series of gaps of nonresponsiveness because many T cell specificities will have been eliminated during the acquisition of self tolerance (Pullen *et al.*, 1988). Finally, T cells require accessory signals via the CD28 molecule, in an analogous manner to the CD40 system for B cells (reviewed by Linsley and Ledbetter, 1993). Stimulation with antigen alone, without the CD28 signal, may induce antigen-specific nonresponsiveness or anergy.

In addition to all the factors listed above, to be immunogenic, a substance must usually be nonself to avoid nonresponse due to self tolerance. Immunogenicity is therefore a property that depends on both the structure and chemical nature of the substance, and on the mechanisms by which lymphocytes are tolerized and activated.

4.1 What Types of Substances are Antigenic?

An extremely wide range of substances may be antigenic. Perhaps the most common antigens are proteins. The great majority of the exposed surface of proteins is immunogenic (Benjamin *et al.*, 1984). For virtually any protein antigen, many different antibodies may bind. The binding site for an antibody or T cell receptor is known as an *antigenic determinant* or *epitope*. The greater the difference in sequence of a protein from self, the more antigenic it is likely to be, and the greater the number of epitopes.

Carbohydrates are often antigenic and capable of eliciting an immune response, particularly if they are polymeric or are attached to a to protein (glycoproteins) or lipid (glycolipids).

In essence, virtually any substance is potentially antigenic. The more it differs from self, the more antigenic it is likely to be. However, the 'rules of the game' are somewhat different for very small molecules such as hormones or vitamins. Tolerance to these is often weak or nonexistent, and antibodies to virtually any hormone or other small compound such as vitamins, drugs, dyes and other synthetic molecules can usually be generated if immunization is carried out with the compound coupled to an immunogenic carrier.

While most of this book concentrates on the production of antibodies, it should be pointed out that a lack of immunogenicity is often highly desirable, for example in artificial joints or heart valves, drugs or intravenous material such as blood transfusions, macromolecular plasma expanders used to treat shock, or antibodies used to neutralize drug overdoses. While there are no hard and fast rules, simple substances such as stainless steel or teflon are often nonimmunogenic, perhaps because they are very insoluble in water and/or because they have little tendency to form complexes with proteins.

4.2 Proteins that are Very Different from Self

Proteins that are not present in the mammalian genome, such as keyhole limpet haemocyanin, are highly immunogenic. This is to be expected because the great majority of peptides (after proteolytic processing by antigen-presenting cells) will be seen as nonself by T cells, and most of the surface of the protein will be perceived as foreign by B cells. Keyhole limpet haemocyanin is therefore a very effective carrier for immunization with smaller haptenic molecules such as peptides, drugs, hormones or vitamins.

When the corresponding gene is present in the recipient genome, the strength of response will depend on the degree of sequence difference between self and the immunizing protein. As a very rough rule of thumb, proteins from two distant mammalian species, such as mice and humans, might be expected to be about 80% identical in their protein sequences. In other words, about 20% of amino acids might differ between these species in a typical protein. Such differences are very easily detected by the immune system, and numerous antigenic determinants would be recognized.

Occasionally, one finds that immunization with foreign antigens may provoke an antiself response, particularly if complete Freund's adjuvant is used. One such example is the production of autoantibodies to the acetylcholine receptor that develop when rats are immunized with the corresponding protein from electric eels (Lennon *et al.*, 1975; Lindstrom *et al.*, 1988). In this case, it is easy to see how the rat T cells would provide help, as the antigen is clearly foreign. The fact that the rat's B cells are able to respond to self acetylcholine receptor is more difficult to explain, and suggests that in this case, B cells have not been irreversibly tolerized to self during their development.

4.3 Proteins that are Very Similar to Self: Immune Response Genes

Some proteins are much more highly conserved than others. For example, histones are among the most highly conserved proteins, and their amino acid sequence varies little between species. As the amino acid sequence of proteins becomes closer and closer to self, they tend to become less and less immunogenic, and gaps in the repertoire become more and more apparent. The limiting of the repertoire owing to tolerance in the case of antigens that are very similar to self reveals another level of control, the histocompatibility-linked immune response genes (Ir genes; Benacerraf and McDevitt, 1972). For example, when mice are immunized with the B subunit of lactate dehydrogenase from pig hearts, some strains respond well while others respond very poorly. The genes that control responsiveness map to the class II region of the mouse MHC (Melchers *et al.*, 1973). It is now clear that Ir genes often reflect the ability of individual peptides to bind to MHC molecules and thus initiate immune responses (Rammensee *et al.*, 1993; Engelhard, 1994a, b).

The genetic factors that govern immune responsiveness are particularly well seen when inbred mice are immunized with proteins from different mouse strains where there are slight allelic differences in the sequence of that protein. Allelic differences between proteins of different individuals within a species are known as *allotypes*. In many cases, allotypic differences were first detected serologically by cross-immunization between strains of mice. In some cases, the structural difference has been shown to be a single amino acid substitution. The immune response to allotypic differences in proteins epitomizes the exquisite ability of the immune system to detect minor differences in protein structure.

For example, the T cell membrane glycoprotein Thy-1 has two allelic forms (Thy-1.1 and Thy-1.2) which differ by a single amino acid (Williams and Gagnon, 1982; Williams, 1989; Reif and Schlesinger, 1989). Immunization of Thy-1.1 mice with T cells from Thy-1.2 mice results in the production of antibodies that recognize the Thy-1.2 allele but not the Thy-1.1 allele, because the recipient lymphocytes are tolerant to the self allele. Although nonspecific T cell help is available for potential anti-Thy-1.1 B cells by virtue of T cell stimulation with the nonself Thy-1.2 antigen, the fact that anti-Thy-1.1 B cells do not respond may be taken as evidence that B cells with this antiself specificity are either non-responsive or deleted.

Interestingly, the very small difference between the self and nonself forms of Thy-1 is also reflected in MHC-linked immune response genes that control the ability to respond to the foreign Thy-1 allele. These genes are almost certainly identical to the A_α and A_β loci of the H-2 complex (Klein and Zaleski, 1989).

Allelic forms of mouse β_2 microglobulin differ in a single amino acid, and yet this difference can be detected by the immune system, as confirmed by the production of monoclonal antibodies (Michaelson *et al.*, 1980; Goding and Walker, 1980; Chorney *et al.*, 1982).

There are also several allelic forms of mouse IgM, which reflect genetic differences in the constant region of the μ chain gene (Warner *et al.*, 1977; Black *et al.*, 1978; Schreier *et al.*, 1986; Schüppel *et al.*, 1987). The difference between two IgM alleles is a single conservative amino acid substitution at codon 222 (lysine versus arginine) out of more than 400 amino acids for the whole constant region (Schreier *et al.*, 1986). This minor difference is sufficient to elicit easily detectable allele-specific antibodies.

Membrane IgD also has multiple allelic forms in the mouse (Goding *et al.*, 1976; Goding, 1977, 1980). One site of variation is on the Fc portion, while another is on the Fd portion, presumably in the C_H1 domain (Kessler *et al.*, 1979; Goding, 1980). Anti-allotype antibodies can be very useful as structural probes, as these antibodies have very high specificity. Antibodies to IgD allotypes provided some of the first highly specific antibodies to IgD and were generated without the need to purify the IgD molecule. This approach of alloimmunization was a key tool in the early analysis of lymphocyte membrane proteins and the classification of lymphocyte subsets, but has now been superseded by the use of monoclonal antibodies. Immunoglobulin allotypes will be discussed in more detail in Chapter 6.

When the amino acid sequences of proteins are very similar to self, the immune response is often very weak or undetectable. In such cases, the use of immunological adjuvants such as Freund's complete adjuvant (see Section 4.10) may make the difference between a detectable and a nondetectable response.

In the case of cell-surface antigens, it has often been observed that immunization with cells from mice that differ only in the protein of interest will not elicit antibodies, but if the donor and recipient also differ for other genes, strong antibodies may be found (Goding *et al.*, 1977; Lake and Douglas, 1978; Zaleski and Gorzynski, 1980; Lake *et al.*, 1989; Klein and Zaleski, 1989). The presence of multiple genetic differences appears to have a strong adjuvant effect. Cross-immunization of immunoglobulin congenic C57BL/6 mice with immunoglobulin heavy chain congenic partner strain C57BL/6.Ig^e (which differs from C57BL/6 only in the immunoglobulin heavy chain locus) will not generate detectable antibodies to IgD (Goding *et al.*, 1977). However, immunization of C57BL/6 mice with CBA mice, in which there are multiple genetic differences, results in very strong antibodies to IgD.

These effects are not very well understood. Both MHC and non-MHC genes have been implicated. The additional determinants required for a vigorous response have been variously called 'helper' or 'carrier' determinants, but because they are probably not directly analogous to the classical meaning of these terms, Klein and Zaleski have proposed the term *acolyte* determinants.

The requirement for acolyte determinants could be important when mice are immunized with mouse cells that have been transfected with human genes, because it is sometimes found that the antibody response to the human gene product is poor unless the transfected cells are derived from a different strain of mouse than the recipient.

A particularly interesting example where the body encounters antigens that are very close to self occurs in cancer, where the neoplastic process results from somatic mutations, translocations and deletions, many of which could be expected to produce 'modified self' proteins that could be detected by the immune system (see also Section 2.6). In a few cases, an immune response can be detected by the presence of cytotoxic T cells or the presence of antitumour antibodies (reviewed by Houghton, 1994; Boon *et al.*, 1994), but the response is usually weak and ineffectual, as demonstrated by the continued growth of the tumour. The same types of considerations regarding weak alloantigens such as Thy-1 presumably apply to tumour antigens. In addition, it would be expected that in most cases, the tumour cells would lack specialized lymphocyte-activating molecules such as the CD28 ligands B7 and CTLA-4. Accordingly, the tumour cells would not be able to provide the necessary second signal, without which antigenic stimulation may lead to specific anergy or nonresponsiveness (Nossal, 1994).

4.4 Proteins that are Identical to Self: Can Self Tolerance be Broken?

Although there are occasional exceptions, immunization with foreign antigens does not usually elicit significant quantities of antibodies to the self form of the immunogen, even if the antigen is given in complete Freund's adjuvant (Berzofsky and Berkower, 1993). The foreign antigen would have numerous epitopes in common with the recipient (as would be revealed by cross-reaction if the antigen were used to immunize a third species of recipient). This general failure to form auto-antibodies by immunization with foreign proteins presumably reflects the fact that lymphocytes with antiself receptors have usually been deleted or anergized by prior contact with self. It follows that deliberate immunization with self is generally unlikely to work.

Does this mean that immunization with self should never elicit the production of auto-antibodies, Ehrlich's *horror autotoxicus*? In some cases, autoimmunization is possible. It is becoming increasingly clear that self tolerance is by no means absolute, particularly if the antigen is given together with complete Freund's adjuvant (see Section 4.10). However, the robustness of self tolerance appears to vary greatly, depending on the individual antigen.

In some cases, the self antigen may be physically sequestered from the immune system, for example in the eye or the testis, and hence the developing lymphocytes have no chance to come into contact with the antigen and thus to become tolerant to it.

Self tolerance may also be weak or lacking for self antigens that develop late in ontogeny. In this case, it is presumed that some lymphocytes may become fully mature before having an opportunity to become 'tolerized' by the relevant self antigen. Akashi and Eishi (1991) immunized rats with autologous tissue homogenates in complete Freund's adjuvant, and found that it was possible to generate auto-antibodies to certain late-developing but not early-developing self antigens. Adams *et al.* (1987) found that expression of transgenic SV40 T antigen early in embryogenesis resulted in much greater tolerance than if expression began later. Alderuccio *et al.* (1993) found that experimental autoimmune gastritis could be prevented by expression of the relevant antigens in the thymus, presumably because of exposure to self at this early stage in lymphocyte development would result in more effective tolerance.

As discussed previously, self may be defined as 'that which is present in the body all the time' (Section 2.4). This concept is dramatically illustrated in the case of 'knockout' mice in which a self gene (tissue plasminogen activator; t-PA) has been deleted by homologous recombination (Declerck *et al.*, 1995). Immunization of the knockout mice with 10 μg of mouse t-PA in complete Freund's adjuvant resulted in a vigorous antibody response and allowed the production of numerous hybridomas secreting auto-antibodies to t-PA, which reacted against a diverse array of epitopes, most of which were highly conserved across species. Apart from its theoretical interest, this strategy could be applied to the generation of antibodies against any self protein that was not essential for survival or a functioning immune system.

It is interesting to note that several antitumour immune responses have been found to be directed against self proteins that are not mutated (reviewed by Houghton, 1994; Boon *et al.*, 1994). One such antigen recognized by cytotoxic T cells was an unmutated gene expressed in mastocytoma cells. Other examples include enzymes such as tyrosinase which are part of the pathway of melanin biosynthesis in normal melanocytes and which are expressed, apparently unmutated, in malignant melanoma cells (Houghton, 1994). In other words, an immune response against a tumour may be directed against molecules that are expressed by the normal cellular counterpart of the tumour and are structurally identical to self. The meaning behind these surprising observations remains to be discovered.

Notwithstanding the above, from a practical point of view, immunization with self proteins, even using complete Freund's adjuvant, will generally fail to elicit significant amounts of antibody.

Where there is a choice, the species of antigen donor and recipient should be as distant as possible. In the cases where this is not possible, or where the protein is extremely highly conserved between species, it may sometimes be possible to break B cell tolerance by coupling the protein to a highly immunogenic carrier by physical cross-linking, or by expressing the genetically engineered antigen as a fusion protein with a foreign antigen such as β-galactosidase from *Escherichia coli* or glutathione *S*-transferase from *Schistosoma japonicum* (Smith and Johnson, 1988).

4.5 Denatured (Unfolded) Proteins

The epitopes recognized by T cells are short peptides in an extended conformation in the MHC groove, whose conformations are likely to bear little resemblance to those in the native protein (Madden *et al.*, 1991; Engelhard, 1994a, b). It follows that the conformation of proteins should be largely irrelevant to T cell activation and self tolerance. Apart from the possible influence of unfolding on the proteolytic processing events, it would be expected that T cells would respond equally well to native and denatured proteins.

In marked contrast, recognition of protein antigen by B cells is critically dependent on its conformation, because of the stereospecific complementarity of the antigen–antibody bond (Laver *et al.*, 1990). Both activation and tolerance in B cells will depend on stereospecific recognition of the conformation of the antigen (Benjamin *et al.*, 1984; Goodnow *et al.*, 1989). It is therefore to be expected, and is indeed observed, that many antibodies to native proteins do not react with denatured forms of the same protein, and vice versa (Arnon, 1973; Crumpton, 1974). For example, rabbit antibodies to the native form of human apolipoprotein A-I were ineffective in recognizing the same protein that had been fixed with formaldehyde. However, when rabbits were immunized with apolipoprotein A-I that had been treated with formalin, the resulting antibodies reacted strongly and specifically with the formalin-fixed antigen (Harrach and Robenek, 1991).

These observations have great practical importance. If the antigen to be studied is in a modified or denatured form, much better results may be obtained if the immunization protocol starts with the same form of antigen that will be used once the antibodies are generated. This strategy seems a rational and effective way of generating antibodies that will work well on fixed tissues, and could also be applied to the production of monoclonal antibodies.

4.5.1 Linear and Conformational Epitopes

It has been proposed that epitopes can be divided into *linear* and *conformational* epitopes (Sela, 1969; Arnon, 1973; Crumpton, 1974; see also Berzofsky and Berkower, 1993 for a critical discussion of this concept). The idea of linear epitopes arose out of studies involving short synthetic peptides. A small portion of polyclonal antibodies to native proteins can be shown to react with short peptides from the same protein, and antibodies to short synthetic peptides can sometimes react with the native protein.

However, it must be emphasized that in all cases where the crystallographic structure of an antigen–antibody complex has been solved, it is found that there are multiple points of contact between the antigen and antibody. These points of contact almost always involve amino acids that are not contiguous in the antigen sequence, but rather come together in the folded protein. *The great majority of antigenic determinants recognized by antibodies and B cells are conformational* (Laver *et al.*, 1990). The concept of linear determinants has much greater relevance for T cells because of the manner in which peptides are bound to MHC molecules.

4.5.2 B and T Cell Tolerance to Denatured Self: Relevance for Antibody Production to Self or Highly Conserved Proteins

Continuing the previous line of argument, self tolerance in T cells must be directed towards self peptides rather than the overall conformation of self proteins. In contrast, B cells could only be tolerant to those conformations of self that have been 'seen' by the developing B cell population (Benjamin *et al.*, 1984; Goodnow *et al.*, 1989). In many cases, one might expect that the developing B cell population would not have encountered denatured forms of self proteins, and thus would not have had the chance to become tolerant to them. As most protein antigens are T dependent, T cell tolerance might suffice to keep autoreactive B cells in check.

This line of argument implies that if one immunizes with a denatured self protein and at the same time provides physically linked T cell help, for example in the form of an engineered bacterial hybrid fusion protein containing the nominal antigen fused to β-galactosidase from *E. coli* or glutathione *S*-transferase from *S. japonicum* (Smith and Johnson, 1988), one might expect to be able to generate highly specific autoantibodies, or antibodies to highly conserved proteins. The main limitation of this approach is that the resulting antibodies would react only with the denatured form of the protein. This may not be such a severe limitation in many experimental situations as the antigen is denatured, for example in histological sections in which the tissue has been fixed, and in Western blots.

Protein antigens expressed in bacteria often fold incorrectly, and antibodies against bacterially expressed proteins may fail to bind to the native protein.

Consider the case of the monoclonal mouse antibody 4H4, which was raised against an 'EF hand' calcium-binding motif from human phosphodiesterase I (Belli *et al.*, 1994). This motif was made in *E. coli* in the vector pGEX (Smith and Johnson, 1988), and expressed as a fusion protein contiguous with glutathione *S*-transferase of *S. japonicum*. Somewhat contrary to expectations, the resulting antibody only reacted with the EF hand when calcium was absent, when the EF hand was presumably unfolded. Although the antibody was highly specific, it reacted equally well with mouse and human phosphodiesterase I, suggesting that the B cells of the recipient mouse were not tolerant to the unfolded EF hand, having never encountered it in the absence of calcium. T cell help would have been provided by the glutathione *S*-transferase. *This approach might be applied to generate antibodies to unfolded forms of any self protein, although it is not yet known if the results are general.*

4.6 Recognition of Short Peptides by T and B Lymphocytes

T cells 'see' antigen in the form of short peptides bound to MHC molecules. The peptide recognized by helper T cells bound to class II MHC proteins is typically about 14 or more amino acids in length, while the peptide recognized by cytotoxic T cells bound to class I molecules is typically 8–9 amino acids in length. As mentioned previously, whether or not T cells will respond to a given peptide will depend on three main factors. First, the peptide must be capable of reaching the MHC groove, whether from outside the cell or by being pumped from the cytoplasm by specific peptide transporter molecules (see Monaco, 1992). Second, the peptide must be capable of stable binding to MHC molecules (Madden *et al.*, 1991; Engelhard, 1994a, b). Third, the peptide must be encountered by T cells with appropriate receptors for the peptide:MHC complex, and which are capable of responding. For any given protein antigen, one may expect there to be a large number of peptides that could potentially be recognized by T cells (Germain and Margulies, 1993; Engelhard, 1994a, b).

In the great majority of instances, typical protein antigens will produce sufficient peptides after processing by cells to result in at least some that will stimulate T cells. However, for a given individual short peptide, T cells may or may not respond, depending on whether or not the conditions mentioned above have been met.

Provided that there is sufficient T cell help available, B cells are also capable of responding to short peptides. However, immunization with short peptides alone, in the absence of carrier, is unlikely to generate significant amounts of antibody because many short peptides will not be noticed by T cells. For reliable production of antipeptide antibodies, the peptide must be physically linked to an immunogenic carrier protein (see Clarke *et al.*, 1987, for an example). The question of whether such antibodies are likely to recognize the native protein from which the peptide was derived is addressed in the next section.

4.7 Antibodies Against Short Peptides Sometimes Recognize the Native Protein

Immunization with short peptides coupled to immunogenic protein carriers will sometimes result in antibodies that react with the protein from which the peptide was derived (Walter *et al.*, 1980; Lerner, 1984). This finding has proven extremely useful, particularly for the identification and characterization of proteins whose sequence has been predicted from newly cloned genes or cDNAs (Doolittle, 1986).

The reaction of antipeptide antibodies with native proteins raises several conceptual and practical issues. Peptides of less than about 50 amino acids usually have little stability of conformation because there is insufficient folding energy. *However, all antibodies, even those to short peptides, are highly stereospecific.* It follows that the great majority of such antipeptide antibodies would recognize the peptide in a conformation that is different to that seen in the native protein. Is the claim that a high percentage of antipeptide antibodies will recognize the native protein really true?

Not all antipeptide antibodies react with the native protein. As the literature generally publishes only positive results, it may have given a misleading impression. Many unpublished failures have occurred. Antipeptide antibodies are thought to be more likely to react with the native protein if the peptides are derived from the amino- or carboxy-terminus of the protein (Doolittle, 1986). Hydrophilic peptides with charged amino acids may also be favoured because charged residues usually lie on the surface of proteins. Other factors that have been suggested to favour reactivity of antipeptide sera with the native protein are when the peptide is derived from a region of the protein with high mobility (Tainer *et al.*, 1984), or the presence of amphipathic helices (Berzofsky and Berkower, 1993). Unfortunately, most of these predictive procedures are of rather uncertain reliability, as are the computer algorithms that are based on them.

All the claims for a high frequency of reactivity of antipeptide antibodies to native proteins are based on the assumption that the test antigen was truly native. In at least some cases, this is unlikely to be true, particularly for solid-phase assays such as enzyme-linked immunosorbent assay (ELISA), in which the nominally native antigen is bound to plastic plates by an unknown mechanism. There is abundant evidence that a large proportion of protein antigen bound to plastic ELISA plates is denatured (Smith and Wilson, 1986, and references therein). Careful quantitative studies of antibodies bound to plastic plates have shown that up to 90% of the bound protein is denatured (Butler *et al.*, 1993). It is very likely that this result would also apply to other proteins.

That antipeptide antibodies often fail to recognize the native protein should not be surprising. As pointed out by Benjamin *et al.* (1984) and Laver *et al.* (1990), the great majority of antigenic determinants are conformational, and in virtually every case where crystallographic data are available the antibody binds to amino acids that are not contiguous in the amino acid sequence. Indeed, the recent solving of the crystal structure of a complex between an antibody and a peptide from a myohemerythrin helical complex has shown that the conformation of the peptide is in the form of a β-turn when bound to antibody, but is an α-helix in the native protein (Stanfield *et al.*, 1990). There is no way that this antibody could bind to the native protein.

Despite these limitations, antipeptide antibodies can be extremely useful for probing Western blots and histological sections, or indeed for any situation in which the antigen is denatured, or in which it can be denatured without perturbing the experimental system. Antipeptide antibodies should be used with very great caution in situations in which the native structure of the antigen is important, such as immunoprecipitation in nonionic detergents. Such experiments may often fail because the antibody does not react with the native protein.

4.8 Carbohydrates and Lipids

Carbohydrate polymers have been known to be antigenic since the beginnings of the study of immunology. They are commonly encountered in pathogenic organisms, such as pneumococci, and it is to be expected that there has been substantial evolutionary pressure to produce effective responses against them (reviewed by Berzofsky and Berkower, 1993). As a general rule, low molecular weight carbohydrates are not immunogenic, but may become so after coupling to an immunogenic carrier. The carbohydrate moieties of glycoproteins and glycolipids are also immunogenic. Other important carbohydrate antigens include the ABO blood group substances, which are glycolipid in nature (Yamamoto *et al.*, 1990). For background information on carbohydrate antigens, the reader is referred to the reviews of Wiegandt (1985), Sharon and Lis (1989, 1993), Karlsson (1991) and Drickamer and Carver (1992)

The manner in which carbohydrate antigens stimulate the immune system is somewhat different to that of proteins. A classical example of a carbohydrate antigen is bacterial lipopolysaccharide (LPS). LPS has intrinsic mitogenicity for B cells, owing to its lipid A moiety. It also has potent adjuvant properties, and is capable of inducing endotoxic shock, which may be fatal. LPS binds to a serum protein known as LPS-binding protein (LBP; Schumann *et al.*, 1990), which then binds to the CD14 molecule on the surface of monocytes and neutrophils (Wright *et al.*, 1990; Ferrero *et al.*, 1993), activating them and causing the release of cytokines including tumour necrosis factor α (TNF-α), a primary mediator of endotoxic shock (Beutler and Cerami, 1988; Bone, 1991). LPS also has potent immunological adjuvant activity.

Although many carbohydrate antigens are capable of eliciting very high titre antibodies, and these may be of the IgG class, the question of whether T cells can recognize carbohydrates has been controversial. Paradoxically, one of the reasons for this controversy has been the progress in understanding the role of the histocompatibility antigens in presenting antigen to T cells. As the mechanism of peptide binding to MHC proteins became clear, it became less clear how carbohydrates could bind. However, there is now a considerable and growing body of evidence that some T cells can recognize carbohydrate or glycolipid antigens (Litvin *et al.*, 1981; Moll *et al.*, 1989; Michaëlsson *et al.*, 1994; Haurum *et al.*, 1994; Tanaka *et al.*, 1994; Beckman *et al.*, 1994; Bendelac, 1995; Sieling *et al.*, 1995). In many such cases, recognition by T cells is not MHC restricted, or is restricted by the so-called non-classical (class Ib) MHC molecules. In some cases the relevant T cells have been shown to bear γδ receptors for antigen (Strominger, 1989; Hedrick, 1992; Parham; 1994; Klein and O'hUigin, 1994; Beckman *et al.*, 1994).

In addition to their intrinsic interest and importance, carbohydrate antigens can be the cause of experimental artefacts. For example, the carbohydrate moiety on one glycoprotein may be identical in structure to that on an unrelated glycoprotein, and be the source of serological cross-reaction between them. It has recently been shown that antibodies to galactosyl determinants may bind to agarose, a commonly used support matrix for affinity chromatography (Osborn *et al.*, 1994). If the presence of such antibodies is not recognized, they could be the source of misleading observations.

The fact that lipids are hydrophobic and tend to form micelles and membranes may make them less accessible to the immune system than more hydrophilic molecules such as carbohydrates and proteins. The degree to which the immune system can respond to lipids has been somewhat controversial, although as early as the 1940s, there was considerable evidence that sterols can be immunogenic when attached to large carrier proteins (Landsteiner, 1962; Aniagolu *et al.*, 1995 and references therein). Similarly, antibodies to steroid hormones may be raised by immunization with these molecules bound to immunogenic protein carriers, and these antibodies have been widely used for radioimmunoassay.

There seems to be no reason in principle why the immune system should not see the more hydrophilic portions of lipid-containing molecules, such as hydroxyl, phosphate or carbohydrate groups (Bendelac, 1995; Sieling *et al.*, 1995). The extent to which the more hydrophobic portions of such amphipathic molecules are seen by the immune system is currently rather unclear. Antibodies to phospholipids such as cardiolipin are commonly said to be found in patients with systemic lupus erythematosus, but their specificity may be directed to protein:lipid complexes rather than the lipids in isolation (Roubey, 1994; Harris *et al.*, 1995 and references therein).

4.9 Haptens: Small Molecules, Such as Drugs, Hormones and Synthetic Compounds

It has been known since the days of Ehrlich that under appropriate conditions, the immune system is capable of making highly specific antibodies to virtually any dye, drug or other compound that a chemist can synthesize. This amazing flexibility of responsiveness testifies to the enormous diversity of the repertoire of the immune system.

As mentioned at the beginning of this chapter, these low molecular weight compounds are not usually capable of initiating an immune response on their own, but are capable of eliciting very strong responses when coupled to an immunogenic carrier or 'schlepper' (Yiddish). Landsteiner (1921) coined the term *hapten* (from the Greek, meaning to touch, to grasp or to fasten) to describe this phenomenon. Functionally, a hapten is virtually the equivalent of an epitope. The surface of protein antigens can be considered to be the equivalent of innumerable haptens.

Can low molecular weight self-compounds act as haptens if coupled to immunogenic carriers? In general, the answer appears to be yes. The immune system appears to be 'ignorant' of small self molecules such as steroids, thyroxine, adrenaline, vitamins and low molecular weight metabolites, presumably because they are not present in sufficient concentrations to tolerize. There is little evidence that T cells can recognize such compounds and it is hard to see how they could cross-link B cell receptors. Provided that T cell help is provided by an immunogenic carrier, it would appear that virtually any low molecular weight compound could act as a hapten and lead to the formation of antibodies.

In addition to their theoretical importance, haptens have great practical significance. It is possible to make antibodies to virtually any low molecular weight compound by coupling it to an immunogenic carrier. Such antibodies are frequently used to measure the concentration of hormones, drugs and metabolites by radioimmunoassay or ELISA.

The unintentional coupling of drugs, dyes or other compounds to proteins may sometimes be harmful. The new antigenic determinants created by the coupling may extend beyond the haptenic group itself, and may stimulate both T cells and B cells. Hypersensitivity or allergic reactions to nickel, penicillin or other drugs and dyes may occur by this mechanism.

4.10 Adjuvants

An adjuvant is a substance which augments immune responses in a non-specific manner (Fenichel and Chirigos, 1984; Ada, 1993; Audibert and Lise, 1993; Gupta *et al.*, 1993). The most commonly used adjuvants are *Freund's complete adjuvant* (a water-in-oil emulsion in which killed and dried *Mycobacterium bovis* bacteria are suspended in the oil phase) and *Freund's incomplete adjuvant* (omitting the bacteria) (reviewed by Cox and Coulter, 1992; Stewart-Tull, 1995).

Other adjuvants include aluminium compounds such as aluminium hydroxide gel and alums such as potassium alum (K_2SO_4 Al_2SO_4) which strongly adsorb protein antigens from solution to form a precipitate. Alum-precipitated proteins are often administered together with killed *Bordetella pertussis* organisms (whooping cough vaccine) (Munoz, 1964). Many other strategies for increasing immunogenicity have been explored, including the covalent coupling of protein to lipid (Hopp, 1984).

In recent years, there have been numerous new adjuvants, including pluronic block polymers that give a large surface area, such as TiterMax™, SAF-1 (Syntex Corporation) which combines pluronic polymers with an MDP derivative, and DETOX™ (Ribi ImmunoChem Research), which contains monophosphoryl lipid A, a mycobacterial cell wall skeleton, and squalene (reviewed by Ada, 1993; Allison and Byars, 1991; Cox and Coulter, 1992). It has been shown that small polymeric lipid–amphipathic viral glycoprotein aggregates (iscoms; immunostimulatory complexes) based on the lipid Quil A (the active component of saponin) are particularly effective in increasing the immune response (Morein *et al.*, 1984, 1987). Grubhofer (1995) described a new adjuvant formulation based on *N*-acetylglucosaminyl-*N*-acetylmuramyl-L-alanyl-D-isoglutamine, which is claimed to be particularly effective and non-toxic. It is available commercially as Gerbu Adjuvant (Gerbu Biotechnki GmbH, Gaiberg, Germany).

The main problem with adjuvants is that the most potent ones tend to have strong inflammatory properties and it is not clear whether these inflammatory properties can be separated from adjuvanticity (see Allison *et al.*, 1991; Gupta *et al.*, 1993; Stewart-Tull, 1995). Promotion of inflammation may be intrinsic to adjuvanticity. The claims that the more recent adjuvants are less toxic remain to be tested in a rigorous way. In my hands, some of the newer agents appear more toxic than Freund's adjuvant, and are often less effective.

Freund's and other adjuvants that are in common use in research laboratories are considered too toxic for human use. There is currently a massive search for better and safer adjuvants, particularly for use with vaccines in humans (Ada, 1993; Hui, 1994). The only adjuvant approved for human use is alum, but its potency as an adjuvant is mediocre (Ada, 1993).

The mechanisms of action of adjuvants are complex and poorly understood (Freund, 1956; Allison and Byars, 1991; Ada, 1993; Audibert and Lise, 1993; Stewart-Tull, 1995). They probably include slow, prolonged release of antigen in a highly aggregated form, together with pharmacologically active substances from the bacterial additives, which stimulate lymphocytes directly or indirectly via the induction of secretion of cytokines. These immunostimulatory substances include muramyl dipeptide from *Mycobacterium bovis* (Ellouz *et al.*, 1974; Chedid *et al.*, 1976), which among other actions may stimulate the release of IL-1 (Bahr *et al.*, 1987), 'pertussigen' from *B. pertussis* (Munoz and Bergman, 1977; Munoz and Sewell, 1984; Sewell *et al.*, 1984), lipid A from bacterial lipopolysaccharide (reviewed by Hui, 1994) which stimulates the release of TNF-α, and γ-inulin or combinations of γ-inulin and alum (Cooper and Steele, 1988, 1991; Cooper *et al.*, 1991).

Adjuvants can have major effects on the types of cytokines produced, which can in turn influence both the amount and the predominant class of antibody produced (see Audibert and Lise, 1993, and Section 3.3.10). Complete Freund's adjuvant is said to be a potent stimulator of Th1 cells, which produce IFN-γ and IL-2, and tend to help production of IgG2a, while alum is said to favour a Th2 response, in which the T cells produce IL-4, 5, 10 and 13, and tend to favour switching to IgG1 and IgE (Grun and Maurer, 1989; Mosmann and Coffman, 1989; Audibert and Lise, 1993). There is some evidence that *B. pertussis* promotes IgE responses (Hirashima *et al.*, 1981).

However, these general tendencies may not always apply for a particular antigen, and they should be taken as guidelines only. While it is said that complete Freund's adjuvant favours Th1 cells and IgG2a responses, the production of monoclonal antibodies to *Toxoplasma gondii* gave exactly the opposite result. Immunization with parasites without adjuvant provided many hybridomas making IgG2a (Handman *et al.*, 1980), while immunization in complete Freund's adjuvant provided many IgG1 hybridomas (Johnson *et al.*, 1981).

Many cytokines can probably act as adjuvants, and some recent vaccination strategies have involved genetically engineered constructs that include expression of cytokines as well as the nominal antigen (Flexner *et al.*, 1987; Ramshaw *et al.*, 1987; Ramsay *et al.*, 1994). Even if the vectors incorporating IL-2 have not fulfilled their promise, constructs using more recently discovered cytokines may turn out to be more effective.

4.11 Summary

As a general rule, the greater the phylogenetic distance between the antigen and the recipient, the more vigorous the immune response. Responses against highly conserved mammalian proteins are often weak and mainly IgM owing to lack of stimulation of T cells. Not all proteins are equally immunogenic and it is frequently found that very minor impurities in an antigen may evoke strong antibody responses.

As a broad generalization, arrays of repeating determinants or a high state of aggregation favour a strong antibody response. For the production of monoclonal antibodies, cell-surface antigens are often sufficiently immunogenic without the use of adjuvants, particularly if 'acolyte determinants' are present (see Section 4.3).

Water-soluble proteins, particularly if they are closely related in structure to their mouse homologues and/or if they are monomeric and not aggregated, may evoke feeble responses unless adjuvants are used. For soluble antigens, complete Freund's adjuvant is still the best choice. It remains the gold standard by which other adjuvants are judged (Ada, 1993). In my opinion, it is unproven that the newer adjuvants are either less toxic or more effective.

Practical details of adjuvant use for production of antibodies for research purposes are discussed in Sections 8.2.1 and 8.2.2.

References

Ada, G. L. (1993) Vaccines. *In* 'Fundamental Immunology' (Paul, W. E., ed.), 3rd edn, pp. 1309–1352. Raven Press, New York.

Adams, T., E., Alpert, S. and Hanahan, D. (1987) Non-tolerance and autoantibodies to a transgenic self antigen expressed in pancreatic β cells. *Nature* **325**, 223–228.

Akashi, T. and Eishi, Y. (1991) Developmental expression of autoimmune target antigens during organogenesis. *Immunology* **74**, 524–532.

Alderuccio, F., Toh, B-H., Tan, S-S., Gleeson, P. A. and van Driel, I. R. (1993) An autoimmune disease with multiple molecular targets abrogated by the transgenic expression of a single autoantigen in the thymus. *J. Exp. Med.* **178**, 1365–1375.

Allison, A. C. and Byars, N. E. (1991) Immunological adjuvants: desirable properties and side effects. *Mol. Immunol.* **28**, 279–284.

Aniagolu, J., Swartz, G. M., Dijkstra, J., Madsen, J. W., Raney, J. J. and Green, S. (1995) Analysis of anticholesterol antibodies using hydrophobic membranes. *J. Immunol. Methods* **182**, 85–92.

Arnon, R. (1973) Immunochemistry of enzymes. *In* 'The Antigens' (M. Sela, ed.), Vol. 1, pp. 88–159. Academic Press, New York.

Audibert, F. M. and Lise, L. D. (1993) Adjuvants: current status, clinical perspectives and future prospects. *Immunol. Today* **14**, 281–284.

Bahr, G. M., Chedid, L. A. and Behbehani, K. (1987) Induction, *in vivo* and *in vitro*, of macrophage membrane interleukin-1 by adjuvant-active synthetic muramyl peptides. *Cell. Immunol.* **107**, 443–454.

Beckman, E. M., Porcelli, S. A., Morita, C. T., Behar, S. M., Furlong, S. T. and Brenner, M. B. (1994) Recognition of a lipid antigen by CD1-restricted $\alpha\beta^+$ T cells. *Nature* **372**, 691–694.

Belli, S. I., Sali, A. and Goding, J. W. (1994) Divalent cations stabilize the conformation of plasma cell membrane glycoprotein PC-1 (alkaline phosphodiesterase I). *Biochem. J.* **304**, 75–80.

Benacerraf, B. and McDevitt, H. O. (1972) Histocompatibility-linked immune response genes. *Science* **175**, 273–279.

Bendelac, A. (1995) CD1: Presenting unusual antigens to unusual T lymphocytes. *Science* **269**, 185–186.

Benjamin, D. C., Berzofsky, J. A., East, I. J., Gurd, F. R. N., Hannum, C., Leach, S. J., Margoliash, E., Michael, J. G., Miller, A., Prager, E. M., Reichlin, M., Sercarz, E. E., Smith-Gill, S. J., Todd, P. E. and Wilson, A. C. (1984) The antigenic structure of proteins: a reappraisal. *Annu. Rev. Immunol.* **2**, 67–101.

Berzofsky, J. A. and Berkower, I. J. (1993) Immunogenicity and antigen structure. *In* 'Fundamental Immunology' (W. E. Paul, ed.), 3rd edn, pp 235–282. Raven Press, New York.

Beutler, B. and Cerami, A. (1988) The common mediator of shock, cachexia and tumor necrosis. *Adv. Immunol.* **42**, 213–232.

Black, S. J., Goding, J. W., Gutman, G. A., Herzenberg, L. A., Loken, M. R., Osborne, B. A., van der Loo, W. and Warner, N. L. 1978. Immunoglobulin iso-antigens (allotypes) in the mouse. V. Characterization of IgM allotypes. *Immunogenetics* **7**: 213–230.

Bone, R. C. (1991) The pathogenesis of sepsis. *Ann. Int. Med.* **115**, 457–469.

Boon, T., Cerottini, J-C., Van den Eynde, B., van der Bruggen, P. and van Pel, A. (1994) Tumor antigens recognised by T lymphocytes. *Annu. Rev. Immunol.* **12**, 337–366.

Butler, J. E., Ni, L., Brown, W. R., Joshi, K. S., Chang, J., Rosenberg, B. and Voss, E. W. (1993) The immunochemistry of sandwich ELISAs. VI. Greater than 90% of monoclonal and 75% of polyclonal anti-fluorescyl capture antibodies (CAbs) are denatured by passive adsorption. *Mol. Immunol.* **30**, 1165–1175.

Chedid, L., Audibert, F., Lefrancier, P., Choay, J. and Lederer, E. (1976) Modulation of the immune response by a synthetic adjuvant and analogs. *Proc. Natl Acad. Sci. USA* **73**, 2472–2475.

Chorney, M., Shen, F. W., Michaelson, J., Taylor, B. and Boyse, E. A. (1982) Monoclonal antibody to an alloantigenic determinant on beta 2 microglobulin (B2M). *Immunogenetics* **16**, 91–93.

Clarke, B. E., Newton, S. E., Carrol, A. R., Francis, M. J., Appleyard, G., Syred, A. D., Highfield, P. E., Rowlands, D. J. and Brown, F. (1987) Greatly improved immunogenicity of a peptide epitope by fusion to hepatitis B core protein. *Nature* **330**, 381–384.

Cooper, P. D. and Steel, E. J. (1991) Algammulin, a new vaccine adjuvant comprising gamma inulin particles containing alum; preparation and in vitro properties. *Vaccine* **9**, 351–357.

Cooper, P. D. and Steele, E. J. (1988) The adjuvanticity of gamma inulin. *Immunol. Cell. Biol.* **66**, 345–352.

Cooper, P. D., McComb, C. and Steele, E. J. (1991) The adjuvanticity of algammulin, a new vaccine adjuvant. *Vaccine* **9**, 408–415.

Cox, J. C. and Coulter, A. R. (1992) Advances in adjuvant technology and application. *In* 'Animal Parasite Control Utilizing Biotechnology' (W. K. Yong, ed.), pp. 49–112. CRC Press, Boca Raton, Florida.

Crumpton, M. J. (1974) Protein antigens: The molecular bases of antigenicity and immunogenicity. *In* 'The Antigens' (M. Sela, ed.), Vol. 2, pp. 1–78. Academic Press, New York.

Declerck, P. J., Carmeliet, P., Verstreken, M., De Cock, F. and Collen, D. (1995) Generation of monoclonal antibodies against autologous proteins in gene-inactivated mice. *J. Biol. Chem.* **270**, 8397–8400.

Doolittle, R. F. (1986) 'Of Urfs and Orfs. A Primer on how to Analyze Derived Amino Acid Sequences'. University Science Books, Mill Valley, California.

Drickamer, K. and Carver, J. (1992) Carbohydrates and glycoconjugates: upwardly mobile sugars gain status of information-bearing molecules. *Curr. Opin. Struct. Biol.* **2**, 653–654.

Ellouz, F., Adam, A., Ciorbaru, R. and Lederer, E. (1974) Minimal structural requirements of adjuvant activity of bacterial peptidoglycan derivatives. *Biochem. Biophys. Res. Commun.* **59**, 1317–1325.

Engelhard, V. H. (1994a) Structure of peptides associated with class I and class II MHC molecules. *Annu. Rev. Immunol.* **12**, 181–208.

Engelhard, V. H. (1994b) How cells present antigens. *Sci. American* **271**, 44–51.

Fenichel, R. L. and Chirigos, M. A. (eds) (1984) 'Immune Modulation Agents and Their Mechanisms'. Marcel Dekker, New York.

Ferrero, A., Tsuberi, B. Z., Tesio, L., Rong, G. W., Haziot, A. and Goyert, S. M. (1993) Transgenic mice expressing human CD14 are hypersensitive to lipopolysaccharide. *Proc. Natl Acad. Sci. USA* **90**, 2380–2384.

Flexner, H., Hugin, A. and Moss, B. (1987) Prevention of vaccinia virus infection in immunodeficient mice by vector-directed IL-2 expression. *Nature* **330**, 259–262.

Freund, J. (1956) The mode of action of immunologic adjuvants. *Adv. Tuberc. Res.* **7**, 130–148.

Germain, R. N. and Margulies, D. H. (1993) The biochemistry and cell biology of antigen processing and presentation. *Annu. Rev. Immunol.* **11**, 403–450.

Goding, J. W. (1977) Allotypes of IgM and IgD receptors in the mouse: a probe for lymphocyte differentiation. *Contemp. Top. Immunobiol.* **8**: 203–243.

Goding, J. W. (1980) Structural studies of lymphocyte surface IgD. *J. Immunol.* **124**, 2082–2088.

Goding, J. W. and Walker, I. (1980) Allelic forms of β_2-microglobulin in the mouse. *Proc. Natl Acad. Sci. USA*. **73**: 7395–7399.

Goding, J. W., Warr, G. W. and Warner, N. L. (1976) Genetic polymorphism of IgD-like cell surface immunoglobulin in the mouse. *Proc. Natl Acad. Sci. USA* **73**, 1305–1309.

Goding, J. W., Scott, D. W. and Layton, J. E. (1977) Genetics, cellular expression and function of IgD and IgM receptors. *Immun. Rev.* **12**, 152–186.

Goodnow, C. C., Crosbie, J., Adelstein, S., Lavoie, T. B., Smith-Gill, S. J., Mason, D. Y., Jorgensen, H., Brink, R. A., Pritchard-Briscoe, H., Loughnan, M., Loblay, R. H., Trent, R. J. and Basten, A. (1989) Clonal silencing of self-reactive B lymphocytes in a transgenic mouse model. *Cold Spring Harbor Symp. Quant Biol.* **LIV**, 907–920.

Grubhofer, N. (1995) An adjuvant formulation based on *N*-acetylglucosaminyl-*N*-acetylmuramyl-L-alanyl-D-isoglutamine with dimethyldioctadecylammonium chloride and zinc-L-proline complex as synergists. *Immunol. Lett.* **44**, 19–24.

Grun, L. L, and Maurer, P. H. (1989) Different T helper subsets elicited in mice utilizing two different adjuvant vehicles: The role of endogenous interleukin-1 in proliferative responses. *Cell. Immunol.* **121**, 134–145.

Gupta, R. K., Relyveld, E. H., Lindblad, E. B., Bizzini, B., Ben-Efraim, S. and Gupta, C. K. (1993) Adjuvants—a balance between toxicity and adjuvanticity. *Vaccine* **11**, 293–306.

Handman, E., Goding, J. W. and Remington, J. (1980) Detection and characterization of membrane antigens of *Toxoplasma gondii*. *J. Immunol.* **124**, 2578–2583.

Harrach, B. and Robenek, H. (1991) Immunocytochemical localization of apolipoprotein A-I using polyclonal antibodies raised against the formaldehyde-modified antigen. *J. Microscopy* **161**, 97–108.

Harris, E. N., Goldsmith, G., Picrangeli, S., Gharavi, A. and Branch, W. (1995) Phospholipid binding antibodies warrant continued investigation. *Blood* **85**, 2276–2277.

Haurum, J. S., Arsequell, G., Lellouch, A. C., Wong, S. Y. C., Dwek, R. A., McMichael, A. J. and Elliott, T. (1994) Recognition of carbohydrate by major histocompatibility complex class I-restricted, glycopeptide-specific cytotoxic T lymphocytes. *J. Exp. Med.* **180**, 739–744.

Hedrick, S. M. (1992) Dawn of the hunt for nonclassical MHC function. *Cell* **70**, 177–180.

Hirashima, M., Yodoi, J. and Ishizaka, K. (1981) Regulatory role of IgE-binding factors from rat T lymphocytes. V. Formation of IgE-potentiating factor by T lymphocytes treated with *Bordetella pertussis* vaccine. *J. Immunol.* **126**, 838–842.

Hopp, T. P. (1984) Immunogenicity of a synthetic HBsAg peptide; enhancement by conjugation to a fatty acid carrier. *Mol. Immunol.* **21**, 13–16.

Houghton, A. N. (1994) Cancer antigens: Immune recognition of self and altered self. *J. Exp. Med.* **180**, 1–4.

Hui, G. S. (1994) Liposomes, muramyl dipeptide derivatives, and nontoxic lipid A derivatives as adjuvants for human malaria vaccines. *Am. J. Trop. Med. Hyg.* **50** (Suppl. 4), 41–51.

Johnson, A. M., McNamara, P. J., Neoh, S. H., McDonald, P. J. and Zola, H. (1981) Hybridomas secreting monoclonal antibody to *Toxoplasma gondii*. *Aust. J. Exp. Biol. Med. Sci.* **59**, 303–306.

Karlsson, K-A. (1991) Glycobiology: A growing field for drug design. *Trends Pharmacol. Sci.* **12**, 265–272.

Kessler, S. W., Woods, V. L., Finkelman, F. D. and Scher, I. (1979) Membrane orientation and location of multiple distinct allotypic determinants of mouse lymphocyte IgD. *J. Immunol.* **123**, 2772–2778.

Klein, J. and O'hUigin, C. (1994) The conundrum of nonclassical major histocompatibility complex genes. *Proc. Natl Acad. Sci. USA* **91**, 6251–6252.

Klein, J. and Zaleski, M. (1989) Genetic control of the immune response to the Thy-1 antigens—ten years later. *In* 'Cell Surface Antigen Thy-1. Immunology, Neurology, and Therapeutic Applications' (A. E. Reif and M. Schlesinger, eds), pp. 335–366. Marcel Dekker, New York.

Lake, P. and Douglas, T. C. (1978) Recognition and genetic control of helper determinants for cell surface antigen Thy-1. *Nature* **275**, 220–222.

Lake, P., Mitchison, N. A., Clark, E. A., Khorshidi, M., Nakashima, I., Bromberg, J. S., Brunswich, M. R., Szensky, T., Sainis, K. B., Sunshine, G. H., Vavila-Castilo, L.,

Woody, J. N. and Lebwohl, D. (1989) The regulation of antibody responses to antigens of the cell surface: Studies with Thy-1 and H-2 antigens. *In* 'Cell Surface Antigen Thy-1. Immunology, Neurology, and Therapeutic Applications' (A. E. Reif and M. Schlesinger, eds), pp. 367–394. Marcel Dekker, New York.

Landsteiner, K. (1921) Uber heterogenetisches Antigen und Hapten XV. *Mitteilung uber Antigene. Biochem. Z.* **119**, 294–306.

Landsteiner, K. (1962) The Specificity of Serological Reactions. Originally printed by Harvard University Press, 1945; reprinted by Dover Publications, New York, 1962. pp. 110–113.

Laver, W. G., Air, G. M., Webster, R. G. and Smith-Gill, S. J. (1990) Epitopes on protein antigens: Misconceptions and realities. *Cell* **61**, 553–556.

Lennon, V. A., Lindstrom, J. M. and Seybold, M. E. (1975) Experimental autoimmune myasthenia: A model of myasthenia gravis in rats and guinea pigs. *J. Exp. Med.* **141**, 1365–1375.

Lerner, R. A. (1984) Antibodies of predetermined specificity in biology and medicine. *Adv. Immunol.* **36**, 1–44.

Lindstrom, J., Shelton, D. and Fujii, Y. (1988) Myasthenia gravis. *Adv. Immunol.* **42**, 233–284.

Linsley, P. S. and Ledbetter, J. A. (1993) The role of the CD28 receptor during T cell responses to antigen. *Ann. Rev. Immunol.* **11**, 191–212.

Litvin, N. L., Benacerraf, B. and Germain, R. N. (1981) B-lymphocyte response to trinitrophenyl-conjugated Ficoll: Requirement for T lymphocytes and Ia-bearing adherent cells. *Proc. Natl Acad. Sci. USA* **78**, 5113–5117.

Madden, D. R., Gorga, J. C., Strominger, J. L. and Wiley, D. C. (1991) The structure of HLA-B27 reveals nonamer self-peptides bound in an extended conformation. *Nature* **353**, 321–325.

Melchers, I., Rajewsky, K. and Shreffler, D. C. (1973) Ir-LDH_B: Map position and functional analysis. *Eur. J. Immunol.* **3**, 754–761.

Michaelson, J., Rothenberg, E. and Boyse, E. A. (1980) Genetic polymorphism of murine β-2 microglobulin detected biochemically. *Immunogenetics* **11**, 93–95.

Michaëlsson, E., Malmström, V., Reis, S., Engström, A., Burkhardt, H. and Holmdahl, R. (1994) T cell recognition of carbohydrates on type II collagen. *J. Exp. Med.* **180**, 745–749.

Moll, H., Mitchell, G. F., McConville, M. J. and Handman, E. (1989) Evidence for T cell recognition in mice of a purified lipophosphoglycan from *Leishmania major*. *Infect. Immun.* **57**, 3349–3356.

Monaco, J. J. (1992) Antigen presentation. Not so groovy after all? *Curr. Opin. Immunol.* **2**, 433–435.

Morein, B., Sundquist, B., Hoglund, S., Dalsgaard, K. and Osterhaus, A. (1984) Iscom, a novel structure for antigenic presentation of membrane proteins from enveloped viruses. *Nature* **308**, 457–460.

Morein, B., Lovgren, K., Hoglund, S. and Sundquist, B. (1987) The ISCOM: an immunostimulating complex. *Immunol. Today* **8**, 333–337.

Mosmann, T. R. and Coffman, R. L. (1989) Th1 and Th2 cells: Different patterns of lymphokine expression lead to different functional properties. *Annu. Rev. Immunol.* **7**, 145–173.

Munoz, J. (1964) Effect of bacteria and bacterial products on the immune response. *Adv. Immunol.* **4**, 397–440.

Munoz, J. J. and Bergman, R. K. (1977) '*Bordetella pertussis*. Immunological and Other Biological Activities'. Marcel Dekker, New York.

Munoz, J. J. and Sewell, W. A. (1984) Effect of pertussigen on inflammation caused by Freund's adjuvant. *Infect. Immun.* **44**, 637–641.

Nossal, G. J. V. (1994) Negative selection of lymphocytes. *Cell* **76**, 229–239.

Osborn, A., Kelleher, M., Handman, E. and Goding, J. W. (1994) Antigalactosyl antibodies that react with unmodified agarose; a potential source of artefacts in immunoaffinity chromatography. *Anal. Biochem.* **217**, 181–184.

Parham, P. (1994) Chewing the fat. *Nature* **372**, 615–616.

Pullen, A. M., Marrack, P. and Kappler, J. W. (1988) The T-cell repertoire is heavily influenced by tolerance to polymorphic self-antigens. *Nature* **335**, 796–801.

Rammensee, H-G., Falk, K. and Rotzschke, O. (1993) MHC molecules as peptide receptors. *Curr. Opin. Immunol.* **5**, 35–44.

Ramsay, A. J., Husband, A. J., Ramshaw, I. A., Bao, S., Matthaei, K. I., Koehler, G. and Kopf, M. (1994) The role of interleukin-6 in mucosal IgA antibody responses *in vivo*. *Science* **264**, 561–563.

Ramshaw, I. A., Andrew, M. E., Phillips, S. M., Boyle, D. B. and Coupar, B. E. H. (1987) Recovery of immunodeficient mice from a vaccinia virus/IL-2 recombinant infection. *Nature* **329**, 545–546.

Reif, A. E, and Schlesinger, M. (1989) 'Cell Surface Antigen Thy-1. Immunology, Neurology, and Therapeutic Applications'. Marcel Dekker, New York.

Roubey, R. A. S. (1994) Autoantibodies to phosphlipid-binding plasma proteins: A new view of lupus anticoagulants and other 'antiphospholipid: autoantibodies. *Blood* **84**, 2854–2867.

Schreier, P. H., Quester, S. and Bothwell, A. (1986) Allotypic differences in murine μ genes. *Nucl. Acid. Res.* **14**, 2381–2389.

Schumann, R. R., Leong, S. R., Flaggs, G. W., Gray, P. W., Wright, S. D., Mathison, J. C., Tobias, P. S. and Ulevitch, R. J. (1990) Structure and function of lipopolysaccharide binding protein. *Science* **249**, 1429–1431.

Schüppel, R., Wilke, J. and Weiler, E. (1987) Monoclonal anti-allotype antibody towards BALB/c IgM. Analysis of specificity and site of a V-C crossover in recombinant strain BALB-Ig-V^a/Igh-C^b. *Eur. J. Immunol.* **17**, 739–741.

Sela, M. (1969) Antigenicity: Some molecular aspects. *Science* **166**, 1365–11374.

Sewell, W. A., Munoz, J. J., Scollay, R. and Vadas, M. A. (1984) Studies on the mechanism of enhancement of delayed-type hypersensitivity by pertussigen. *J. Immunol.* **133**, 1716–1722.

Sharon, N. and Lis, H. (1989) Lectins as cell recognition molecules. *Science* **246**, 227–234.

Sharon, N. and Lis, H. (1993) Carbohydrates in cell recogntion. *Sci. American* **268**, 74–81.

Sieling, P. A., Chatterjee, D., Porcelli, S. A., Prigozy, T. I., Mazzaccaro, R. J., Soriano, T., Bloom, B. R., Brenner, M. B., Kronenberg, M., Brennan, P. J. and Modlin, R. L. (1995) CD1-restricted T cell recognition of microbial lipoglycan antigens. *Science* **269**, 227–230.

Smith, A. D. and Wilson, J. E. (1986) A modified ELISA that selectively detects monoclonal antibodies recognizing native antigen. *J. Immunol. Methods* **94**, 31–35.

Smith, D. B. and Johnson, K. S. (1988) Single-step purification of polypeptides expressed in *Escherichia coli* as fusions with glutathione *S*-transferase. *Gene* **67**, 31–40.

Stanfield, R. L., Fieser, T. M., Lerner, R. A. and Wilson, I. A. (1990) Crystal structures of an antibody to a peptide and its complex with peptide antigen at 2.8 Å. *Science* **248**, 712–719.

Stewart-Tull, D. E. S. (1995) 'The Theory and Practical Application of Adjuvants'. John Wiley and Sons, New York.

Strominger, J. L. (1989) The γδ T cell receptor and class Ib MHC-related proteins: enigmatic molecules of immune recognition. *Cell* **57**, 895–898.

Tainer, J. A., Getzoff, E. G., Alexander, H., Houghten, R. A., Olson, A. J., Lerner, R. A. and Hendrickson, W. A. (1984) The reactivity of antipeptide antibodies is a function of the atomic mobility of sites in a protein. *Nature* **312**, 127–134.

Tanaka, Y., Sano, S., Nieves, E., de Libero, G., Rosa, D., Modlin, R. L., Brenner, M. B., Bloom, B. R. and Morita, C. T. (1994) Nonpeptide ligands for human γδ T cells. *Proc. Natl Acad. Sci. USA* **91**, 8175–8179.

Walter, G., Scheidmann, K. H., Carbone, A., Laudano, A. P. and Doolittle, R. F. (1980) Antibodies specific for the carboxy-and amino terminal regions of simian virus 40 large tumour antigen. *Proc. Natl Acad. Sci. USA* **77**, 5197–5200.

Warner, N. L., Goding, J. W., Gutman, G., Warr, G. W., Herzenberg, L. A., Osborne, B. A., van der Loo, W., Black, S. J. and Loken, M. R. (1977) Allotypes of mouse IgM immunoglobulin. *Nature* **265**, 447–449.

Weigandt, H. (1985) (ed.) 'Glycolipids. New Comprehensive Biochemistry, Vol. 10' (A. Neuberger. and L. L. M. van Deenen, gen. eds). Elsevier, Amsterdam.

Williams, A. F. (1989) The structure of Thy-1 antigen. *In* 'Cell Surface Antigen Thy-1. Immunology, Neurology, and Therapeutic Applications' (A. E. Reif and M. M. Schlesinger, eds), pp. 49–69. Marcel Dekker, New York.

Williams, A. F. and Gagnon, J. (1982) Neuronal cell Thy-1 glycoprotein: homology with immunoglobulin. *Science* **216**, 696–703.

Wright, S. D., Ramos, R. A., Tobias, P. S., Ulevitch, R. J. and Mathison, J. C. (1990) CD14, a receptor for complexes of lipopolysaccharide (LPS) and LPS binding protein. *Science* **249**, 1431–1433.

Yamamoto, Y., Clausen, H., White, T., Marken, J. and Hakomori, S. (1990) Molecular basis of the histo-blood group ABO system. *Nature* **345**, 229–233.

Zaleski, M. B. and Gorzynski, T. J. (1980) Non-H-2 antigens serving as carrier determinants in a primary anti-Thy-1 response in mice. *Immunol. Commun.* **9**, 309–317.

5 Antibody Structure and Function

5.1 Structure of Antibodies

As a general rule, antibodies are symmetrical molecules made up to two identical glycosylated heavy chains of M_r (relative molecular mass, or molecular weight) 50 000–75 000, and two identical nonglycosylated light chains of M_r ≈25 000. The heavy chains are joined by disulfide bonds to each other, and each light chain is joined by a disulfide bond to one heavy chain, making up a basic subunit of two heavy chains and two light chains (H_2L_2; Fig. 5.1). Proteins which have the general structural features of antibodies, but which do not have known antigen-binding properties, are known as *immunoglobulins*. Although disulfide bonding between chains is characteristic of immunoglobulins, the exact number and location of such bonds is somewhat variable (Nisonoff *et al.*, 1975; Carayannopoulos and Capra, 1993).

Immunoglobulins may be grouped into five main *classes*: IgM, IgD, IgG, IgE and IgA. Individual classes of immunoglobulins have distinctive structural and biological properties (Table 5.1). The class of an immunoglobulin molecule is determined by its heavy chains. Thus IgM, IgD, IgG, IgE and IgA possess μ, δ, γ, ε, and α heavy chains respectively.

While IgM and IgD have single genes, the remaining classes can often be subdivided into subclasses, each with its own constant region gene. It appears likely that the various Cγ genes evolved quite differently since the time of mammalian speciation because their number and relationships vary greatly between species (see Hayashida *et al.*, 1984; Gough and Cory, 1986). For example, there are four subclasses of IgG in mouse, rat and human, but only one in the rabbit (Brüggemann, 1987). Similarly, there are two genes for Cα in the human, one in the mouse, and at least 13 Cα genes in the rabbit

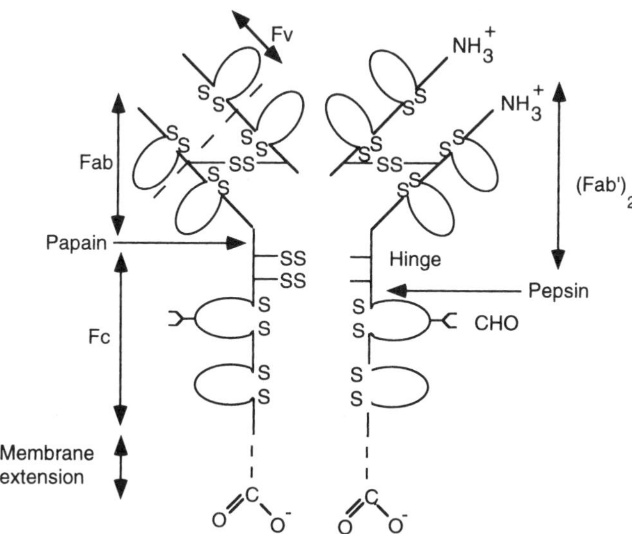

Fig. 5.1. Structure of an IgG molecule. Each chain is made up of a series of homology units of approximately 110 amino acids. The sites of proteolytic cleavage usually lie between the homology units. Membrane forms possess a hydrophobic C-terminal extension, but are otherwise identical in sequence to the secretory forms.

(Brüggemann, 1987; Burnette *et al.*, 1989; Lai *et al.*, 1989). Similarly, there is a single Cε gene in the mouse and rat, but there are two genes for Cε in the human, although one is a nonfunctional pseudogene (Lai *et al.*, 1989). [The committee of immunoglobulin nomenclature of the World Health Organization (1973) has recommended that all the letters for the various subclasses be written on the same line, thus IgG1 and not IgG_1].

There are two types of light chains, κ and λ. In the mouse, rat and human, there is a single Cκ gene, but the number of λ genes varies greatly between species, with four in the mouse and four functional λ genes and two λ pseudogenes in humans (Lai *et al.*, 1989; Max, 1993). With only rare exceptions, each lymphocyte is committed to the production of either κ or λ light chains and, accordingly, each immunoglobulin class contains molecules with either κ or λ light chains (see Giachino *et al.*, 1995 for recently discovered exceptions to this rule).

In the mouse and rat, over 95% of light chains are κ. In the mouse, there is only one κ light chain class but, paradoxically, there are four λ chain subclasses (λ_I–λ_{IV}). In humans the ratio of κ:λ bearing immunoglobulins in serum is about 60:40.

Both κ and λ light chains appear to be able to take part in all biological functions of antibodies. The light chains play an important part in determining the specificity of antibodies, but all known 'effector' functions of antibodies (e.g. complement-mediated lysis of cells) are determined by the heavy

Table 5.1 Properties of mouse immunoglobulin classes (isotypes)

	IgM	IgD	IgG	IgE	IgA
Concentration in serum	0.1–1.0 mg/ml	1–10 μg/ml	3–20 mg/ml	0.1–1 μg/ml	1–3 mg/ml
M_r	900 000	180 000	160 000	190 000	170 000–500 000
Heavy chains	μ	δ	γ	ε	α
Heavy chain M_r					
Secretory	80 000	65 000	55 000	80 000	70 000
Membrane	84 000	70–75 000	65 000	?	
Light chains	κ or λ	κ or λ	κ or λ	κ or λ	κ or λ
Subclasses	1	1	4	1	1
Stoichiometry (secreted form)	(μ2 κ2)5, (μ2 κ2)6 (μ λ2)5, (μ2 λ2)6	(δ2 λ2)2 (δ2 λ2)	γ2 κ2 γ2 λ2	ε2 κ2 ε2 λ2	(α2 κ2)1–3 (α2 λ2)1–3
Stoichiometry (receptor form)	μ2 κ2 μ2 λ2	δ2 κ2, δ1 κ1 δ2 λ2, δ1 λ1	γ2 κ2 γ2 λ2	?	α2 κ2 α2 λ2
Carbohydrate	9–12%	12–15%	2–3%	12%	7–11%
Prosthetic groups (secreted form)	J Chain	–	–	–	J Chain Secretory piece
Complement-mediated cytotoxicity	+	–	G1 weak or - others +	–	–
Placental passage	–	–	+ (All subclasses)	–	–
Protein A binding	–	–	+ (See text)	–	–
Protein G binding	–	–	+ (See text)	–	–
E_{280}0.1% 1cm	Absorbance at 280nm largely reflects tryptophan content, and will vary depending on variable region sequence. Typical values are 1.3–1.5. For a rough estimate, assume a value of 1.4.				
Serum half-life	1 day	Short (probably less than 1 day)	G1, G2a, G3 4 days G2b 2 days	Short (probably less than 1 day)	1 day
Special properties	First antibody in most responses	Major class on lymphocyte surface. Minor class in serum	Major class in serum. IgG-bearing lymphocytes are rare.	Minor class in serum. Binds to mast cells. Responsible for allergic reactions, histamine and serotonin release.	Major class in secretions (e.g. tears, saliva, bile, gut)

chains. In some special circumstances one light chain may predominate, such as the existence of an unusually suitable V_λ gene and the absence of a suitable $V\kappa$ gene.

Each light and heavy chain is made up of a series of *homology units* of *c.* 110 amino acids (Fig. 5.1). The amino acid sequences of individual homology units are sufficiently similar to suggest that they arose by gene duplication from a common primordial gene. Each homology unit characteristically contains one intrachain disulfide bond between cysteine residues situated about 20 amino acids from each end.

Each homology unit is folded into a *domain*, which is a compact, globular structure containing large amounts of β-pleated sheet. This motif has been termed the *immunoglobulin fold* by Poljak *et al.* (1973). Individual domains are rather resistant to proteolytic attack (Fig. 5.1). In contrast, the short linking sequences joining individual domains are often susceptible to proteolysis.

Some classes of antibody contain an extra 'hinge' region (Fig 5.1) which has a more open structure and a more flexible conformation, and is especially susceptible to proteolysis. The flexibility of the hinge probably has biological value, as it allows the antibody to alter the spacing of its variable regions to accommodate polymeric antigens with differing spacing between repeating epitopes. The recent solving of the structure of a complete IgG antibody including the hinge has provided our first glimpse of the conformation of this important region (Harris *et al.*, 1992; see also Brekke *et al.*, 1995, for a discussion on the role of the hinge).

The *N*-terminal amino acid sequence of the light and heavy chains varies greatly between molecules, and is therefore known as the *variable* (V) region. The *N*-terminal homology units of heavy and light chains are designated V_H and V_L respectively. Within each variable region lie three *hypervariable* segments. The V_H and V_L domains are folded in such a way that brings the hypervariable regions together to create an *antigen-combining site*, the specificity of which is determined by both heavy and light chains. In general, each antibody molecule (H_2L_2) has two identical antigen-combining sites. Some antibodies (IgM and IgA) also have higher order polymeric forms $(H_2L_2)_n$, with correspondingly increased numbers of combining sites. The mechanisms of generation of the diverse combining sites will be discussed in a later section of this chapter.

Even after reduction of all interchain disulfide bonds, the heavy and light chains of immunoglobulin molecules remain tightly associated with each other. The reduced chains can only be separated after denaturation in urea, guanidine or acid, indicating the presence of strong noncovalent interactions between chains. X-ray analysis shows that the V_H and V_L domains are in intimate contact with each other, as are C_L and C_H1, and pairs of C_H3 (reviewed by Padlan, 1994). Conversely, interaction between pairs of C_H2 do not occur. 'Vertical' interactions between domains (e.g. between C_H2 and C_H3) are generally weaker.

The affinity of interacting domains is such a prominent feature of antibody structure that some authors define a domain as a *pair* of corresponding homology units folded together (Fig. 5.1). However, the term domain is more commonly used to describe the globular structure made up by a single homology unit. The term domain is also used in a looser way which is virtually synonymous with homology unit sequence.

5.1.1 Nature of the Antigen–Antibody Bond

In recent years, many crystal structures of antigen–antibody complexes have been solved, and we now have a detailed knowledge of the the antigen–antibody bond (Padlan, 1994). The variable region module is made up of a *framework* of a V_H–V_L pair which varies little between individual antibodies or between species, with six hypervariable loops protruding from one end (three from each light chain and three from each heavy chain; Fig. 5.2). These loops, which form the antigen-combining site, are highly variable in sequence, length and conformation, as would be expected if antibodies were to recognize a diverse range of different antigens.

The antigen-combining site thus forms a convoluted surface which is complementary to the antigen to which it binds. In some cases, the combining site consists of a cavity, while in others there are protruding loops in a more 'male' configuration. The combining site often has a somewhat hydrophobic character, which would be expected to make it more sticky. It tends to have numerous amino acids which are capable of forming hydrogen bonds (e.g.

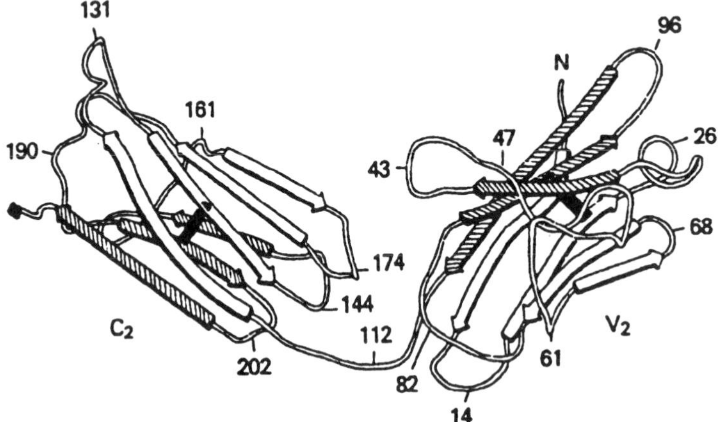

Fig. 5.2. Structure of the antibody-combining site of an antibody light chain. The variable region is to the right and the constant domain to the left. The three hypervariable loops (complementarity-determining regions) protrude to the right, and are marked with amino acids 26 (CDR1), 68 (CDR2) and 96 (CDR3). Reprinted with permission from Edmundson et al. (1975), Biochemistry ***14****, 3953–3961. Copyright 1975, American Chemical Society.*

tyrosine, serine, threonine), and a relative paucity of aliphatic hydrophobic amino acids, which are incapable of forming hydrogen bonds. Charged amino acids contribute to many antigen–antibody interactions because of their ability to form salt bridges, but these bonds usually comprise only a minor component of the total binding energy.

The bond between antigen and antibody usually involves the hydrophobic effect (reduction in free energy by the bringing together of two hydrophobic surfaces, removing them from water), multiple weak noncovalent interactions (hydrogen bonding, van der Waals interactions) and salt bridges. The surface area buried by the interaction with antigen varies greatly, but may be typically 100–1000 $Å^2$ (Padlan, 1994). Typical dissociation constants for the binding of antigen are 10^{-5}–10^{-10} M. Dissociation constants of more than about 10^{-5} M are difficult to measure and probably of little biological importance.

There has been considerable debate concerning whether or not there is a significant conformational change upon binding of antibody to antigen. Is the interaction more like a lock and key or a handshake (induced fit)? There is now an increasing body of evidence that significant conformational changes can occur (Colman, 1991; Davies and Chacko, 1993; Wilson and Stanfield, 1993).

Foote and Milstein (1994) have recently shown that the widely held assumption that the amino acid sequence of a given antibody is sufficient to specify a unique conformation may not be correct. An individual antibody-combining site may have several different isomeric conformations.

Antigen–antibody bonds are usually stable over a pH range of 4–9 and a wide range of salt concentrations (0–1.0 M NaCl), although occasional monoclonal antibodies may be dissociated from their antigen under these mild conditions (Herrmann and Mescher, 1979). By judicious choice of screening procedures, it should be possible to isolate monoclonal antibodies which are able to be disrupted by very minor changes. Dissociation of antigen–antibody bonds more usually requires the use of strong denaturing conditions, such as a pH of less than 3.0 or greater than 10.0, or high concentrations of thiocyanate (3.5 M), guanidine-hydrocholoride (6 M) or urea (9 M). Usually, but not always, antibody activity is restored once the denaturing conditions are removed. Dissociation of antigen–antibody interactions is discussed in detail in Chapter 11.

5.1.2 Membrane Immunoglobulins

The heavy chains of membrane bound receptor forms of antibody are identical to the secreted forms over most of their sequence, but differ at their *C* terminus. Secretory immunoglobulins have a hydrophilic *C* terminus, while membrane immunoglobulins have a hydrophobic *C* terminus which is responsible for membrane attachment (Rogers and Wall, 1984).

The *C*-terminal membrane extension can be divided into three segments.

The first is a highly charged *c.* 15 amino-acid sequence rich in acidic amino acids, and is followed by an uninterrupted sequence of *c.* 25 uncharged and predominantly hydrophobic amino acids, which presumably spans the lipid bilayer. Finally, there is a basic triplet (Lys-Val-Lys), which probably marks the entry into the cytoplasm. Membrane IgM and IgD terminate at this point (Rogers *et al.*, 1980: Cheng *et al.*, 1982) while membrane IgG has an additional cytoplasmic tail of *c.* 28 amino acids (Tyler *et al.*, 1982).

5.2 Proteolytic Fragmentation of Immunoglobulins

Controlled proteolysis of immunoglobulins often generates large fragments, with the cleavage sites usually lying between domains.

Proteolytic attack of IgG by papain at neutral pH cleaves at the hinge, generating two identical Fab fragments, each containing one intact light chain disulfide-bonded to a fragment of the heavy chain (Fd) containing the V_H and C_H1 domains (Fig. 5.1). The Fab fragments each contain one antigen-combining site. The remaining portion of the IgG molecule consists of a dimer of C_H2 and C_H3, and is called Fc because, in some cases, it can be crystallized. The Fc portion is homogeneous, and contains no antigen-binding activity.

In a few very special cases, it is possible to generate an even smaller antigen-binding fragment (Fv) by proteolysis (Fig. 5.1). The Fv fragment contains only the noncovalently associated V_H and V_L domains (Inbar *et al.*, 1972; Kakimoto and Onoue, 1974). F_v fragments are not easy to make by proteolysis, but have assumed a much greater importance in recent times because they can be made by genetic engineering, and are well suited to expression in bacteria because they are small and not glycosylated (see Zdanov *et al.*, 1994).

Another proteolytic reaction of major importance is the attack of antibodies by pepsin at a pH around 4. Like papain, pepsin cleaves IgG at the hinge, but the pepsin cleavage site lies on the carboxyl side of the inter-heavy chain disulfide bond(s) (Fig. 5.1). The divalent antigen-binding fragment so produced is known as $F(ab')_2$. The heavy chain fragment in $F(ab')_2$ is a little longer than the papain Fd fragment, and is known as Fd′. Pepsin cleavage usually results in the breakdown of the C_H2 domain into small peptides, although under certain circumstances, a fragment approximating a dimer of C_H3 (pFc′) may be recovered.

Controlled proteolysis of immunoglobulins is useful in many practical situations, and also illustrates some of the basic principles and nomenclature of antibody structure. However, the rate and nature of fragmentation depends greatly on the species and class of antibody. The classical recipes which work well for human or rabbit IgG may fail completely when applied to the mouse or rat, or to classes other than IgG. We will return to this important subject in Chapter 9.

5.3 Properties of the Individual Immunoglobulin Classes

The physical and biological properties of the individual classes of immunoglobulin are summarized in Table 5.1. Much useful information concerning the structure, purification and properties of individual antibody classes can be found in the volume edited by Di Sabato *et al.* (1985).

5.3.1 IgM

IgM is phylogenetically the oldest immunoglobulin, and is also the first to appear in ontogeny. The μ heavy chain has four constant region domains (Kehry *et al.*, 1979, 1982). Homology considerations suggest that the 'extra' domain over IgG is $C_\mu 2$, which is located between $C_\mu 1$ and the first inter-heavy chain disulfide bonds. IgM lacks a well-defined hinge, although the preferred site of proteolytic attack is usually the region of the $C_\mu 2$ domain (Shimizu *et al.*, 1974). Overall, IgM is rather resistant to proteolysis. Fragmentation of IgM will be discussed in Chapter 9.

In mammals, secretory IgM consists of a pentamer (M_r 9×10^5) or hexamer (M_r 1.18×10^6) of $\mu_2\lambda_2$ or $\mu_2\kappa_2$ subunits (monomers) with an M_r of 180 000 (Davis *et al.*, 1988, 1989; Davis and Shulman, 1989; Randall *et al.*, 1992; Brewer *et al.*, 1994a, b). The multimeric nature of IgM has great biological value. It helps overcome the relatively low affinity of individual combining sites that have not yet been subject to somatic mutation (Gearhart *et al.*, 1981) and facilitates its ability to activate complement. The hexameric form of IgM is up to 20 times more efficient at activating complement than the more well-known pentameric form (Davis *et al.*, 1988). Like polymeric IgA, IgM is also transported into bodily secretions via the receptor for polymeric immunoglobulin (see Section 5.3.5).

The membrane receptor form of IgM is a monomer in which the heavy chains possess a hydrophobic *C*-terminal extension (Rogers *et al.*, 1980), and may also acquire covalent lipid (Pillai and Baltimore, 1987). IgM receptors are associated with several proteins in the membrane, including Ig-α, Ig-β and others, without which membrane IgM cannot be transported to the cell surface (Wienands and Reth, 1992; Reth, 1994). These accessory chains are involved in signal transduction (Cambier *et al.*, 1994; Reth, 1994; Terashima *et al.*, 1994).

The arrangement of disulfide bonds between subunits of polymeric IgM is still not completely clear, and may not be the same in all IgM molecules (see Davis *et al.*, 1988, 1989; Davis and Shulman, 1989; Brewer *et al.*, 1994a, b). Older views that the only disulfide bonds between monomeric subunits of mouse IgM are situated at the penultimate cysteine 575 are probably not correct. Elimination of this cysteine by site-directed mutagenesis still allows polymerization into hexameric IgM, indicating that cysteine 414 can link monomers (Davis *et al.*, 1988).

IgM polymerization occurs early in the secretory pathway, most likely in the endoplasmic reticulum, and occurs in a stepwise manner, with non-covalent assembly preceding the formation of disulfide bonds between monomeric subunits. IgM does not proceed along the secretory pathway until assembly is complete (Brewer *et al.*, 1994a, b).

Pentameric but not hexameric IgM also contains a single J chain, an acidic non-immunoglobulin polypeptide of M_r *c.* 15 000, which is rich in carbohydrate and cysteine residues. It is believed to be disulfide bonded to the Cys 575 residues of two adjacent IgM monomers. Recent work has cast some doubt on the exact stoichiometry, and it is possible that the J chain can form disulfide bonds to other cysteines (Davis and Shulman, 1989; Randall *et al.*, 1992).

The role of J chain is still not well understood. Previous suggestions that J chain is essential for assembly or secretion of IgM have recently been challenged (Randall *et al.*, 1992). In the absence of J chain, secreted IgM is predominantly in the hexameric form, but when J chain is expressed, IgM is secreted mainly as pentamers, demonstrating that J chain regulates the structure and function of secreted IgM (Randall *et al.*, 1992; see also Niles *et al.*, 1995). J chain was found to have little effect on the rate of IgM secretion, but had a major effect on whether it was in the pentameric or hexameric form.

IgM constitutes roughly 5% of total serum immunoglobulins, although the levels vary depending on the degree of antigenic stimulation. Catabolism of IgM is rapid. The serum half-life in the mouse is approximately 1 day (Spiegelberg, 1974), and about 2 days in the rat (Bazin *et al.*, 1990a).

5.3.2 IgD

In marked contrast to all the other immunoglobulins classes, IgD is rarely secreted, and its main function appears to be as a receptor for antigen (Blattner and Tucker, 1984). The murine δ chain has an M_r of *c.* 68 000, and contains only two constant region domains (Cδ1 and Cδ3) and an extended hinge. The rat δ chain has a similar structure (Sire *et al.*, 1982; Bazin *et al.*, 1990a). The human δ chain is somewhat larger, containing three constant region domains (Cδ1, Cδ2 and Cδ3), and also has an extended hinge (Takahashi *et al.*, 1982).

The great majority of mature B lymphocytes have IgD on their surface, usually in combination with IgM, and on any one B cell the two classes of receptor have the same specificity for antigen (Goding and Layton, 1976). Immature B lymphocytes possess only IgM on their surface, and IgD is acquired with maturation. Most splenic B cells have approximately 10 times as much IgD as IgM on their surface (Havran *et al.*, 1984). After stimulation of the B cell with mitogen, IgD is rapidly lost (Bourgois *et al.*, 1977). This orderly appearance and disappearance suggest a unique physiological role for IgD, but to date none has been found. Any defects in immune response in mice in which the IgD gene has been inactivated are subtle (Nitschke *et al.*, 1993; Roes and Rajewsky, 1993).

Unlike IgM, IgD is highly susceptible to proteolytic cleavage at the hinge (Vitetta and Uhr, 1975; Goding, 1980; Takahashi *et al.*, 1982). Membrane IgM and IgD both contain a hydrophobic *C*-terminal segment which ends in Lys-Val-Lys. The terminal tripeptide presumably marks the entry into the cytoplasm. The amino acid sequences of the transmembrane segments and cytoplasmic tails of IgM and IgD are highly homologous, but there are some differences in the associated signal-transducing proteins (Kim *et al.*, 1994c; Terashima *et al.*, 1994). Unlike membrane IgM, IgD is capable of cell surface expression in the absence of Ig-α and Ig-β proteins in a variant-transfected myeloma cell line, but is then found to be attached to the membrane via a phosphatidylinosityl anchor (Wienands and Reth, 1992). It is not yet known whether these observations have any physiological significance in normal B cells.

The role of membrane IgD remains something of a mystery. Older beliefs that the acquisition of IgD signalled a change in the B cell towards resistance to tolerance induction are not easily reconciled with the apparently normal B-cell tolerance in mice in which the IgD gene has been inactivated (Nitschke *et al.*, 1993; Roes and Rajewsky, 1993). Brink *et al.* (1992) found that the only overt difference between the behaviour of membrane IgM and IgD was a greater degree of downregulation of IgD after induction of clonal anergy by antigen. The search for a clear biological function for membrane IgD continues.

Secretory forms of IgD also exist, and very rarely IgD myelomas are found (Finkelman *et al.*, 1981). In contrast to membrane IgD, the secretory form has a hydrophilic *C* terminus (Takahashi *et al.*, 1982; Cheng *et al.*, 1982). The levels of secretory IgD in serum are very low (of the order of 1–10 μg/ml). Secretory IgD does not appear to fix complement, cross the placenta, cause mast cell degranulation, or localize in secretions. Like membrane IgD, its role is completely unknown.

There is some variation in inter-heavy-chain disulfide bonds in murine IgD. The membrane form sometimes lacks an inter-heavy-chain disulfide bond (Pollock *et al.*, 1980) while the secretory form appears to be a disulfide-bonded dimer of the basic δ_2L_2 structure (Finkelman *et al.*, 1981).

5.3.3 IgG

IgG is by far the most abundant immunoglobulin in serum. It consists of two heavy chains of M_r of *c.* 160 000. The constant regions of γ chains have three regular domains (Cγ1, Cγ2 and Cγ3) and a short hinge.

Human, mouse and rat each have four IgG subclasses: G1–4 in humans;

G1, G2a, G2b and G3 in the mouse; G1 G2a, G2b and G2c in the rat. The names of the IgG subclasses reflect the history of their discovery, and the name of one IgG subclass in one species does not necessarily reflect a structural or functional similarity with the subclass of the same name in another species. The four human IgG subclasses are very closely related to each other (typically greater than 90% amino acid identity except for IgG3, which has an unusually large hinge region), but are much less closely related to the IgG subclasses in rodents (Hamilton, 1994).

The four IgG subclasses have heavy chains with virtually identical molecular weights, and cannot be distinguished with certainty by electrophoretic mobility in agarose or cellulose acetate under nondenaturing conditions (Section 5.5). In general, γ1 antibodies have a 'faster' (i.e. more anodal) mobility than γ2, although there is considerable overlap. It is virtually impossible to separate polyclonal mouse or rat IgG subclasses by classical procedures such as ion exchange chromatography or zone electrophoresis. The heavy chains of mouse IgG2b run as a double band on SDS–polyacrylamide gel electrophoresis (Köhler *et al.*, 1978; Parham, 1983; Lo *et al.*, 1984). The structural basis for this is now known to be the partial *O*-glycosylation of threonine 221A (see Kim *et al.*, 1994a, b; Yamaguchi *et al.*, 1995).

Despite their virtually identical size and overlapping electrophoretic mobilities, the mouse γ1, γ2a, γ2b and γ3 subclasses have rather dissimilar amino acid sequences, and except for γ2a and γ2b show little serological cross-reaction. They are so different that some authors have considered them to be different classes (named IgF, IgG, IgH and IgI, respectively; see Potter and Lieberman, 1967). This nomenclature is now obsolete.

The C_H1 homology units of mouse γ1, γ2a and γ2b are *c.* 85% homologous. The C_H2 domains of γ2a and γ2b are 94% homologous, but only 67% homologous to γ1, while the C_H3 domains of γ2a and γ2b are only 61% homologous to each other and *c.* 55% homologous to γ1 (Sikorav *et al.*, 1980).

Hinge region homology between mouse IgG subclasses varies from 18% (γ1:γ2b) to 64% (γ2a:γ2b). Although there is a correlation between hinge region flexibility and ability to fix complement (Oi *et al.*, 1984), it is now clear that this is not a requirement for complement activation (Brekke *et al.*, 1995). The most important sequence for the binding and activation of the C1q component of complement lies in the C_H2 domain. Activation of complement requires this region, an intact disulfide bond between the heavy chains in the hinge region, the presence of *N*-linked carbohydrate, and paired C_H3 domains (Brekke *et al.*, 1995).

There has been some uncertainty regarding the properties of the IgG subclasses, particularly in relation to interaction with complement. Part of the confusion probably resulted from the use of artificially aggregated myeloma proteins to assess complement fixation, rather than 'real' antibody binding and cytotoxicity assays (see Bazin, 1990a; Hirayama *et al.*, 1982; Brekke *et al.*, 1995). Table 5.2 summarizes current knowledge.

The relationships between the various IgG subclasses in mice and rats have

Table 5.2 Properties of mouse and rat IgG Subclasses

Species	Subclass	Complement mediated Cytotoxicity	Binding to Protein A-Sepharose*	Binding to Protein G-Sepharose‡	Comments
Mouse	Gl	–	pH > 8 (weak)†	++	Fast electrophoretic mobility. Major subclass.
	G2a	++	pH > 4.5	++	Major subclass.
	G2b	+++	pH > 3.5	++	Tends to precipitate in low salt concentrations.
	G3	+	pH > 4.5	+	Minor subclass; associated with anticarbohydrate activity. Tends to precipitate in low salt concentrations.
Rat	G1	–	–	++	Minor subclass. Sometimes binds to protein A. Similar to mouse IgG1
	G2a	±	±	+	Major subclass.
	G2b	+	-	+	Major subclass.
	G2c	+	+	++	Minor subclass; associated with anticarbohydrate activity. Tends to precipitate in low salt concentrations. Similar to mouse IgG3

* See Ey et al, 1978, Malm, 1987.

† Binding of mouse IgG1 to protein A-Sepharose is strong in a buffer of pH >8.5 and containing > 1 M NaCl.

‡ Binding to protein G-Sepharose 4 Fast Flow or protein G-Superose (Pharmacia) is considerably stronger than binding to protein G alone. See Akerström *et al.* (1985). For many species, binding of IgG to protein G occurs equally strongly between pH 6 and 9, and at physiological ionic strength. However, it should be noted that Pharmacia recommend the use of low ionic strength buffers with protein G-Superose and protein G-Sepharose 4 Fast Flow for mouse and rat IgG. Binding of mouse IgG is in 20 mM sodium phosphate at pH 7.0, while binding of rat IgG is in 2 mM sodium phosphate, pH 7.0 (see Pharmacia literature). It may be presumed that the choice of low ionic strength buffers is not accidental, and perhaps contributes to strengthening of binding by nonspecific electrostatic effects. Elution from protein G-Sepharose is in 0.1 M glycine-hydrochloride, pH 2.7.

now been established (Table 5.2; Bazin *et al.*, 1990a). Rat IgG2a and IgG1 are very similar to each other and to mouse IgG1. Mouse IgG2a and IgG2b are very similar to each other, with the main differences being in the C_H3 domain, and are very similar to rat IgG2b. Rat IgG2c is similar to mouse IgG3 and human IgG2 (Der Balian *et al.*, 1980; Nahm *et al.*, 1980; Bazin *et al.*, 1990a).

Mouse IgG3 and its rat homologue IgG2c have some properties that set them apart from the other IgG subclasses (Greenspan and Cooper, 1992). These subclasses tend to predominate in responses to carbohydrate antigens, particularly bacterial polysaccharides. They show a marked tendency towards self-aggregation, are often cryoglobulins (they precipitate in the cold) or euglobulins (precipitate at low ionic strength), and tend to aggregate upon freezing. Greenspan and Cooper (1992) summarized the evidence and found that the reason for these properties may be related to a unique tendency for antibodies of this subclass to have noncovalent interactions between the Fc regions of adjacent molecules, most likely through the C_H3 domains. Mouse IgG2b is also somewhat prone to aggregation.

5.3.4 Binding of IgG to Fc Receptors on Mammalian Cells

One very important property of IgG is the binding of its Fc portion to 'Fc receptors' present on many cell types (Hulett and Hogarth, 1994). The binding of immune complexes to Fc receptors or the antigen-induced cross-linking of immunoglobulin bound to Fc receptors is often followed by a series of biological consequences, including cell activation or degranulation, and/or transport of the complex into the cell (endocytosis) or across the cell to the opposite side of an epithelial surface (transcytosis). In addition, the binding of antibodies and immune complexes to Fc receptors can generate a large number of experimental artefacts.

There are four main types of Fc receptor for IgG; FcγRI, FcγRII, FcγRIII and FcγRn, which differ in their affinity for IgG, their subclass and species specificity and their cellular distribution. Of these, FcγRI is the only one that binds monomeric IgG with high affinity; it is known as the high-affinity Fcγ receptor. FcγRII and FcγRIII have much lower affinity, but may bind IgG aggregates with higher affinity than monomers.

In the mouse, these Fc receptors are encoded by single genes, but in humans, there are multiple genes for most Fc receptor types. In addition, there is added diversity because of alternate mRNA splicing patterns, so the overall picture is complex (Hulett and Hogarth, 1994). The properties, binding specificity and distribution of Fcγ receptors is shown in Table 5.3. Because it is common to use mouse monoclonal antibodies on human cells, Table 5.3 also lists the specificity of the human Fcγ receptors for the various mouse IgG subclasses.

The Fcγn receptor is unique in that it is present only in newborn mice and rats, and transports IgG from maternal milk across the intestinal epithelium and into the bloodstream. This receptor is unusual in that the binding of IgG is pH dependent, with strong binding at pH 6.5 in the gut, but negligible binding at pH 7.4 on the blood side of the epithelium, allowing its release (Simister, 1990; Hulett and Hogarth, 1994).

Binding of antibody to Fc receptors is important from a practical viewpoint because it may cause artefactual nonspecific binding to tissues. This artefact has probably been responsible for many misleading observations in the literature (Vitetta and Uhr, 1975; Winchester *et al.*, 1975). Controls for Fc receptor binding should include use of a control antibody of the *same subclass* but different specificity to the active antibody. It is wise to deaggregate IgG antibodies by high-speed centrifugation (100 000 ***g***) just before each use. The

Table 5.3 Properties of Fcγ Receptors

Receptor type	Affinity for IgG	IgG classes that bind	Cellular distribution
hFcγRI	10^{-8}–10^{-9} M	hIgG2,3 hIgG4 (lower affinity) mIgG2a, mIgG3	Monocytes, macrophages; inducible on eosinophils and neutrophils with IFN-γ.
mFcγRI	10^{-7}–10^{-8} M	mIgG2a	Macrophages, monocytes; may be similar to human.
hFcγRII	<10^{-7} M	Essentially only binds IgG complexes; hIgG1, hIgG3 hIgG2, hIgG4 (weak) mIgG1, mIgG2a,b Human FcγRII has multiple isoforms with differing specificity.	All leukocytes, macrophages, Langerhans cells, platelets, B cells, some T cells; absent from NK cells.
mFcγRII	<10^{-7} M	mIgG1,2a,2b, mIgE; does not bind mIgG3	Similar to human. Some isoforms predominate in B cells, others in monocytes & macrophages.
hFcγRIII (multiple isoforms)	<10^{-7} M	hIgG2,4 (GPI-linked form) weak hIgG1,3 (GPI-linked form) Specificity of transmembrane form not yet known. All mouse IgG except IgG2b?	GPI-linked form on neutrophils, TM form on macrophages and NK cells Upregulated by cytokines (see Hulett and Hogarth, 1994 for details)
mFcγRIII (single isoform)		mIgG1, 2a, 2b mIgE (low affinity)	
FcRn (rat and mouse only)	10^{-8} M	IgG (binds at pH 6.4; releases at pH 7.4)	Intestinal epithelial cells, yolk sac

Beckmann Airfuge and the Beckman TL-100 benchtop ultracentrifuge are ideal for this purpose, because only 5–10 minutes at 100 000 ***g*** are required. Alternatively, Fab or $F(ab')_2$ fragments may be used to prevent binding to Fc receptors (Chapter 9).

5.3.5 Bacterial Fc Receptors: Interaction of IgG with Staphylococcal Protein A and Streptococcal Protein G

Many bacteria have IgG-binding proteins on their surface (Boyle and Reis, 1987; Boyle, 1990). The role of these 'bacterial Fc receptors' in pathogenicity is unclear (see Myhre and Kronvall, 1980), but they are extremely useful tools for the detection and purification of IgG (Goding, 1978; Langone, 1982; Boyle and Reis, 1987; Boyle, 1990). Bacterial Fc receptors have been subdivided into five main types (Woof and Burton, 1990). Of these, the most widely exploited are type I (protein A from *Staphylococcus aureus*) and type III (protein G from group G streptococci).

Staphylococcal protein A binds to IgG subclasses of many species, and has been widely used to detect and purify IgG (Goding, 1978; Langone, 1982; Surolia *et al.*, 1982; Boyle, 1990). Affinity chromatography on protein A-agarose is a very simple and efficient way of purifying IgG, and differential elution from protein A with stepwise pH gradients is capable of separating most mouse IgG subclasses (Ey *et al.*, 1978; Seppälä *et al.*, 1981). The protein A-binding region has been localized to the junction of the C_H2 and C_H3 domains (Lancet *et al.*, 1978). Binding is predominantly hydrophobic, but with contributions by polar residues (Deisenhofer, 1981).

There is considerable variation in the ability of different subclasses of mouse and rat IgG to bind to protein A (see Chapter 9). Protein A binds mouse IgG2a, IgG2b and IgG3 strongly, but binds mouse and rat IgG1 weakly and does not bind rat IgG2a or rat IgG2b at all (Bazin *et al.*, 1990b). Binding of mouse IgG1 (and presumably rat IgG1) is greatly strengthened by the use of buffers combining alkaline pH (typically 8.5; see Ey *et al.*, 1978) and high ionic strength (up to 1 M NaCl; see Östlund, 1986; Malm, 1987). Purification of IgG by affinity chromatography on protein A-agarose is discussed in detail in Chapter 9.

It has been shown that preparations of protein A are mitogenic for lymphocytes, but this activity is not present in protein A prepared by recombinant DNA technology, and appears to be due to contaminating staphylococcal enterotoxins, which can act as 'superantigens' (Schrezenmeier and Fleischer, 1987; Fleischer, 1989; Marrack and Kappler, 1990; but see also Barrett, 1990).

The selectivity of staphylococcal protein A towards certain IgG subclasses has sometimes been a disadvantage, and another bacterial immunoglobulin-binding molecule, protein G of streptococci, has recently been characterized (Björck and Kronvall, 1984; Åkerström *et al.*, 1985; Fahnestock, 1987; Stone *et al.*, 1989; Björck and Åkerström, 1990; Fahnestock *et al.*, 1990; Myhre, 1990). Protein G binds to at least two regions on the γ heavy chain, one on the Fd portion and one on the Fc (Stone *et al.*, 1989; Derrick and Wigley, 1992). The structure of a complex of an Fab fragment of mouse IgG1 with protein G has recently been determined (Derrick and Wigley, 1992). It appears likely that protein G can bind to antibody Fd regions without interfering with antigen binding. The structural basis of the binding of protein G to the Fc portion of IgG is not yet known.

Protein G has a different and somewhat broader selectivity towards immunoglobulin subclasses from different species, and forms a very useful complement to protein A (Åckerström *et al.*, 1985; Table 5.2). Protein G binds all mouse IgG subclasses, and binds mouse IgG1 much more strongly than protein A. The affinity of binding of rat IgG to protein G is said to be about 10 times less than that for other species (Björck and Åckerström, 1990), but is still adequate for affinity chromatography. Protein G also binds rat IgG2a and IgG2c strongly, and rat IgG1 and IgG2b somewhat less strongly (Nilsson *et al.*, 1982; Nilson *et al.*, 1986).

Although the optimal binding pH for free protein G to IgG is said to be about 5 (see Åkerström and Björck, 1986; Björck and Åkerström, 1990), the manufacturer of protein G-Sepharose 4 Fast Flow and protein G-Superose (Pharmacia) recommends binding at pH 7.0, and elution at pH 2.7 (see below).

It should also be noted that the strength of binding of protein G to certain antibody classes and species is stronger when the protein G is coupled to agarose beads. Presumably, extra binding energy is conferred by a nonspecific component. This component is probably electrostatic. Some literature recommends the use of low-salt buffers (20 mM sodium phosphate, pH 7.0 for mouse IgG, and 2 mM sodium phosphate pH 7.0 for rat IgG; Pharmacia, 1995). It may be presumed that the choice of such low ionic strength buffers is not accidental. One suspects that they were chosen because higher ionic strength buffers weakened the nonspecific electrostatic component. The use of such low ionic strength buffers is certainly not obligatory, and it is possible to get good recovery of mouse and rat IgG using buffers with physiological salt concentration. Because the optimal binding of IgG to protein G is at a pH of about 5 (Åckerström and Björck, 1986), it may be best to bind at a slightly acidic pH. The use of low ionic strength buffers may cause precipitation of some antibodies that have euglobulin properties (especially some mouse IgG2b and IgG3 and some rat IgG2c; see Sections 5.3.3 and 9.2.2). The practical use of protein G for IgG purification will be discussed in more detail in Chapter 9.

In addition to binding IgG, protein G binds albumin via a distinct region on the protein G molecule (Åckerström *et al.*, 1987; Björck *et al.*, 1987). This region has been deleted from the protein G that is commercially available from Pharmacia (Protein G Sepharose 4 Fast Flow and Protein G Superose).

Genetically engineered hybrid molecules between protein A and protein G are commercially available coupled to agarose beads, and combine the best features of each molecule (eg protein A/G; Pierce).

5.3.6 IgE

IgE is a quantitatively very minor immunoglobulin class (serum concentration *c.*0.3 μg/ml). It consists of two light chains and two ε chains of M_r *c.*70 000

(Liu *et al.*, 1980) consisting of four constant region domains (Dorrington and Bennich, 1978; Ishida *et al.*, 1982). Switching from IgM to IgE requires the cytokine IL-4, and mice in which the IL-4 gene is disrupted are unable to make IgE (Kuhn *et al.*, 1991).

IgE is responsible for hypersensitivity and allergic reactions which are mediated by binding of IgE to mast cells, followed by degranulation upon antigenic stimulation. Binding of IgE to mast cells occurs via a specific receptor for Fc(ε), and the signal for degranulation involves antigen-mediated cross-linking of these receptors (Kinet and Metzger, 1990).

It has always been assumed that the nuisance value of IgE in causing allergy, asthma and occasionally anaphylaxis must be balanced by some important survival value. It has long been suspected that IgE plays a role in defence against parasites such as schistosomes (Capron and Capron, 1994; Gounni *et al.*, 1994). The powerful technology of making knockout mice offers a chance to test this idea. Deletion of the α chain of the high-affinity receptor for IgE from the germline of mice abolishes anaphylaxis (Dombrowicz *et al.*, 1993). It will be interesting to see if these mice have decreased defences against schistosomes and other parasites.

IgE myelomas are very rare in man, and to date none have been reported in the mouse. However, a number of rat IgE myelomas have been described (Bazin *et al.*, 1973). IgE-secreting hybridomas with specificity against ovalbumin and dinitrophenyl have been constructed (Böttcher *et al.*, 1978; Liu *et al.*, 1980; Eshhar *et al.*, 1980).

5.3.7 IgA

IgA is the main immunoglobulin of body fluids, such as saliva, tears, colostrum and intestinal secretions (Tomasi and Grey, 1972; Mestecky and McGhee, 1987), and is also present at low concentrations in serum. The molecular weight of the α heavy chain is approximately 68 000–75 000. The range is due to carbohydrate heterogeneity. There are three constant region homology units (Tucker *et al.*, 1981). A major role for interleukin-6 in IgA production is suggested by the fact that mice in which the IL-6 gene is disrupted have a severe deficiency in IgA responses (Ramsay *et al.*, 1994).

In serum, IgA consists mainly of an α_2L_2 tetramer of M_r *c.* 180 000, but it is also present as disulfide-bonded polymers of this basic structure (Nisonoff *et al.*, 1975). IgA myeloma proteins show similar heterogeneity. The disulfide bond between polymer subunits is easily reduced. In IgA myelomas of BALB/c mice, the light chains are not disulfide bonded to the heavy chains. The disposition of the disulfide bonds in mouse IgA is somewhat variable (Stanisz *et al.*, 1983), and will be discussed in Chapter 6.

In contrast to the serum form, IgA in secretory fluids mostly occurs as a disulfide-bonded dimer of the basic α_2L_2 structure. Secretory IgA also contains J chain disulfide bonded to the Fc piece (Koshland, 1975, 1985), and a second nonimmunoglobulin component known as secretory piece or secretory component. Secretory component is attached to α chains by strong non-covalent forces, and also by disulfide bonding in some species (Underdown *et al.*, 1977). Secretory component is a proteolytic fragment of the receptor for polymeric immunoglobulin (pIg) on epithelial cells. The intact pIg receptor contains five extracellular domains, which are strikingly homologous to each other and to antibody variable regions (Mostov *et al.*, 1984, Mostov, 1994).

Transport of IgA across epithelia involves binding to the pIg receptor, endocytosis via coated vesicles, transport through the cell, and exocytosis at the luminal side of the cell with part of the receptor (the secretory piece) still attached (Underdown, 1990; Mostov, 1994). In addition to its transport function, secretory piece may protect IgA from proteolysis (Underdown and Dorrington, 1974). The receptor for polymeric immunoglobulin is expressed at high levels in hepatocytes in rats, mice and chickens, but not in humans (Underdown, 1990).

The mechanism of binding of IgM and polymeric IgA to the pIg receptor is not well characterized (see Mostov, 1994). Earlier suggestions that it may be mediated via J chain (Brandtzaeg and Prydz, 1984; Brandtzaeg, 1985) are still somewhat controversial (Underdown, 1990).

IgA is encoded by a single gene in rats and mice, but has two subclasses (IgA1 and IgA2) in man (Lai *et al.*, 1989) and at least 13 Cα genes in rabbits (Burnett *et al.*, 1989). IgA2 constitutes about 10% of IgA in serum but about 50% in secretory fluids (Grey *et al.*, 1968).

IgA does not mediate complement-dependent cell lysis by the direct pathway. Its main role is probably to neutralize viruses and bacteria, and to mediate defences in bodily secretions and the lumen of the gut. It may serve as a major route of excretion of antigen via the bile in rodents and chickens, but probably not in man (Underdown, 1990). The increased valency of secretory IgA polymers may aid viral neutralization, and its great resistance to proteolysis probably improves its survival in the gut (Underdown and Dorrington, 1974).

5.4 Biosynthesis and Assembly of Immunoglobulins

The biosynthesis of immunoglobulins follow the same general pathway as other secretory and membrane glycoproteins (reviewed by Wall and Kuehl, 1983; see also Brewer *et al.*, 1994a, b). Nascent chains are formed on polyibrosomes bound to the cytoplasmic surface of the endoplasmic reticulum, and their *N*-terminal hydrophobic 'leader sequence' guides them through the membrane and into the lumen. As the nascent chain appears on the luminal side its leader sequence is proteolytically removed and glycosylation occurs

(Tartakoff and Vassalli, 1979). Membrane and secretory forms are assembled in the same intracellular compartment (Goding, 1982).

Carbohydrate is added to the nascent heavy chains via preformed dolichol phosphate donor, and is initially of the high mannose endoglycosidase H-sensitive type containing *N*-acetyl glucosamine and mannose, but lacking galactose and sialic acid (Tartakoff and Vassalli, 1979; Goding and Herzenberg, 1980; Wall and Kuehl, 1983; Brewer *et al.*, 1994a). In most cases, the carbohydrate of immunoglobulin is attached to asparagine. Glycosylation of serine and threonine is uncommon in immunoglobulins, but is known to occur in IgA and IgD. Mouse IgG2b may possess an *O*-linked glycan on one chain (Parham, 1983; Kim *et al.*, 1994a, b; Yamaguchi *et al.*, 1995). The light chains are usually not glycosylated.

Shortly after synthesis, pairing of heavy and light chains commences. The pathway for assembly of IgG is relatively simple. In some cases the order of assembly is H → HH → HHL → LHHL, while in others it is H → HL → LHHL (Baumal and Scharff, 1973; Bergman and Kuehl, 1979). Some lymphocytes make membrane IgG as well as large amounts of secretory IgG. In these cells, the law of mass action together with the fact that membrane and secretory chains are made in the same compartment results in the formation of aysmmetrical membrane IgG molecules containing one membrane chain and one secretory chain (Goding, 1982).

Polymerization of IgM is considerably more complex. It occurs in the endoplasmic reticulum (Tartakoff and Vassalli, 1979; Brewer *et al.*, 1994a), and IgM does not move to the Golgi until assembly of polymers is completed.

In recent years, it has become apparent that the folding and assembly of multimeric secretory and membrane proteins is not spontaneous, but requires the assistance of numerous other proteins (reviewed by Hurtley and Helenius, 1989; Gething and Sambrook, 1992). Intermediates in glycoprotein synthesis, including immunoglobulin heavy chains, are transiently bound to a 78 kDa protein called BiP, which is a molecular chaperone and a member of the stress-70 family. Release from BiP requires ATP. The role of BiP appears to be to prevent aggregation while the protein makes multiple attempts to fold. Once folding and/or assembly is achieved, release occurs. A second endoplasmic reticulum protein of apparent molecular mass 90 kDa, termed *calnexin*, also binds to glycoprotein-folding intermediates (Ou *et al.*, 1993).

Most immunoglobulin-secreting cells make an excess of light chains, and these are usually secreted (Parkhouse, 1971; Goding and Herzenberg, 1980). This may have some biological logic. The production of large amounts of isolated heavy chains without light chains is thought to be toxic to the cell, possibly because they compete for the binding to BiP, which may impair its role in other essential cellular processes (see Köhler, 1980; Argon *et al.*, 1983; Gething and Sambrook, 1992).

In contrast to the cytoplasm, the environment of the lumen of the endoplasmic reticulum is oxidizing (Hwang *et al.*, 1992). However, the formation of disulfide bonds is not spontaneous, and appears to require assistance via protein disulfide isomerase and possibly other enzymes (see Freedman, 1989; Bardwell and Beckwith, 1993; Brewer *et al.*, 1994a).

Following initial glycosylation and assembly, immunoglobulins are transported to the Golgi apparatus, via a vesicular transport system (Rothman and Orci, 1992; Söllner *et al.*, 1993). In the Golgi, the carbohydrate chains are trimmed and galactose and sialic acid added, producing modified carbohydrates which are generally resistant to endoglycosidase H (Tartakoff and Vassalli, 1979; Goding and Herzenberg, 1980; Brewer *et al.*, 1994a).

As is the case for all glycoproteins, glycosylation of immunoglobulins probably serves multiple functions (see Varki, 1993, and Fiedler and Simons, 1995 for discussion). It has long been suspected that *N*-linked glycans might serve as signals for intracellular trafficking, but the evidence for this has been somewhat weak and contradictory. However, recent work suggests that this question requires re-examination (see Fiedler and Simons, 1995). Glycosylation may also help folding and cover hydrophobic patches. Inhibition of *N*-linked glycosylation by tunicamycin has quite variable effects, which depend on the class and individual properties of each antibody (Wall and Kuehl, 1983). Treatment of cells with tunicamycin reduces IgM secretion but does not affect IgG secretion (Blatt and Haimovitch, 1981). Glycosylation is essential for activation of complement (Brekke *et al.*, 1995), but this may merely reflect the need for proper folding.

5.5 Serum Electrophoresis

Serum is a complex mixture of several hundred proteins. The total protein concentration is about 70 mg/ml. By far the most abundant single protein is albumin (~40 mg/ml). When analysed by electrophoresis in agarose or cellulose acetate at pH 8.3–8.6, albumin is visible as a single intense band at the anodal end (Fig. 5.3), consistent with its small size and acidic nature (pI ~4.9). Most of the remaining bands in normal serum electrophoresis consist of mixtures of several different components, although some are identifiable as distinct proteins (Jeppsson *et al.*, 1979; Janik, 1984; Killingsworth, 1985; Killingsworth and Warren, 1986).

Proteins with the general structure of antibodies are referred to as immunoglobulins or the virtually obsolete term *gamma globulins*. They migrate as a broad band situated close to the origin (Fig. 5.3). Depending on the pH and the degree of electro-endosmosis in the supporting matrix, they may move slightly towards the cathode, remain near the origin, or move slightly towards the anode. Typical immunoglobulin levels in normal mouse serum range from 1–10 mg/ml, depending on the degree of antigenic stimulation. In specific-pathogen-free mice, the immunoglobulin levels may be virtually undetectable.

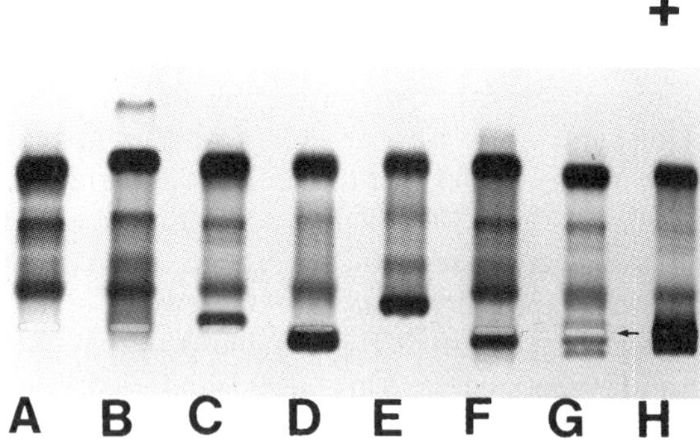

Fig. 5.3. Agarose gel electrophoresis of serum from mice. The anode is at the top, and the origin is marked with an arrow. (A) Specific-pathogen-free mice. Note the virtual absence of γ globulins. (B) Mice infected with the parasite M. corti. *These mice have very high levels of polyclonal IgG1, and raised levels of a protein that migrates more rapidly than albumin. (C) Mice bearing the IgG1 myeloma MOPC-21. (D) Mice bearing the IgG2a myeloma MOPC-173. (E) Mice bearing the IgG2b myeloma MPC-11. (F) Mice bearing the IgG3 myeloma Y5606. (G) Mice bearing the hybridoma MAR-18.5 (Lanier* et al., *1982), which contains γ1 chains from MOPC-21 and γ2a chains from spleen. Note the three spikes in the γ globulin region, due to γ1/γ1, γ1/γ2a and γ2a/γ2a hybrid molecules. (H): Mice bearing the IgM myeloma HPC-76.*

As a general rule, IgG1 and IgA antibodies have a more anodal ('faster') mobility than IgG2, but there is considerable overlap. IgM antibodies tend to stay close to the origin because of retardation by sieving effects.

Serum electrophoresis is a simple and extremely useful way of characterizing the presence and quantity of monoclonal antibodies, and will be referred to frequently in this book.

References

Åkerström, B. and Björck, L. (1986) A physico-chemical study of protein G—a molecule with unique IgG-binding properties. *J. Biol. Chem.* **261**, 10240–10247.

Åkerström, B., Brodin, T., Reis, K. and Björck, L. (1985) Protein G: A powerful tool for binding and detection of monoclonal and polyclonal antibodies. *J. Immunol.* **135**, 2589–2592.

Åkerström, B., Nielsen, E. and Björck, L. (1987) Definition of the IgG-and albumin-binding regions of streptococcal protein G. *J. Biol. Chem.* **262**, 13388–13391.

Argon, Y., Burrone, O. R. and Milstein C. (1983) Molecular characterization of a non-secreting myeloma mutant. *Eur. J. Immunol.* **13**, 301–305.

Bardwell, J. C. A. and Beckwith, J. (1993) The bonds that tie: catalyzed disulfide bond formation. *Cell* **74**, 769–771.

Barrett, D. J. (1990) Lymphocyte stimulation by bacterial Fc receptors. *In* 'Bacterial Immunoglobulin-binding Proteins' (M. D. P. Boyle, ed.), pp. 279–293. Academic Press, San Diego.

Baumal, R. and Scharff, M. D. (1973) Synthesis, assembly and secretion of mouse immunoglobulin. *Transplant Rev.* **14**, 163–183.

Bazin, H., Beckers, A., Deckers, C. and Moriame, M. (1973) Transplantable immunoglobulin-secreting tumors in rats. V. monoclonal immunoglobulins secreted by 250 ileocecal immunocytomas in LOU/WsI rats. *J. Natl Cancer Inst.* **51**, 1359–1361.

Bazin, H., Rousseaux, J., Rousseaux-Prevost, R., Platteau, B., Querinjean, P., Malache, J. M. and Delaunay, T. (1990a) Rat Immunoglobulins. *In* 'Rat Hybridomas and Rat Monoclonal Antibodies' (H. Bazin, ed.). CRC Press, Boca Raton.

Bazin, H., Malache, J. M., Nisol, F. and Delaunay, T. (1990b) Purification of rat monoclonal antibodies from ascitic fluid, serum or culture supernatant. *In* 'Rat Hybridomas and Rat Monoclonal Antibodies' (H. Bazin, ed.). CRC Press, Boca Raton.

Bergman, L. W. and Kuehl, W. M. (1979) Formation of intermolecular disulfide bonds on nascent immunoglobulin polypeptides. *J. Biol. Chem.* **254**, 5690–5694.

Björck, L. and Åkerström, B. (1990) Streptococcal protein G. *In* 'Bacterial Immunoglobulin-Binding Proteins'. (M. D. P. Boyle, ed.), pp. 113–126. Academic Press New York.

Björck, L. and Kronvall, G. (1984) Purification and some properties of streptococcal protein G., a novel IgG-binding reagent. *J. Immunol.* **133**, 969–973.

Björck, L., Kastern, W., Lindahl, G. and Widebäck, K. (1987) Streptococcal protein G, expressed by streptococci or by *Escherichia coli*, has separate binding sites for human albumin and IgG. *Mol. Immunol.* **24**, 2047–2053.

Blatt, C. and Haimovitch, J. (1981) The selective effect of tunicamycin on the secretion of IgM and IgG produced by the same cells. *Eur. J. Immunol.* **11**, 65–66.

Blattner, F. R. and Tucker, P. W. (1984) The molecular biology of immunoglobulin D. *Nature*, **307**, 417–422.

Bourgois, A., Kitajima, K., Hunter, I. R. and Askonas, B. A. (1977) Surface immunoglobulins of lipopolysaccharide stimulated spleen cells. The behaviour of IgM, IgD and IgG. *Eur. J. Immunol.* **7**, 151–153.

Boyle, M. D. P. (1990) (ed.) Bacterial immunoglobulin-binding proteins. Academic Press, San Diego.

Boyle, M. D. P. and Reis, K. J. (1987) Bacterial Fc receptors. *Bio/Technology* **5**, 697–703.

Böttcher, I., Hämmerling, G. and Kapp, J. F. (1978) Continuous production of monoclonal IgE antibodies with known allergenic specificity by a hybrid cell line. *Nature* **275**, 761–762.

Brandtzaeg, P. (1985) Role of J chain and secretory component in receptor mediated glandular and hepatic transport of immunoglobulins in man. *Scand. J. Immunol.* **22**, 111–146.

Brandtzaeg, P. and Prydz, H. (1984) Direct evidence for an integrated function of J chain and secretory component in epithelial transport of immunoglobulins. *Nature*, **311**, 71–73.

Brekke, O. H., Michaelsen, T. E. and Sandlie, I. (1995) The structural requirements for complement activation by IgG: does it hinge on the hinge? *Immunol. Today* **16**, 85–90.

Brewer, J. W., Randall, T. D., Parkhouse, R. M. E. and Corley, R. B. (1994a) Mechanism and subcellular localization of secretory IgM polymer assembly. *J. Biol. Chem.* **269**, 17338–17348.

Brewer, J. W., Randall, T. D., Parkhouse, R. M. E. and Corley, R. B. (1994b) IgM hexamers? *Immunol. Today* **15**, 165–168.

Brink, R., Goodnow, C. C., Crosbie, J., Adams, E., Eris, J., Mason, D. Y., Hartley, S. B.

and Basten, A. (1992) Immunoglobulin M and D antigen receptors are both capable of mediating B lymphocyte activation, deletion or anergy after interaction with specific antigen. *J. Exp. Med.* **176**, 991–1005.

Brüggemann, M. (1987) Genes encoding immunoglobulin constant regions. *In* 'Molecular Genetics of Immunoglobulin' (F. Calabi and M. S. Neuberger, eds), pp. 51–80. Elsevier, Amsterdam.

Burnett, R. C., Hanly, W. C., Zhai, S. and Knight, K. L. (1989) The IgA heavy chain gene family in rabbit: cloning and sequence analysis of 13 Cα genes. *EMBO J.* **8**, 4041–4047.

Cambier, J. C., Pleiman, C. M. and Clark, M. R. (1994) Signal transduction by the B cell receptor and its coreceptors. *Annu. Rev. Immunol.* **12**, 457–486.

Capron, M. and Capron, A. (1994) Immunoglobulin E and effector cells in schistosomiasis. *Science* **264**, 1876–1877.

Carayannopoulos, L. and Capra, J. D. (1993) Immunoglobulins: Structure and Function. In 'Fundamental Immunology' (Paul, W. E., ed) 3rd edn, pp. 283–314.

Cheng, H-L., Blattner, F. R., Fitzmaurice, L., Mushinski, J. F. and Tucker, P. W. (1982) Structure of genes for membrane and secreted murine IgD heavy chains. *Nature* **296**, 410–415.

Colman, P. M. (1991) Antigen–antigen receptor interactions. *Curr. Opin. Struct. Biol.* **1**, 232–236.

Davies, D. R. and Chacko, S. (1993) Antibody structure. *Acc. Chem. Res.* **26**, 421–427.

Davis, A. and Shulman, M. J. (1989) IgM—Molecular requirements for its assembly and function. *Immunol. Today* **10**, 118–128.

Davis, A. C., Roux, K. H. and Shulman, M. J. (1988) On the structure of polymeric IgM. *Eur J. Immunol.* **18**, 1001–1008.

Davis, A. C., Roux, K. H., Pursey, J. and Shulman, M. J. (1989) Intermolecular disulfide bonding in IgM: effects of replacing cysteine residues in the μ heavy chain. *EMBO J.* **8**, 2519–2526.

Deisenhofer, J. (1981) Crystallographic refinement and atomic models of a human Fc fragment and its complex with fragment B of Protein A from *Staphylococcus aureus* at 2.9- and 2.8Å resolution. *Biochemistry* **20**, 2361–2370.

Der Balian, G. P., Slack, J., Clevinger, B. L., Bazin, H. and Davie, J. M. (1980) Antigenic similarities of rat and mouse IgG subclasses associated with anti-carbohydrate specificities. *Immunogenetics* **7**, 199–203.

Derrick, J. P. and Wigley, D. B. (1992) Crystal structure of a streptococcal protein G domain bound to an Fab fragment. *Nature* **359**, 752–754.

Di Sabato, G., Langone, J. J. and Van Vunakis, H. (1985) Immunochemical Techniques. Section 1. Purification and Characterization of Serum Immunoglobulins. *Methods Enzymol.* **116**, 1–200.

Dombrowicz, D., Flamand, V., Brigman, K. K., Koller, B. H. and Kinet, J-P. (1993) Abolition of anaphylaxis by targeted disruption of the high affinity immunoglobulin E receptor α chain gene. *Cell* **75**, 969–976.

Dorrington, K. J. and Bennich, H. H. (1978) Structure-function relationships for human immunoglobulin *E. Immunol. Rev.* **41**, 3–25.

Edmundson, A. B., Ely, K. R., Abola, E. E., Schiffer, M. and Panagiotopoulos, N. (1975) Rotational allomerism and divergent evolution of domains in immunoglobulin light chains. *Biochemistry* **14**, 3953–3961.

Eshhar, Z., Ofarim, M. and Waks, T. (1980) Generation of hybridomas secreting murine reaginic antibodies of anti-DNP specificity. *J. Immunol.* **124**, 775–780.

Ey, P., Prowse, S. J. and Jenkin, C. R. (1978) Isolation of pure IgG1, IgG2a and IgG2b immunoglobulins from mouse serum using protein A-Sepharose. *Immunochemistry* **15**, 429–436.

Fahnestock, S. L. R. (1987) Cloned streptococcal protein G genes. *Trends Biotechnol.* **5**, 79–83.

Fahnestock, S. L. R., Alexander, P., Filpula, D. and Nagle, J. (1990) Structure and evolution of the streptococcal genes encoding protein G. *In* 'Bacterial Immunoglobulin-Binding Proteins' (M. D. P. Boyle, ed), pp. 133–148. Academic Press, New York.

Fiedler, K. and Simons, K. (1995) The role of *N*-glycans in the secretory pathway. *Cell* **81**, 309–312.

Finkelman, F. D., Kessler, S. W., Mushinski, J. F. and Potter, M. (1981) IgD-secreting murine plasmacytomas: identification and partial characterization of two IgD myeloma proteins. *J. Immunol.* **126**, 680–687.

Fleischer, B. (1989) Bacterial toxins as probes for the T-cell antigen receptor. *Immunol. Today*. **10**, 262–264.

Foote, J. and Milstein, C. (1994) Conformational isomerism and the diversity of antibodies. *Proc. Natl Acad. Sci. USA* **19**, 10370–10374.

Freedman, R. B. (1989) Protein disulfide isomerase: multiple roles in the modification of nascent secretory proteins. *Cell* **57**, 1069–1072.

Gearhart, P. J., Johnson, N. D., Douglas, R. and Hood, L. (1981) IgG antibodies to phosphorylcholine exhibit more diversity than their IgM counterparts. *Nature* **291**, 29–34.

Gething, M-J. and Sambrook, J. (1992) Protein folding in the cell. *Nature* **355**, 33–45.

Giachino, C., Padovan, E. and Lanzavecchia, A. (1995) $\mu^+\kappa^+$ dual receptor B cells are present in the human peripheral repertoire. *J. Exp. Med.* **181**, 1245–1250.

Goding, J. W. (1978) Use of staphylococcal protein A as an immunological reagent. *J. Immunol. Methods* **20**, 241–253.

Goding, J. W. (1980) Structural studies of murine lymphocyte surface IgD. *J. Immunol.* **124**, 2082–2088.

Goding, J. W. (1982) Asymmetrical surface IgG on MOPC-21 plasmacytoma cells contains one membrane heavy chain and one secretory heavy chain. *J. Immunol.* **128**, 22416–2421.

Goding, J. W. and Layton, J. E. (1976) Antigen-induced co-capping of IgM and IgD-like receptors on murine B cells. *J. Exp. Med.* **144**, 852–857.

Goding, J. W. and Herzenberg, L. A. (1980) Biosynthesis of lymphocyte surface IgD in the mouse. *J. Immunol.* **124**, 2540–2547.

Gough, N. M. and Cory, S. (1986) The murine immunoglobulin heavy chain constant region locus. *In* 'Handbook of Experimental Immunology' (D. M. Weir, L. A. Herzenberg, C. Blackwell. and L. A. Herzenberg, eds), Vol. 3, 4th edn, 88.1–88.26. Blackwell, Oxford.

Gounni, A. S., Lamkhloued, B., Ochial, K., Tanaka, Y., Delaporte, E., Capron, A., Kinet, J-P, and Capron, M. (1994) High-affinity IgE receptor on eosinophils is involved in defence against parasites. *Nature* **367**, 183–186.

Greenspan, N. S. and Cooper, L. J. N. (1992) Intermolecular cooperativity: a clue to why mice have IgG3? *Immunol. Today* **13**, 164–168.

Grey, H. M., Abel, C. A., Yount, W. J. and Kunkel, H. G. (1968) A subclass of human γA-globulins (γA2) which lacks the disulfide bonds linking heavy and light chains. *J. Exp. Med.* **128**, 1223–1236.

Hamilton, R. G. (1994) 'The Human IgG Subclasses'. Calbiochem--Novabiochem International, San Diego.

Harris, L. J., Larson, S. B., Hasel, K. W., Day, J., Greenwood, A. and McPherson, A. (1992) The three-dimensional structure of an intact monoclonal antibody for canine lymphoma. *Nature* **360**, 369–372.

Havran, W. L., Di Giusto, D. L. and Cambier, J. C. (1984) mIgM:mIgD ratios on B cells: mean mIgD expression exceeds mIgM by 10-fold on most splenic B cells. *J. Immunol.* **132**, 1712–1716.

Hayashida, H., Miyata, T., Yamawaki-Kataoka, Y., Honjo, T., Wels, J. and Blattner, F. (1984) Concerted evolution of the mouse immunoglobulin gamma chain genes. *EMBO J.* **3**, 2047–2053.

Herrmann, S. H. and Mescher, M. F. (1979) Purification of the H-2K^k molecule of the murine major histocompatibility complex. *J. Biol. Chem.* **254**, 8713–8716.

Hirayama, N., Hirano, T., Köhler, G., Kurata, A., Okumura, K. and Ovary, Z. (1982) Biological activities of antitrinitrophenyl and antidinitrophenyl mouse monoclonal antibodies. *Proc. Natl Acad. Sci. USA* **79**, 613–615.

Hulett, M. D. and Hogarth, P. M. (1994) Molecular basis of Fc receptor function. *Adv. Immunol.* **57**, 1–127.

Hurtley, S. M. and Helenius, A. (1989) Protein oligomerization in the endoplasmic reticulum. *Annu. Rev. Cell Biol.* **5**, 277–307.

Hwang, C., Sinskey, A. J. and Lodish, H. F. (1992) Oxidised redox state of glutathione in the endoplasmic reticulum. *Science* **257**, 1496–1502.

Inbar, D., Hochman, Y. and Givol, D. (1972) Localization of antibody-combining sites within the variable portions of heavy and light chains. *Proc. Natl Acad. Sci. USA* **69**, 2659–2662.

Ishida, N., Ueda, S., Hayashida, H., Miyata, T. and Honjo, T. (1982) The nucleotide sequence of the mouse immunoglobulin epsilon gene: comparison with the human epsilon gene sequence. *EMBO J.* **1**, 1117–1123.

Janik, B. (1984) 'High Resolution Electrophoresis and Immunofixation of Serum Proteins on Cellulosic Media'. Gelman Sciences, Inc, Ann Arbor, Michigan.

Jeppsson, J-O., Laurell, C-B. and Franzen, B. (1979) Agarose gel electrophoresis. *Clin. Chem.* **25**, 629–638.

Kakimoto, K. and Onoue, K. (1974) Characterization of the F_v fragment isolated from a human immunoglobulin M. *J. Immunol.* **112**, 1373–1382.

Kehry, M., Sibley, C., Fuhrman, J., Schilling, J. and Hood, L. E. (1979) Amino acid sequence of a mouse immunoglobulin μ chain. *Proc. Natl Acad. Sci. USA* **76**, 2932–2936.

Kehry, M., Fuhrman, J. S., Schilling, J. W., Rogers, J., Sibley, C. H. and Hood, L. E. (1982) Complete amino acid sequence of a mouse μ chain: homology among heavy chain constant region domains. *Biochemistry* **21**, 5415–5424.

Killingsworth, L. M. (1985) 'High Resolution Protein Electrophoresis. A Clinical Overview with Case Studies'. Helena Laboratories, Beaumont, Texas.

Killingsworth, L. M. and Warren, B. M. (1986) 'Immunofixation. For the Identification of Monoclonal Gammopathies'. Helena Laboratories, Beaumont, Texas.

Kim, H., Matsunaga, C., Yoshino, A., Kato, K. and Arata, Y. (1994a) Dynamical structure of the hinge region of immunoglobulin G as studied by ^{13}C nuclear magnetic resonance spectroscopy. *J. Mol. Biol.* **236**, 300–309.

Kim, H., Yamaguchi, Y., Masuda, K., Matsunaga, C., Yamamoto, K., Irimura, T., Takahashi, N., Kato, K. and Arata, Y. (1994b) O-glycosylation in hinge region of mouse immunoglobulin G2b. *J. Biol. Chem.* **269**, 12345–12350.

Kim, K-M., Adachi, T.k, Nielsen, P. J., Terashima, M., Lamers, M. C., Köhler, G. and Reth, M. (1994c) Two new proteins preferentially associated with membrane immunoglobulin D. *EMBO J.* **13**, 3793–3800.

Kinet, J-P. and Metzger, H. (1990) Genes, structure and actions of the high-affinity Fc receptor for immunoglobulin E. *In* Fc Receptors and the Action of Antibodies (H. Metzger, ed.), pp. 239–259. American Society for Microbiology, St Paul.

Koshland, M. E. (1975) Structure and function of the J chain. *Adv. Immunol.* **20**, 41–69.

Koshland, M. E. (1985) The coming of age of the immunoglobulin J chain. *Annu. Rev. Immunol.* **3**, 425–453.

Köhler, G. (1980) Immunoglobulin chain loss in hybridoma lines. *Proc. Natl Acad. Sci. USA* **77**, 2197–2199.

Köhler, G., Hengartner, H. and Shulman, M. J. (1978) Immunoglobulin production by lymphocyte hybridomas. *Eur. J. Immunol.* **8**, 82–88.

Kuhn, R., Rajewsky, K. and Muller, W. (1991) Generation and analysis of interleukin-4 deficient mice. *Science* **254**, 707–710.

Lai, E., Wilson, R. K. and Hood, L. E. (1989) Physical maps of the mouse and human immunoglobulin-like loci. *Adv. Immunol.* **46**, 1–59

Lancet, D., Isenman, D., Sjödahl, J., Sjöquist, J. and Pecht, I. (1978) Interaction between staphylococcal protein A and immunoglobulin domains. *Biochem. Biophys. Res. Commun.* **85**, 608–614.

Langone, J. J. (1982) Protein A of *Staphylococcus aureus* and related immunoglobulin receptors produced by streptococci and pneumonococci. *Adv. Immunol.* **32**, 157–252.

Lanier, L. L., Gutman, G. A., Lewis, D. E., Griswold, S. T. and Warner, N. L. (1982) Monoclonal antibodies against rat immunoglobulin kappa chains. *Hybridoma* **1**, 125–131.

Liu, F-T., Bohn, J. W., Ferry, E. L., Yamamoto, H., Molinaro, C. A., Sherman, L. A., Kliman, N. R. and Katz, D. H. (1980) Monoclonal dinitrophenyl-specific murine IgE antibody: preparation, isolation and characterization. *J. Immunol.* **124**, 2728–2737.

Lo, M. M. S., Tong, T. Y., Conrad, M. K., Strittmatter, S. M., Hester, L. D. and Snyder, S. H. (1984) Monoclonal antibody production by receptor-mediated electrically induced cell fusion. *Nature* **310**, 792–794.

Malm, B. (1987) A method suitable for the isolation of monoclonal antibodies from large volumes of serum-containing hybridoma cell culture supernatants. *J. Immunol. Methods* **104**, 103–109.

Marrack, P. and Kappler, J. (1990) The staphylococcal enterotoxins and their relatives. *Science* **248**, 705–711.

Max, E. (1993) Immunoglobulins. Molecular genetics. *In* 'Fundamental Immunology' (W. E. Paul, ed.), 3rd edn, pp. 315–382. Raven Press, New York.

Mestecky, J. and McGhee, J. R. (1987) Immunoglobulin A (IgA): molecular and cellular interactions involved in IgA biosynthesis and immune response. *Adv. Immunol.* **40**, 153–245.

Mostov, K. E., Friedlander, M. and Blobel, G. (1984) The receptor for trans-epithelial transport of IgA and IgM contains multiple immunoglobulin-like domains. *Nature*, **308**, 37–43.

Mostov, K. E. (1994) Transepithelial transport of immunoglobulins. *Annu. Rev. Immunol.* **12**, 63–84.

Myhre, E. B. (1990) Interaction of bacterial immunoglobulin receptors with sites in the Fab region. *In* 'Bacterial Immunoglobulin-Binding Proteins' (M. D. P. Boyle, ed.), pp. 243–256. Academic Press, New York.

Myhre, E. B. and Kronvall, G. (1980) Demonstration of specific binding sites for human serum albumin in group C and G streptococci. *Infect. Immun.* **27**, 6–14.

Niles, M. J., Matsuuchi, L. and Koshland, M. E. (1995) Polymer IgM assembly and secretion in lymphoid and nonlymphoid cell lines: Evidence that J chain is required for pentamer IgM synthesis. *Proc. Natl Acad. Sci. USA* **92**, 2884–2888.

Nilson, B., Björck, L. and Åkerström, B. (1986) Detection and purification of rat and goat immunoglobulin G antibodies using protein G-based solid phase radioimmunoassays. *J. Immunol. Methods* **91**, 275–281.

Nilson, R., Myhre, E. B., Kronvall, G. and Sjögren, H. O. (1982) Fractionation of rat IgG subclasses and screening for IgG Fc-binding to bacteria. *Mol. Immunol.* **19**, 119–126.

Nisonoff, A., Hopper, J. E. and Spring, S. B. (1975) 'The Antibody Molecule'. Academic Press, New York.

Nitschke, L., Kosco, M. H., Köhler, G. and Lamers, M. C. (1993) Immunoglobulin D-deficient mice can mount normal immune responses to thymus-independent and -dependent antigens. *Proc. Natl Acad. Sci. USA* **90**, 1887–1891.

Ou, W-J., Cameron, P. H., Thomas, D. Y. and Bergeron, J. J. M. (1993) Association of folding intermediates of glycoproteins with calnexin during protein maturation. *Nature* **364**, 771–776.

Östlund, C. (1986) Large-scale purification of monoclonal antibodies. *Trends Biotechnol.* **4** (November), 288–293.

Padlan, E. (1994) Anatomy of the antibody molecule. *Mol. Immunol.* **31**, 169–217.

Parham, P. (1983) On the fragmentation of monoclonal IgG1, IgG2a, and IgG2b from BALB/c mice. *J. Immunol.* **131**, 2895–2902.

Parkhouse, R. M. E. (1971) Immunoglobulin M Biosynthesis: Production of intermediates and excess light chain in mouse myeloma MOPC-104E. *Biochem. J.* **123**, 633–641.

Pharmacia (1995) 'The World of Pharmacia Biotech '95', p. 369. Pharmacia, Uppsala.

Pillai, S. and Baltimore, D. (1987) Myristoylation and the post-translational acquisition of hydrophobicity by the membrane immunoglobulin heavy-chain polypeptide in B lymphocytes. *Proc. Natl Acad. Sci. USA* **84**, 7654–7658.

Poljak, R. J., Amzel,L. M., Avey, H. P., Chen, B. L., Phizackerley, R. P. and Sau., F. (1973) Three-dimensional structure of the Fab' fragment of a human immunoglobulin at 2.8Å resolution. *Proc. Natl Acad. Sci. USA* **70**, 3305–3310.

Pollock, R. R., Dorf, M. E. and Mescher, M. F. (1980) Genetic control of murine IgD structural heterogeneity. *Proc. Natl Acad. Sci. USA* **77**, 4256–4259.

Potter, M. and Lieberman, R. (1967) Genetics of immunoglobulin in the mouse. *Adv. Immunol.* **7**, 92–145.

Ramsay, A. J., Husband, A. J., Ramshaw, I. A., Bao, S., Matthaei, K. I., Koehler, G. and Kopf, M. (1994) The role of interleukin-6 in mucosal IgA antibody responses *in vivo*. *Science* **264**, 561–563.

Randall, T. D., Brewer, J. W. and Corley, R. B. (1992) Direct evidence that J chain regulates the polymeric structure of IgM in antibody-secreting B cells. *J. Biol. Chem.* **267**, 18002–18007.

Reth, M. (1994) B cell antigen receptors. *Curr. Opin. in Immunol.* **6**, 3–8.

Roes, J. and Rajewsky, K. (1993) Immunoglobulin D (IgD)-deficient mice reveal an auxiliary receptor function for IgD in antigen-mediated recruitment of B cells. *J. Exp. Med.* **177**, 45–55.

Rogers, J. and Wall, R. (1984) Immunoglobulin RNA rearrangements in B lymphocyte differentiation. *Adv. Immunol.* **35**, 39–59.

Rogers, J., Early, P., Carter, C., Calame, K., Bond, M., Hood, L. and Wall, R. (1980) Two mRNAs with different 3′ ends encode membrane-bound and secreted forms of immunoglobulin μ chain. *Cell* **20**, 303–312.

Rothman, J. E. and Orci, L. (1992) Molecular dissection of the secretory pathway. *Nature* **355**, 409–415.

Schrezenmeier, H. and Fleischer, B. (1987) Mitogenic activity of staphylococcal protein A is due to contaminating staphylococcal enterotoxins. *J. Immunol. Methods* **105**, 133–137.

Shimizu, A., Watanabe, S., Yamamura, Y. and Putnam, F. W. (1974) Tryptic digestion of immunoglobulin M in urea: conformational lability of the middle part of the molecule. *Immunochemistry* **11**, 719–727.

Sikorav, J-L., Auffray, C. and Rougeon, F. (1980) Structure of the constant and 3′

untranslated regions of the murine Balb/c γ2a heavy chain messenger RNA. *Nucleic Acids Res.* **8**, 3143–3155.

Simister, N. E. (1990) Transport of monomeric antibodies across epithelia. *In* 'Fc Receptors and the Action of Antibodies' (H. Metzger, ed.), pp. 57–73. American Society for Microbiology, Washington DC.

Sire, J., Auffray, C. and Jordan, B. R. (1982) Rat immunoglobulin delta heavy chain gene: nucleotide sequences derived from cloned cDNA. *Gene* **20**, 377–386.

Söllner, T., Whiteheart, S. W., Brunner, M., Erdjument-Bromage, H., Geromanos, S., Tempst, P. and Rothman, J. E. (1993) SNAP receptors implicated in vesicle targeting and fusion. *Nature* **362**, 318–324.

Spiegelberg, H. L. (1974) Biological activities of immunoglobulins of different classes and subclasses. *Adv. Immunol.* **19**, 259–294.

Stanisz, A. M., Lieberman, R., Kaplan, A. and Davie, J. M. (1983) IgA polymorphism in mice: NZB and BALB/c mice produce two forms of IgA. *Molec. Immunol.* **20**, 983–988.

Stone, G. C., Sjobring, U., Björck, L., Sjöquist, J., Barber, C. V. and Nardella, F. A. (1989) The Fc binding site for streptococcal protein G is in the C gamma 2-C gamma 3 interface region of IgG and is related to the sites that bind staphylococcal protein A and human rheumatoid factors. *J. Immunol.* **143**, 565–570.

Surolia, A., Pain, D. and Khan, M. I. (1982) Protein A: nature's universal anti-antibody. *Trends Biochem. Sci.* **7**, 74–76.

Takahashi, N., Tataert, D., Debuire, B., Lin, L-C. and Putnam, F. W. (1982) Complete amino acid sequence of the δ heavy chain of human immunoglobulin D. *Proc. Natl Acad. Sci. USA* **79**, 2850–2854.

Tartakoff, A. and Vassalli, P. (1979) Plasma cell immunoglobulin M molecules. Their biosynthesis, assembly and intracellular transport. *J. Cell. Biol.* **83**, 284–299.

Terashima, M., Kim, K-M., Adachi, T., Nielsen, P. J., Reth, M., Köhler, G. and Lamers, M. C. (1994) The IgM antigen receptor of B lymphocytes is associated with prohibitin and a prohibitin-related protein. *EMBO J.* **13**, 3782–3792.

Tomasi, T. B. and Grey, H. M. (1972) Structure and function of immunoglobulin A. *Prog. Allergy* **16**, 81–185.

Tucker, P. W., Slighton, J. L. and Blattner, F. R. (1981) Mouse IgA heavy chain gene sequence: Implications for evolution of immunoglobulin hinge exons. *Proc. Natl Acad. Sci. USA* **78**, 7684–7688.

Tyler, B. M., Cowman, S. D., Gerondakis, S. D., Adams, J. M. and Bernard, O. (1982) Messenger RNA for surface immunoglobulin γ chains encodes a highly conserved transmembrane sequence and a 28-residue intracellular domain. *Proc. Natl Acad. Sci. USA* **79**, 2008–2012.

Underdown, B. J. and Dorrington, K. J. (1974) Studies on the structural and conformational basis for the relative resistance of serum and secretory immunoglobulin A to proteolysis. *J. Immunol.* **112**, 949–959.

Underdown, B. J. (1990) Transcytosis by the receptor for polymeric immunoglobulin. *In* 'Fc Receptors and the Action of Antibodies' (H. Metzger, ed.), pp. 74–93. American Society for Microbiology, Washington DC.

Underdown, B. J., DeRose, J. and Plaut, A. (1977) Disulfide bonding of secretory component to a single monomer subunit in human secretory IgA. *J. Immunol.* **118**, 1816–1821.

Varki, A. (1993) Biological roles of oligosaccharides: all of the theories are correct. *Glycobiology* **3**, 97–130.

Vitetta, E. S. and Uhr, J. W. (1975) Immunoglobulin-receptors revisited. *Science* **189**, 964–969.

Wall, R. and Kuehl, M. (1983) Biosynthesis and regulation of immunoglobulins. *Annu. Rev. Immunol.* **1**, 393–422.

Wienands, J. and Reth, M. (1992) Glycosyl-phosphatidylinositol linkage as a mechanism for cell-surface expression of immunoglobulin D. *Nature* **356**, 246–248.

Wilson, I. A. and Stanfield, R. L. (1993) Antibody–antigen interactions. *Curr. Opin. Struct. Biol.* **3**, 113–118.

Winchester, R. J., Fu, S. M., Hoffman, T. and Kunkel, H. G. (1975) IgG on lymphocyte surfaces: technical problems and significance of a third cell population. *J. Immunol.* **114**, 1210–1212.

Woof, J. M. and Burton, D. R. (1990) The nature and interaction of bacterial Fc receptors and IgG. *In* 'Bacterial Immunoglobulin-binding Proteins' (M. D. P. Boyle, ed.), Vol. 1, pp 305–316. Academic Press, San Diego.

World Health Organization (1973) Nomenclature of human immunoglobulins, *Bull. WHO* **48**, 373.

Yamaguchi, Y., Kim, H., Kato, K., Masuda, K., Shimada, I. and Arata, Y. (1995) Proteolytic fragmentation with high specificity of mouse immunoglobulin G. *J. Immunol. Methods* **181**, 259–267.

Zdanov, A., Li, Y., Bundle, D. R., Deng, S-J., MacKenzie, C. R., Narang, S. A., Young, N. M. and Cygler, M. (1994) Structure of a single-chain antibody variable domain (Fv) fragment complexed with a carbohydrate antigen at 1.7Å resolution. *Proc. Natl Acad. Sci. USA* **91**, 6423–6427.

6 Genetics of Antibodies

The genetics of antibodies presented a fascinating problem for immunologists (Kindt and Capra, 1984; Max, 1993). How was it possible for an animal to make perhaps 10^{10} antibodies when the total genome contains only about 10^5 genes? The amino acid sequences of antibodies suggested that there were numerous variable region genes, yet the constant regions behaved as if they were single genes. How could a clone of lymphocytes switch antibody class and yet retain antibody specificity? Do the membrane forms of antibodies differ in structure from secretory antibodies? If so, how? If not, how are they held in the membrane?

The answers to these questions came slowly at first, and then with great speed with the development of recombinant DNA technology. In 1965, Dreyer and Bennett made the radical proposal that there were multiple genes for the variable regions but only single copies of the constant region genes, and that the DNA rearranged during development of lymphocytes. The proposal was particularly daring because at that time it was thought that the DNA content of all somatic cells was identical. This visionary model has now been proven correct in its essentials.

In the early days of the study of antibodies, it was only possible to look at the products of the genes, but not the genes themselves. A great deal of speculation took place based on the protein structure, but definitive answers to these questions had to wait until the arrival of recombinant DNA technology, when it became possible to examine the genes directly. We now have a satisfying overall picture, although many details are still not understood. In view of the remarkable diversity of antibodies, together with their enormous survival value, it is hardly surprising that complex and unprecedented genetic mechanisms were found to be involved.

Before discussing the structure and rearrangement of antibody genes, we will discuss some of the knowledge derived from the study of the genetic variation of antibodies at the protein level, which led to a broad outline of the arrangement and grouping of antibody genes. Although the study of antibodies by recombinant DNA has been extremely powerful, much of the earlier information derived from studying the variation of antibodies at the protein level remains of lasting value because it provides tools for following the origin of antibodies and cells in many experimental situations.

6.1 Immunoglobulin Allotypes

The products of allelic forms of the same gene are known as *allotypes*. Allotypes are now known for all mouse heavy chains (Stall, 1995). There are no known serologically detectable allotypes of mouse κ chains. The constant regions of λ1 chains of mice show a single amino acid difference between BALB/c and SJL mice (Arp *et al.*, 1982). In the rat, the κ chains show extensive polymorphism, and variants have also been described for α, γ2b and γ2c chains (Gutman, 1986; Bazin *et al.*, 1990).

Immunoglobulin allotypes are important for several reasons. They allow genetic linkage studies to be carried out, and the origin of immunoglobulins to be determined in cell transfer experiments. They were central in demonstrating allelic exclusion (see later), in showing that variable and constant region genes were linked, and in showing that all the heavy chain genes are linked to each other. However, a detailed account of the esoteric world of immunoglobulin allotypes is beyond the scope of this book. A comprehensive review of mouse immunoglobulin allotypes is given by Stall (1995).

The heavy chain locus in the mouse is designated *Igh*, and the constant region loci are designated *Igh-C*. The genes are named in order of discovery *Igh-1* to *Igh-8* (Table 6.1). (Note that the order of the genes on the chromosome is completely different; see Fig. 6.1.) Alleles at each locus are designated by superscripts (e.g. *Igh-1*^{b}). The combination of *Igh-C* alleles in a particular mouse strain is known as a *haplotype*. In many cases the haplotype designation is the same letter as the allele for that strain at the *Igh-1* (γ2a) locus (Table 6.2).

The structural basis for immunoglobulin allotypes is now well understood. The variation appears to fall into two categories. In some cases, there are single point mutations, such as for mouse IgM (Schreier *et al.*, 1986). These are known as 'simple' allotypes. Some allotypes are classified as 'complex', because there are multiple amino acid substitutions at several different regions of the chain (Sheppard and Gutman, 1981a, b; Ollo and Rougeon, 1983). In the former case, the variation can be explained by simple point mutations, but in the case of allotypes with multiple substitutions between alleles, more complex genetic mechanisms such as gene conversion have probably been responsible (Sheppard and Gutman, 1981a, b; Ollo and Rougeon, 1983).

Table 6.1 Locus symbols for mouse immunoglobulin heavy chains

Constant region	Igh-C
Constant region loci	Igh-1 (γ2a)
	Igh-2 (α)
	Igh-3 (γ2b)
	Igh-4 (γ1)
	Igh-5 (δ)
	Igh-6 (μ)
	Igh-7 (ε)
	Igh-8 (γ3)

Table 6.2 Distribution of alleles of the Igh-C loci in the Igh-C haplotypes

Haplotype	Prototype Strain	Igh-6 μ	Igh-5 δ	Igh-8* γ3	Igh-4 γ1	Igh-3 γ2b	Igh-1 γ2a	Igh-7 ε	Igh-2 α
a	BALB/c	*a*	*a*	*a*	*a*	*a*	*a*	*a*	*a*
b	C57BL10/J	*b*	*b*	*a*	*b*	*b*	*b*	*b*	*b*
c	DBA/2	*a*	*a*	*a*	*a*	*a*	*c*	?	*c*
d	AKR/J	*n(d)*‡	*a*	*a*	*d*	*d*	*d*	*a*	*d*
e	A/J	*e*	*e*	*a*	*a*	*e*	*e*	*a*	*d*
f	CE/J	*a*	*a*	*a*	*a*	*f*	*f*	?	*f*
g	RIII/J	*a*	*a*	*a*	*a*	*g*	*g*	?	*g*
h	SEA/J	*a*	*a*	*a*	*a*	*a*	*h*	?	*a*
j	CBA/H	*a*	*a*	*a*	*a*	*a*	*j*	*a*	*a*
k	KH-1†	?	?	*a*	*a*	*a*	*k*	?	*c*
l	KH-2†	?	?	*a*	*a*	*a*	*l*	?	*c*
m	Ky†	?	?	*a*	*b*	*b*	*m*	?	*b*
n	NZB	*n(d)*	*a*	*a*	*d*	*e*	*e*	?	*d*
o	AL/N	*e*	*e*	*a*	*a*	*d*	*d*	?	?
p	SWR/J	?	?	?	?	*f*	*c*	?	?

* Allelic forms of γ3 chains have only been found in wild mice and the SPE strain of *Mus spretus* (Huang *et al.*, 1984; Amor *et al.*, 1984).

† Haplotypes derived from wild mice.

‡ This allele was first identified in NZB mice and is thus given the designation Igh-6n. However, recent studies indicate that it is common to both AKR and NZB and should be designated Igh-6d.

I am grateful to Dr Alan Stall for providing me with updated information for this table. For further information, see Stall (1995).

Allotypic variation can affect many properties of antibodies. *Heterologous antisera (e.g. rabbit anti-mouse IgG) may have a pronounced bias towards a particular haplotype.* For example, one such antiserum reacted about 20-fold more strongly against IgG2a of the *a* haplotype than the *b* haplotype (Stall, 1995). If this difference is not appreciated, it could potentially cause serious errors in experimental measurements of antibody levels. Allotypic variation can also affect physical properties such as solubility, binding to staphylococcal protein A (Seppälä *et al.*, 1981) and electrophoretic mobility (Herzenberg *et al.*, 1967). Mouse IgG of the *b* allotype (i.e. from C57BL and related strains) is less soluble than IgG of the *a* allotype (from BALB/c).

In BALB/c mice, the light chains of myeloma IgA are disulfide-bonded to each other and not to the heavy chains, while the conventional IgA structure is seen in myelomas from NZB mice (Warner and Marchalonis, 1972). A similar genetic variation is seen in human IgA2 (Grey *et al.*, 1968). The different arrangement of disulfide bonds seen in BALB/c and NZB IgA myeloma proteins could have reflected the known allotypic difference, but it has been shown that each strain produces both forms of IgA, and that in each strain both forms share the serological allotypic characteristics of that strain (Stanisz *et al.*, 1983). It would therefore appear that the myeloma proteins are not necessarily typical of the structure of all IgA molecules produced by these strains. These mouse strains have a single α chain gene (Lai *et al.*, 1989), so the dimorphism in mouse IgA structure is probably not a consequence of the known allotypic differences in α chain (see Stanisz *et al.*, 1983, for further discussion).

By a process of repeated backcrossing, a number of 'allotype congenic' partner strains of mice have been produced, in which the heavy chain locus derived from one strain is transferred to the other strain (see Stall, 1995). These pairs are histocompatible, and allow the transfer of lymphoid cells between them. The use of anti-allotype antibodies can then allow the identification and quantification of cells and antibodies derived from the donor and the recipient.

Allotypes of membrane immunoglobulin are particularly useful in tracing the origin of lymphocytes in cell transfer experiments. Allelic variation has been described for membrane IgD (Goding *et al.*, 1976) and IgM (Black *et al.*, 1978), and numerous monoclonal antibodies have been described that recognize allotypic determinants membrane immunoglobulin (Oi *et al.*, 1978; Stall and Loken, 1984; Schüppel *et al.*, 1987; reviewed by Stall, 1995).

6.2 Antibody Genes

6.2.1 Three Clusters of Genes

Antibody genes exist in three clusters; κ light chains, λ light chains, and heavy chains (Max, 1993). In the mouse, the κ group lies on chromosome 6, the λ genes on chromosome 16, and the heavy chain group on chromosome 12. In each case, the genes for the variable region are linked to the constant region genes.

In recent years, studies using recombinant DNA technologies have given an extremely detailed picture of mouse immunoglobulin genes (reviewed by Max, 1993). A simplified picture is given in Fig. 6.1. The overall organization of the DNA is best thought of as a series of discrete coding segments separated by variable distances.

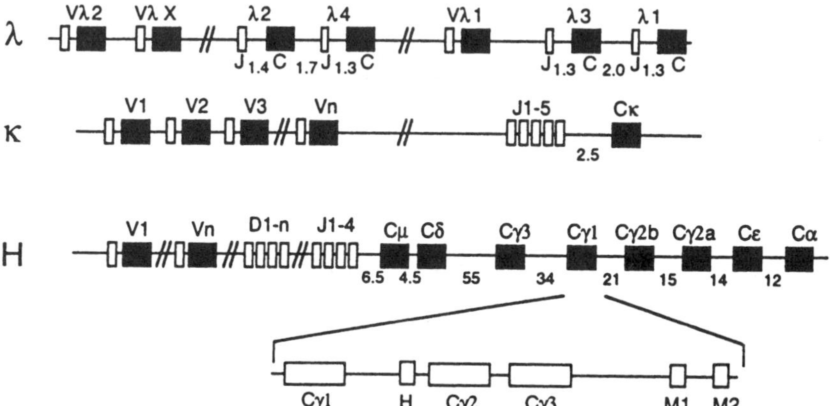

Fig 6.1. Organization of immunoglobulin genes in the mouse. There are three clusters, encoding λ light chains (chromosome 16), κ light chains (chromosome 6) and heavy chains (chromosome 12). The light chain genes contain segments encoding the hydrophobic leader (signal) sequence, the variable region, the joining (J) region, and the constant region. The heavy chain complex possesses in addition a number of 'D' (diversity) segments between V and J. The constant region domains of the heavy chains are encoded by discrete segments (open boxes). Downstream from the last constant domain of each heavy chain lie two segments (M1 and M2) which encode a hydrophobic C-terminal extension present in membrane immunoglobulin.

6.2.2 The κ Complex

The mouse κ complex has only one constant region gene, but has a few hundred variable region genes. Each variable region (V) gene consists of two separate coding segments, one encoding the N-terminal leader (signal) sequence, and the other encoding the body of the variable region (Fig. 6.1). The variable region genes are situated a large distance upstream (i.e. on the 5′ side) of the constant region genes. (By convention, DNA sequences are written the same way as RNA, with the 5′ end to the left.)

Approximately 2500 bases upstream from the κ constant region gene is a cluster of five 'J' or joining genes, each of some 39 nucleotides, which encode the last few amino acids of the variable region. J3 is probably a pseudogene and unable to be expressed because of a faulty splice site. In lymphoid cells expressing κ light chains, one V_κ gene is joined directly onto one J gene, and the intervening DNA is excised (Fig 6.2). The J genes downstream from the chosen J are retained, but the upstream J genes are deleted.

6.2.3 The λ Complex

In many respects, the mouse λ complex is the simplest. There are now known to be four λ constant region genes (λ1–λ4), but paradoxically there are only three

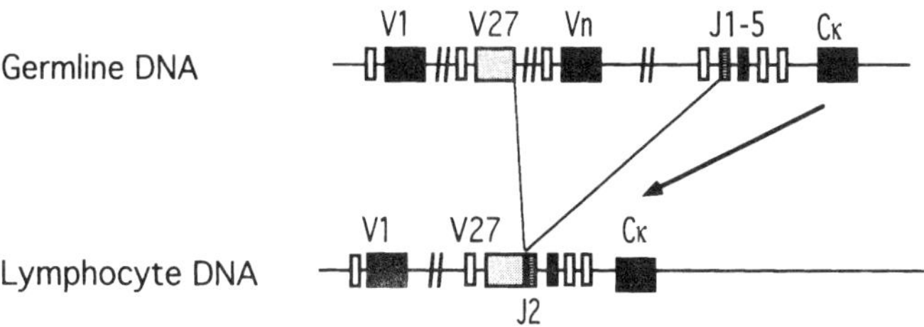

Fig 6.2. Rearrangements of DNA encoding κ light chains in the mouse. In the germline, there are several hundred variable region genes, each with its own leader sequence slightly upstream, five J genes and one constant region gene. Expression of light chains in lymphocytes involves movement of a particular V gene such that it becomes contiguous with a particular J gene. The intervening DNA (containing V and J genes) is deleted, but the J genes and DNA between the chosen J and C_κ remain unchanged. (They are subsequently treated like an intron and removed by RNA splicing.) In this diagram, the J2 gene is expressed, and the J1 gene is deleted. The J3 gene (black) is probably a nonfunctional pseudogene because of a faulty splice site.

variable region genes (Vλ1, Vλ2 and Vx; Fig. 6.1). Vx is rarely used, and the J–λ4 locus is not expressed owing to a faulty splice site. For κ and heavy chains, each variable region gene has its own leader sequence separated from the main portion. Approximately 1000 bases upstream from each λ constant region gene is a single J gene. A functional λ chain gene is formed in a similar way to the κ genes. In lymphoid cells expressing λ chains, one V_λ gene is joined onto a J gene and the intervening DNA excised. There is evidence that there may be additional copies of the lambda locus in some strains of wild mice (Max, 1993).

6.2.4 The Surrogate Light Chain Locus (V_{preB}/λ5)

An additional λ-like gene is situated on chromosome 16, which encodes the 'surrogate light chain' expressed in pre-B cells. This locus, which does not rearrange, consists of two genes that encode a λ-like constant domain known as λ5 or ι (iota) and two variable-like domains known as V_{preB1} and V_{preB2} or ω (omega) (Pillai and Baltimore, 1988; Melchers *et al.*, 1993, 1994). The λ5 polypeptide is disulfide-bonded to membrane μ chains in pre-B cells, while the V_{preB} domain is attached noncovalently. The surrogate light chain appears to serve to form an IgM-like complex on the surface of pre-B cells which signals that a productive heavy chain rearrangement has taken place, and gives a signal to the cell to commence light chain rearrangement (Tsubata *et al.*, 1992).

6.2.5 The Heavy Chain Complex

The heavy chain complex is the most complex in its organization (Fig. 6.3). There are a few hundred variable region genes, four J genes, and one constant region gene for each immunoglobulin class. The pool of V_H genes is shared by all heavy chain classes. Each constant region gene is divided into a series of segments, each coding for one domain or functional polypeptide segment. The hinge region of γ and δ chains are encoded by discrete segments, while the hinge region of α chains is encoded as a 5′ extension of the $C_\alpha 2$ segment.

In additional to the J genes, there are about 12 'diversity' (D) genes, each of some 10–20 nucleotides. The D segments are situated between the V_H and J_H genes (Fig. 6.3), and encode much of the third hypervariable region of the heavy chains.

A functional heavy chain gene is formed by joining a particular V_H gene to a particular D gene, and by joining that D gene to a particular J gene. As is the case for κ, the J genes downstream from the chosen J gene are retained (Fig. 6.3).

6.2.6 Immunoglobulin Gene Enhancers

Enhancers are genetic regulatory sequences that act to increase the rate of transcription of a gene in a tissue-specific manner. They may be situated at the 5′ end of the gene, in an intron in the middle of the gene, or at the 3′ end of the gene, and may act over very long distances, possibly by 'looping back' of the DNA to a region close to the start of transcription.

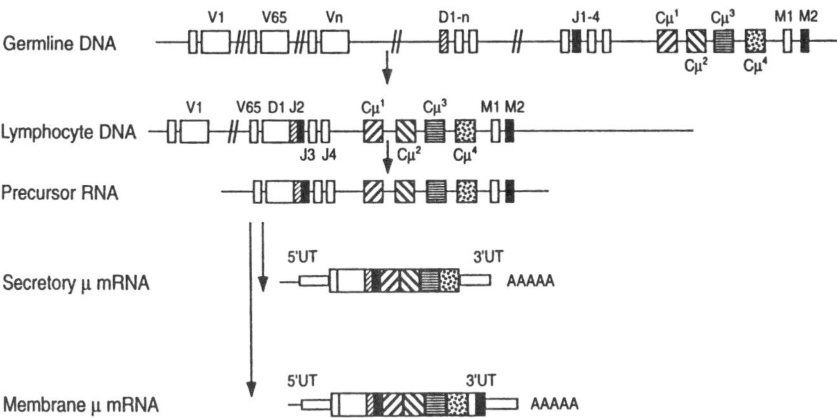

Fig 6.3. DNA rearrangements in heavy chain genes, and subsequent RNA processing. Commitment of stem cells to the B lymphocyte series involves somatic recombination making an individual V gene contiguous with a D gene, and the D gene contiguous with a J gene. However, the J gene does not become contiguous with a constant region gene. Transcription into precursor RNA is apparently a faithful copy of the rearranged DNA. The RNA is then spliced to remove all intervening sequences, but untranslated sequences persist at both ends. Finally, the mRNA is cleaved and a poly A tail is added. Membrane and secretory heavy chains are directed by separate RNA species, which result from alternate splicing at the 3′ end. M1 and M2 represent the exons encoding the hydrophobic membrane C-terminus of the μ heavy chain. In this illustration, the V_H65, D1 and J2 genes are expressed.

Each variable region gene has its own promoter at its 5′ end, creating a puzzle concerning why only the most *C*-proximal V gene is expressed. It is now known that high-level expression of a V gene depends on the presence of enhancer sequences, which were first found in the introns between J_H and C_μ (or to be more precise, between J_H and S_μ) and in the intron between J_κ and C_κ.

It was somewhat disappointing to find that the discovery of these enhancers did not lead to levels of expression of antibody genes that were comparable to those seen in plasmacytomas. More recently, enhancer sequences have been found 3′ of Cα, Cκ, Cλ1 and Cλ4. These 3′ enhancers are much more powerful in their action than the enhancers in the introns on the 5′ side of the constant region genes (reviewed by Max, 1993). Only the 3′ enhancer, and not the intron enhancer, is necessary for high-level expression of κ genes in hybridomas. The 3′ enhancer also appears essential for somatic hypermutation and affinity maturation (Betz *et al.*, 1994).

6.3 Sequential Rearrangement and Expression of Antibody Genes in the Developing B Cell

Expression of antibody genes in the developing B cell occurs in a progressive way. Lymphocytes have the ability to test for the accomplishment of a productive rearrangement (as indicated by production of a full-length immunoglobulin chain), which delivers a signal to activate the next stage. Failure to generate a functional product causes developmental arrest and cell death by apoptosis. The B cell also appears to be able to sense when a fully functional IgM molecule is expressed on its membrane; this signals the cell to cease further rearrangements.

The heavy chain locus is the first to rearrange, initially by D-J joining and followed by V–D joining. Production of a functional μ chain seems to be sensed by its expression on the membrane physically associated with the surrogate light chain, and this suppresses further heavy chain rearrangement and activates light chain rearrangement. If one μ allele rearranges productively, further μ rearrangements are suppressed, while a nonproductive rearrangement permits the cell to attempt to rearrange the other μ allele. If neither μ allele is rearranged productively the cell dies via apoptosis. If the first attempt at μ rearrangement was successful, the other μ allele will be in the germline configuration. It may be assumed that if one μ allele is rearranged productively and the other nonproductively, the nonproductive allele rearranged first, and the productive allele represents a successful second attempt.

Successful μ rearrangement allows the cell to move on to the next stage, when attempts are made to rearrange the light chain genes. Because there are two different light chain types (κ and λ), and each has two alleles, the developing B cell has at least four chances to rearrange productively. It is presumed that shortly after succeeding at this task, the recombination machinery for light chains is turned off, with one allele of one light chain rearranged productively, and the remainder either rearranged nonproductively (failed attempts) or in the germline (not attempted because of shutdown of rearrangement owing to successful production of a light chain).

Early theories that the λ locus only rearranges after attempts have failed at producing a productive κ rearrangement are no longer favoured. It now appears that the probability of κ rearrangement is higher than λ, perhaps owing to the larger number of κ genes, and as a result the κ locus usually but not always rearranges first (Max, 1993, and references therein).

6.3.1 Mechanisms of Monospecificity, Commitment to One Light Chain Type, and Allelic Exclusion

The fact that the antibody gene rearrangements are orderly and monitored by the cell for productive expression ensures that the great majority of cells express a single allelic form of the heavy chain and a single allelic form of a single light chain type (allelic exclusion). Allelic exclusion appears to be a result of the sensing of a productive heavy chain rearrangement via membrane immunoglobulin expression, and consequent cessation of further heavy chain rearrangements, because it is abolished by deletion of the membrane exon of IgM (Kitamura and Rajewsky, 1992). These effects must also be able to be mediated by membrane IgD, because transgenic expression of membrane IgD causes suppression of VDJ rearrangement and thereby allelic exclusion, and activation of light chain rearrangement (Iglesias *et al.*, 1987).

If a lymphocyte expresses more than one heavy chain class (e.g. IgM and IgD), they are always encoded by the same chromosome (*haplotype exclusion*; Herzenberg *et al.*, 1977; Stall, 1995). However, a lymphocyte may express either maternal and paternal light chains in conjunction with either maternal or paternal heavy chains.

The commitment of an individual B cell to a given heavy chain haplotype and a given V_H and V_L gene and a given light chain (κ or λ) is preserved even after cell division and switching of heavy chain class and somatic hypermutation and affinity maturation (see Section 3.3.9).

6.4 RNA Processing

Transcription begins a short distance 5′ to the initiation codon (ATG) of the active V gene, and continues though the various D, J and constant region segments. The final stopping point is not known, and may vary with individual circumstances. The initial long nuclear transcript is processed until all the sequences that intervene between coding segments are excised (Fig. 6.3). Segments of untranslated sequence persist at the 5′ and 3′ ends. The poly A addition signal (AAUAAA) lies 20–30 bases from the start of the poly A tail of the final messenger RNA (mRNA). The poly A tail is not encoded in the DNA but is added enzymically after cleavage of the mRNA. It appears to stabilize the mRNA.

In addition to excision of intervening sequences, heavy chain RNA may undergo differential splicing at its 3′ end to generate distinct mRNAs, one coding for the membrane form and the other coding for the secretory form (Fig. 6.3).

6.5 Generation of Antibody Diversity

The diversity of antibody-combining sites depends on a large number of genetic mechanisms (Tonegawa, 1983). There are a few-hundred variable region genes for heavy chains and a similar number for κ light chains. As discussed previously, the degree of diversity of λ light chain genes in the mouse is extremely limited. The random association of p light chains with q heavy chains potentially generates $p \times q$ antibody specificities. If p and q both are around 100, we have 10 000 combining sites. This mechanism is known as *combinational association*.

The introduction of D and J segments introduces a very large additional amount of diversity, because they lie in or close to the third hypervariable region and thus contribute to the specificity of the antigen-combining site. Variation in the precise splicing points generates additional diversity. Additional diversity arises as a result of variable trimming of the 5′ and 3′ ends of the D segments of heavy chains by an as yet uncharacterized exonuclease, followed by nontemplate-directed filling in with random nucleotides, probably via terminal transferase, to generate 'N' regions. Finally, a large amount of additional diversity is generated by somatic mutation (Tonegawa, 1983). This is particularly apparent in IgG antibodies, and accounts for the gradual rise in antibody affinity seen after immunization (Gearhart *et al*., 1981; Küppers *et al.*, 1993). The number of different antibodies that may be generated is therefore virtually infinite.

6.6 Genetics of Immunoglobulin Class Switches

Most mature but unstimulated B lymphocytes express both IgM and IgD on their surface, and the specificity of these two receptors is the same (Goding and Layton, 1976). Coexpression of IgM and IgD is probably achieved by differential splicing of a long primary transcript through the μ and δ genes (see Fig. 6.3).

Following antigenic stimulation, B cells mature into antibody-secreting cells. Production of immunoglobulin secretory mRNA is increased several-

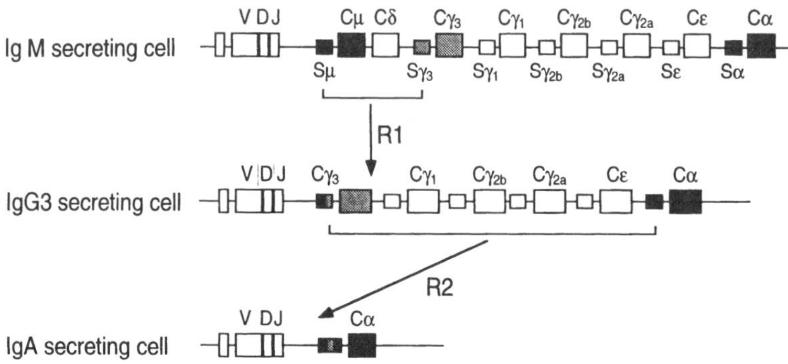

Fig. 6.4. Immunoglobulin class switching in lymphocytes. On the 5′ side of each constant region gene lies a switch (S) sequence. Recombination in the switch sequences mediates class switching, which is virtually always in the order of the genes and irreversible. In this illustration, the recombination event, R1, results in switching from IgM to IgG3, and a second recombination event, R2, results in switching from IgG3 to IgA.

hundredfold, and membrane mRNA synthesis reduced. The first antibody-secreting cells in most immune responses secrete IgM. IgD-secreting cells are rare at all stages in the immune response, perhaps because the Cδ gene lacks the switching mechanisms available to other heavy chain genes (see below). Days or weeks after the initial response, there is a shift towards secretion of other classes, particularly IgG.

The switch from IgM to IgG, IgA or IgE secretion involves further recombination events which translocate a particular V gene onto successive C genes, via short highly repetitive 'switch' sequences located on the 5′ side of each constant region gene (Fig. 6.4). The intervening DNA is removed.

B cell class switching seems to require three stimuli; antigen, engagement of CD40 with its ligand, and cytokines (see Sections 3.3.7–3.3.10). A typical mature B cell expressing IgM and IgD receptors has the potential to switch to any other subclass, but switching is in general a one-way event. Once a cell has switched to IgA, it loses the potential to express IgG or IgE (Fig. 6.4). The light chain type and antibody specificity remain the same during class switches.

6.7 Affinity Maturation and Somatic Mutation in B Cells

As discussed previously (Sections 3.3.7–3.3.10), the switch from IgM to IgG and other classes is accompanied by rounds of cell division and somatic mutation of antibody variable region genes, which, together with selection of the highest affinity cells by antigen on follicular dendritic cells, results in the survival and expression of clones secreting higher and higher affinity antibody.

The genetic mechanism of this somatic hypermutation is still poorly understood. It involves only the variable region and adjacent intron, does not involve the constant region, and occurs independently from selection by antigen (Sharpe *et al.*, 1991). Both the κ intron and the κ 3′ enhancer are essential, but the promoter regions are not (Betz *et al.*, 1994). The mutations consist entirely of single base substitutions, and no insertions or deletions are seen. The mutations tend to cluster around 'hot spots' such as the serine-31 codon and the first complementarity-determining region. Transitions (substitution of one purine for the other or one pyrimidine for the other) are favoured over transversions (substitution of a purine for a pyrimidine or vice versa, such that the purine–pyrimidine axis is reversed) (Betz *et al.*, 1993).

6.8 Immunoglobulin Genes and B Cell Neoplasia

During the analysis of rearrangements of the immunoglobulin heavy chain region in mouse plasmacytomas, it was noticed that there was frequently a rearrangement that involved an incoming sequence that was not part of the immunoglobulin locus on chromosome 12. This locus was shown to be the oncogene *c-myc*, which seems to be a transcriptional factor that controls cellular proliferation and survival and is located on chromosome 15 (reviewed by Cory, 1986). The aberrant rearrangement generally involves the alleles on the chromosome that is not productively rearranged for immunoglobulin synthesis.

Cytogenetically, a reciprocal 12:15 translocation is seen (Showe and Croce, 1987). It is generally presumed, but not proven, that the translocation occurs as an aberration of the normal breakage and rejoining mechanism involved in generation of a functional antibody gene. The joining region at the translocation lacks the degree of precision seen in physiological rearrangements.

It is now apparent that the great majority of mouse and rat plasmacytomas, and also most Burkitt's lymphomas in humans, involve a chromosomal translocation between the *myc* locus and the immunoglobulin heavy chain locus. The coding region of the *myc* gene is not altered by this translocation, but the regulation of the *myc* gene comes under the control of the immunoglobulin locus, resulting in deregulation of expression.

Transgenic mice in which the translocated *c-myc* gene under the control of the E-μ enhancer develop a very high incidence of lymphoid tumours, demonstrating that the translocation is causally involved in the neoplastic transformation of plasma cells. However, the fact that the transgenic mice are not born with the tumours indicates that other genetic events are required for the development of the fully malignant state (Cory, 1986).

The nature of mouse plasmacytomas is discussed in more detail in Section 7.1.

References

Amor, M., Bonhomme, F., Guenet, J-L., Petter, F. and Cazenave, P-A. (1984) Polymorphism of heavy chain immunoglobulin isotypes in the *Mus* subgenus. I. Limited polymorphism revealed by antibodies raised in SPE wild derived inbred strain. *Immunogenetics* **20**, 577–581.

Arp, B., McMullen, M. D. and Storb, U. (1982) Sequences of immunoglobulin λ1 genes in a λ1 defective mouse strain. *Nature* **298**, 184–187.

Bazin, H., Rousseaux, J., Rousseaux-Prevost, R., Platteau, B., Querinjean, P., Malache, J. M. and Delaunay, T. (1990) Rat Immunoglobulins. *In* 'Rat Hybridomas and Rat Monoclonal Antibodies' (H. Bazin, ed.), pp. 5–42. CRC Press, Boca Raton.

Betz, A. G., Milstein, C., Gonzalez-Fernandez, A., Pannell, R., Larson, T. and Neuberger, M. S. (1994) Elements regulating somatic hypermutation of an immunoglobulin κ gene: critical role for the intron enhancer/matrix attachment region. *Cell* **77**, 239–248.

Betz, A. G., Rada, C., Pannell, R., Milstein, C. and Neuberger, M. S. (1993) Passenger transgenes reveal intrinsic specificity of the antibody hypermutation mechanism: Clustering, polarity, and specific hot spots. *Proc. Natl Acad. Sci. USA* **90**, 2385–2388.

Black, S. J., Goding, J. W., Gutman, G. A., Herzenberg, L. A., Loken, M., Osborne, B., van der Loo, W. and Warner, N. (1978) Immunoglobulin isoantigens (allotypes) in the mouse. (V) Characterization of IgM allotypes. *Immunogenetics* **7**, 213–230.

Cory, S. (1986) Activation of cellular oncogenes in hemopoietic cells by chromosomal translocation. *Adv. Cancer. Res.* **47**, 189–234.

Dreyer, W. N. and Bennett, J. C. (1965) The molecular basis of antibody formation: a paradox. *Proc. Natl Acad. Sci. USA* **54**, 864–869.

Gearhart, P. J., Johnson, N. D., Douglas, R. and Hood, L. (1981) IgG antibodies to phosphorylcholine exhibit more diversity than their IgM counterparts. *Nature* **291**, 29–34.

Goding, J. W. and Layton, J. E. (1976) Antigen-induced co-capping of IgM and IgD-like receptors on murine B cells. *J. Exp. Med.* **144**, 852–857.

Goding, J. W., Warr, G. W. and Warner, N. L. (1976) Genetic polymorphism of IgD-like cell surface immunoglobulin in the mouse. *Proc. Natl Acad. Sci. USA* **73**, 1305–1309.

Grey, H. M., Abel, C. A., Yount, W. J. and Kunkel, H. G. (1968) A subclass of human γA-globulins (γA2) which lacks the disulfide bonds linking heavy and light chains. *J. Exp. Med.* **128**, 1223–1236.

Gutman, G. A. (1986) Rat immunoglobulin allotypes. *In* 'Handbook of Experimental Immunology' 4th edn. (D. M. Weir, L. A. Herzenberg, C. Blackwell and L. A. Herzenberg, eds.) Volume 3, pp. 98.1–98.10. Blackwell, Oxford.

Herzenberg, L. A., Minna, J. D. and Herzenberg, L. A. (1967) The chromosome region for immunoglobulin heavy chains in the mouse: allelic electrophoretic mobility differences and allotype suppression. *Cold Spring Harbor Symp. Quant. Biol.* **32**, 181–186.

Herzenberg, L. A., Black, S. J., Loken, M. R., Okumura, K., van der Loo, W., Osborne, B. A., Hewgill, DL., Goding, J. W., Gutman, G. and Warner, N. L. (1977) Surface markers and functional relationships on cells involved in murine B cell differentiation. *Cold Spring Harbor Symp. Quant. Biol.* **41**, 33–45.

Huang, C-H., Huang, H-J. S. and Lee, S-C. (1984) Detection of immunoglobulin heavy chain IgG3 polymorphism in wild mice with xenogeneic monoclonal antibodies. *Immunogenetics* **20**, 565–575.

Iglesias, A., Lamers, M. and Köhler, G. (1987) Expression of immunoglobulin δ chain causes allelic exclusion in transgenic mice. *Nature* **330**, 482–484.

Kindt, T. J. and Capra, J. D. (1984) 'The Antibody Enigma'. Plenum Press, New York.

Kitamura, D. and Rajewsky, K. (1992) Targeted disruption of μ chain membrane exon causes loss of heavy-chain allelic exclusion. *Nature* **356**, 154–156.

Küppers, R., Zhao, M., Hansmann, M-L. and Rajewsky, K. (1993) Tracing B cell development in human germinal centres by molecular analysis of single cells picked from histological sections. *EMBO J.* **12**, 4955–4967.

Lai, E., Wilson, R. K. and Hood, L. E. (1989) Physical maps of the mouse and human immunoglobulin-like loci. *Adv. Immunol.* **46**, 1–59

Max, E. (1993) Immunoglobulins. Molecular Genetics. *In* 'Fundamental Immunology' 3rd edn. (W. E. Paul ed.), pp. 315–382. Raven Press, New York.

Melchers, F., Karasuyama, H., Haasner, D., Bauer, S., Kudo, A., Sakaguchi, N., Jameson, B. and Rolink, A. (1993) The surrogate light chain in B-cell development. *Immunol. Today* **14**, 60–68.

Melchers, F., Haasner, D., Grawunder, U., Kalberer, C., Karasuyama, H., Winkler, T. and Rolink, A. G. (1994) Roles of IgH and L chains and of surrogate H and L chains in the development of cells of the B lymphocyte lineage. *Annu. Rev. Immunol.* **12**, 209–225.

Oi, V., Jones, P., Goding, J. W., Herzenberg, L. A. and Herzenberg, L. A. (1978) Properties of monoclonal antibodies to mouse Ig allotypes, H-2 and Ia antigens. *Curr. Top. Microbiol. Immunol.* **81**, 115–129.

Ollo, R. and Rougeon, F. (1983) Gene conversion and polymorphism: Generation of mouse immunoglobulin γ2a chain alleles by differential gene conversion by γ2b chain gene. *Cell* **32**, 515–523.

Pillai, S. and Baltimore, D. (1988) The ω and ι surrogate immunoglobulin light chains. *Curr. Topics. Microbiol. Immunol.* **137**, 136–139.

Schreier, P. H., Quester, S. and Bothwell, A. (1986) Allotypic differences in murine μ genes. *Nucl. Acids Res.* **14**, 2381–2389.

Schüppel, R., Wilke, J. and Weiler, E. (1987) Monoclonal anti-allotype antibody towards BALB/c IgM. Analysis of specificity and site of a V-C crossover in recombinant strain BALB-Ig-V^a/Igh-C^b. *Eur. J. Immunol.* **17**, 739–741.

Seppälä, I., Sarvas, H., Péterfy, F. and Mäkelä, O. (1981) The four subclasses of IgG can be isolated from mouse serum by using protein A-Sepharose. *Scand. J. Immunol.* **14**, 335–342.

Sharpe, M., Milstein, C., Jarvis, J. M. and Neuberger, M. S. (1991) Somatic hypermutation of immunoglobulin κ may depend on sequences 3′ of Cκ and occurs on passenger transgenes. *EMBO J.* **10**, 2139–2145.

Sheppard, H. W. and Gutman, G. A. (1981a) Complex allotype of rat kappa chains are coded for by structural alleles. *Nature* **293**, 669–671.

Sheppard, H. W. and Gutman, G. A. (1981b) Allelic forms of rat kappa chain genes: Evidence for strong selection at the level of nucleotide sequence. *Proc. Natl Acad. Sci. USA* **78**, 7064–7068.

Showe, L. C. and Croce, C. M. (1987) The role of chromosomal translocations in B- and T-cell neoplasia. *Annu. Rev. Immunol.* **5**, 253–277.

Stall, A. M. and Loken, M. R. (1984) Allotypic specificities of murine IgD and IgM recognized by monoclonal antibodies. *J. Immunol.* **132**, 787–795.

Stall, A. M. (1995) Mouse immunoglobulin allotypes. *In* 'Handbook of Experimental Immunology' 5th edn. (D. M. Weir, L. A. Herzenberg, C. Blackwell and L. A. Herzenberg), Volume 3. Blackwell, Oxford. (In press).

Stanisz, A. M., Lieberman, R., Kaplan, A. and Davie, J. M. (1983) IgA polymorphism in mice: NZB and BALB/c mice produce two forms of IgA. *Mol. Immunol.* **20**, 983–988.

Tonegawa, S. (1983) Somatic generation of antibody diversity. *Nature* **302**, 575–581.

Tsubata, T., Tsubata, R. and Reth, M. (1992) Crosslinking of the cell surface immunoglobulin (μ-surrogate light chain complex) on pre-B cells induces activation of V gene rearrangement at the immunoglobulin κ locus. *Int. Immunol.* **4**, 637–641.

Warner, N. L. and Marchalonis, J. J. (1972) Structural differences in mouse IgA myeloma proteins of different allotypes. *J. Immunol.*, **109**, 657–661.

7 Introduction to Monoclonal Antibodies

In order to appreciate the revolutionary nature of monoclonal antibodies, let us consider a typical immune response. Most antigens will consist of a mixture of molecules, and the antibody response will be directed against many different regions (determinants) on each molecule. Some regions may provoke a more vigorous response than others. There is no such thing as absolute purity, and even if the antigen is very pure, it is frequently found that very minor impurities in the antigen may provoke strong antibody responses. If one is unlucky, one may find that a minor contaminant is much more immunogenic than the desired antigen (Fig. 7.1).

Monoclonal antibodies overcome all these problems, allowing extreme specificity, not just for the immunizing protein but for a particular region on it, and making the degree of purity of the antigen irrelevant (Köhler and Milstein, 1975, 1976). By immortalizing single clones of antibody-forming cells, the immune response is thus dissected into its component parts. Instead of dealing with an extremely heterogeneous mixture of antibodies in polyclonal antisera, one can work with a homogeneous, chemically defined entity. Once the desired clone is isolated, there is no need to purify more antigen, and the specificity will be maintained essentially indefinitely, with no variation between animals. Accordingly, monoclonal antibodies have provided the solution to many biological problems.

Towards the end of this chapter, we will see that monoclonal antibodies have created some new problems. Sometimes, the specificity of monoclonal antibodies can be too narrow for a particular purpose. Conversely, like all antibodies, a monoclonal antibody will usually react with antigens that are closely related in structure to the desired antigen. What is less commonly appreciated is that occasionally a monoclonal antibody may react with other antigens that

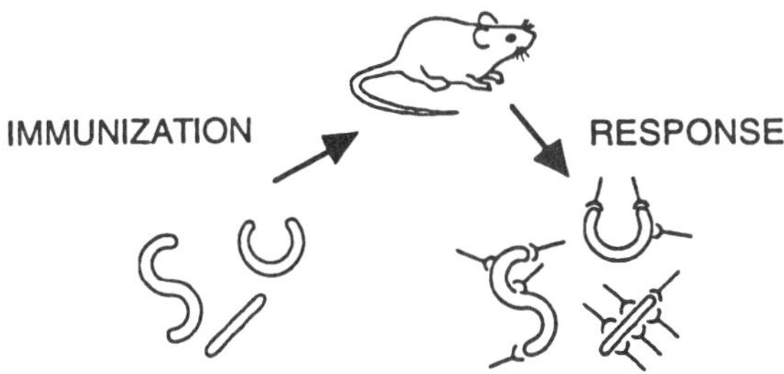

Fig. 7.1. A typical immune response. Antigen consists of a mixture of molecules, and the antibody response is directed against many different regions (determinants or epitopes) on each molecule. However, some regions provoke a more vigorous response than others.

bear no obvious structural relationship to the desired antigen (Lane and Koprowski, 1982; Ghosh and Campbell, 1986). These limitations are not serious, and provided they are kept in mind, the power of monoclonal antibodies may be exploited in ways that are almost infinitely flexible.

Köhler and Milstein's revolutionary work was built on the foundations laid by many other discoveries, particularly the study of mouse plasmacytomas and their adaptation to tissue culture, and the development of methods for cell fusion and selection of hybrids. We will now turn to these important background topics, and then return to the production of monoclonal antibodies.

7.1 Myeloma (Plasmacytoma)

The tumour caused by malignantly transformed antibody-secreting cells is known as *plasmacytoma* or *myeloma*. (While pathologists may make a morphological distinction between these terms, for our purposes they may be regarded as biologically indentical.) Spontaneous myelomas are rare in mice (Potter, 1972), but are quite common in the LOU/C strain of rats (Bazin *et al.*, 1973; Bazin, 1990; Bazin *et al.*, 1990b). Myeloma occasionally occurs in other species and is quite common in man.

In 1959, it was accidentally discovered that peritoneal irritants could cause development of myelomas in BALB/c mice (Merwin and Redmon, 1963). Subsequently, it was found that mineral oil or pristane were potent inducers of myeloma in BALB/c mice (Potter and Boyce, 1962; Potter, 1972). The plasmacytomas made by Potter using this approach have the prefix MOPC- (for mineral oil plasmacytoma). Approximately 40% of BALB/c mice will develop myeloma within 1 year of a series of three intraperitoneal injections of mineral oil (Warner, 1975). The only other laboratory strain of mouse where plasmacytomas can be induced is NZB. F_1 hybrids between NZB and BALB/c are also susceptible (Warner, 1975).

The development of plasmacytomas in mice is usually associated with a chromosome translocation between chromosome 15, which bears the oncogene *c-myc*, and the heavy chain locus on chromosome 12. The translocation causes abnormal and constitutive expression of the *c-myc* gene, and is an important step in the malignant transformation of the cell. In a few cases, a variant translocation involves the κ locus on chromosome 6 instead of the heavy chain on chromosome 12 (reviewed by Cory, 1986). While there is compelling evidence that this translocation is causally involved in the pathogenesis of mouse and rat plasmacytomas and Burkitt's lymphoma in humans, it is not sufficient to generate the fully malignant phenotype, which requires additional genetic events, most of which are still unknown.

The development of a large number of well-characterized mouse myelomas (Potter, 1972, 1977) has been responsible for much of our current knowledge of immunoglobulin structure, biosynthesis and genetics. The mouse myelomas that have been studied number in the thousands.

The immunoglobulin classes found in mouse and rat myelomas do not precisely reflect the abundance of classes in serum. BALB/c myelomas are 30–50% IgA, 30% IgG and 3% IgM, while NZB myelomas are 20% IgA, 50% IgG and 2% IgM (Hood *et al.*, 1976). Two IgD myelomas have been found in the mouse (Finkelman *et al.*, 1981), but to date no IgE myelomas. In the rat, the corresponding figures are 5% IgA, 46% IgG, 3% IgM, 1% IgD and 45% IgE (Bazin *et al.*, 1973, 1978, 1990; Bazin, 1990a, b).

In the great majority of cases, the antigen-binding specificity of myeloma proteins is not known, and attempts to generate antigen-specific myelomas by intense antigenic stimulation prior to tumour induction have failed. However, extensive screening programmes have turned up a number of antigen-binding myelomas, and these have been very useful in studies of the antigen-combining site (reviewed by Potter, 1977). The most common specificities have been for carbohydrate antigens and phosphoryl choline, presumably reflecting antigenic stimulation by microorganisms in the gut.

The rate of synthesis of immunoglobulins by plasmacytomas is quite high, and may be typically 20–30% of total protein synthesis (Wall and Kuehl, 1983). The synthesis of light chains is usually somewhat greater than that of heavy chains, possibly because isolated heavy chains may be toxic to the cell (Köhler, 1980; Argon *et al.*, 1983; Wall and Kuehl, 1983). A plausible mechanism for this toxicity might be the overloading of the chaperonin BiP, which binds unfolded and partly assembled proteins in the endoplasmic reticulum (see Gething and Sambrook, 1992). The excess light chains are usually secreted. Provided that the cells do not die and release their cytoplasmic contents during labelling with radioactive amino acids, immunoglobulin will be the major labelled protein in myeloma supernatants (see Section 9.6.3).

7.1.1 Adaptation of Mouse and Rat Plasmacytoma Cells to Continuous Tissue Culture

Many attempts have been made to produce continuous lines of plasmacytomas grown in tissue culture, but all failed until the work of Horibata and Harris (1970). The reasons for the difficulty in adapting mouse plasmacytomas to culture are still not completely clear, but may relate to the fact that most primary mouse lymphoid tumours require mercaptoethanol in the culture medium for growth, and this was not known in the early 1970s (A. W. Harris, personal communication). The success of Horibata and Harris in adapting these mouse lines to tissue culture presumably depended on the growth of mutants that lacked the requirement for mercaptoethanol. The spontaneously occurring plasmacytomas of LOU/C rats are much easier to adapt to tissue culture (Bazin *et al.*, 1990). The first such lines were described by Burtonboy *et al.* in 1973. They did not require mercaptoethanol.

In addition to the *myc* translocation and the need for mercaptoethanol in culture, many plasmacytomas require growth factors. This was first shown by Metcalf (1973, 1974), who found that colonies of nontissue-culture adapted mouse plasmacytomas could be grown in soft agar provided that conditioned medium was added. With hindsight, we can guess that the active ingredient studied by Metcalf in these pioneering experiments was probably what is now known as interleukin-6 (IL-6). Many plasmacytomas, particularly those of human or rat origin, require IL-6 for growth *in vitro* (Bazin and Lemieux, 1987, 1989; Shimizu *et al.*, 1989; van Snick, 1990; Harris *et al.*, 1992). It may be significant that two most common mouse plasmacytomas that are grown in tissue culture, respectively, have mutations in the IL-6 receptor (MOPC-21; Sugita *et al.*, 1990) or constitutively produce IL-6 (MPC-11) which may act in an autocrine manner (Blankenstein *et al.*, 1990).

IL-6 has been seen in many different situations, such as the plasmacytoma growth factor secreted constitutively by the macrophage cell line P388D1 (Nordan and Potter, 1986) and by the mouse plasmacytoma line MPC-11 (Blankenstein *et al.*, 1990). It has often been given different names, such as interferon β_2 (IFNβ_2), interleukin HP-1 (IL-HP1), B cell stimulating factor-2 (BSF-2), plasmacytoma growth factor (PCT-GF), hybridoma growth factor (HGF), CDF, and 26K factor (see van Snick, 1990). By common agreement it is now universally designated IL-6. The role of IL-6 and other growth factors for myelomas and hybridomas is also discussed in Section 8.9.3.

Primary mouse plasmacytomas also appear to require factors elicited by pristane (the active component of mineral oil used to induce the tumours), and this may reflect the need for contact with stromal cells (Degrassi *et al.*, 1993). With repeated passage of the tumours *in vivo*, the requirement for stromal cells lessens, and adaptation of the cells to tissue culture becomes easier (Degrassi *et al.*, 1993).

The early passages of cultured plasmacytomas tended to be intolerant of dilution and grew poorly when small numbers of cells were present. After multiple tissue culture generations, they now grow very well, even when diluted to the level of single cells.

The tissue culture adapted version of MOPC-21 (renamed P3K or P3) produced by Horibata and Harris (1970) was given to Milstein and was used in subsequent work on formation of hybrids. Milstein was also given a rat plasmacytoma line named S210 from Bazin's laboratory (Bazin *et al.*, 1990). An azaguanine-resistant clone of S210 was isolated in Milstein's laboratory and named 210RCY3Ag1, and subsequently renamed Y3.Ag1.2.3. In retrospect, it was lucky that the MOPC-21 line was used for the fusion experiments. For reasons that are still not well understood, it fuses with spleen cells far more readily than any other cultured mouse plasmacytoma line.

7.1.2 Chain Loss and Mutations in Myeloma Cells

Compared with other gene products, there is a rather high frequency of variants and deletions in immunoglobulin synthesis by cultured myeloma cells (Adetugbo *et al.*, 1977). The loss of heavy chain production is particularly common (Cotton *et al.*, 1973; Scharff, 1974). In one case, loss of the heavy chain was shown to be caused by a 50 base pair deletion in the V_H gene (Thammana, 1994). Isolated loss of light chain production without loss of the heavy chain is extremely rare, perhaps indicating that such cells cannot survive (Köhler, 1980; Argon *et al.*, 1983; see Section 7.1). In almost all the cases where heavy chains are secreted alone (heavy chain disease), there has been a substantial deletion of part of the heavy chain (Franklin and Frangione, 1971).

Cook and Scharff (1977) found a variant frequency of 0.2–2% in the phosphorylchlorine-binding myeloma S107, although this figure is much larger than seen by other workers (Adetugbo *et al.*, 1977). More recent work has shown that S107 differs from many other plasmacytomas in that V region mutations occur constitutively at a high rate, even though it represents a late stage of B cell differentiation. It may therefore be a very useful model cell for the study of somatic hypermutation (Zhu *et al.*, 1995).

7.1.3 Formation of Cell Hybrids by Chemically Induced Fusion

When cells are treated with Sendai virus or a high concentration of polyethylene glycol (PEG), their membranes fuse and multinucleate cells called *heterokaryons* are formed (Ringertz and Savage, 1976; Abbott and Povey, 1995). At the next cell division, the nuclei of heterokaryons fuse, and the daughter cells possess a more or less equal share of the genetic material. The mechanism of fusion is still poorly understood (see Zimmerberg *et al.*, 1993).

It is possible that both the PEG itself and an unidentified contaminant in the PEG are required for fusion (Wojciezsyn *et al.*, 1983).

For reasons that are poorly understood, the resulting hybrid cells are not genetically stable and there is a strong tendency for loss of chromosomes. This loss is not completely random. Depending on the species and perhaps on the individual cell types, there is usually a preferential loss of chromosomes from one or other cell (Ringertz and Savage, 1976; Abbott and Povey, 1995). By comparing the retained functions with the retained chromosomes, it has been possible to map a large number of genes to individual chromosomes.

Cell fusion may also be accomplished by a brief, intense direct current electrical pulse (Hewish and Werkmeister, 1989; Werkmeister *et al.*, 1991).

7.1.4 Selection Procedures for Hybrids

When a cell mixture is subjected to reagents that promote fusion, the fusion events are poorly controlled. In addition to A–B fusions, it is to be expected that many fusions will be A–A or B–B, or even higher multiples of these. Thus, if it is desired to produce a long-term hybrid cell line from two cell types, a selection procedure is required.

The most common selection procedure is that devised by Littlefield in 1964. Littlefield's procedure depends on the fact that when the main biosynthetic pathway for guanosine is blocked by the folic acid antagonist aminopterin, there is an alternative 'salvage' pathway in which the nucleotide metabolites hypoxanthine or guanine are converted to guanosine monophosphate via the enzyme *hypoxanthine guanine phosphoribosyl transferase* (HGPRT or HPRT; Fig. 7.2). Cells lacking HGPRT die in a medium containing hypoxanthine, aminopterin and thymidine (HAT medium) because both the main and the salvage pathways are blocked. However, an HGPRT$^-$ cell can be made to grow in HAT medium if it is provided with the missing enzyme by fusion with an HGPRT$^+$ cell.

Selection of HGPRT$^-$ cells is performed by use of the toxic base analogues 8-azaguanine or 6-thioguanine, which are incorporated into DNA via HGPRT. Because the salvage pathway is not normally essential for cell survival, mutants that lack HGPRT will continue growing, while cells that possess HGPRT will die.

Selection of HGPRT$^-$ variants is usually easy because the enzyme is coded for by the X chromosome, and there is normally only one active copy of the X chromosome per cell (Lyon, 1961). Selection of variants lacking thymidine kinase (an autosomal enzyme) is more difficult, because two simultaneous rare events are required. (This simplified description does not take into account the many complexities of the HGPRT locus (Caskey and Kruh, 1979; Stout and Caskey, 1988; Cariello and Skopek, 1993), nor does it consider the fact that most cultured myeloma cells are subtetraploid rather than diploid (Ohno *et al.*, 1979).)

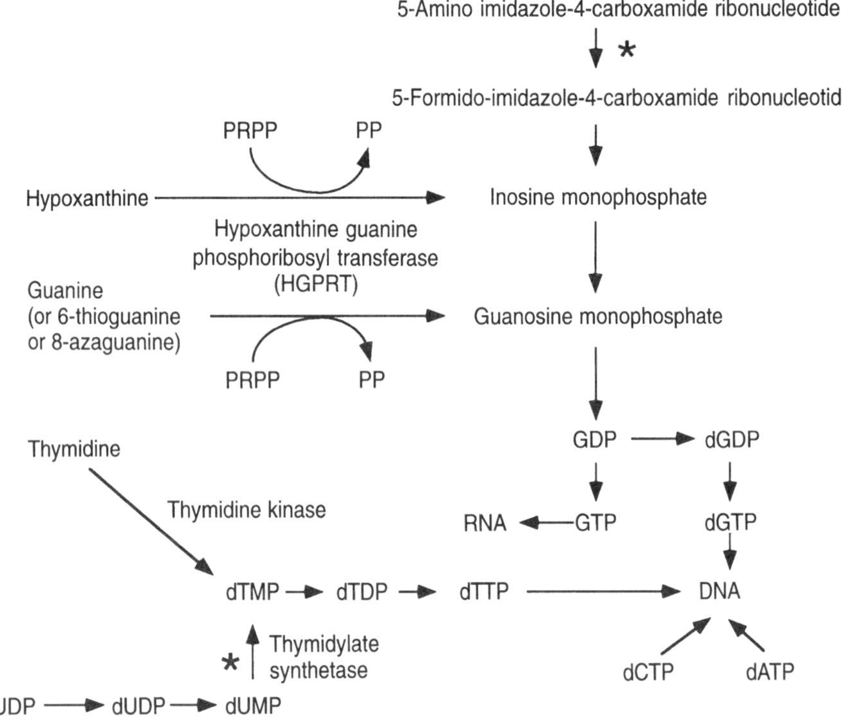

Fig. 7.2. Metabolic pathways relevant to hybrid selection in medium containing hypoxanthine, aminopterin and thymidine (HAT medium). When the main synthetic pathways are blocked with the folic acid analogue aminopterin (*), the cell must depend on the 'salvage' enzymes HGPRT and thymidine kinase. HGPRT$^-$ cells can be selected by growth in medium containing the toxic base analogues 6-thioguanine or 8-azaguanine, which are incorporated into the cell via HGPRT. Only HGPRT$^-$ cells survive. HGPRT$^-$ cells cannot grow in HAT medium unless they are fused with HGPRT$^+$ cells.

7.2 Monoclonal Antibodies of Predefined Specificity Produced by Hybridomas

For decades, immunologists have sought ways of producing homogeneous antibodies of defined specificity (Krause, 1970). Some limited successes were achieved by screening myeloma proteins for antigen binding (see Potter, 1977), but a general method did not become available until 1975, when Köhler and Milstein used cell hybridization to produce continuous cell lines secreting antibody of predefined specificity (see Köhler and Milstein, 1975, 1976).

Milstein had been interested in the genetic control of antibody synthesis. In a study designed to examine the basis of allelic exclusion, Cotton and Milstein (1973) constructed hybrids between rat and mouse myeloma cells, and found that the synthesis of both species of immunoglobulin was retained, but hybrid polypeptides (indicating that V-C joining took place at the mRNA level) were not seen. This result was consistent with the idea that V-C joining took place at the DNA level. However, mixed IgG molecules were observed that were made up of various combinations of the rat and mouse polypeptides (Cotton and Milstein, 1973).

In 1975, Köhler and Milstein took the bold step of extending the previous experiments by fusing a HAT-sensitive variant of MOPC-21 myeloma cells with spleen cells from mice immunized with sheep red cells. The fusion was mediated by Sendai virus, and hybrids were selected by growth in medium containing hypoxanthine, aminopterin and thymidine (HAT medium; see Section 7.1.4). It was known that normal spleen cells could only survive a few days in culture, but it was hoped that they would 'complement' the missing hypoxanthine guanosine phosphoribosyl transferase (HGPRT) in the myeloma cells, and that the myeloma cells would provide the 'immortality' needed for continuous culture (Fig. 7.3).

The experiment worked exactly as planned, and a number of cloned hybrid lines secreting anti-sheep erythrocyte antibodies were produced. The cell lines were capable of forming tumours when injected into mice. These tumours are now known as hybridomas. The serum of myeloma or hybridoma-bearing

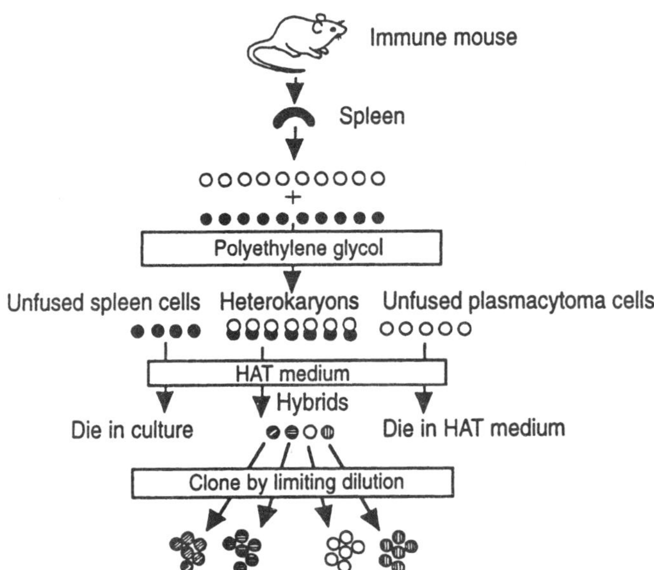

Fig. 7.3. Production of hybridomas. Spleen cells from immune mice are fused with HGPRT⁻ myeloma (plasmacytoma) cells using polyethylene glycol. The binucleate fusion products are known as heterokaryons. *At the next division, the nuclei fuse, generating hybrid cells, which grow in HAT medium. Unfused myeloma cells die in HAT medium (see Fig. 7.2), and unfused spleen cells can only survive for a few days in culture. Hybrids are tested for production of antibody of the desired specificity, and cloned by limiting dilution.*

mice contains large amounts of homogeneous antibody. The use of Sendai virus has been superseded by PEG (Pontecorvo, 1976; Galfrè and Milstein, 1981), but except for this modification, the basic procedure is essentially unchanged.

The implications of Köhler and Milstein's discovery were revolutionary. For the first time, unlimited quantities of absolutely specific and uniform antibodies recognizing only one antigenic site could be produced, even if the immunizing antigen was weakly immunogenic or grossly impure. All that was needed was an appropriate way of screening for antibody with the desired properties.

7.2.1 Cloning of Hybridomas

Because of the high probability of chromosome loss in the hybrids, and to ensure that the antibodies were indeed monoclonal, it was essential to clone the hybrid cell lines. Köhler and Milstein used soft agar cultures to grow discrete colonies, and detected clones secreting anti-erythrocyte antibodies by overlaying with sheep red cells and complement. Colonies secreting anti-erythrocyte antibodies were surrounded by a zone of haemolysis.

Cloning in soft agar has the disadvantage that unless an overlay technique is available, colonies need to be plucked out and regrown in liquid culture prior to testing for antibody production. For this reason, most workers now prefer to clone by limiting dilution. If cells are grown in small numbers, the fraction of wells with growth should follow the Poisson distribution (see Lefkovits and Waldmann, 1979; Coller and Coller, 1983):

$$f(0) = e^{-\lambda}$$

where $f(0)$ is the fraction of wells with no growth, and λ is the *average* number of clones per well.

If $\lambda = 1, f(0) = 0.37$. In other words, to obtain a reasonable probability that wells with growth contain single clones, more than 37% of wells should have no growth. This analysis assumes a cloning efficiency of 100%, and that there is no clumping of cells.

Hybridoma lines should be cloned at least twice to make absolutely certain that each is a true clone because the Poisson distribution specifies only probabilities and not certainties (Coller and Coller, 1983). If there are 37% of wells with no growth in the first cloning, it is necessary to have less than 32% with no growth in the second cloning to achieve a 95% probability of clonality, and more than 94% with no growth to achieve a 99% probability of clonality (Coller and Coller, 1983).

Repeated subcloning is also important because of the relatively high probability of growth of nonproducer variants owing to chromosome loss. After two cycles of cloning, the rate of chromosome loss is small, although the risk of overgrowth by nonproducer cells is never completely eliminated.

7.2.2 Nature of the Fusing Cell in Spleen

In a certain sense, the original experiments worked *too* well. Although only about 1% of spleen cells actively secrete immunoglobulin, about 10% of the hybrid cell lines secreted antibody. There may be two explanations. First, there seems to be a preference for myeloma cells to fuse with dividing B cells. Consistent with this idea, it appears that the spleen cells that fuse with myeloma cells are larger than average and have recently undergone antigen stimulation and blast transformation (Andersson and Melchers, 1978). Moreover, Paslay and Roozen (1981) found that the peak hybridoma production corresponded to the peak of proliferation, which preceded the peak of production of antibody-secreting cells and the peak of serum antibodies. The other possible explanation may be that myeloma cells have the ability to activate nonsecreting B cells to rapid secretion (Eshhar *et al.*, 1979).

It is probable that the myeloma cells also fuse with T cells and other non-B cells in the spleen. However, fusion of cells of different lineages usually results in extinction of differentiated function, and it is unlikely that these fusions would result in antibody-secreting hybrids (Köhler *et al.*, 1977; Boshart *et al.*, 1993).

7.2.3 Nonproducer Variants of Myeloma Cells for Fusion

The original cell line used by Köhler and Milstein secreted IgGl with κ light chains. As expected from the earlier experiments involving rat–mouse fusions (Cotton and Milstein, 1973), mixed molecules of IgG containing light and heavy chains from both MOPC-21 and the spleen cells were found. Milstein and Köhler proposed that the MOPC-21 heavy chain be designated G and its light chain K, while the heavy chain of the spleen cell should be designated H and the light chain L. The product of the myeloma would then be GK, the spleen cell HL, and the mixed molecules HLGK.

If the association of chains were random, only a small minority of IgG molecules would be derived entirely from the spleen. A combining site with the myeloma light or heavy chains (or both) would not be expected to bind antigen. Thus, many secreted IgG molecules would be completely inactive, while some would possess only one active antigen-binding site. Only a minority of molecules would have two functional combining sites (Figs 5.2 and 7.4). Fusion of two different antibody-forming cells allows the formation of mixed antibodies of dual specificity (Milstein and Cuello, 1983).

In 1976, Köhler *et al.* (1976) described a variant of MOPC-21 (P3-NS1-Ag4-1; abbreviated NS-1) which lacked heavy chain synthesis. This defect is now known to be due to a 50 base pair deletion in the V_H framework region (Thammana, 1994). Although NS-1 continued to make κ light chains, they were degraded intracellularly and were not secreted. Fusion of NS-1 with spleen cells resulted in the production of active antibody-secreting hybrids.

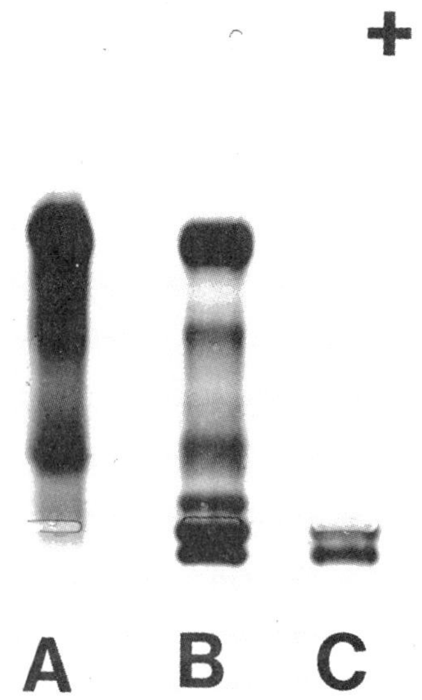

Fig. 7.4. Agarose electrophoresis of serum from normal mice (A); from mice bearing the hybridoma MAR-18.5 (see Fig. 5.3 lane B); and MAR-18.5 antibody purified by affinity chromatography on its antigen (lane C). Note that in lane B there are three spikes in the γ globulin region, but only two of these contain material that binds antigen (lane C). The most anodal spike has both heavy chains from MOPC-21 (γ1/γ1), the middle spike has one MOPC-21 γ1 chain and one spleen γ2a chain, while the most cathodal spike has both heavy chains from the spleen (γ2a/γ2a). Because of their size, the heavy chains dominate the charge of the IgG molecule, and hence its mobility.

The immunoglobulins secreted by NS-1 bore only spleen cell heavy chain, but had light chains derived from both spleen and MOPC-21 (i.e. HLK). In other words, the MOPC-21 light chains were 'rescued' by binding to the introduced heavy chains, suggesting that they did not fold correctly on their own.

If the light and heavy chain pairing were random and the rates of synthesis of spleen and MOPC-21 light chains equal, one would expect 25% of molecules to possess only the spleen cell heavy chains (and therefore have two functional combining sites), 50% to have one spleen cell light chain and one MOPC-21 light chain (one functional site) and 25% to have both light chains from MOPC-21 (no functional combining sites). The use of NS-1 as fusion partner thus allows about 75% of the secreted immunoglobulin to have antibody activity.

Subsequently, several cell lines have been produced which synthesize neither heavy nor light chains, but which allow the production of antibody-secreting hybridomas when fused with spleen cells. These include Sp 2/0-Ag-14 (Shulman *et al.*, 1978), X63-Ag8.653 (Kearney *et al.*, 1979) and NS0/1

(Galfrè and Milstein, 1981). Of these, X63-Ag8.653 is probably the cell of choice because it is widely available, has a high fusion frequency and is easy to grow.

Several rat lines have been shown to be suitable for fusion. Y3-Ag1.2.3 (abbreviated Y3) is an azaguanine resistant derivative of the LOU/C strain myeloma R2310.RCY3 (Bazin *et al.*, 1990), and secretes κ chains but does not express heavy chains (Galfrè *et al.*, 1979). Fusion of spleen cells with Y3 usually results in HLK antibodies.

Bazin (1982, 1990a) has developed a total nonproducer rat line suitable for fusion (IR 983F) (Bazin, 1990). A total nonproducer line, YB2/0, has also been derived from a fusion between Y3 myeloma cells and spleen cells from an AO rat (Kilmartin *et al.*, 1982; Galfrè and Milstein, 1981).

7.2.4 Strategies for Production of Human Monoclonal Antibodies

Monoclonal antibodies of human origin would be extremely useful for many purposes, particularly for administration to humans for tumour immunotherapy, imaging, and neutralization of toxins, viruses and drug overdoses. The administration of mouse monoclonal antibodies to humans almost always results in the production of human antimouse antibodies (HAMA), which limits their effectiveness and can cause serious side-effects such as serum sickness and even anaphylaxis. Despite intense effort, human monoclonal antibodies have not been easy to produce (see James and Bell, 1987; Thompson, 1988). Almost every year, a paper or review is published that states that the problem has been very difficult, but is now solved. Let us look at the difficulties, and then explore current progress.

The first problem has been the lack of a suitable fusion partner. This problem is perhaps a little less surprising if one bears in mind the very limited number of mouse myeloma lines that work well for making mouse monoclonal antibodies. Almost everyone still uses a derivative of the original MOPC-21 line made by Horibata and Harris in 1970. Most of the so-called 'myeloma' cell lines of human origin have proved to be Epstein–Barr virus-transformed B lymphoblastoid lines, which have low levels of immunoglobulin production, grow slowly in large clumps, and seem to be somewhat unstable in long-term culture as hybrids. While many attempts have been made to adapt human myelomas to tissue culture, few have succeeded. The knowledge that many human myelomas are dependent on IL-6, either from other cells or via an autocrine route (Shimizu *et al.*, 1989) has had some promise, but so far general solution has not been found (James and Bell, 1987; Thompson, 1988). Fusion of human cells with mouse lines such as X63-Ag8.653 have had some success, and there is a feeling that the instability of interspecies hybrids may have been somewhat exaggerated. With repeated subcloning, some hybrids made in this way are said to be quite stable (Thompson, 1988).

The second problem is the source of human B cells. The only easily available cells are from peripheral blood, and these are generally nondividing and unsuitable for fusion. If fused, they tend to give low-affinity IgM antibodies. For ethical reasons, it is difficult to immunize humans to most antigens.

Many potential solutions have been proposed. The first is to 'humanize' mouse antibodies by grafting their complementarity-determining regions (CDRs) onto human immunoglobulin backbones (Borrebaeck, 1992; Mayforth, 1993; Winter, 1993; Owens and Young, 1994). While elegant and simple in theory, this approach has many pitfalls. It requires a careful assessment of the three-dimensional structure of the antibody, in order to make sure that the CDRs fit the new background with a minimum of distortion. In many cases, the affinity of the resulting antibody is reduced. Even if such antibodies are successfully engineered, there is still some antimouse response when they are injected into humans, albeit a greatly reduced one. Finally, this approach is extremely labour-intensive, and requires highly specialized expertise in genetic engineering.

Another potential approach is to activate the human B cells *in vitro*. This is not as easy as it appears. Banchereau and colleagues have developed a B cell culture system in which the cells are stimulated by IL-4 and CD40, using anti-CD40 antibodies and a mouse cell line expressing the human FcγII receptor (Darveau *et al.*, 1993). After activation *in vitro*, the B cells were fused with the mouse cell line X63-Ag8.653, and many clones secreting human immunoglobulin were produced. Unfortunately, this paper only tested for immunoglobulin production and not for specific antibody, so it is difficult to tell whether it will fulfil its promise for production of human monoclonal antibodies.

Yet another approach exploited the reconstitution of mice with severe combined immunodeficiency (SCID; Bosma *et al.*, 1983) using human immune system (Mosier *et al.*, 1988). While this system seemed promising at first, it has not been generally adopted.

Yet another approach has been to make combinatorial libraries of human antibody genes encoding Fv fragments and express them in *Escherichia coli* (Chiswell and McCafferty, 1992; Nissim *et al.*, 1994). The Fv fragment of antibodies is not usually glycosylated, and is small and therefore easier to fold than the whole molecule, so this approach has much promise. The main problem at present is to create a library with sufficient diversity to ensure that there will be some combinations with sufficiently high affinity for any antigen. The most promising version of this technology involves phage 'display' systems in which the antibody CDRs are inserted into the gene III coat protein of *fd* phage, which allows simultaneous screening and selection for the phage carrying the appropriate genes (Nissim *et al.*, 1994; see Chapter 14). Future improvements to this system may include mutagenesis and various other strategies to improve the affinity of the selected clone (Giesow, 1992).

Recently, transgenic mice have been produced in which the endogenous mouse immunoglobulin genes have been knocked out and a set of human immunoglobulin genes inserted (Green *et al.*, 1994; Wagner *et al.*, 1994). For technical reasons, only a relatively small number of human immunoglobulin genes were inserted, and the antibody responses obtained after immunization were rather variable in quantity and affinity. This approach appears extremely promising, and may make all other approaches obsolete because it should be possible to perform fusions and make monoclonal antibodies in exactly the same way as for mouse antibodies.

7.3 Differences Between Conventional and Monoclonal Serology

The invention of hybridoma antibodies has done much to put serology on a firm scientific basis. The old uncertainties of specificity and reproducibility have been replaced by the promise of unlimited supplies of standardized, monospecific antibodies. Terms like titre and avidity have become virtually obsolete. We can now talk about mass and affinity of antibody in a very precise way. However, the successful use of monoclonal antibodies requires a firm understanding of the differences between conventional and monoclonal serology.

7.3.1 Cross-reactions Due to Structural Relatedness Between Antigens

Cross-reaction may be defined as the reaction of an antiserum against an antigen molecule not present in the immunizing preparation. It is usually a manifestation of structural similarities between the immunizing antigen and the cross-reacting antigen (Fig. 7.5). An obvious practical point concerning cross-reaction is that it places some limits on the specificity of antibodies.

Consider an antiserum containing antibodies against the carbohydrate

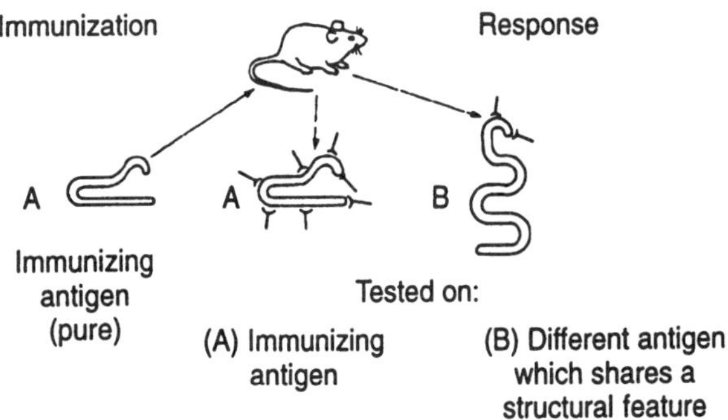

Fig. 7.5. Cross-reaction due to a shared structural feature between the immunizing antigen (A) and the test antigen (B).

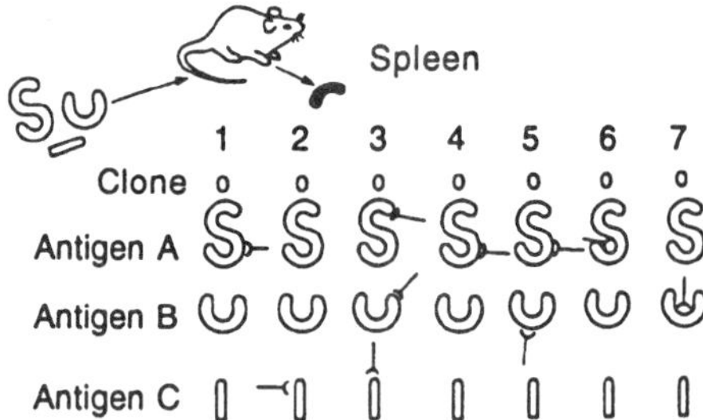

Fig. 7.6 Cross-reaction at the level of the individual clone. Clones 1, 2, 4, 6 and 7 show no cross-reaction, while clones 3 and 5 cross-react because of a shared structural feature in antigens A, B and C. Note that this structural feature does not need to be absolutely identical in A, B and C. The affinities of binding of clone 3 to A, B and C may be quite different.

moieties of a particular glycoprotein. Carbohydrate structures are widely shared between different glycoproteins, and such an antiserum is thus very likely to cross-react with other glycoproteins, even if the polypeptide portions are totally unrelated. In other words, antibodies may have lectin-like properties owing to carbohydrate cross-reactions.

The proportion of antibodies in a polyclonal antiserum which will cross-react in a given situation will depend on a number of factors. Particularly important is the degree of structural relatedness between the two antigens. Thus, a sheep antimouse IgG antiserum will cross-react very extensively with rat IgG, with perhaps half or more of the total mass of antibodies binding both antigens. (As far as the sheep is concerned, a mouse looks very much like a rat.) Conversely, cross-reaction of the same antiserum with human IgG may be weak or undetectable. (The sheep would generally have little trouble distinguishing a mouse from a human.) The cross-reacting antibody subpopulation may be removed from an antiserum by absorption with the cross-reacting antigen coupled to agarose beads, implying that within the total antibody population there exists a cross-reacting subset and a noncross-reacting subset.

Now consider the situation at the level of the individual clone. Figure 7.6 shows a simplified model of the immune response of a mouse to three antigens. Clones 1 and 4 recognize a determinant which is only present on antigen A. However, clone 5 recognizes a determinant on the same portion of A as clones 1 and 4, but it also recognizes a structure on antigen B which is *not* recognized by clones 1 and 4. By definition, then, the determinant recognized by clone 5 must be different to that recognized by clones 1 and 4.

It is crucial to understand that the determinant on antigen B which is recognized by clone 5 is not necessarily identical with the determinant recognized by the same antibody on antigen A. It is quite possible to have a similar, but non-identical determinant that allows detectable binding, perhaps of lower affinity (see Birnbaum and Kourilsky, 1981, for an example). Depending on the desired application, the specificity of clone 1 and 4 or of clone 5 may be preferred.

7.3.2 Cross-reactions Due to Multiple Specificity of Individual Clones

Although antibodies are highly specific, they cannot perform miracles. The combining site of an antibody is merely a highly complex surface that can bind other structures which have a complementary shape. As mentioned previously (see Section 5.2.1), the antigen–antibody interaction is probably more like a handshake than a lock and key; the surfaces may mould to fit each other. In addition, the primary structure of the variable regions may not necessarily specify a unique conformation (Foote and Milstein, 1994). It is quite possible for a homogeneous monoclonal antibody to recognize two or more antigens that are not related in any obvious way. This result was predicted by Talmage in 1959, and evidence was found by Richards in 1975.

The problem of unexpected cross-reactions of monoclonal antibodies has been reviewed by Lane and Koprowski (1982), and by Ghosh and Campbell (1986). Three main interactions between different antigens and a given antibody can be envisaged. *Group I* interactions represent the binding to identical structures in the different antigens. *Group II* interactions represent binding to a closely related structure. *Group III* interactions represent binding to an apparently unrelated structure, which happens by chance to form a sufficient number of interactions with the antibody to produce measurable binding. Group III interactions are expected and usually observed to be of low affinity, but are strengthened by multivalent binding. In other words, they are unlikely to be detected when monovalent Fab′ fragments are tested; they will sometimes be seen with intact IgG. They are most common with IgM, because the large number of binding sites greatly increases the total strength of binding (avidity; see Fig. 2.4). As a general rule, monoclonal antibodies of the IgM class often have unexpected cross-reactivities, and are best avoided unless there is absolutely no choice.

There are now many examples where a monoclonal antibody can bind two apparently quite unrelated antigens (Lane and Koprowski, 1982; Ghosh and Campbell, 1986). For example, a rat IgM monoclonal antibody to mouse Thy-1 also bound to a determinant on the variable region of κ light chains of the mouse myeloma TEPC-15 (Pillemer and Weissman, 1981). Similarly, some monoclonal antibodies cross-react with two cytoskeletal proteins, tropomyosin and vimentin (Blose *et al.*, 1982).

The interactions between specificity, multivalency and cross-reaction have been elegantly described in a series of papers by Parham. These papers should be referred to for more detail (Ways and Parham, 1983; Parham, 1984a, b).

One should not be too surprised at the concept of multispecificity of monoclonal antibodies. A stable antigen–antibody complex will result whenever there is sufficient number and strength of short-range interactions, and this might be achieved in many different ways. Considering the enormous number of different biological macromolecules, it is to be expected that occasionally there will be high-affinity interactions between functionally unrelated molecules. For example, tubulin has a high affinity for lactoperoxidase (Rousset and Wolff, 1980) and actin has a high affinity for deoxyribonuclease I (Hitchcock, 1980).

These types of odd cross-reactions are seldom, if ever, seen in polyclonal sera. The specificity of a conventional polyclonal antiserum is the result of a consensus of thousands of different clones, and the likelihood is that cross-reactions due to multispecificity will be random, and therefore diluted out. Thus, an antiserum against an absolutely pure antigen A may contain clones that react with A, C, P, Z, other clones that react with A, B, D, E, and other clones that react with A, Q, R, Y, etc. The only common reactivity is that against A, and the other specificities will be too dilute to be detected. This effect may give added specificity to polyclonal responses, as predicted by Talmage in 1959.

The specificity of a given monoclonal antibody is therefore unlikely to be absolute. Given a large enough panel of unrelated test antigens, sooner or later a cross-reaction will be found. Unexpected cross-reactions as cited by Lane and Koprowski (1982) are uncommon in IgG antibodies, but more frequent in IgM. It is important to realize that they can and do occur.

7.3.3 Sometimes, a Monoclonal Antibody May Be Too Specific

The specificity of conventional polyclonal antisera depends on a consensus of hundreds of thousands of clonal products, which bind to antigenic determinants covering most or all of the external surface of the antigen (Benjamin *et al*., 1984). As a result, small changes in the structure of the antigen owing to genetic polymorphism, heterogeneity of glycosylation or slight denaturation will usually have little effect on polyclonal antibody binding. Similarly, a subset of antibodies from a polyclonal antiserum will usually bind to antigens which have been modified or denatured (Burnette, 1981).

In contrast, monoclonal antibodies usually bind to one single unique site on the antigen molecule. If for any reason this site is altered, the antibody may or may not continue to bind. Whether this is seen as a problem or an advantage will obviously depend on the individual circumstances. If the monoclonal antibody is used in a radioimmunoassay for a human serum protein, a minor genetic polymorphism (known or unknown) in that protein could cause gross errors. Similarly, if monoclonal antibodies were used for classification of microorganisms, they might not give exactly the same reactivity patterns as conventional antisera.

It is therefore essential that monoclonal antibodies be tested and characterized in the assay system in which they are to be used.

In future years, one may anticipate that commercial preparations of antibodies might be comprised of pools of several different clonal products, such that their nominal specificity will be maintained in all circumstances.

7.3.4 Affinity of Polyclonal and Monoclonal Antibodies

Most polyclonal antisera contain antibodies of a wide range of affinities, with dissociation constants from about 10^{-6} M to 10^{-9} M (Mason and Williams, 1980; Ways and Parham, 1983; Parham, 1984a, b; Dower *et al.*, 1984). Antibodies with affinities of less than about 10^{-6} M are not usually detected by the standard methods, and the higher affinity antibodies tend to dominate in most practical situations. These very high-affinity antibodies may comprise only a small fraction of the total. It is therefore to be expected that extremely high-affinity monoclonal antibodies will be less common than lower affinity ones.

The homogeneity of monoclonal antibodies means that each clonal product will have a well-defined affinity. The use of antibodies for affinity chromatography or immunofluorescence usually involves extensive washing to remove unbound antibody. If a monoclonal antibody has a low affinity, excessive washing may dissociate the antigen–antibody complexes.

The extreme heterogeneity of polyclonal antibodies has a number of other important practical consequences. It allows an antibody population to be covalently modified (e.g. with fluorochromes or ^{125}I) with the expectation that although some antibodies will no longer bind, enough reactivity will be preserved to allow the experiment to proceed. Binding may still be detected even if more than 95% of the antibodies are destroyed.

Monoclonal antibodies may behave quite differently under these conditions. In some cases, it may be possible to alter antigen or antibody quite extensively without destruction of binding. In others, it may be found that seemingly minor modifications to antibody or antigen may abolish binding completely (Nussenzweig *et al.*, 1982).

7.3.5 Kinetics of Binding of Monoclonal Antibodies

Individual monoclonal antibodies vary greatly in the kinetics of binding to antigen (Mason and Williams, 1980; Ways and Parham, 1983; Parham, 1984a, b; Dower *et al.*, 1984). The binding of individual monoclonal antibodies to the cell surface may reach saturation in as little as 15 min, or as long as 90 min (J. Goding, unpublished data). It seems clear that the rate of association is sometimes limited by more than simple diffusion. These results might be understood in terms of the need for an individual antigenic determinant that is recognized by a given monoclonal antibody to be in a particular (transient) conformation before the antibody can bind. Such slow 'on' rates might be expected to be highly dependent on temperature.

The highly individual kinetics of binding of monoclonal antibodies to their antigen makes it strongly advisable to keep incubation times constant from one experiment to another. Failure to do so may lead to errors in quantification and poor reproducibility in cases where the kinetics of binding are slow.

7.3.6 Differences Due to Failure of Cross-linking

Many important immunological assays depend on cross-linking of antigen by antibody. Monoclonal antibodies would be expected to cause cross-linking only if the relevant antigen had two or more antibody-binding sites. Thus, monoclonal antibodies would only be expected to form precipitates in gel diffusion (Ouchterlony) assays when the antigen is dimeric or multimeric. Because of their multivalency, IgM antibodies are more likely to cause cell agglutination than IgG. Whether or not a monoclonal antibody will agglutinate cells will also depend on the density and orientation of the relevant antigenic sites. Binding *per se* will not guarantee agglutination.

The activation of complement also depends on the close proximity of the Fc portions of antibodies, and it is to be expected that some monoclonal IgG antibodies may fail to activate complement. Neuberger and Rajewsky (1981) studied the relative effectiveness of monoclonal mouse antibodies in initiating complement-mediated lysis of erythrocytes. IgM, IgG2a, IgG2b and IgG3 were cytotoxic, while IgD and IgA were not. IgG1 was weakly cytotoxic in some cases. Howard *et al.* (1979) found that certain mixtures of monoclonal antibodies to different determinants were highly cytotoxic, while the individual antibodies on their own were weakly cytotoxic or not cytotoxic at all.

7.3.7 Individual Monoclonal Antibodies Have Distinct Physical and Chemical Properties

The binding of polyclonal antibodies to their antigen is usually stable over a wide range of pH (*c.*4–9) and salt concentration (0–1 M NaCl). In contrast, some monoclonal antibodies may be very susceptible to minor changes. Herrmann and Mescher (1979) found that the monoclonal antibody 11–4 bound its antigen at pH 7.0 and at salt concentrations of less than 100 mM, but at pH 8.0 and 200 mM NaCl binding was virtually abolished.

Ease of disruption of an antigen–antibody complex by changing the pH or ionic strength or adding denaturants is a property that is distinct from affinity. It is, in principle, quite possible to have a very high-affinity interaction that is easily disrupted by a minor change in physical conditions. For example, if a particular antibody had very high-affinity binding which depended critically on the ionization of a histidine residue, slight changes of pH could cause drastic loss of this affinity by modifying the charge on this residue.

The sensitivity of some monoclonal antibodies to environmental changes is not necessarily a disadvantage. Indeed, it may facilitate affinity chromatography by allowing very gentle elution conditions. Conversely, some monoclonal antibodies may require very harsh conditions (e.g. pH 2.0, 3.5 M KSCN or 7 M guanidine-hydrochloride) before the antigen–antibody bond is disrupted (see Chapter 11).

The desired affinity of a monoclonal antibody, its binding to modified or denatured antigen, and its sensitivity to environmental conditions may all be taken into account when designing hybridoma screening protocols. If high-affinity antibodies or rapid 'on' kinetics are desired, one might keep the incubation times short. Similarly, if it is vital that the antibody recognize the denatured antigen, or the antigens in a particular ionic environment, these conditions might be used during the screening assay.

In addition to their highly individual binding specificity and affinity, monoclonal antibodies may have highly individual physical or chemical properties unrelated to antigen binding. As mentioned earlier in this chapter, IgM, mouse IgG2b and IgG3, and rat IgG2c often precipitate in solutions of low ionic strength (euglobulin precipitation). The proteins will usually (but not always) redissolve upon raising the salt concentration. Euglobulin precipitation is often exploited in the purification of monoclonal antibodies (see Section 9.2.2).

Occasionally, monoclonal antibodies will precipitate in the cold, and redissolve upon warming. Immunoglobulins with this property are known as 'cryoglobulins' (reviewed by Grey and Köhler, 1973; see also Brekke, 1995). The precise structural features that cause precipitation in the cold are not clear. If a monoclonal antibody is a cryoglobulin its usefulness may be greatly diminished because many immunological procedures are carried out in the cold. Unexpected cryoglobulin properties of a monoclonal antibody could result in significant experimental artefacts.

References

Abbott, C. and Povey, S. (1995) 'Somatic Cell Hybrids'. IRL Press, Oxford.

Adetugbo, K., Milstein, C. and Secher, D. S. (1977) Molecular analysis of spontaneous somatic mutants. *Nature* **265**, 299–304.

Andersson, J. and Melchers, F. (1978) The antibody repertoire of hybrid cell lines obtained by fusion of X63–Ag8 myeloma cells with mitogen-activated B-cell blasts. *Curr. Top. Microbiol. Immunol.* **81**, 130–139.

Argon, Y., Burrone, O. R. and Milstein C. (1983) Molecular characterization of a non-secreting myeloma mutant. *Eur. J. Immunol.* **13**, 301–305.

Bazin, H. (1982) Production of rat monoclonal antibodies with the LOU rat non-secreting IR983F myeloma cell line. *In* 'Protides of the Biological Fluids' (H. Peeters, ed.), 29th Colloquium, pp. 615–618. Pergamon Press, Oxford.

Bazin, H. (1990a) 'Rat Hybridomas and Rat Monoclonal Antibodies'. CRC Press, Boca Raton.

Bazin, H. (1990b) The LOUVAIN (LOU) rats. *In* 'Rat Hybridomas and Rat Monoclonal Antibodies' (H. Bazin, ed.), pp. 43–52. CRC Press, Boca Raton.

Bazin, H. and Lemieux, R. (1987) Role of the macrophage-derived hybridoma growth factor in the *in vitro* and *in vivo* proliferation of newly formed B cell hybridomas. *J. Immunol.* **139**, 780–787.

Bazin, H. and Lemieux, R. (1989) Increased proportion of B cell hybridomas secreting monoclonal antibodies of desired specificity in cultures containing macrophage-derived growth factor (IL-6) *J. Immunol. Methods* **116**, 245–249.

Bazin, H., Beckers, A., Deckers, C. and Moriame, M. (1973) Transplantable immunoglobulin-secreting tumors in rats. V. Monoclonal immunoglobulins secreted by 250 ileocecal immunocytomas in LOU/WsI rats. *J. Natl Cancer Inst.* **51**, 1359–1361.

Bazin, H., Beckers, A., Urbain-Vansanten, G., Pauwels, R., Bruyns, C., Tilkin, A. F., Platteau, B. and Urbain, J. (1978) Transplantable IgD immunoglobulin-secreting tumors in the rat. *J. Immunol.* **121**, 2077–2082.

Bazin, H., Pear, W. S., Klein, G. and Sümegi, J. (1990a) Rat immunocytomas (IR). *In* 'Rat Hybridomas and Rat Monoclonal Antibodies' (H. Bazin, ed.), pp. 53–68. CRC Press, Boca Raton.

Benjamin, D. C., Berzofsky, J. A., East, I. J., Gurd, F. R. N., Hannum, C., Leach, S. J., Margoliash, E., Michael, J. G., Miller, A., Prager, E. M., Reichlin, M., Sercarz, E. E., Smith-Gill, S. J., Todd, P. E. and Wilson, A. C. (1984) The antigenic structure of proteins: a reappraisal. *Annu. Rev. Immunol.* **2**, 67–101.

Birnbaum, D. and Kourilsky, F. M. (1981) Differences in the cell binding affinity of a crossreactive monoclonal anti-Ia alloantibody in mice of different H-2 haplotypes. *Eur. J. Immunol.* **11**, 734–738.

Blankenstein, T., Qin, Z., Li, W. and Diamantstein, T. (1990) DNA rearrangement and constitutive expression of the interleukin 6 gene in a mouse plasmacytoma. *J. Exp. Med.* **171**, 965–970.

Blose, S. H., Matsumura, F. and Lin, J. J. C. (1982) Structure of vimentin 10-nm filaments probed with a monoclonal antibody that recognizes a common antigenic determinant on vimentin and tropomyosin. *Cold Spring Harbor Symp. Quant. Biol.* **46**, 455–463.

Borrebaeck, C. A. K (1992) 'Antibody Engineering—A Practical Guide'. W. H. Freeman, San Francisco.

Boshart, M., Nitsch, D. and Schütz, G (1993) Extinction of gene expression in somatic cell hybrids—a reflection of important regulatory mechanisms? *Trends Genet.* **9**, 240–245

Bosma, G. C., Custer, R. P. and Bosma, M. J. (1983) A severe combined immunodeficiency mutation in the mouse. *Nature* **301**, 527–530.

Brekke, O. H., Michaelsen, T. E. and Sandlie, I. (1995) The structural requirements for complement activation by IgG: does it hinge on the hinge? *Immunol. Today* **16**, 85–90.

Burnette, W. N. (1981) 'Western blotting': Electrophoretic transfer of proteins from sodium dodecyl sulfate-polyacrylamide gels to unmodified nitrocellulose and radiographic detection with antibody and radioiodinated protein A. *Anal. Biochem.* **112**, 195–203.

Burtonboy, G., Bazin, H., Deckers, C., Beckers, A., Lamy, M. and Heremans, J. F. (1973) Transplantable immunoglobulin-secreting tumors in rats. 3. Establishment of immunoglobulin-secreting cell lines from LOU-Ws1 strain rats. *Eur. J. Cancer* **9**, 259–262.

Cariello, N. F. and Skopek, T. R. (1993) *In vivo* mutation at the human HPRT locus. *Trend. Genet.* **9**, 322–326.

Caskey, C. T. and Kruh, G. D. (1979) The HPRT locus. *Cell* **16**, 1–9.

Chiswell, D. J. and McCafferty, J. (1992) Phage antibodies: will new 'coliclonal' antibodies replace monoclonal antibodies? *Trends Biotechnol.* **10**, 80–83.

Coller, H. A. and Coller, B. S. (1983) Statistical analysis of repetitive subcloning by the limiting dilution technique with a view toward ensuring hybridoma monoclonality. *Hybridoma* **2**, 91–96.

Cook, W. D. and Scharff, M. D. (1977) Antigen-binding mutants of mouse myeloma cells. *Proc. Natl Acad. Sci. USA* **74**, 5687–5691.

Cory, S. (1986) Activation of cellular oncogenes in hemopoietic cells by chromosomal translocation. *Adv. Cancer. Res.* **47**, 189–234.

Cotton, R. G. H. and Milstein, C. (1973) Fusion of two immunoglobulin-producing myeloma cells. *Nature* **244**, 42–43.

Cotton, R. G. H., Secher, D. S. and Milstein, C. (1973) Somatic mutation and origin of antibody diversity. Clonal variability of the immunoglobulin produced by MOPC-21 cells in culture. *Eur. J. Immunol.* **3**, 135–140.

Darveau, A., Chevrier, M-C., Neron, S., Delage, R. and Lemieux, R. (1993) Efficient preparation of human monoclonal antibody-secreting heterohybridomas using peripheral B lymphocytes cultured in the CD40 system. *J. Immunol. Methods* **159**, 139–143.

Degrassi, A., Hilbert, D. M., Rudikoff, S., Anderson, A. O., Potter, M. and Coon, H. G. (1993) *In vitro* culture of primary plasmacytomas requires stromal cell feeder layers. *Proc. Natl Acad. Sci. USA* **90**, 2060–2064.

Dower, S. K., Ozato, K. and Segal, D. M. (1984) The interaction of monoclonal antibodies with MHC class I antigens on mouse spleen cells. I. Analysis of the mechanism of binding. *J. Immunol.* **132**, 751–758.

Eshhar, Z., Blatt, C., Bergman, Y. and Haimovitch, J. (1979) Induction of secretion of IgM from cells of the B cell line 38C-13 by somatic cell hybridisation. *J. Immunol.* **122**, 2430–2434.

Finkelman, F. D., Kessler, S. W., Mushinski, J. F. and Potter, M. (1981) IgD-secreting murine plasmacytomas: identification and partial characterization of two IgD myeloma proteins. *J. Immunol.* **126**, 680–687.

Foote, J. and Milstein, C. (1994) Conformational isomerism and the diversity of antibodies. *Proc. Natl Acad. Sci. USA* **19**, 10370–10374.

Franklin, E. C. and Frangione, B. (1971) The molecular defect in a protein (CRA) found in γl heavy chain disease, and its genetic implications. *Proc. Natl Acad. Sci. USA* **68**, 187–191.

Galfrè, G. and Milstein, C. (1981) Preparation of monoclonal antibodies: strategies and procedures. *Methods Enzymol.* **73**, 3–46.

Galfrè, G., Milstein, C. and Wright, B. (1979) Rat × rat hybrid myelomas and a monoclonal anti-Fd portion of mouse IgG. *Nature* **277**, 131–133.

Gething, M-J. and Sambrook, J. (1992) Protein folding in the cell. *Nature* **355**, 33–45.

Ghosh, S. and Campbell, A. M. (1986) Multispecific monoclonal antibodies. *Immunol. Today* **7**, 217–222.

Giesow, M. J. (1992) Antibody engineering—successful affinity maturation in vitro. *Trends Biotechnol.* **10**, 299–301.

Green, L. L., Hardy, M. C., Maynard-Currie, C. E., Tsuda, H., Louie, D. M., Mendez, M. J., Abderrahim, H., Noguchi, M., Smith, D. H., Zeng, Y., David, N. E., Sasai, H., Garza, D., Brenner, D. G., Hales, J. F., McGuiness, R. P., Capon, D. J., Klapholz, S. and Jakobovits, A. (1994) Antigen-specific human monoclonal antibodies from mice engineered with human Ig heavy and light chain YACs. *Nature Genetics* **7**, 13–21.

Grey, H. M. and Kohler, P. F. (1973) Cryoimmunoglobulins. *Semin. Hematol.* **10**, 87–112.

Harris, J. F., Hawley, R. G. and Crawford-Sharpe, G. C. (1992) Increased frequency of both total and specific monoclonal antibody producing hybridomas using a fusion partner that constitutively expresses recombinant IL-6. *J. Immunol. Methods* **148**, 199–207.

Herrmann, S. H. and Mescher, M. F. (1979) Purification of the H-2K^k molecule of the murine major histocompatibility complex. *J. Biol. Chem.* **254**, 8713–8716.

Hewish, D. R. and Werkmeister, J. W. (1989) The use of an electroporation apparatus for the production of murine hybridomas. *J. Immunol. Methods* **120**, 285–289.

Hitchcock, S. E. (1980) Actin-deoxyribonuclease I interaction. *J. Biol. Chem.* **255**, 5668–5673.

Hood, L., Loh, E., Hubert, J., Barstard, P., Eaton, B., Early, P., Fuhrman, J., Johnson, N., Kronenberg, M. and Schilling, J. (1976) The structure and genetics of mouse immunogloublins: an analysis of NZB myeloma proteins and sets of BALB/c myeloma proteins binding particular haptens. *Cold Spring Harb. Symp. Quant. Biol.* **41**, 817–836.

Horibata, K. and Harris, A. W. (1970) Mouse myelomas and lymphomas in culture. *Exp. Cell Res.* **60**, 61–77.

Howard, J. C., Butcher, G. W., Galfrè, G., Milstein, C. and Milstein, C. P. (1979) Monoclonal antibodies as tools to analyse the serological and genetic complexities of major transplantation antigens. *Immunol. Rev.* **47**, 139–174.

James, K. and Bell, G. T. (1987) Human monoclonal antibody production. Current status and future prospects. *J. Immunol. Methods* **100**, 5–40.

Kearney, J. F., Radbruch, A., Liesegang, B. and Rajewsky, K. (1979) A new mouse myeloma cell line which has lost immunoglobulin expression but permits the construction of antibody in secreting hybrid cell lines. *J. Immunol.* **123**, 1548–1550.

Kilmartin, J. V., Wright, B. and Milstein, C. (1982) Rat monoclonal antibodies derived by using a new non-secreting rat cell line. *J. Cell Biol.* **93**, 576–582.

Köhler, G. (1980) Immunoglobulin chain loss in hybridoma cell lines. *Proc. Natl Acad. Sci. USA* **77**, 2197–2199.

Köhler, G. and Milstein, C. (1975) Continuous cultures of fused cells secreting antibody of predefined specificity. *Nature* **256**, 495–497.

Köhler, G. and Milstein, C. (1976) Derivation of specific antibody-producing tissue culture and tumor lines by cell fusion. *Eur. J. Immunol.* **6**, 511–519.

Köhler, G., Howe, S. C. and Milstein, C. (1976) Fusion between immunoglobulin-secreting and non-secreting myeloma cell lines. *Eur J. Immunol.* **6**, 292–295.

Köhler, G., Pearson, T. and Milstein, C. (1977) Fusion of T and B cells. *Somat. Cell. Genet.* **3**, 303–312.

Krause, R. M. (1970) The search for antibodies with molecular uniformity. *Adv. Immunol.* **12**, 1–56.

Lane, D. and Koprowski, H. (1982) Molecular recognition and the future of monoclonal antibodies. *Nature* **296**, 200–202.

Lefkovits, I. and Waldmann, H. (1979) 'Limiting Dilution Analysis of Cells in the Immune System.' Cambridge University Press, Cambridge.

Littlefield, J. W. (1964) Selection of hybrids from matings of fibroblasts *in vitro* and their presumed recombinants. *Science* **145**, 709–710.

Lyon, M. F. (1961) Gene action in the X chromosome of the mouse. (*Mus. musculus* L.) *Nature* **190**, 372–373.

Mason, D. W. and Williams, A. F. (1980) The kinetics of antibody binding to membrane antigens in solution and at the cell surface. *Biochem. J.* **187**, 1–20.

Mayforth, R. D. (1993) 'Designing Antibodies'. Academic Press, San Diego.

Merwin, R. M. and Redmon, L. W. (1963) Induction of plasma cell tumors and sarcomas in mice by diffusion chambers placed in the peritoneal cavity. *J. Natl. Cancer Inst.* **31**, 997–1017.

Metcalf, D. (1973) Colony formation in agar by murine plasmacytoma cells: Potentiation by hemopoietic cells and serum. *J. Cell. Physiol.* **81**, 397–410.

Metcalf, D. (1974) The serum factor stimulating colony formation in vitro by murine plasmacytoma cells: Response to antigens and mineral oil. *J. Immunol.* **113**, 235–243.

Milstein, C. and Cuello, A. C. (1983) Hybrid hybridomas and their use in immunocytochemistry. *Nature* **305**, 537–540.

Mosier, D. E., Gulizia, R. J., Baird, S. M. and Wilson, D. B. (1988) Transfer of a functional human immune system to mice with severe combined immunodeficiency. *Nature* **335**, 256–259.

Neuberger, M. S. and Rajewsky, K. (1981) Activation of mouse complement by monoclonal mouse antibodies. *Eur. J. Immunol.* **11**, 1012–1016.

Nissim, A., Hoogenboom, H., Tomlinson, I. M., Flynn, G., Midgley, C., Lane, D. and Winter, G. (1994) Antibody fragments from a 'single pot' phage display library as immunochemical reagents. *EMBO J.* **13**, 692–698.

Nordan, R. P. and Potter, M. (1986) A macrophage-derived factor required by plasmacytomas for survival and proliferation *in vitro*. *Science* **233**, 566–569.

Nussenzweig, M. C., Steinman, R. M., Witmer, M. D. and Gutchinov, B. (1982) A monoclonal antibody specific for mouse dendritic cells. *Proc. Natl Acad. Sci. USA* **79**, 161–165.

Ohno, S., Babonits, M., Wiener, F., Spira, J., Klein, G. and Potter, M. (1979) Nonrandom chromosome changes involving the Ig gene-carrying chromosomes 12 and 6 in pristane-induced mouse plasmacytomas. *Cell* **18**, 1001–1007.

Owens, R. J. and Young, R. J. (1994) The genetic engineering of monoclonal antibodies. *J. Immunol. Methods* **168**, 149–165.

Parham, P. (1984a) Changes in conformation with loss of alloantigenic determinants of a histocompatibility antigen (HLA-B7) induced by monoclonal antibodies. *J. Immunol.* **132**, 2975–2983.

Parham, P. (1984b) The binding of monoclonal antibodies to cell surface molecules. Quantitative analysis of the reactions and cross-reactions of an antibody (MB40.3) with four HLA-B molecules. *J. Biol. Chem.* **259**, 13077–13083.

Paslay, J. and Roozen, K. (1981) The effect of B cell stimulation on hybridoma formation. *In* 'Monoclonal antibodies and T-cell hybridomas, perspectives and technical advances' (G. J. Hämmerling, U. Hämmerling and J. F. Kearney, eds), pp. 551–559. Elsevier/North-Holland Biomedical Press, Amsterdam.

Pillemer, E. and Weissman, I. L. (1981) A monoclonal antibody that detects a V_κ-TEPC-15 idiotypic determinant cross-reactive with a Thy-1 determinant. *J. Exp. Med.* **153**, 1068–1079.

Pontecorvo, G. (1976) Production of indefinitely multiplying mammalian somatic cell hybrids by polyethylene glycol (PEG) treatment. *Somatic Cell Genet.* **1**, 397–400.

Potter, M. (1972) Immunoglobulin-producing tumors and myeloma proteins of mice. *Physiol. Rev.* **52**, 631–719.

Potter, M. (1977) Antigen-binding myeloma proteins of mice. *Adv. Immunol.* **25**, 141–211.

Potter, M. and Boyce, C. R. (1962) Induction of plasma cell neoplasms in strain BALB/c mice with mineral oil adjuvants. *Nature* **193**, 1086–1087.

Richards, F. F., Konigsberg, W. H., Rosenstern, R. W. and Varga, J. M. (1975) On the specificity of antibodies. *Science* **187**, 130–137.

Ringertz, N. R. and Savage, R. E. (1976) 'Cell Hybrids'. Academic Press, New York.

Rousset, B. and Wolff, J. (1980) Lactoperoxidase-tubulin interactions. *J. Biol. Chem.* **255**, 2514–2523.

Shimizu, S., Yoshioka, B., Hirose, Y., Sugai, S., Tachibana, J. and Konda, S. (1989) Establishment of two interleukin 6 (B cell stimulatory factor 2/interferon β_2)-dependent human bone marrow-derived myeloma cell lines. *J. Exp. Med.* **169**, 339–344.

Shulman, M., Wilde, C. D. and Köhler, G. (1978) A better cell line for making hybridomas secreting specific antibodies. *Nature* **276**, 269–270.

Stout, J. T. and Caskey, C. T. (1988) The Lesch-Nyhan syndrome: clinical, molecular and genetic aspects. *Trends Genet.* **4**, 175–178.

Sugita, T., Totsuka, T., Saito,M., Yamasaki, K., Taga, T., Hirano, T. and Kishimoto, T. (1990) Functional murine interleukin 6 receptor with the intracisternal A particle gene product at its cytoplasmic domain. *J. Exp. Med.* **171**, 2001–2009.

Talmage, D. W. (1959) Immunological specificity. *Science* **129**, 1643–1648.

Thammana, P. (1994) Characterization of a deletion in the P3 myeloma heavy chain V region sequence in the non-secreting cell line NS1. *Mol. Immunol.* **31**, 77–78.

Thompson, K. M. (1988) Human monoclonal antibodies. *Immunol. Today* **9**, 113–117.

van Snick, J. (1990) Interleukin-6: an overview. *Annu. Rev. Immunol.* **8**, 253–278.

Wagner, S. D., Williams, G. T., Larson, T., Neuberger, M. S., Kitamura, D., Rajewsky,K., Xian, J. and Brüggermann, M. (1994) Antibodies generated from human immunoglobulin miniloci in transgenic mice. *Nucl. Acids Res.* **22**, 1389–1393.

Wall, R. and Kuehl, M. (1983) Biosynthesis and regulation of immunoglobulins. *Annu. Rev. Immunol.* **1**, 393–422.

Warner, N. L. (1975) Autoimmunity and the pathogenesis of plasma cell tumor induction in NZB inbred and hybrid mice. *Immunogenetics* **2**, 1–20.

Ways, J. P. and Parham, P. (1983) The binding of monoclonal antibodies to cell-surface molecules. A quantitative analysis with immunoglobulin G against two alloantigenic determinants of the human transplantation antigen HLA-A2. *Biochem J.* **216**, 423–432.

Werkmeister, J. A., Tebb, T. A., Kirkpatrick, A. and Shukla, D. D. (1991) The use of peptide-mediated electrofusion to select monoclonal antibodies against specific and homologous regions of the potyvirus coat protein. *J. Immunol. Methods* **143**, 151–157.

Winter, G. (1993) Immunological techniques. *Curr. Op. Immunol.* **5**, 253–255.

Wojciezsyn, J. W., Schlegel, R. A., Limley-Sapanski, K. and Jacobson, K. A. (1983) Studies on the mechanism of polyethylene glycol-mediated cell fusion using fluorescent membrane and cytoplasmic probes. *J. Cell. Biol.* **96**, 151–159.

Zhu, M., Rabinowitz, J. L., Green, N. S., Kobrin, B. J. and Scharff, M. D. (1995) A well-differentiated B-cell line is permissive for somatic mutation of a transfected immunoglobulin heavy-chain gene. *Proc. Natl Acad. Sci. USA* **92**, 2810–2814.

Zimmerberg, J., Vogel, S. S. and Chernomordik, L. V. (1993) Mechanisms of membrane fusion. *Annu. Rev. Biomol. Struct.* **22**, 433–466.

8 Production of Monoclonal Antibodies

Although the technology of hybridoma production is now firmly established, there are a large number of steps involved, and each of these may be carried out in many different ways. The diversity of published approaches reflects both individual biological problems and previous experience. The methods also vary in convenience, speed, reliability and expense. There is no one right approach and ultimately each investigator must choose and adapt the published strategies to individual needs. An appreciation of the variables and compromises will help minimize the effort required.

Much useful information on general aspects of tissue culture will be found in the volume edited by Jakoby and Pastan (1979) and a monograph by Freshney (1993).

8.1 Choice of Normal Lymphocyte Donor

As outlined in Chapter 4, strong antibody responses are most easily obtained when the immunizing antigen differs structurally from components of the body of the recipient. This is usually achieved by immunizing across a species barrier. For example, mice will make large amounts of antibody to IgG from chicken, human, rabbit or rat, but will not normally make antibody against mouse IgG unless the donor IgG structural genes differ from those of the recipient.

When planning the immunization protocol, it is worthwhile thinking about the ways in which the antibodies are to be used. As mentioned in Section 4.5, there is often little serological cross-reaction between native and denatured protein. If the monoclonal antibodies are to be used on formalin-fixed tissues, for example, it may be very helpful to immunize with formalin-treated antigen and screen in a way that detects the relevant antibodies (see Harrach and Robenek, 1991 for an example).

In most situations, the choice of species to be immunized will be limited to the mouse and rat. The mouse is usually preferred as recipient because of its convenience, the availability of suitable myeloma lines, and its rapid breeding (gestation time 21 days; first mating at 6–8 weeks).

However, if it is desired to make antibodies against a mouse protein for which there are no known genetic variants, immunization of mice is unlikely to be successful. In such cases, rats may be used as recipients. The other advantage of rats is their larger size, resulting in higher volumes of serum per animal. Moreover, the breeding time of rats is not much longer than that of mice (Galfrè *et al.*, 1979; Galfrè and Milstein, 1981).

All the currently available mouse myeloma cell lines used for hybridoma production are of BALB/c origin, and unless there are compelling reasons to the contrary (e.g. inability of BALB/c to respond to the antigen), BALB/c mice should be used as recipients for the initial immunization. The resulting hybridomas may then be grown in BALB/c mice.

On the other hand, if C57BL mice are used as recipients, the resulting hybridomas will be rejected by C57BL mice because of the BALB/c histocompatibility antigens of the myeloma cells. Such hybridomas could only be grown in C57BL × BALB/c F_1 mice. Hybridomas from outbred animals will usually fail to grow as tumours in any recipient, because they are unlikely to be histocompatible. Occasionally, the need for histocompatibility can be overcome by injecting relatively large numbers (10^7) of hybridoma cells into BALB/c mice that have been sublethally irradiated (400–500 R (4–5 Gy)).

All available rat myelomas suitable for fusion are of LOU/C origin (Bazin, 1990a, b; Bazin *et al.*, 1990). These tumour lines will also grow in LOU/M rats, which are better breeders, but which do not have a high incidence of spontaneous myeloma (Bazin *et al.*, 1973, 1990). If, on the other hand, DA rats are used as recipients for antigen, the resulting hybridomas could only be grown in LOU × DAF_1 rats (Kilmartin *et al.*, 1982).

Several investigators have successfully made hybridomas by fusing spleen cells from rats with myeloma cells from mice (Ledbetter and Herzenberg, 1979; McKearn *et al.*, 1980; Springer, 1980; Kincade *et al.*, 1981). These hybrids appear to be as stable as mouse–mouse hybrids, but in general they can only be grown in tissue culture. Occasionally, such hybrids will grow in 'nude' (athymic) mice, but these mice are difficult to breed and maintain in good health. Rat–mouse hybrids will sometimes grow in mice or rats which have been sublethally irradiated (400–500 R).

8.2 Immunization Protocol

Protocols for immunization of animals vary widely. The first injection of antigen should be given in a highly aggregated form because soluble monomeric proteins in their native state are poorly immunogenic and tend to induce tolerance (Dresser, 1962). Subsequent injections may be either soluble or aggregated.

When the antigen is a water-soluble protein, the use of an adjuvant is usually essential (See Section 4.10). By far the most widely used adjuvant is Freund's complete adjuvant. Details of its use will be given in Section 8.2.1. Alternatively, the antigen may be precipitated on alum, with or without the addition of killed *Bordetella pertussis* organisms; this is generally much less effective. Recent advances in adjuvant technology have been reviewed by Cox and Coulter (1992) and by Stewart-Tull (1995).

As noted in Section 4.10, the use of adjuvant may influence the class of antibody produced, but no hard and fast rules can be given. Handman and Remington (1980) found that mice which are chronically infected with *Toxoplasma gondii* make substantial quantities of IgG2 and IgG3 antibodies, but virtually no IgG1. Hybridomas made from the spleens of these mice had a similar class distribution to the serum antibody. In contrast, Johnson *et al.* (1981) immunized mice with the same parasite in Freund's adjuvant, and obtained a substantial IgG1 response and many IgG1 hybridomas. The antigens recognized in these two cases appeared to be the same proteins, but it is not known whether the same determinants were involved.

If the antigen is of very low molecular weight, or is poorly immunogenic for any other reason, strong responses may often be elicited by coupling it to an immunogenic carrier molecule (see Chapter 4).

8.2.1 Preparation of Antigen in Freund's Adjuvant

Vials of Freund's complete adjuvant are available from many suppliers. It is important that the vial be shaken to resuspend the mycobacteria prior to use. It is traditional to use equal volumes of antigen (in aqueous solution) and adjuvant, although it has been suggested that the formation of water-in-oil emulsion (rather than oil-in-water, which has little adjuvant effect) is more reliable (infallible) if 2–4 volumes of oil are used for each volume of aqueous immunogen (Hurn and Chantler, 1980).

The antigen solution, in buffered saline or water, is sucked into a 5 ml glass syringe with a Luer lock attachment. A second glass syringe is filled with Freund's complete adjuvant. Any excess air is expelled, and the syringes connected together via a disposable plastic three-way stopcock (Top Surgical, Japan). The aqueous phase is injected into the oil, and then the mixture is passed rapidly back and forth between the syringes a few dozen times.

The emulsion should be tested by allowing a drop to fall onto the surface of a beaker of water. The first drop will often spread, but subsequent drops should remain as discrete globules. The emulsion should be thick and creamy. Rapid dispersion of the drops is indicative of an oil-in-water emulsion. The presence of detergents such as sodium dodecyl sulfate, Triton X-100 or Nonidet P-40 in the antigen solution may prevent proper emulsification. Detergents may be removed by acetone precipitation of the antigen (Hager and Burgess, 1980).

Freund's adjuvant should be prepared with care. Accidental injection into the hand may result in permanently stiff or useless fingers, and hypersensitivity reactions can be very severe (Chapel and August, 1976). Glass syringes are recommended, because the plunger of plastic disposable syringes tends to swell and stiffen in oil. Failure to use Luer-lock fittings will usually result in the couplings coming undone, and the resultant widespread spraying of adjuvant may cause permanent eye damage.

The site of injection into the recipient is not of crucial importance. Intradermal injections often cause large painful ulcers, and the popular intra foot-pad route is also inhumane and probably unnecessary. Adequate priming can usually be achieved by intraperitoneal, subcutaneous, or intramuscular routes. The spleen is often enlarged after intraperitoneal injection of Freund's adjuvant, and intraperitoneal injection gives excellent results for hybridoma production. A typical volume of emulsion is 50–200 μl per mouse, and perhaps a little more for rats. Although the older literature suggests that antigen doses of 1–50 mg are necessary, more recent experience suggests that 1–50 μg are often adequate.

It is customary to give booster immunizations in incomplete Freund's adjuvant (i.e. without the tubercle bacilli) to prevent possible hypersensitivity reactions to the bacteria. However, boosting with antigen subcutaneously or intraperitoneally in phosphate-buffered saline (PBS) is often perfectly adequate. Typically, 1–3 boosters may be given at intervals of 2–8 weeks. The final boost is best given 2–4 days prior to fusion, and is often given in aqueous form rather than in adjuvant. Some authors recommend that the final boost be given intravenously (e.g. Stähli *et al.*, 1980), but in my experience this often results in the animal dying of anaphylaxis within minutes. Intraperitoneal boosting is safer, and probably nearly as effective.

In summary, a reasonable schedule would involve priming with 10 μg antigen emulsified in 200 μl complete Freund's adjuvant given intraperitoneally, and boosting twice with the same dose in PBS or Freund's incomplete adjuvant at 3–4-weekly intervals. A final intraperitoneal boost of the same or somewhat higher dose in aqueous solution may be given (see Stähli *et al.*, 1980), and the fusion performed 3–4 days later. Both the spleen and the regional draining lymph nodes may be used for fusion (Kearney *et al.*, 1979). Lymph nodes are somewhat deficient in macrophages, and should usually be cultured together with feeder cells (see Section 8.9.3). The syringes should be thoroughly washed in several changes of acetone, followed by standard wash procedure in detergent.

8.2.2 Preparation of Antigen Adsorbed to Alum

An alternative adjuvant system involves precipitation of protein antigen adsorbed to alum. The following protocol is adapted from that described by Hudson and Hay (1976):

(1) Make up the antigen solution in 10 ml 0.2 M aluminium potassium sulfate (alum; $AlK(SO_4)_2.12\ H_2O$) in water. The pH will be about 3.0.
(2) Add 4.5 ml of 1.0 M $NaHCO_3$ (pH ≈8.0). Mix. A white precipitate should form. Leave for 15 min at room temperature.
(3) Centrifuge (300 *g*, 15 min). Discard supernate, and wash the precipitate three times in phosphate-buffered saline, pH 7.4.
(4) Resuspend the insolubilized protein to the desired concentration.
(5) Some authors add killed *Bordetella pertussis* vaccine (10^{10} organisms/ml; see Bradley and Shiigi, 1980). Note that the adjuvanticity of *B. pertussis* may vary markedly between batches.
(6) Inject the suspension containing 20–200 μg antigen, subcutaneously or intraperitoneally, and boost at 3–4 week intervals in the same way, until the antibody titre is strong. A final boost should be given 3–4 days prior to fusion, and is probably best given in water-soluble form intraperitoneally.

8.2.3 Immunization with Intact Cells

Intact living cells are highly immunogenic, and do not usually require the use of adjuvants. As noted in Section 4.3, it is sometimes found that if the genetic difference between self and the injected cells is very small, the response may be poor, owing to lack of what has been called *acolyte determinants*. This situation may arise if the cells are from a closely related strain of mouse, or if the cells are syngeneic with the recipient but transfected with a gene from another species, such as human. In these cases, a much stronger response may often be achieved by choosing a cell line or mouse strain such that there are multiple genetic differences. It will usually be possible to discriminate against the production of antibodies to the undesired determinants. For example, if transfected cells are used as the immunogen, one can perform duplicate screening on transfected and untransfected cells. One can be reasonably confident that antibodies that react only with the transfected cells are directed to the product of the transfected gene.

Cells should be washed 3–4 times in phosphate-buffered saline to remove serum components, and injected intraperitoneally. A typical dose is 10^7 cells, although any dose from 2×10^6 to 5×10^7 will probably be just as good. Recipients may be boosted at intervals of 3–8 weeks, and fusion performed 2–4 days after the last boost. There is some evidence that better yields of active hybridomas are obtained if the recipients are 'rested' for 4–8 weeks between the penultimate and final boost (Oi and Herzenberg, 1980).

Long immunization schedules are probably unnecessary, and may even be counter-productive. Trucco *et al.* (1978) obtained many interesting hybrids by fusing 4 days after a single injection of human lymphoblastoid cells.

8.2.4 *In Vitro* Immunization

It has been shown that it is possible to immunize mouse lymphocytes *in vitro* prior to fusion (Luben and Mohler, 1980; Luben *et al.*, 1982). This procedure has not been widely adopted. It appears somewhat unreliable, and tends to provide IgM antibodies, which are difficult to purify and fragment, and may show unexpected cross-reactions with unrelated antigens (see Section 7.3.2).

8.3 Choice of Myeloma Cell Line for Fusion

There are now numerous myeloma cells which have been used successfully for the generation of hybridomas. However, not all cultured myelomas are capable of forming hybridomas, and occasionally a subline of a previously satisfactory line may fail to work. If this happens, it is worthwhile testing for mycoplasma contamination; see Section 8.16.2. A new subline should be obtained from a laboratory where it is being used successfully. The advantages of 'nonproducer' myelomas as fusion partners have been discussed in the previous chapter (Section 7.2.3). Such lines are now widely available, and should be used routinely.

8.3.1 Mouse Myelomas for Fusion Partners

By far the most widely used cell lines for hybridoma production are descendants of MOPC-21 which have been selected for HAT sensitivity and loss of endogenous immunoglobulin heavy chains (e.g. P3-NS1-Ag 4–1; abbreviated NS-1) or loss of both heavy and light chains (e.g. X63-Ag8.653 or NS0/1). NS-1 has been very popular for some time, but the production of an inactive light chain is undesirable.

For anyone starting a project to make monoclonal antibodies, it is strongly recommended that a total nonproducer line be used, such as X63-Ag8.653 (Kearney *et al.*, 1979) or Sp2/0-Ag-14 (abbreviated Sp2), which is a total nonproducer variant selected from a hybridoma involving fusion of MOPC-21 and BALB/c spleen cells.

A variant of Sp2 termed F0 has been produced by Fazekas and Scheidegger (1980). F0 cells are claimed to have particularly rapid growth, high fusion frequency and good cloning efficiency, but they have not been widely used.

Other mouse lines that have been used by small numbers of workers include variants of the BALB/c myelomas MPC-11 (Yelton *et al.*, 1978) and S-194 (Trowbridge, 1978) but they do not appear to offer any significant advantages over NS-1, X63-Ag8.653 or NS0/1, and their use is not recommended.

8.3.2 Rat Myelomas

All of the currently available rat myelomas which have been used for hybridoma production are derived from the LOU/C strain of rats (Bazin *et al.*, 1973, 1990; Bazin, 1982). These include Y3-Ag 1.2.3 (abbreviated Y3) which is an azaguanine-resistant derivative of the myeloma R210.RCY3, and secretes κ chains (Burtonboy *et al.*, 1973; Galfrè *et al.*, 1979).

As mentioned above, it is strongly recommended that a total nonproducer line be used for fusion. The rat line IR983F (Bazin, 1982; Bazin *et al.*, 1990) is probably the line of choice. It will grow in LOU/C or LOU/M rats. A second total nonproducer line YB2/0 (Galfrè and Milstein, 1981; Kilmartin *et al.*, 1982) was derived from a fusion between Y3 myeloma cells and spleen cells from an AO rat. It has the significant disadvantage that it must be grown in LOU/C × AO or LOU/M × AO F_1 rats.

The production of rat hybridomas has been very straightforward in some laboratories, but has given a great deal of trouble in others (see Bazin, 1990 for discussion). Those who have failed to produce rat hybrids have often been very successful in producing mouse hybrids. It has been quite difficult to understand the reason for these different experiences.

When problems growing rat hybrids are encountered, the usual observation is that the hybrids grow very well until just about the time when they are ready for antibody screening, and then within a day or two, or often even overnight, the hybrids all die. Sometimes, they survive to the point of screening, but die suddenly when diluted into a larger volume. The suspicion has arisen that the cells die because they run out of a growth factor, and attention has focused on interleukin-6 (IL-6), because of the known dependence of some plasmacytomas and hybridomas on this cytokine. There is now quite good evidence that many, or even most, rat hybridomas have a requirement for IL-6 (Bazin and Lemieux, 1987, 1989), and it is recommended that it be added to the cultures routinely. A convenient source of IL-6 is the supernatant from the macrophage line P388D1 (Nordan and Potter, 1986) or possibly MPC-11 (Blankenstein *et al.*, 1990). A derivative of Sp2/0 cells has been produced that is transfected with recombinant IL-6 and expresses it constitutively; this line has increased yield of hybridomas when compared with untransfected Sp2/0 cells (Harris *et al.*, 1992). It would be interesting to see whether transfection of rat myeloma lines with IL-6 would make more robust fusion partners.

However, it seems fairly clear that IL-6 is not the complete solution, and an additional factor is required for the reliable growth of rat hybrids. The experimental detail that appears to distinguish those scientists who find the production of rat hybrids easy and those that do not appears to be cell density. Successful laboratories usually plate out the hybrids at high cell density (more than 10^6/ml), and often add extra 'feeder' cells such as peritoneal cells (20 000/0.2 ml well), spleen cells, or thymocytes (10^6/0.2 ml well) (see Ravoet and Bazin, 1990). This may point to the requirement of the hybrids for another factor in addition to IL-6. Exhaustive questioning of those who are successful in producing rat hybrids has not revealed any other variable.

It is highly recommended that for rat hybrids, the cell density be kept very high, especially in the early stages of production. The cells should be diluted very gradually (perhaps only 1:2 in the early phases), and spleen cells, peritoneal cells or thymocytes should be used as feeders during dilution and expansion of the early cultures. One has the impression that once the hybrids are established, they are more robust and less dependent on cell density.

Rat hybrids may also be more intolerant of variation in pH, and attention should be paid to adequate buffering and CO_2 control.

8.3.3 Human Myeloma Cell Lines for Fusion

In contrast to mouse and rat myelomas, it has been extraordinarily difficult to adapt human myelomas to continuous growth as tissue culture lines. Over the years, several reports have claimed success, but in many cases the lines so produced have turned out to be lymphoblastoid lines derived from Epstein–Barr virus (EBV) transformed normal B cells.

Lines such as RPMI-8226 (Nilsson, 1971), H My-2 (Edwards *et al.*, 1982), IM9 (Everson *et al.*, 1973) and GM-1500 (Croce *et al.*, 1980) are all probably EBV-transformed lymphoblastoid lines, even though most of them arose from patients with myeloma. For a line to be acceptable as a true myeloma, it should secrete large amounts of immunoglobulin of the same idiotype and heavy and light chain isotype as the paraprotein in the patient's serum. EBV-transformed lymphoblastoid lines have low levels of secretion, grow in large clumps, and possess the EBV nuclear antigen (Reedman and Klein, 1973). In a few cases, EBV-transformed cell lines have been found suitable for hybridoma production (Croce *et al.*, 1980; Edwards *et al.*, 1982), although many unpublished failures have occurred.

A number of cell lines have been produced that appear to be true myelomas. The U-266 line was grown from a patient with an IgE(κ) myeloma (Nilsson *et al.*, 1970), and continued to secrete IgE after adaptation to continuous tissue culture growth. Many later sublines of U-266 have lost production of the ε heavy chain. A HAT-sensitive variant of U-266 (SK-007) has been produced by Olssen and Kaplan (1980) and was claimed to be suitable for fusion with normal human B cells, resulting in antibody-secreting hybridomas, but is rarely used now. A second human myelomas line (Karpas 707) has been described (Karpas *et al.*, 1982a, b), and is also rarely used. More recently, several other strategies and human cell lines have been claimed to be suitable for the production of human monoclonal antibodies (Kudo *et al.*, 1991; Darveau *et al.*, 1993, Shammah *et al.*, 1993). In view of the many disappointments from earlier lines, one can only adopt a somewhat sceptical 'wait and see' attitude to any new human myeloma cell line. Sooner or later, the problem must be solved.

8.4 Equipment Required for Fusion and Growth of Hybridomas

The production of hybridomas requires the following equipment:

(1) Biohazard or laminar flow hood. This is not absolutely essential, and a 'still air' box to protect from airborne organisms will suffice. While there is no documented evidence that the procedures for the production of mouse hybridomas are harmful to humans, it is now considered preferable to use biohazard hoods which do not blow air over the cultures and into the face of the operator.

The use of biohazard hoods should be regarded as mandatory for any tissue culture work using human cells or any other material of human origin. As a general rule, the use of hypodermic needles should be avoided when using human material, and great care must be taken to guard against infection with hepatitis or human immunodeficiency viruses.

(2) CO_2 incubator. It is standard practice to buffer the cultures with CO_2 (7–10%) and sodium bicarbonate (2.0–3.7 g/l). The CO_2 regulation must be reliable, but need not be complex or expensive. The older simple ball-type metering devices are inexpensive and trouble-free, but are now regarded as very old-fashioned. The more complex and expensive electronic CO_2 regulators that are now standard are becoming more reliable.

In my opinion, it is best to avoid incubators with inbuilt fans, although most modern incubators have them. The problem with fans is that they are usually inaccessible and impossible to clean, although some manufacturers have made the fans both accessible and autoclavable. If an incubator with a fan becomes contaminated with moulds, the spores are blown into the cultures. Incubators with smooth unobstructed interiors that are easily dismantled and cleaned are preferable.

If there is an air intake at the top of the incubator, it is a good idea to cover it with fine gauze or permeable surgical dressing to prevent moulds and dust from dropping into the incubator. The tissue culture room should have an ultraviolet germicidal lamp, and it should be switched on whenever no one is in the room. Care should be taken to ensure that the ultraviolet light cannot be switched on when the room is occupied, as ultraviolet light can cause damage to the eyes and skin.

Temperature regulation and uniformity are important. The temperature gauge on most incubators cannot be trusted, and the internal temperature should be measured frequently with a thermometer in a beaker of water. It should be 37 ± 0.5°C. It is important that the incubator be adequately humidified to prevent drying of cultures.

(3) Liquid nitrogen facility. This is absolutely essential for storage of cells, and must be organized so that there is no possibility of the nitrogen running out.

(4) Inverted microscope, preferably with phase optics. This is absolutely essential for monitoring growth of cells.

(5) 37°C Water bath. This must be cleaned frequently, as it is a potent source of infection for cultures.

(6) Centrifuge. This does not need to be refrigerated.

(7) Sterile pipettes. Disposable plastic pipettes are convenient, but very expensive. Alternatively, if glass pipettes are used, a hot-air sterilizing oven will be required. The tissue culture room should be kept clean, tidy and uncluttered to minimize the chance of contamination.

(8) Gamma counter. This is desirable for screening assays, but not absolutely essential, because it can be replaced by autoradiography or nonisotopic assays such as enzyme-linked immunoassay.

(9) Others. Animal holding facilities, small forceps, scissors, tissue culture ware, standard microscope, haemocytometers.

(10) A note concerning yeast contamination. It has been observed that laboratory workers who make their own bread or pastry sometimes have great difficulty with contamination of tissue culture by yeasts, presumably owing to carry-over from the kitchen to the laboratory.

8.5 Preparation of Spleen Cells

Instruments required are medium-sized scissors, fine scissors and fine forceps. They may be adequately sterilized by boiling in water for 5 min or by autoclaving.

The animal may be killed by cervical dislocation or asphyxiation in CO_2. (The former is instantaneous, and therefore preferable). The animal should then be swabbed with 70% ethanol, the superficial skin pinched up over the left side of the abdomen, and a small cut made over the spleen. The skin is then torn back, revealing the abdominal muscles, through which the spleen will be visible. From this point on, sterile technique must be used.

The abdominal wall over the spleen is pinched up with fine forceps, and a small incision made with fine scissors, taking care to avoid the gut. The spleen is gently delivered through the incision, holding it by its mesentery to avoid crushing it, and is released by cutting the mesentery and placed into a sterile Petri dish.

A single-cell suspension is then made by gently teasing the spleen through a sterile sieve into culture medium, using fine forceps. The medium should be at room temperature. Alternatively, the spleen may be cut into a few pieces with scissors, and the cells gently released into medium by pressure between the frosted ends of two sterile glass microscope slides.

The cells are then harvested by centrifugation (400 *g* for 5 min), washed twice in medium, and counted. Erythrocytes may be removed from an aliquot of cells for counting by using 3% acetic acid as diluent. Typically, a mouse spleen will yield *c.* 10^8 nucleated cells, and a rat spleen about 5–10 times as much.

8.6 Preparation of Myeloma Cells

8.6.1 Culture Medium

The two most popular media are Dulbecco's modified Eagle's medium (DME) or RPMI-1640, which is very similar. The main differences are that RPMI-1640 has no pyruvate, and DME has no asparagine. Neither of these ingredients is essential for hybridomas.

Most culture media are buffered with HCO_3^-/CO_2. Typically, one might use 3.4 g $NaHCO_3$/l, in equilibrium with 10% CO_2 in air to give the desired pH. Supplementary buffering may be useful, especially with very small volume cultures, and can be obtained by addition of HEPES up to 20 mM final concentration.

It is strongly recommended that the tissue culture medium be stored protected from direct sunlight or room fluorescent light, because light generates highly toxic photoproducts (Stoien and Wang, 1974; Wang, 1976; Griffin *et al.*, 1981).

Glutamine is somewhat unstable, and some investigators add extra (2 mM final concentration). It is doubtful whether this is necessary. Glutamine is stored as a concentrated solution at –20°C. The main problem with the instability of glutamine is that its decomposition is accompanied by the release of ammonia, which is very toxic to cells (Heeneman *et al.*, 1993). Culture media should be stored at 4°C and not kept for long periods for this reason. Decomposition of glutamine is much more rapid at 20°C or 37°C (Heeneman *et al.*, 1993).

It is customary to add 100 μg/ml of streptomycin and penicillin, although if tissue culture technique is good, antibiotics should not be necessary. The addition of pyruvate (1.0 mM) and 2-mercaptoethanol (5×10^{-5} M) is optional; there is no evidence that they make any difference to hybridoma production using the standard myeloma cell lines.

8.6.2 Fetal Calf Serum

As a general rule, fetal calf serum (10–15%) must be added to the medium. The quality of fetal calf serum has greatly improved in recent years, and extensive testing of many different batches is now usually unnecessary. None the less, the serum chosen should be capable of supporting the growth of the myeloma cells at 1 cell per well, and this test should be carried out on each new batch. When a good batch is found, a bulk order (5–20 l) should be placed. Serum is stable for at least 1–2 years, and probably much longer, when frozen at –20°C.

Some investigators have used horse serum or newborn calf serum, but in general the best results are obtained with fetal calf serum. The precise ingredients in serum which make cells grow are still poorly understood. Of particular importance are the iron transport protein transferrin, hormones (e.g. insulin, thyroxine) and possibly trace elements such as selenium. Some commercial suppliers market medium supplements that reduce the requirement for serum, but it is still not really feasible to use serum-free medium for the routine production of hybridomas.

8.6.3 Growth of Myeloma Cells

It is important to maintain a number of vials of myeloma cells in liquid nitrogen as insurance against infection, genetic drift and a host of other disasters which can occur in the laboratory. When vials are removed from this stock, the cells should be grown in bulk, and fresh vials replaced as soon as possible.

The myeloma cells are maintained in the laboratory in exponential growth in medium with 10% fetal calf serum. A convenient way to do this is to have a series of small Petri dishes, each holding 5 ml medium. It is not necessary to use special tissue culture grade plastic dishes for myeloma cells. Standard bacteriological Petri dishes are adequate, and much less expensive.

Cells are seeded at serial tenfold dilutions (six dishes are ample). Once a week they are passaged from a dish in which the cells are fairly dense but not overgrown. Typical doubling times are 14–16 h.

If the cells are in exponential growth, the doubling time (t_D) may be calculated from the expression

$$t_D = \frac{0.693t}{\log_e \frac{N}{N_0}}$$

Where t = elapsed time, N_0 = starting number of cells, N = final number of cells, and e = 2.718.

Similarly, if the doubling time is known, and a certain number of cells are required, one may arrange the initial number of cells and the culture time according to the formula

$$\frac{N}{N_0} = e^{\frac{0.69t}{t_D}}$$

Understanding exponential growth is crucial for success in tissue culture.

Healthy growing cells divide every 8–48 h, but notwithstanding claims to the contrary, the doubling times of mammalian cells in culture are virtually never less than 10 h. Exponential growth means that the cells will appear to be growing extremely slowly when dilute, but rapidly when concentrated.

8.6.4 Maximum Cell Density

As the cells become very dense, they start to look unhealthy, and viability drops. This happens quite suddenly; if the cultures are too dense on one day, by the next day most of the cells will be dead.

Most mouse myeloma cells will grow to a density of approximately 10^6/ml in roller bottles (rotate at 3 r.p.m.), and somewhat lower density (3×10^5/ml) in stationary culture. Rat myeloma cells and hybrids involving rat cells tend to adhere loosely to the walls of tissue culture vessels. They do not usually grow well in roller bottles. Cells may also be grown in 'spinner' flasks (Galfrè and Milstein, 1981).

8.7 Preparation of HAT and HT Medium

HAT and HT medium are prepared by addition of concentrated stock solutions to standard medium. Concentrated stocks are now available commercially (e.g. Boehringer, Difco), or can be made up in the laboratory as described below.

For HT concentrate, dissolve 136 mg hypoxanthine and 39 mg thymidine in 100 ml deionized distilled water warmed to 70–80°C. Sterilize by filtration, and store in 1.0 ml aliquots at –20°C.

To prepare HAT concentrate, take 100 ml of 100 × HT medium as above, and add 1.8 mg aminopterin. If the aminopterin does not dissolve readily, add a few drops of 1.0 M NaOH. Sterilize by filtration, and store as for HT. *Aminopterin is light sensitive, and may deteriorate, resulting in failure of HAT selection and growth of unfused myeloma cells. It is a good idea to test the HT and HAT concentrates for toxicity and effectiveness prior to use. The myeloma cells should grow normally in HT medium, but should all die within 1–3 days in HAT medium.* As a final test, the HAT medium should allow the growth of established hybridoma lines. (It is conceivable that an established hybrid line grown in normal medium could regain HAT sensitivity by loss of the X chromosome derived from the spleen cell fusion partner.)

8.8 Fusion Protocol: Important Variables

The first hybridomas were made using Sendai virus as the fusing agent, but virtually all hybridomas are now produced with polyethylene glycol (PEG). PEG is commercially available and inexpensive, and its use results in a higher fusion frequency and greater reproducibility. PEG-induced fusion of myeloma cells was first described by Galfrè *et al.* (1977). Although a number of minor modifications have been described since then, the procedure has not been changed in its essentials. The most important variables are:

(1) *The concentration of PEG.* Below about 30% PEG, very few hybrids are formed. Above 50% PEG, toxicity becomes overwhelming. If 40–50% PEG is used, dilution of PEG after fusion must be slow.
(2) *The purity of the PEG.* It appears likely that a contaminant in the PEG is essential for cell fusion (Wojciezsyn *et al.*, 1983). None the less, some batches of PEG are excessively toxic to cells, and cannot be used. If in doubt, try a new batch of PEG from a different source.
(3) *The pH of the PEG mixture.* Sharon *et al.* (1980) have shown that the fusion frequency is highly dependent on pH. Maximal numbers of hybridomas were obtained at pH 8.0–8.2.
(4) *The duration of exposure to PEG.* The fusion frequency, but also toxicity, increases with the time of exposure to PEG. Lower concentrations of PEG (30–35%) can be tolerated for longer times (e.g. 7 min) than higher concentrations (50% for no more than 1–2 min, see Gefter *et al.*, 1977).

Factors which appear to be of minor importance include temperature (20–37°C), cell numbers, the molecular weight of the PEG, the ratio of spleen cells to myeloma cells, and the presence or absence of serum or erythrocytes during fusion (Zola and Brooks, 1982). Norwood *et al.* (1976) have suggested that the addition of 15% (v/v) dimethyl sulfoxide to 42% (w/v) PEG results in better fusion, but the difference is probably small (Fazekas and Scheidegger, 1980).

It should be noted that there is always a certain percentage of experiments in which the fusions do not work, and a degree of variability in the yield of hybrids. It is therefore essential to perform a number of fusions. Sometimes, success will result from the first fusion. In other cases, it may be necessary to perform as many as 5–10 fusions before the desired hybrid is obtained.

8.8.1 Fusion Protocol: Procedure and Practice

This procedure is that used by Joan Curtis and Emanuela Handman, and is very reliable.

PEG (10 g), molecular weight 1500 (BDH cat. no. 29575) is autoclaved in a glass bottle. After allowing to cool, but while the PEG is still liquid, 10 ml of sterile Dulbecco's modified Eagle's medium (without serum) is added and thoroughly mixed. The pH should be slightly alkaline (pink, not orange or purple).

The spleen is taken from a mouse that has been immunized and boosted 3–4 days prior to the day of fusion. Thymuses for feeder cells are taken from

4–6-week-old mice, made into cell suspensions and washed once. Medium is Dulbecco's modified Eagle's medium. Before starting, put all reagents in a 37°C incubator or water bath. Put aside a separate 50 ml tube of medium without FCS in the 37°C incubator for the dilution after the fusion. The myeloma cells are diluted into 200 ml medium 2 days before the fusion. It is essential that they be in logarithmic growth, and the cell concentration on the day of fusion should not be more than about 10^5/ml. It is not essential to perform cell counts, because the fusion is very reliable as long as the size of the cell pellets for spleen and myeloma are about the same.

(1) Remove one spleen from mouse in sterile hood and make into a single cell suspension in medium without FCS. Spin down the myeloma cells (200 ml) at the same time. Wash cells three times, spinning at 400–700 g for 7 minutes each time. If necessary, adjust the amount of myeloma cells so that the size of the pellet is about the same as that for the spleen cells.
(2) Combine the spleen and myeloma cells into a single 50 ml tube for the final wash. Keep cells at 37°C while fusion is taking place.
(3) Remove all liquid from the tube, making sure that the pellet is dry. Place the tube in a 37°C water bath for 2 min.
(4) Gently stir the cell pellet with a sterile bacteriological loop (no added medium).
(5) Add 1 ml of 50% PEG to the cells over the next 30 s by running it down the side of the tube and letting it run into the cells. Then, with the same pipette, stir the mixture for another 30 s. Do not pipette up and down; stir only.
(6) Over the next 60 s, add a total of 7 ml of the fresh medium without FCS. Add 1 ml over the first 15 s, 2 ml over the next 30 s, and 4 ml over the next 45 s, until all the 7 ml is added over a period of about 1 min.
(7) At this stage, the cells will be in small clumps. Stir but do not pipette at this stage. Once the 7 ml have been added, the remaining medium without FCS can be added rapidly with stirring to a final volume of 50 ml over 1–2 min. Clumps of cells may still be visible after the tube is full, but this is not a cause for concern.
(8) Spin fused cells at 400–700 *g* for 5 min.

During this time, make up the final culture medium (100 ml DME plus 10% FCS, plus 1 ml of 200 mM glutamine and 1 ml of 100 × HAT concentrate, and cells from 1–2 thymuses from 4–6-week-old mice). If the spleen is large, you may want to increase the volume of medium so that you do not get too many colonies in each well.

(9) Resuspend the fused cells in the culture medium, making sure that all the clumps are broken down by gently mixing. Do not vortex.
(10) Plate out cells into the 96-well flat-bottom plates, at 200 μl/well. Each tray will take about 20 ml; for a typical fusion, use 4–6 trays. If the spleen

was large, as it often is if the immunization used Freund's adjuvant, it may be advisable to put the cells into more plates to avoid getting too many colonies in each well.

Plating out can be done easily by pouring the cells into a large sterile Petri dish and using an 8-channel Titertek pipette to dispense them into the 96-well plates. Put the plates into a 37°C incubator. The next day, check plates to make sure that there has been no microbial contamination.

(11) Leave the plates in the incubator for 6–8 days, and then check for the growth of colonies. Mark the top of the plate to indicate how many colonies in each well (e.g. 1 dot per colony). Wells that are positive for antibody should probably be cloned quickly if they have more than one clone visible.

(12) After about 10–12 days, very carefully remove half of the old medium and replace with fresh medium, still containing HAT. Repeat this every few days.

(13) Test each well for antibody production.

8.8.2 The Protocol of Gefter *et al.*

Gefter *et al.* (1977) have shown that good fusion frequencies are obtained with less variation if the PEG concentration is lowered to 30–35%, and the cells exposed for 7–9 min at room temperature. The mixture of spleen and myeloma cells is centrifuged in serum-free medium, and the supernate removed. The cell pellet is then loosened by tapping the tube, and a solution of 30–35% PEG (molecular weight 1000; 1–5 ml) is added at room temperature. The cells are then centrifuged at 200 ***g*** for 2 min and left for an additional 5–7 min at room temperature. They are then diluted in a large volume of medium without serum, harvested by centrifugation and cultured. The dilution step may be rapid.

8.8.3 Electrically Induced Fusion

Induction of fusion with PEG is essentially random, and the fraction of hybrids of the desired specificity is therefore generally low. It is also possible to induce cell fusion by a brief but intense electrical pulse, which will selectively fuse cells that are touching each other. If the spleen cell–myeloma contact were mediated by an antigen bridge, it would seem logical that the frequency of antigen-specific hybrids might be greatly increased. There is some evidence for this proposal (Lo *et al.*, 1984; Wojchowski and Sytkowski, 1986; Stenger *et al.*, 1988; Hewish and Werkmeister, 1989; Werkmeister *et al.*, 1991). While the earlier methods for cell fusion by electrical pulses were very complicated, they have now been greatly simplified. Electroporation apparatus is now available in many laboratories, and the method may become more popular.

The bridge between myeloma and antigen-binding cell involves the biotin–avidin system (see Section 10.3). The myeloma cells are directly conjugated with biotin, while the spleen cells are incubated with biotinylated antigen and washed, so that only specific antigen-binding cells retain the antigen. The biotinylated myeloma cells and the spleen cells with bound biotinylated antigen are then bridged by adding streptavidin, which is a tetrameric protein with four biotin-binding sites. Finally, the conjugates are subjected to a brief electrical pulse to induce fusion, and then plated out in HAT medium. The practical procedure is fairly flexible, and the following is suggested as a guide.

Myeloma cells are biotinylated as follows. Wash 10^7–10^8 myeloma cells three times in PBS, and resuspend in 1.0 ml PBS. To a separate tube containing 1.0 ml PBS, add 50 μl freshly prepared biotin succinimide ester (1.0 mg/ml in DMSO), and mix. Without delay, combine the contents of the two 1.0 ml tubes and mix gently. Leave on ice for 1 h. Wash three times in PBS to remove unbound biotin, and resuspend in 1.0 ml PBS. The wash medium should contain 50 μg/ml DNAse I to digest DNA released from cells killed by procedure.

When working with biotin succinimide ester, it is important that the solution should not contain any nucleophiles such as azide, Tris or any other amines. The biotin succinimide ester must be kept dry in a dessicator and only dissolved in DMSO immediately prior to use, as it has a lifetime of only a few minutes in aqueous solution. DMSO usually contains significant amounts of water. It cannot be assumed that biotin succinimide ester will be stable in DMSO for more than a few minutes unless special precautions are taken to keep the DMSO anhydrous.

The antigen is coupled with biotin as follows. Dialyse the protein against 0.1 M $NaHCO_3$ overnight, and then dilute to 1.0 mg/ml in the same buffer, avoiding azide, amines or other nucleophiles. Make up the biotin succinimide ester in DMSO at 1.0 mg/ml immediately prior to use, and add 120 μl of biotin solution per ml of protein. Leave at room temperature for 1 h, and then dialyse overnight against PBS to remove unbound biotin (see also Section 10.3 for further details of the biotin–avidin system).

Hold the spleen cells (typically 10^8) with biotinylated antigen (concentration is not critical, try 100 μg/ml) on ice for 1–3 h to allow it to bind, then wash cells three times in PBS and resuspend in 1.0 ml PBS. Add 50 μl of 1.0 mg/ml streptavidin in PBS and leave on ice for a further 30 min to allow it to bind to the biotinylated antigen, then wash the spleen cells twice in PBS, and resuspend in 1 ml PBS.

Fusion is carried out as follows. Add 1 ml biotinylated myeloma cells to 1 ml spleen cells, centrifuge at 200 ***g*** for 5 min, and incubate the pellet for 1 h at 37°C. Wash the cells and resuspend in 2 ml low ionic strength isotonic medium (0.213 g/l Na_2HPO_4, 0.068 g/l KH_2PO_4, 93.1 g/l sucrose, pH 7.2, 300 mosmol), and transfer 400 μl aliquots into electroporation cuvettes.

Fusion is induced by two pulses of the Bio-Rad Gene Pulser, at 400 V using 1 μF capacitance. The cells are then grown in HAT medium as for polyethylene glycol-mediated fusions (see Werkmeister *et al.*, 1991). It may be worth experimenting with the conditions of fusion by varying the capacitance and voltage. *If the cells are fused in PBS or culture medium rather than in low ionic strength buffer as above, it would probably be necessary to increase the capacitance 10–100-fold to achieve the same time constant, because of the much lower resistance (i.e. higher conductivity) of the buffer.*

The advantages of electrofusion include economy in terms of amount of antigen, the routine production of high-affinity antibodies, and a reduction in the amount of cell farming and screening. The disadvantages include increased complexity and the need for special equipment. To date, it has not been widely adopted, but it is worth considering.

8.9 Early Growth

Some investigators plate the cells into normal medium (containing 10–15% FCS) for 1 day prior to adding HAT (e.g. Galfrè and Milstein, 1981; Oi and Herzenberg, 1980), while many others plate directly into HAT (Hämmerling *et al.*, 1981; Zola and Brooks, 1982). There seems to be no clear reason for delaying the introduction of HAT selection, and direct plating into HAT is now becoming common practice.

8.9.1 Large or Small Cultures?

The other main area in which there is a good deal of variation is in the size of the initial cultures. Galfrè and Milstein (1981) recommended plating a total of 10^8 spleen cells plus the myeloma cells into 43 × 1.0 ml Linbro plates, together with 10^5 normal spleen cells per ml (feeder cells). The latter are probably unnecessary, as each well already contains 2×10^6 spleen cells. Oi and Herzenberg (1980) have used a slightly different protocol, starting with a total of 3×10^8 cells and plating into 300 × 100 μl wells (i.e. 10^6 cells per 100 μl). Depending on the frequency of hybridoma formation (typically 1 clone per 10^5–10^7 input cells), each well may have none, one or several clones. Plating at relatively low numbers of cells per well has the advantage that 'positive' wells will probably contain monoclonal antibodies from the start. Thus, screening assays which are searching for antibody of a *particular specificity* (e.g. detecting a subpopulation of cells) will not be confused by the presence of more than one clonal product.

Most workers in the field now plate out the fusion into 96-well plates. It is a good idea to leave the outermost row of wells empty, because they are most prone to infection by airborne microorganisms.

8.9.2 Feeding the Cultures

Another area where some variation exists is in the feeding of cultures after fusion. Feeding is carried out by removal of about half the culture medium by suction, followed by its replacement with fresh medium. Feeding gradually dilutes out any antibody made by normal (unfused) antibody-secreting cells, which may remain alive for up to a week.

Feeding also removes waste products and replenishes nutrients. Some authors have recommended feeding every 3–4 days (e.g. Oi and Herzenberg, 1980), but in this case the feeding also served the purpose of gradually phasing in the HAT selection. If the hybrids are plated immediately into HAT, it is usually sufficient to feed only after the first 7 days (Zola and Brooks, 1982). Subsequent feeding should be guided by the rate of cell growth. If the cultures become dense and the medium turns yellow, it may be necessary to feed more frequently.

8.9.3 Feeder Cells

Lymphoid cells often grow poorly or die when grown at low density. The reasons are still not well understood, but may relate to requirements for growth factors, and possibly also toxic products from the tissue culture vessels. Choice of the particular batch of FCS may make a big difference.

Although it is rarely necessary on a routine basis, the growth of small numbers of cells is sometimes improved by incubating the tissue culture ware with medium overnight at 37°C, and then replacing it with fresh medium before the cells are added. It is presumed that something toxic in the plastic is being washed out.

These problems can be overcome (at least partly) by culturing together with a slow-growing or nongrowing population of cells. These cells are usually termed 'feeders', implying that they make something needed for growth. If culture conditions are optimal, feeder cells may make little difference. They do no harm however, and may reduce variation between experiments (Fazekas and Scheidegger, 1980). The current concept of the action of feeder cells is that they supply growth factors, particularly IL-6, but they are very likely also to supply others (see Section 8.3.2). If the fusion mixture is plated out at high cell density, added feeders may make little or no difference because the spleen cells used for fusion also act as feeders.

As noted previously, the presence of feeder cells may make a crucial difference between success and failure of production of rat hybrids.

Commonly used feeder cells include thymocytes (typically 10^6/0.2 ml well; Lernhardt *et al.*, 1978; Oi and Herzenberg, 1980), normal spleen cells (10^5/ml; Galfrè and Milstein, 1981) or peritoneal cells (Hengartner *et al.*, 1978; typically 2×10^4/0.2 ml). Peritoneal cells are harvested by washing out the peritoneal cavity with sterile saline, using a syringe and needle and taking care to avoid puncturing the gut. Roughly half are lymphocytes and half are macrophages. If the mice are from specific pathogen-free colonies, yields will be 3–5 × 10^6 cells per mouse. 'Conventional' mice will yield up to 10 times as many cells.

Peritoneal cells from rats are collected by anaesthetizing the rat, injecting 20 ml of sterile culture medium into the peritoneal cavity, and gently massaging the abdomen. After 2–5 min, the rat is killed and the medium aspirated. Typically, one may recover about 2 × 10^7 cells (Ravoet and Bazin, 1990).

The feeder cells do not have to be histocompatible with the hybrids. They do not even need to be from the same species. Rat thymocytes seem to function well for mouse–mouse hybrids.

8.9.4 Phasing out HAT Selection

After 1 week in HAT medium, it is safe to assume that all the parental myeloma cells will be dead, and that any growing cells will be hybrids. At this time, HAT selection may be terminated. It is generally recommended that the hybrids be cultured in HT medium for a few days prior to using normal medium. Theoretical considerations suggest that this step should be carried out carefully.

The inhibition constant (K_i) of aminopterin for dihydrofolate reductase is less than 10^{-9} M (Calabresi and Chabner, 1990), and the concentration of aminopterin in HAT medium is 4 × 10^{-7} M. *The aminopterin concentration must therefore drop by more than 400-fold before the enzyme can regain activity.* Unlike hypoxanthine and thymidine, aminopterin is metabolized extremely slowly. In man, the main route of elimination of the closely related drug methotrexate is via urinary excretion; catabolism does not appear to occur to a significant degree (Calabresi and Chabner, 1990). *The main route of elimination of aminopterin from the cultures must therefore be by dilution.* Accordingly, it would be prudent to remove as much of the HAT medium as possible prior to feeding with HT medium, and to repeat the process several times prior to using ordinary medium.

Zola and Brooks (1982) found that some hybrids died when taken out of HT medium, and suggested that the cells be grown in HT medium permanently. It may be that the very slow catabolism of aminopterin, together with its extremely high affinity for dihydrofolate reductase, resulted in unexpectedly slow recovery of enzyme activity. The ingredients for HT medium are inexpensive, and the long-term growth in HT would not result in much extra work, so their proposal should be considered.

8.10 Screening Assays

The choice of an appropriate screening assay is one of the most important parts of hybridoma production. In principle, any method that is capable of detecting antibody of the desired specificity may be used. The practical constraints on screening assays are reliability, speed, cost and labour. Many hundreds or thousands of assays will have to be performed during initial screening, cloning, expansion, recloning and bulk culture. Unless the assays can be done in large numbers, and at reasonable cost and effort, the production of hybridomas is likely to fail.

The screening assay should be set up and 'debugged' well before the fusion is started because there will not be enough time to eliminate problems later. A trial bleed from the immune spleen donor may be used as a source of antibody for positive controls in the screening assay, but there is little correlation between the antibody levels in mice and success in producing monoclonal antibodies.

The required 'dynamic range' of the assay (i.e. the ratio of the strongest signal to the weakest that can be detected above background) will depend on whether or not the test antigen is a pure substance. For example, if it is desired to make hybridomas against a purified protein, 100% of the test antigen will be relevant, and a simple system that gives an unambiguous yes/no answer will be adequate. However, if the desired antigen is a minor cell-surface protein, the assay system may need to be capable of detecting very small signals, and should have a dynamic range of at least 10:1 (preferably 100:1).

Choice of screening assays is also influenced by the type of antibody required. Classes of antibody that fix complement might be chosen by using an assay based on cytotoxicity. If protein A-binding antibodies are desired, this will be guaranteed by incorporating protein A in the screening assay. If antibodies are desired which react with formalin-fixed or denatured proteins, the immunization and screening assay should use this form of antigen (Harrach and Robenek, 1991).

8.10.1 Removal of Supernatants

The first step in any assay will involve sterile removal of the supernatants. This may be accomplished by using sterile Pasteur pipettes, but if hundreds of supernatants must be tested, this will be found to be very slow. There are now a variety of alternative procedures which differ in ease, complexity and cost. In my experience, the more complex and sophisticated devices are usually unreliable and cumbersome.

One of the best and cheapest procedures is to use standard laboratory pipettes with disposable conical tips (e.g. Eppendorf, Gilson, Finnpipette, Oxford). The lower part of the pipette should be wiped with 70% ethanol prior to use. Many companies now supply autoclavable and reusable boxes for plastic tips. A convenient autoclavable reusable box for sterilizing 96 tips at one time is made by Treff (Treff AG, CH 9113, cat. no. 9606, Degersheim, Switzerland). Each tip may then be placed onto the pipette without contact with the fingers.

Another alternative is the use of multi-channel pipettes, such as those made by Flow Laboratories (8-channel 100 μl Titertek multi-channel pipette, cat. no. 77-844-00).

8.10.2 Labour-saving Devices for Screening Assays

Screening assays constitute one of the most time-consuming and labour-intensive parts of hybridoma production, so it is important to eliminate any unnecessary steps and to automate those steps that cannot be eliminated. It is neither necessary nor desirable to spend thousands of dollars on complex and sophisticated machines, because they are usually unreliable and insufficiently flexible to meet changing needs. However, a few hundred dollars spent on a repetitive dispenser will greatly facilitate the assays. The Eppendorf Multipette 4780 is ideal for such purposes, because it is inexpensive, robust and reliable, and can be adjusted to deliver a large range of volumes. Sterile disposable tips are available for feeding cultures.

The simultaneous washing of 8–12 wells may be accomplished by the use of the Nunc Immuno Wash or the Skatron Mini-Microwash. These devices simultaneously deliver and aspirate wash fluid, and are reasonably priced. The Drummond Model 590 Microdispenser with microtest manifold and repeating dispenser is also relatively inexpensive, but it does not allow simultaneous delivery and aspiration.

8.10.3 Solid-phase Radioimmunoassay—Soluble Protein Antigen

The solid-phase radioimmunoassay (Catt and Tregear, 1967; Fig. 8.1) is a very simple, sensitive and precise assay for hybridoma antibodies. It is easy to set up and is capable of a very low background and a large dynamic range. A detailed account is given by Tsu and Herzenberg (1980). It is easily adapted to give a colour readout using enzyme-labelled second antibody, which is becoming increasingly popular (see Section 8.10.5).

The assay is based on the fact that polyvinyl surfaces will tightly adsorb nanogram amounts of most proteins. The antigen-containing solution is simply pipetted into the tube, left there for a few hours, and the unbound material washed out. Any remaining protein-binding sites are saturated by a large excess of irrelevant protein, which is usually bovine serum albumin (BSA). It should be noted that a high proportion of the antigen may be denatured by adsorption to plastic but that some native protein usually remains (Smith and Wilson, 1986; Butler *et al.*, 1993).

The antibody-containing solution is then added, left for 30–60 min to react with the antigen, and any unbound material washed out. Finally, a 'revealing' reagent is added. This is usually ^{125}I-labelled affinity-purified anti-

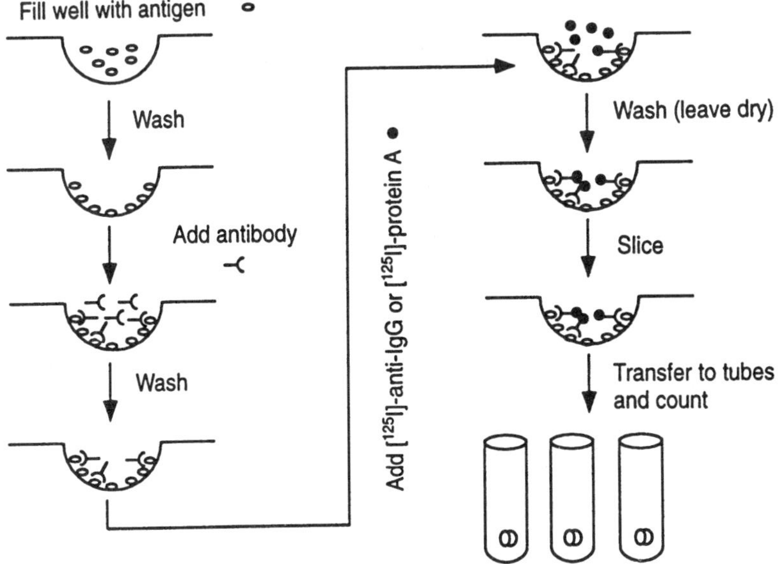

Fig. 8.1. Solid-phase radioimmunoassay for hybridoma screening. Note that this assay is easily adapted to enzymic colour readout (ELISA) rather than the use of radioactivity (see Fig. 8.2).

immunoglobulin or ^{125}I-labelled staphylococcal protein A, which binds specifically to IgG (Sections 5.3.5 and 9.3.3). After 30–60 min, excess revealing reagent is removed, and the wells are washed and counted in a gamma scintillation counter. If a scintillation counter is not available, detection may be by autoradiography with intensifying screens (Parkhouse and Guarnotta, 1978; Nowinski *et al.*, 1979).

A commonly used assay is based on that described by Pierce and Klinman (1976). All steps may be carried out at room temperature. Ninety-six well polyvinyl plates (cat. no. 001–010–2101, Dynatech Laboratories) are coated with 20–50 μl of antigen (10–50 μg/ml) in PBS, pH 7.4, for at least 30–60 min. (Although the optimal pH for coating may vary for each protein, in most cases adequate results are obtained at neutral pH.) The antigen should have no added carrier protein, and detergent will inhibit adsorption. The supernate is removed, but it may be reused several times, as only a small fraction binds. The plate is then washed three times in RIA buffer (PBS plus 0.1% NaN_3 and 1% BSA). Plates are then incubated with culture supernate (20–100 μl) for 1 h. After a further three washes, [^{125}I]-anti-immunoglobulin (10 μCi/μg) or [^{125}I]-protein A (40 μCi/μg) is added (10 000 c.p.m. per well are sufficient). After a further hour, plates are washed three more times and the wells removed for counting by cutting with scissors or slicing with a hot wire (see Tsu and Herzenberg, 1980, for details). The fumes are toxic, and slicing must be carried out in a fume cupboard. The assay is very flexible, and all concentrations and times given above may be varied by at least twofold with only minor changes in results.

It has recently been shown that nonfat powdered milk may be substituted for BSA in buffers and to block plates (Johnson *et al.*, 1984). These authors used 5% (w/v) milk in PBS plus 0.01% Antifoam A (Sigma), but similar results may be obtained with as little as 0.5% milk, and the Antifoam is not essential. The substitution of nonfat powdered milk for BSA results in a major saving in cost, and the results are at least as good, and often better.

If the antigen is pure, a typical positive well might have 300 c.p.m., while negative wells should have less than 100 c.p.m. Even if the antigen is impure, the background should not rise significantly, but positive signals will be smaller. Provided the background is low and reproducible, wells with 2–3 times background may be considered positive.

Some thought should be given to the choice of revealing reagent. The use of [^{125}I]-protein A has certain advantages, and also some disadvantages (Goding 1978, 1980). In the mouse, IgG2a, IgG2b and IgG3 bind at neutral pH, while IgG1 will bind only weakly at pH >8 (Ey *et al.*, 1978). Binding of protein A to mouse IgG1 may be greatly strengthened by a combination of high pH (8.5) and high salt (1.0 M). In general, mouse IgM, IgA, IgD and IgE do not bind. Very few rat IgG molecules bind staphylococcal protein A (Ledbetter and Herzenberg, 1979). Streptococcal protein G will detect most of the rat IgG subclasses (see Section 5.3.5).

One often obtains ample numbers of positive wells from mouse fusions using protein A as the revealing reagent. Its use also guarantees that the antibodies may be subsequently purified by affinity chromatography on protein A–agarose. IgG antibodies are more stable and easier to handle than other classes. Some mouse IgM antibodies bind nonspecifically to the plates, resulting in false positive reactions (Goding, 1980). The use of protein A eliminates this problem by failing to detect IgM. Unless there are special reasons for wanting classes other than IgG2 or IgG3, protein A is often the reagent of choice for mouse antibodies. Sometimes, however, it may be found that the predominant response is IgG1, IgM or IgA. This is particularly likely for IgG1. In these cases, protein G (Section 5.3.5) or anti-immunoglobulin antibodies must be used instead.

If anti-immunoglobulin antibodies are used, they should be affinity-purified to ensure that the majority of counts can bind. Non-affinity-purified antibodies will usually have less than 10% bindable counts, with consequent poor signal:noise ratio.

In general, it is preferable to use antibodies that do not react with light chains or detect IgM, as IgM antibodies are often cross-reactive in unpredictable ways (Section 7.3.2). They often bind nonspecifically to plastic even in the presence of competing protein, and are of low affinity, unstable and difficult to purify.

Many anti-immunoglobulin antisera have substantial anti-light chain activity, and will therefore detect all antibody classes and subclasses. However, some display a marked preference for individual classes, and it is unwise to assume that a randomly chosen antiserum will detect all classes with equal efficiency. If all antibody classes must be detected, the reagent of choice is an antiserum against the Fab fragments of normal polyclonal IgG. Such antisera usually detect all classes with approximately equal efficiency.

8.10.4 Solid-phase Radioimmunoassay—Cell Surface and Viral Antigens

The assay described in Section 8.10.3 is easily adapted to cell surface antigens (Oi *et al.*, 1978) or viruses (Nowinski *et al.*, 1979). One important difference is that if living cells are used, they should be kept cold to prevent antibody-induced 'capping', endocytosis and shedding of antigen–antibody complexes (Taylor *et al.*, 1971). The polyvinyl 96-well plates must be kept on a bed of ice during the assay, and the centrifuge must be refrigerated.

Cell suspensions are prepared by standard techniques and washed twice to remove soluble or loosely adsorbed material. In certain cases, adherent cells such as fibroblasts may be grown in the assay dish. Typically 10^4–10^6 cells are used per well. All steps are performed in RIA buffer (Section 8.10.3). It is advisable to precoat the trays to saturate protein-binding sites on the plastic.

The cells are held at 4°C for 30–90 min with 20–100 μl hybridoma supernatant. The trays are then centrifuged in tray carriers (200–400 ***g***, 2–5 min), and the supernatants removed by gentle vacuum aspiration from the edges of the wells. Cells are then washed 2–3 times by the addition of RIA buffer and centrifugation. (Resuspension is achieved by simply pipetting onto the cell pellet.) The cells are then resuspended in 50–100 μl [^{125}I]-anti-immunoglobulin or [^{125}I]-protein A (Section 8.10.3). After a further 30–90 min, cells are washed again (× 2–3), and the wells cut and counted (Section 8.10.3). The background should be no more than 100–200 c.p.m., and positive wells may contain *c.* 500–3000 c.p.m.

The cellular-binding assay can be modified to avoid the tedious and time-consuming centrifugations. Viruses or crude cellular membrane preparations may be firmly coated onto polyvinyl plates by simple adsorption, and washes may be performed without centrifugation (Nowinski *et al.*, 1979; Howard *et al.*, 1980).

Intact cells in serum-free medium will sometimes adhere to the plastic with sufficient tenacity to allow washing without centrifugation. In most cases, however, it is necessary to attach them by chemical means. Stocker and Heusser (1979) described an assay in which cells are attached to the plate with glutaraldehyde. The chemistry of this reaction is somewhat obscure, and it was not demonstrated that the glutaraldehyde was necessary. None the less, the method works well in the form in which it was published (see also Kincade *et al.*, 1981).

Cells are added to the plates (10^6/well) in protein-free PBS. The plates are then centrifuged at 50–100 ***g*** for 5 min and, without disturbing the cell layer, immersed in a 1-l glass beaker containing freshly prepared 0.25% glutaraldehyde in PBS at 4°C. After 5 min, the plates are washed in RIA buffer. The remainder of the assay is essentially as described above, but omitting the centrifugations.

If glutaraldehyde treatment alone is not sufficient to bind the cells firmly to the plates, adequate adhesion is usually obtained by precoating the plates with poly-L-lysine (10–50 μg/ml in distilled water) prior to adding the cells in protein-free PBS. Kennett (1980) recommends treating the plates with 0.5% glutaraldehyde for 15 min after the cells have been added, but this is probably only necessary for certain cell types. Regardless of how the cells are attached to the plate, all subsequent steps must include carrier protein (0.1% BSA) to saturate any remaining protein-binding sites.

If the cells are fixed with glutaraldehyde, the plates may be stored for several months at 4°C without apparent deterioration. However, many antigenic determinants may be destroyed by glutaraldehyde treatment (Gatti *et al.*, 1974; Kincade *et al.*, 1981).

8.10.5 Assays for Antibodies to Glycolipid Antigens

The solid-phase assay has also been adapted for glycolipid antigens (Smolarsky, 1980). The glycolipids are dissolved in ethanol and added to the wells of polyvinyl plates. The ethanol is then evaporated by a stream of nitrogen followed by high vacuum for 5 min, and the plates washed three times in PBS containing 0.3% gelatin and 1 mM ethylenediamine tetra-acetic acid (EDTA). The remainder of the assay is similar to those previously described. Other assays for antibodies to glycolipids have been described by Gray (1979), Young *et al.* (1979), and Handman and Jarvis (1985).

8.10.6 Enzyme-linked Immunosorbent Assays (ELISA)

The assays described in the previous sections are easily adapted to ELISA readout (Engvall and Pesce, 1978; Voller *et al.*, 1979; Engvall, 1980; Maggio, 1980). The revealing agent is simply conjugated with an enzyme (peroxidase, alkaline phosphatase or β-galactosidase) rather than ^{125}I. After washing away any unbound material, the bound enzyme is revealed by addition of substrate which undergoes a colour change (Fig. 8.2). As discussed earlier, a substantial proportion of any protein antigen adsorbed to ELISA trays is likely to be denatured, although some native antigen is likely to remain (Smith and Wilson, 1986; Butler *et al.*, 1993).

The advantages of ELISA readout are many. No isotopes or scintillation

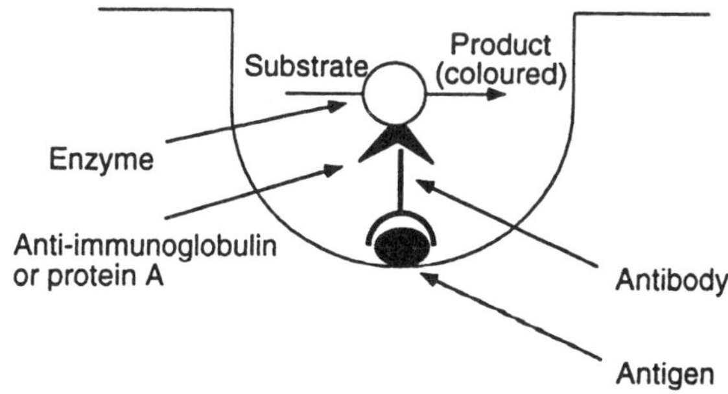

Fig. 8.2. The enzyme-linked immunosorbent assay (ELISA).

counters are required, and the readout may be made by eye. The tedious cutting up of trays, loading into tubes, and loading and unloading the counter are all avoided. If a quantitative readout is needed (unnecessary for hybridoma screening) automated reading devices are available which will scan 96 wells in a minute or so, compared with 1–2 h for scintillation counting. Finally, the reagents for ELISA are inexpensive, and have a long shelf life. Enzyme-conjugated anti-immunoglobulin or protein A or protein G are available commercially from many suppliers.

The most commonly used enzymes are horseradish peroxidase and alkaline phosphatase. Both are capable of giving good results, provided that no endogenous enzyme is present. This may be a serious problem in cell-binding assays. Macrophages and myeloid cells possess high levels of endogenous peroxidase, and B lymphocytes have high levels of alkaline phosphatase (Culvenor *et al.*, 1981; Garcia-Rozas *et al.*, 1982). Thus, these enzymes cannot be used if screening is performed on cells from spleen or lymph node; even thymus cell suspensions will have an unacceptable background. Attempts to destroy endogenous peroxidase require harsh chemical treatment which may destroy antigenic determinants (Farr and Nakane, 1981). However, peroxidase has been used very effectively (Douillard *et al.*, 1980), especially when cultured tumour lines were used as targets. Only one of 23 lines tested had significant endogenous peroxidase activity (Posner *et al.*, 1982). One ingenious way to overcome endogenous peroxidase is to use antibodies coupled with glucose oxidase to generate hydrogen peroxide from glucose. A set amount of peroxidase and colour reagent is then added to detect the H_2O_2. Thus, endogenous peroxidases will not alter the background (D. Fayle, personal communication).

The alkaline phosphatase in leukocytes can be inhibited by the drug levamisole, while alkaline phosphatase from bovine gut is not (see Ponder and Wilkinson, 1981). It may therefore be possible to assay for antibodies to lymphoid cells using alkaline phosphatase-conjugated antibodies if levamisole (1 mM final concentration) is included in the substrate.

The bacterial enzyme β-galactosidase has occasionally been used, and has the advantage that it is not present in mammalian cells (O'Sullivan *et al.*, 1979; Kincade *et al.*, 1981). It has very low sensitivity, and is therefore not recommended.

Although the ELISA is ideal for assays in which the antigen is relatively pure, it has not yet been widely accepted for cell-surface antigens. In addition to the problem of endogenous enzymes, cell-binding ELISA assays suffer from a limited dynamic range. The strongest signals are seldom more than 10 times background, and the background is not as low or as uniform as in RIA. When antibodies against low-concentration membrane proteins are sought, RIA will give much clearer discrimination between positives and negatives.

ELISA assays are usually quite simple to set up. The plate is coated for a few hours with protein antigen (typically 1–10 μg/ml in PBS containing no detergent or extraneous protein), and then all remaining nonspecific protein binding sites are saturated with 1–5% w/v bovine serum albumin in PBS, which is also used as wash and dilution buffer. Hybridoma supernatants are generally assayed undiluted, typically 50–100 μl per well for 1 hour, followed by several washes and then typically 100 μl per well of peroxidase-conjugated anti-mouse immunoglobulin for 1 h, followed by extensive washing and addition of substrate.

Troubleshooting ELISA assays

Occasionally one may encounter high 'background' readings in ELISA assays, especially when using serum rather than hybridoma supernatants, and when using rabbit antibodies. This background is revealed by high colour readings even if the plate is not coated with antigen. The reason for this is not completely clear, but may possibly relate to the marked propensity for IgM to bind nonspecifically to the plate, even after blocking with albumin (see Section 10.8.3). This problem may sometimes be overcome by adding to the diluent solutions normal serum (e.g., 1% v/v) of the same species as the second antibody (e.g., if the second antibody is sheep anti-rabbit IgG, add normal sheep serum to all buffers). This may dramatically lower the background, possibly by allowing sheep IgM to bind to the plate instead of rabbit IgM. If one uses peroxidase-conjugated second antibodies, it is best to avoid the use of powdered milk for blocking or in buffers, as it often has a high level of peroxidase activity.

Other problems occasionally encountered in ELISA systems is variability between the efficiency of binding of proteins to different brands of plastic trays. Dynatech Microtiter Plates (cat. no. 001-010-2101) give excellent results (Dynatech Laboratories, Inc., 14340 Sullyfield Circle, Chantilly, Virginia 22021, USA).

8.10.7 A Simple Method for Conjugating Horseradish Peroxidase to Antibodies or Staphylococcal Protein A

The method to be described is based on that of Wilson and Nakane (1978) but has been greatly simplified. It probably results in a large degree of polymerization, but this does not seem to cause problems and may contribute to its excellent sensitivity. The protocol produces excellent conjugates for ELISA assays and immunohistochemistry (Harahap and Goding, 1988; see Chapter 13). A somewhat more refined but slightly more complex procedure is described by Tijssen and Kurstak (1984).

The method is based on the oxidation of the carbohydrate moieties of horseradish peroxidase (HRPO) by periodate ions, and the subsequent reaction of the resulting aldehyde groups with amino groups of proteins to form Schiff bases. The Schiff bases are stabilized by reduction with sodium borohydride.

pH 4.0 acetate buffer

Add 0.3 g sodium acetate (anhydrous) to 100 ml H_2O. Add 960 μl glacial acetic acid. Mix, and check pH. Adjust if necessary by adding acetic acid or NaOH (± 0.2 pH units is satisfactory).

pH 9.5 carbonate buffer

Add 17.2 g $NaHCO_3$ plus 8.6 g Na_2CO_3 to 1 l H_2O. Make up fresh on day of use. The pH should be close enough without adjustment (± 0.2 pH units).

Practical procedure

(1) Dialyse affinity-purified IgG (1.0 ml of 5.0 mg/ml) against 1 litre pH 9.5 buffer overnight at 4°C.
(2) Weigh out 10 mg horseradish peroxidase (Sigma type VI, cat. no. P8375); dissolve in 1.0 ml H_2O.
(3) Add 200 μl pH 4 buffer. Mix.
(4) Add 200 μl 0.2 M $NaIO_4$ (freshly made up). Solution will turn green. Leave at room temperature for 15 min, protected from light.
(5) Equilibrate a Pharmacia PD-10 column of Sephadex G-25 with 20–30 ml pH 9.5 buffer.
(6) Let column run dry, then load sample and allow to run in.
(7) Add 1.0 ml pH 9.5 buffer. Discard effluent.
(8) Elute 'activated HRPO' with 3.5 ml pH 9.5 buffer.

(9) Add activated HRPO to dialysed IgG.
(10) Leave at room temperature for 60 min.
(11) Add 100 µl $NaBH_4$ (4.0 mg/ml in H_2O).
(N.B. $NaBH_4$ is extremely unstable in water. Store powder in a dessicator, and make up solution immediately before use).
(12) Leave conjugate at room temperature for 1 h, then dialyse against PBS overnight at 4°C.
(13) Add glycerol to conjugates to 50% (v.v). Mix thoroughly, and store at −20°C. The glycerol prevents freezing. The conjugate should be usable at a dilution of more than 1:1000.

For conjugates of HRPO with staphylococcal protein A or protein G, use same procedure but use 4 mg HRPO and 1.0 mg protein A or protein G.

8.10.8 Peroxidase ELISA Assays: Problems, Precautions and Choice of Substrate

The coating of the trays, binding of antibody and washing may be performed as for solid-phase RIA. The carrier protein in the buffers must be free of peroxidase activity (a common contaminant in BSA and milk). Note that azide ions must not be present, as they inhibit peroxidases. Water for buffers should be distilled but not deionized, as the hydrocarbons from deionizing resins often inhibit peroxidase. The conjugate is used at the maximum dilution that gives satisfactory results.

When enzymes with high turnover numbers, such as horseradish peroxidase, are combined with carefully chosen substrates (Al-Kaissi and Mostratos, 1983), the sensitivity of ELISA may equal that of RIA. The assay is carried out in PBS containing H_2O_2 (typically a dilution of 1:3000 from 30% v/v stock, but this may need to be optimized), and a substrate that gives a coloured product. The substrate most often used for peroxidase is *o*-phenylenediamine (OPD), which gives a yellow colour and is one of the most sensitive. An alternative substrate is 3,3′,5,5′ tetramethylbenzidine (TMB), which is marketed as K Blue by Elisa Technologies, 518 Codell Drive, Lexington, KY 40509, USA, and is claimed to be more sensitive and stable than competing brands of the same product, and more sensitive than OPD. Another popular substrate is 2,2′-azino-di-(3-ethylbenzthiazoline sulfonic acid) (ABTS, cat. no. Sigma A-1888), which is produces a stable green colour, but is less sensitive.

Horseradish peroxidase ELISA readouts have the curious property of very high sensitivity for a limited duration. The strong oxidizing effect of the H_2O_2 rapidly inactivates the peroxidase during the reaction, which usually stops almost completely after 30–60 min. This can be seen if one removes the substrate and adds more after the reaction has reached a plateau; little or no further colour will develop. If one uses a lower concentration of peroxide to avoid killing the enzyme, the reaction goes much more slowly and the sensitivity is not as good. At the present, there does not appear to be any way around this problem. If difficulties are encountered, try varying the peroxide concentration.

While 30% H_2O_2 is very stable, hydrogen peroxide is rather unstable when diluted, and dilution should be made immediately before use. If there is any doubt about the H_2O_2, it can be checked by measuring the optical density. For 30% stock diluted 1:1000 in PBS, the optical density at 230 nm should be about 0.7 (Phillips and Morrison, 1970).

Positive wells should be easily visible after 5–60 min at room temperature. The tray should be read at a fixed time (e.g. 30 min), because the colour is not very stable.

The substrate mixture for OPD is made up freshly on the day of use as follows:

Distilled water	100	ml
Citric acid	0.47	g
Na_2HPO_4 (anhydrous)	0.73	g
o-phenylenediamine	40	mg
30% H_2O_2	30	µl.

The volume per well is not critical; 100 µl is typical. The quantitative readout on an ELISA plate reader should be at 492 nm. It is customary to add a 'stopping solution' of 50 µl of 1 M sulfuric or phosphoric acid to make a defined end-point. The acid also intensifies the colour of the product, thus improving the sensitivity.

The substrate mixture for ABTS is made up freshly on the day of use as follows:

Distilled water	100	ml
Citric acid	1.29	g
$Na_2HPO_4.2H_2O$	1.37	g
ABTS	50	mg
30% H_2O_2	30	µl.

The volume per well is not critical; 100 µl is typical. The quantitative readout on an ELISA plate reader should be at 405 nm. The reaction may be stopped by adding 100 µl of 2 mg/ml sodium fluoride in water. The colour is stable for many hours.

8.10.9 Screening by Immunofluorescence

Screening of hybridomas by immunofluorescence has also been widely used (Ledbetter and Herzenberg, 1979; see also Stitz *et al.*, 1988). The sensitivity of immunofluorescence appears to be similar to RIA and ELISA.

Immunofluorescence has the definite advantage that one may detect antibodies against subpopulations of cells or subcellular structures. A detailed account of immunofluorescence is given in Chapter 12. This approach is quite rapid, and is well suited to screening hundreds of samples per day. A simple visual assessment takes only a few seconds. The fluorescence-activated sorter can generate a histogram of the staining of 10 000 cells in a minute or two.

8.10.10 Immunofluorescence Screening on Intact Cells in Suspension

Living cells (typically 10^6) expressing the target antigens are held on ice in 100 μl RIA buffer (see Section 8.10.3) with 5–50 μl hybridoma supernatant for 30–90 min, washed twice, and then held for a similar period with fluorescein-conjugated anti-immunoglobulin antibodies. After washing twice more, they are resuspended in a drop of medium, placed on a microscope slide, and covered with a cover slip which is sealed at the edges with nail polish. The cells are examined using a microscope with appropriate illumination and filters (see Chapter 12).

Alternatively, cells may be analysed by the fluorescence-activated cell sorter (Ledbetter and Herzenberg, 1979; Loken and Stall, 1982; Ormerod, 1990; Shapiro, 1994).

8.10.11 Immunofluorescence on Fixed and Permeabilized Cellular Monolayers

Stitz *et al.*(1988) have described a very easy and rapid method to screen large numbers of antibodies. Although the procedure was developed to detect internal cellular determinants, it is also suitable for the detection of membrane antigens. It is easy for one person to test 1000 samples in a day.

(1) Seed 5×10^4 adherent cells/well into 96-well flat-bottom tissue culture plates, 1 day prior to assay.
(2) On day of assay, gently aspirate the medium, using a suction manifold (Drummond cat. no. 3-00-093 or 3-00-094).
(3) Wash the cells once with RIA buffer (PBS/1% BSA/0.1% azide).
(4) To fix and permeabilize the cells, add 100 μl 100% methanol. Leave at room temperature for 10 min, then aspirate methanol, and air dry until no liquid is left (maximum 5 min). Fixed cells can be stored in plates in 200 μl RIA buffer for a few days at 4°C.
(5) Add 200 μl RIA buffer. Leave for 1 h at room temperature to block.
(6) Aspirate liquid and add 50–100 μl hybridoma supernatant, and leave for 1 h at room temperature.
(7) Gently wash cells with RIA buffer three times.

(8) Add 50–100 μl secondary antibody (FITC–antimouse immunoglobulin; diluted about 1:200 in complete tissue culture medium, and incubate for 1 h at room temperature.
(9) Gently wash cells with RIA buffer.
(10) Turn plate upside down and let it air dry for 5–10 min. The wells are scanned with the plate upside down, using a standard fluorescence microscope and a low- or medium-power objective.

8.10.12 Membrane Immunofluorescence on Intact Cellular Monolayers

The procedure is essentially as above, but omits fixation (step 4), and uses complete tissue culture medium containing 10% fetal calf serum. This appears to be essential to prevent the cells from detaching from the plates during washing.

8.10.13 Cytotoxicity Assays

Hybridoma supernates containing complement-fixing antibodies against cell-surface antigens may be detected by cytotoxicity assays. It is customary to use two-step assays in which antibody is bound to the cells, unbound antibody is washed away, complement is added, and the cells are examined for lysis. The two-step assay was originally developed to circumvent 'anti-complementary' activity in mouse serum. It is possible that a one-step assay (i.e. addition of complement together with the antibody) would be adequate for hybridoma screening.

Cells bearing the target antigen (10^5–10^6) are held with the supernate (10–100 μl) on ice for 30–60 minutes to allow antibody binding. They are then washed once or twice, and a source of complement added (usually guinea-pig or rabbit serum). The cells are held at 37°C for 30 min to allow lysis to occur, and then placed on ice until they are scored for lysis.

The source of complement is critical. Complement-containing serum must be processed rapidly, and stored in small aliquots at −70°C. Once thawed, aliquots should be used immediately and any excess discarded. Repeated freeze–thaw cycles will rapidly inactivate complement.

The optimal dilution of the complement must be determined for each individual batch. A certain degree of background lysis (5–10%) is unavoidable, and is largely due to the presence in the complement of naturally occurring antibodies against cells. These may be minimized by selection of serum from individual animals. Serum from baby rabbits is said to be less toxic. Absorption of natural antibodies on cells at 4°C in the presence of EDTA may be performed (Boyse *et al.*, 1970), but usually a dilution of unabsorbed complement can be found at which the background is acceptable and complement activity adequate. Typical dilutions of complement are 1:10 to 1:20.

Assessment of lysis may be performed using 0.2% trypan blue, or by staining with acridine orange and ethidium bromide (Lee *et al.*, 1975; Parks *et al.*, 1979). The acridine orange/ethidium bromide method is vastly superior in accuracy to the older methods. A stock solution of 1 part per million of acridine orange and ethidium bromide is made in PBS, and stored in a foil-wrapped bottle at 4°C. (Both acridine orange and ethidium bromide are potentially carcinogenic, and should be handled with care.) To assess viability, one drop of cells is placed on a glass slide, and one drop of stain solution added, followed by a cover slip. The slide is examined by fluorescence microscopy, using filters for fluorescein. Living cells stain bright green, while dead cells are bright orange.

The complement-mediated cytotoxicity assay has may variants. It may be miniaturized by the use of Terasaki microcytotoxicity plates (Falcon 3034) and Hamilton syringes with repeating dispensers for aliquoting 1 μl and 5 μl volumes. The use of ^{51}Cr-release as a marker of cell death may be used instead of vital stains, but is not very satisfactory with resting lymphocytes, and works best with mitogen-stimulated blasts as targets (McKearn, 1980).

Screening by cytotoxicity is somewhat cumbersome, and has fallen from favour in recent years. None the less, if only cytotoxic antibodies are of interest, the use of the assay will guarantee that they are selected.

8.10.14 Screening by Immunoprecipitation and Polyacrylamide Gel Electrophoresis

It is customary to identify positive hybridoma wells by simple binding assays, and then identify the antigen biochemically. However, Brown *et al.* (1980) have shown that it is feasible to reverse this order and identify the antigen as part of the screening procedure. The method is unlikely to be the first choice in most circumstances, but when antibodies against a particular polypeptide in a complex mixture are sought, it could be considered.

In order to minimize the number of supernates to be tested, two strategies were adopted. First, all supernates were screened for IgG production by testing their ability to compete with [^{125}I]-IgG for binding to protein A-containing staphylococci. About 25% of wells with growth contained IgG. The second strategy involved pooling a series of horizontal rows (1–12), and pooling a series of vertical columns (A–H), with each primary well contributing 10 μl supernatant. The pools were held for 1 h at 0°C with 100–200 μl ^{125}I-labelled lysate of the immunizing cell (*c.* 10^8 c.p.m.), and then 1 mg of heat-killed and formalin-fixed protein A-bearing *Staphyloccus aureus* was added. (One could substitute 50 μl of 10% v/v protein A-Sepharose for the intact bacteria.) The staphyloccoci, to which the immune complexes were bound, were harvested by centrifugation and washed twice. Bound antigen was released by heating the bacteria to 100°C in SDS sample buffer, and identified by SDS–polyacrylamide electrophoresis and autoradiography. If, for example, and M_r 50 000 polypeptide was present only in row pool C and column pool 11, the relevant antibody would be found in primary well C11. The cells in this well could then be cloned and grown in large numbers. A more detailed account of immunoprecipitation analysis can be found in Chapter 10.

8.10.15 Other Assays

A great many other assays are possible. Indeed, any way in which antibodies have been used may be considered. For example, antibodies against hormones could be identified by standard radioimmunoassay techniques in which ^{125}I- or ^{3}H-labelled hormone is precipitated. In many cases, biological assays might be used. For example, antibodies to an enzyme or hormone might be identified by inhibition of its activity (Frackelton and Rotman, 1980; Secher and Burke, 1980). Antibodies which cause passive cutaneous anaphylaxis (PCA) may be identified by subcutaneous injection into rats (Böttcher *et al.*, 1978; Eshhar *et al.*, 1980).

8.11 Cloning

As soon as positive wells are identified, the hybrid cells should be cloned. Cloning is important to reduce the risk of overgrowth by nonproducer cells, and to ensure that the antibodies are truly monoclonal.

Cells may be cloned by growth in soft agar (Coffino *et al.*, 1972; reviewed by Metcalf, 1977). Typically, two layers are used. A firm underlayer, consisting of 0.5% (w/v) agar in culture medium, is allowed to gel. A second 'soft' agar layer (0.3% agar) which contains the cells to be cloned, is added. If feeder cells are used, they may be grown in the dishes beforehand and the culture medium removed just prior to addition of the lower agar layer. Further details are given elsewhere (Köhler, 1979; Goding 1980; Galfrè and Milstein, 1981).

Cloning in soft agar usually requires the additional step of reculturing in liquid medium before antibody production can be assessed. In contrast, cloning by limit dilution (see below) allows direct testing of supernatants. In a few limited cases, however, actively secreting colonies can be identified by haemolytic overlay techniques (Köhler and Milstein, 1975; Köhler, 1979) or formation of immune precipitates surrounding the colonies (Cook and Scharff, 1977).

The advantages of cloning by limiting dilution are so great that detailed protocols for agar cloning will not be given here.

8.11.1 Cloning by Limit Dilution

The theory of cloning by limiting dilution has been described in Section 7.2.1. As the cloning efficiency will seldom be 100% (especially for newly isolated hybrids), cells should be plated at 10, 3 and 0.5 cells per 200 μl well in 96-well plates (30 wells per group). Cloning efficiency will often be improved by the presence of feeder cells (Section 8.9.3). Probably the simplest and most convenient source of feeder cells is mouse or rat thymocytes (10^6/200 μl well), but some investigators use spleen cells or peritoneal cells (about 10^4 per well). Peritoneal cells are particularly rich in macrophages, which are good source of IL-6 and also help clean up cellular debris. They may be particularly desirable for rat hybridomas.

Macroscopic colonies, visible by looking at the under surface of the plate, should become visible 1–2 weeks after cloning is commenced. The group where about half the wells show growth may be assumed to contain single clones. Clones should be assayed as soon as they become visible, but usually survive and grow for at least a week afterwards. In the first cloning, only a minority of wells with growth may be active. Recloning should always be carried out, because cloning by limit dilution does not guarantee clonality.

It is also possible to clone by limiting dilution without performing cell counts. The hybrids are simply plated out in serial twofold dilutions in 200 μl cultures, and the group in which about half the wells have no growth is considered to contain the progeny of single cells. As previously noted, recloning is highly desirable, because the method does not guarantee clonality.

8.12 Large-scale Cultures

Once the active hybrids have been identified and cloned, they must be grown in larger numbers for freezing and antibody production. These steps seldom cause any problems. The cells are gradually moved to larger and larger cultures, taking care to maintain exponential growth. As soon as there are around 10^7 cells, it is wise to set some aside for freezing in liquid nitrogen (see Section 8.13).

It is important to realize that the risk of overgrowth by nonproducer variants never completely ceases, and any stress on the cells (e.g. overgrowth) increases this probability. Adequate frozen stocks are an important insurance against disaster. In addition, it is wise to test the supernatants for antibody at frequent intervals. A falling antibody level is a sign of overgrowth by nonproducers, and is a warning of the need for recloning before it is too late.

Typical antibody levels in the culture supernatant range from 5–50 μg/ml, depending on the individual clone and on cell density. When large amounts of antibody are required, 1-l cultures (e.g. roller bottles) may be set up, gassed with 10% CO_2 and tightly sealed, and allowed to grow until the cells start dying. This will maximize antibody levels.

8.12.1 Newer Methods of Large-scale Culture

Recently, several new methods have been published for the large-scale production of monoclonal antibodies in tissue culture (Pannel and Milstein, 1992; Falkenberg *et al.*, 1995; Jaspert *et al.*, 1995). They generally maintain the hybridoma cells in a small volume, but in contact with a much larger volume of culture medium via a dialysis membrane, and keep the cells in suspension without damaging them, either by the use of an oscillating bubble or tumbling chamber. In this way, the antibodies are allowed to accumulate in a small volume, and cell densities of up to 3×10^7 per ml are possible. The reports indicate a typical antibody concentration of 1–2 mg/ml, which is not much less than that seen in ascites, where the typical antibody levels are 2–10 mg/ml (see Section 8.14).

These methods are likely to gain popularity for many reasons. Of particular importance are the new regulations concerning animal use, which are now valid in most of Europe, where there are restrictions in growth of ascites tumours in mice (maximum weight gain 20%, and only one tapping; Falkenberg *et al.*, 1995). The simplest system is that of Pannel and Milstein (1992), which consists of a roller bottle with a dialysis tube suspended diagonally across it. This apparatus is commercially available (Techne Cambridge Limited, Duxford, Cambridge CB2 4PZ, England). A similar system has been developed by Jaspert *et al.* (1995), and is also commercially available (Geske and Kretzschmar, Parforceheide 15, 14163 Berlin, Germany). The main advantage of this system is that it allows for the continuous supply of gas to the cultures via a permeable membrane, and only requires medium changes every 3–4 days. The system of Falkenberg *et al.*(1995) is also economical and simple, allows the use of a standard CO_2 incubator and does not require complicated pumping devices. It may be anticipated that this apparatus will also become available commercially. The mini PERM Bioreactor (Heraeus Instruments, Germany) is simple and economical and highly recommended.

It can be anticipated that these and similar devices will become increasingly more popular in the next few years, and may eventually supersede the use of ascites tumours for the large-scale production of monoclonal antibodies.

8.13 Storage of Hybrids in Liquid Nitrogen

Hybridoma cells may be frozen in liquid nitrogen and stored for several years with good recovery. Cells may be frozen in sealed glass ampoules, or preferably in NUNC Cryotubes (32 × 12.5, with screw cap; 1077–1A/3–66656; A/S NUNC, Kamstrup-DK-4000, Roskilde, Denmark), which are far more convenient in terms of labelling, recovery of cells and safety.

Aliquots of cells (*c.* 10^7 per vial) are resuspended in 1–2 ml of a mixture of 90% (v/v) fetal calf serum and 10% dimethyl sulfoxide. The DMSO must be added to the FCS and mixed before adding the cells, or else the cells may be killed by high local concentrations of DMSO. Vials should then be placed immediately into a cooling chamber where the rate of cooling is approximately 1°C per minute, particularly in the critical range of 0°C to –30°C. Once the cells reach a temperature of below –60°C, they may be transferred directly into the liquid nitrogen (–196°C) or stored in the vapour phase (–120°C or colder, depending on position and insulation).

Makers of liquid nitrogen containers (e.g. Union Carbide) supply suitable equipment for controlled freezing. The narrow-neck containers may be fitted with 'freezing cones' which possess a rubber ring which can be set to allow the correct rate of freezing. Other types of liquid nitrogen containers have cardboard drawers for storage of the vials in the vapour phase, and simply inserting the vial into a drawer will result in an adequate rate of cooling. It is possible to place a vial in a foam-insulated box inside a –70°C freezer, and transfer the vial to liquid nitrogen a few hours later, but the rate of freezing in such a makeshift apparatus is somewhat uncertain.

Recently, a simple and very economical cell freezing apparatus has been marketed by Nalgene (Cryo 1°C Freezing Container, cat. no. 5100-0001 or 5100-0002), which consists of a simple metal can with foam insulation. It is filled with isopropanol and stored at –70°C. Vials are placed in it and the apparatus is returned to –70°C for at least 4 h without being disturbed. Under these conditions, the vials will cool at about 1°C/min. The vials may then be removed and placed into the liquid nitrogen tank. This apparatus is likely to become very popular.

Storing hybridoma cell lines at –70°C is not recommended because they may retain some unfrozen water and this will permit some chemical reactions to proceed, ultimately resulting in cell death.

Regardless of how the cells are frozen, it is essential to test-thaw a vial to make sure that the cells may be recovered in a viable and sterile state. The vials should be thawed rapidly in a 37°C water bath, taking care to avoid contact of the water with the seal area of the vial. The vial should then be swabbed with 70% ethanol, dried, and opened. The cells should then be immediately recovered by centrifugation, placed in normal medium (RPMI-1640 or Eagle's medium with 10–15% FCS) and placed in culture. It is a good idea to plate the cells in several ten-fold dilutions after thawing, because it is sometimes found that the most concentrated dish grows poorly or not at all.

Even if the cells from the test vial look very healthy under the microscope upon thawing, they should be cultured for a day or two and re-examined, because sometimes a poor freezing will not reveal itself until 1–2 days after thawing. It is not unknown for apparently healthy cells to be all dead the day after thawing.

8.14 Growing of Hybridomas in Animals

The need for histocompatibility for hybridoma growth has already been discussed (see Section 8.1). In most instances, injection of 10^6–10^7 histocompatible hybrid cells into mice or rats will result in tumour formation after 2–4 weeks. Intraperitoneal injection is the preferred route, because it will usually result in the development of ascites in some individuals. The antibody levels in ascites are not greatly different to those in serum, and range from 2–10 mg/ml. However, it is not uncommon to obtain 2–5 ml or more ascites from a single mouse, compared with 0.5–1.0 ml serum. *Serum or ascites containing monoclonal antibodies grown in mice will also contain normal immunoglobulins.* Although this will limit the ultimate degree of purity and specificity that can be obtained, it is not usually a major problem.

The success rate in tumour development, and the probability of ascites formation, are increased by intraperitoneal injection of 0.5 ml pristane (2,6,10,14-tetramethyl pentadecane; Aldrich Chemical Company, Milwaukee, WI, USA) or incomplete Freund's adjuvant a few days prior to injecting the cells (Hoogenraad *et al.*, 1983; Mueller *et al.*, 1986). If tumours fail to appear, sublethal irradiation (350–500 R) prior to injection of the cells may help.

It is important that mice are inspected frequently for tumour growth, and that they are killed painlessly before the tumour reaches a size that causes distress.

Mice may be bled from the retrobulbar sinus (the widely used term retro-orbital sinus is incorrect) or the tail. The blood is allowed to clot, and the tube 'rimmed' with a wooden stick to detach the clot and allow maximal retraction to occur over a period of several hours. The blood is then centrifuged (5000 ***g*** for 5–10 min) and the serum collected. Ascites does not usually clot but it is essential to remove the cells by centrifugation prior to freezing.

Serum or ascites should be examined by electrophoresis on agarose or cellulose acetate strips for the presence of a 'spike' of monoclonal antibody in the gamma globulin region (see Fig. 5.3). It is then stored in small aliquots, preferably at −70°C. It should not be subjected to multiple freeze–thaw cycles, because these may cause denaturation of the antibody, especially IgM. Stability and storage of monoclonal antibodies will be discussed in Chapter 9.

8.15 Mix-ups of Cells

It is extremely easy to contaminate one cell line with another, although most investigators do not believe it until it happens to them. The notorious contamination of cell lines with HeLa cells (Nelson-Rees and Flandermeyer, 1976) should serve as a warning. Only one cell is enough to take over a culture if it grows rapidly.

The risk of contamination is minimized if a few simple precautions are taken. It is preferable to passage only one cell line at a time, keeping the remainder in the incubator. Care should be taken to avoid aerosols. *A pipette that has contacted cells must obviously never be put into a bottle of medium.*

Contamination of cell lines can also occur during passaging of tumours in animals. Instruments should be boiled before use, and cleaned and boiled again after use. Adequate frozen stocks of cells from early passages will help if a mix-up is suspected. Wherever possible, controls should be built into experiments to allow mix-ups to be detected. Serum electrophoresis (see Fig. 5.3) is also useful for detecting mix-ups, because individual monoclonal antibodies have different electrophoretic mobilities.

8.16 Mycoplasma Contamination

8.16.1 Nature of the Problem

Most laboratories around the world working with continuous cell lines have either chronically or periodically suffered from the problem of mycoplasma contamination (Hay *et al.*, 1989; Rottem and Barile, 1993). Any of a large number of species of mycoplasma or the related genus *Acholeplasma* can become chronic contaminants of cell lines without necessarily causing any grossly observable effect on cell growth or morphology. Some random surveys show incidences of infection greater than 50% in U.S. laboratories.

Since such contamination was first recognized some 35 years ago, it has gradually come to be recognized as a significant source of misleading and invalid observations on cultured cells. For example, Vennegoor *et al.* (1982) unintentionally produced monoclonal antibodies to *Mycoplasma hyorhinis* by immunizing with cultured cells that had been contaminated.

There is now a long list of ways in which the organisms interfere with research on cells. The most obvious are biochemical studies, and a number of reported differences between different types of cell lines proved subsequently to be caused by differences between contaminated and noncontaminated cells. Infection has been shown to affect the ability of cells to form hybrids, and it may cause complete failure of hybridoma production. Mycoplasma infection may also affect the cloning efficiency in culture and the ability to form tumours *in vivo*.

The current belief of most workers who specialize in mycoplasma–cell relationships is that the common primary sources of contamination are bovine (fetal calf serum), porcine (possibly from trypsin) and human, but not the tissue of origin of the cells. Many mycoplasmas are small and supple enough to pass through even 0.22 μm filters. Better quality control by the commercial suppliers of fetal calf serum and routine heating of serum (56°C for 1 h) before use seems to be progressively reducing the incidence of fresh contaminations of bovine origin.

Some epidemiological studies of individual tissue culture laboratories suggest that existing lines are the major source of new contaminations. Another potential source might be laboratory workers handling the cultures. Mouth pipetting should never be used for cultured cell lines. Contaminated lines have not been recognized as a biohazard either to laboratory personnel or to mice. They do, however, represent a threat to cell lines that are presumed to be clean, and will earn the laboratory no gratitude if they are sent to other laboratories.

For a laboratory which has a clean set of lines it is important to identify any newly contaminated lines or imported contaminated lines and to eliminate them promptly. A programme of regular testing of existing lines and immediate testing of imports is highly desirable. Complete cure of infections is usually difficult to achieve, and therefore not worth the effort unless there is no alternative.

8.16.2 Detection of Mycoplasma Contamination

Previously, detection of mycoplasma was difficult and unreliable (Uphoff *et al.*, 1992). The advent of the polymerase chain reaction (PCR) has revolutionized the detection of mycoplasma, and is now the method of choice (Rawadi and Dussurget, 1995). The method is easy and extremely sensitive. It requires a thermal cycling machine and these are now common in biological laboratories.

It is generally recommended that cells should be grown in antibiotic-free medium for a week or so prior to testing, to avoid the possibility that the antibiotics reduce the level of contamination to below the limit of detectability without eliminating it.

The following method is based on the procedure of Callaghan and Goding (unpublished data). It is a simplified version of the procedure of Hopert *et al.* (1993), and detects highly conserved mycoplasma ribosomal RNA genes. It is capable of detecting less than five molecules of rRNA gene DNA from *M. arginini*, *M. orale*, *M. hyorhinis*, *M. pneumoniae*, *M. bovine* group PG50, *M. californicum*, *M. gallisepticum*, *M. bovis*, *M. mycoides*, *M. iowae*, *M. fermentans* and *A. laidlawii* (Weisberg *et al.*, 1989).

The DNA is isolated from the cells as follows. The cells are harvested by centrifugation and DNA isolated by InstaGene™ (Bio-Rad). Take 1 ml of cell suspension, and centrifuge at 12 000 r.p.m. in an Eppendorf centrifuge or equivalent for 1 min. Wash the cells once with PBS and adjust the cells to about 3×10^4/ml in autoclaved water. Take 20 μl of this suspension and add 200 μl InstaGene matrix. Incubate at 56°C for 30 min, followed by 8 min at 100°C. Centrifuge at 12 000 r.p.m. for 3 min, and transfer the supernatant containing the DNA to a fresh tube. The DNA can be stored at –20°C indefinitely. Use 20 μl (about 50 cells) per reaction.

A 10 × PCR buffer is usually provided with the *Taq* polymerase. The

Table 8.1 Cocktail for PCR

Solution	For two reactions	For 20 reactions
25 mM $MgCl_2$	6 µl (final concentration 1.5 mM)	60 µl
10 × PCR Buffer (Promega)	10 µl	100 µl
dNTPs (1.25 mM each)	16 µl (final concentration 0.8 mM total)	160 µl
5′ primer*	50 pmol in 1 µl TE†	500 pmol in 10 µl TE
3′ primer*	50 pmol in 1 µl TE	500 pmol in 10 µl TE
Nuclease-free water	26 µl	260 µl

* See Table 8.2 for the sequence of primers.
† TE is 10 mM Tris-HCl, 1 mM EDTA, pH 8.0, made up in sterile water and autoclaved.

Promega 10 × buffer consists of 500 mM KCl, 100 mM Tris-HCl (pH 9.0 at 25°C), and 1% Triton X-100.

A 'cocktail' for PCR can be made up as shown in Table 8.1.

The PCR is set up as follows. To 30 µl PCR cocktail (Tables 8.1 and 8.2), add the cellular DNA in 20 µl water, followed by 1 unit of *Taq* polymerase. Overlay the sample with mineral oil to prevent evaporation and refluxing, which would reduce the temperatures.

The reaction consists of 5 min at 94°C, followed by 55°C for 1.45 min, followed by 40 cycles of extension at 72°C for 3 min, denaturation at 94°C for 45 s and annealing at 55°C for 45 s. At the end, the sample is held at 72°C for 10 min to ensure that all strands are completed (optional).

Analyse 20 µl of the product on a 2% agarose gel in the presence of ethidium bromide, and inspect on an ultraviolet light box (305 nm) and photograph. A positive result will be indicated by a band at about 500 bp.

As for all PCR, it is essential to take stringent precautions to prevent contamination, as the assay is extremely sensitive. The PCR should be set up using aerosol-resistant pipette tips, preferably in a room that is separate from that where the products will be analysed. It is essential to include negative controls, such as DNA from a noncontaminated line and a 'no DNA' control.

8.16.3 Elimination of Mycoplasma

Elimination of mycoplasma is not easy, and should only be attempted if the cell line is particularly valuable. The organisms rapidly acquire resistance to antibiotics. Probably the best reagent currently available for treatment of mycoplasma-contaminated cell lines is Ciprofloxacin, marketed as Mycoplasma Removal Agent (MRA) by ICN Biomedicals, Inc, PO Box 5023, Costa Mesa, California 92626 (Mowles, 1988; Mowles *et al.*, 1989; Gignac *et al.*, 1991, 1992). Ciprofloxacin is a member of the 4-quinolone group of antibiotics, and inhibits DNA gyrase. It is said to be very effective, but a cure cannot be guaranteed.

Table 8.2 Sequences of Mycoplasma 16S ribosomal RNA genes and PCR primers

	STRAIN	5′ END (5′ → 3′)		3′ END (5′ → 3′)		
		1	10	501		523
1	M arginini (521 bp; M24579)	CGCCTGAGTA	GTATGCTCGC	TTCTCGGGTC	TTGTACACAC	CGC
2	M orale (519 bp; M24659)	CGCCTGAGTA	GTATGCTCGC	TTCTCGGGTC	TTGTACACAC	CGC
3	M hyorhinis (517 bp; M24658)	CGCCTGAGTA	GTATGCNNNN	TTCTCGGGTT	TTGTACACAC	CGC
4	M pneumoniae (504 bp; M29061)	CGCCTGGGTA	GTACATTCGC	TTCTCGGGTC	TTGTACACAC	CGC
5	M bovinepg50 (518 bp; M10588)	CGCCTGAGTA	GTATGCTCGC	TTCTCGGGTC	TTGTACACAC	CGC
6	M californicum (518 bp; M24582)	CGCCTGAGTA	GTACGTTCGC	TTCTCGGGTC	TTGTACACAC	CGC
7	M gallisepticum (504 bp; M22441)	TGCCTGAGTA	GTACATTCGC	TTCTCGGGTC	TTGTACACAC	CGC
8	M bovis (518 bp; M94728)	CGCCTGCGTA	GTATGCTCGC	TTCTCGGGTC	TTGTACACAC	CGC
9	M mycoides (518 bp; M23943)	CGCCTGAGTA	GTATGCTCGC	TTCTCGGGTC	TTGTACACAC	CGC
10	M iowae (595 bp; X55271)	CGCCTGGGTA	GTACATTCGC	TTCTCGGGTC	TTGTACACAC	CGC
11	My fermentans (518 bp; M24289)	CGCCTGAGTA	GTACGTTCGC	TTCTCGGGTC	TTGTACACAC	CGC
12	A laidlawii (519 bp; M23932)	CGCCTGAGTA	GTACGTACGC	TTCTCGGGGT	TTGTACACAC	CGC
	Mycoplasma consensus primers	5′TGCCTGGGTA C A C	GTATGCACGC → 3′ CATT	3′ ← AGCCCAA CG	AACATGTGTG	GCG 5′
	Mouse preproinsulin primers	5′ PREPROINSULIN PRIMER CGAGCTCGAGCCTGCCTATCTTTCAGGTC 5′ 3′		3′ PREPROINSULIN PRIMER CGGGATCCTAGTTGCAGTAGTTCTCCAG 5′ 3′		

The *Mycoplasma* consensus primers are based on conserved regions of the 16S ribosomal RNA gene of *Mycoplasma*, and are designed to detect all *Mycoplasma* species. The mouse preproinsulin primers may be used as positive controls on mouse cells, but are not suitable for human cells.

References

Al-Kaissi, E. and Mostratos, A. (1983) Assessment of substrates for horseradish peroxidase in enzyme immunoassay. *J. Immunol. Methods* **58**, 127–132.

Bazin, H. (1982) Production of rat monoclonal antibodies with the LOU rat non-secreting IR983F myeloma cell line. *In* 'Protides of the Biological Fluids' (H. Peeters, ed.), 29th Colloquium, pp. 615–618. Pergamon Press, Oxford.

Bazin, H. (1990a) 'Rat Hybridomas and Rat Monoclonal Antibodies'. CRC Press, Boca Raton.

Bazin, H. (1990b) The LOUVAIN (LOU) rats. *In* 'Rat Hybridomas and Rat Monoclonal Antibodies' (H. Bazin, ed.), pp. 43–52. CRC Press, Boca Raton.

Bazin, H. and Lemieux, R. (1987) Role of the macrophage-derived hybridoma growth factor in the *in vitro* and *in vivo* proliferation of newly formed B cell hybridomas. *J. Immunol.* **139**, 780–787.

Bazin, H. and Lemieux, R. (1989) Increased proportion of B cell hybridomas secreting monoclonal antibodies of desired specificity in cultures containing macrophage-derived growth factor (IL-6) *J. Immunol. Methods* **116**, 245–249.

Bazin, H., Beckers, A., Deckers, C. and Moriame, M. (1973) Transplantable immunoglobulin-secreting tumors in rats. V. Monoclonal immunoglobulins secreted by 250 ileocecal immunocytomas in LOU/WsI rats. *J. Natl Cancer Inst.* **51**, 1359–1361.

Bazin, H., Pear, W. S., Klein, G. and Sümegi, J. (1990) Rat immunocytomas (IR). *In* 'Rat Hybridomas and Rat Monoclonal Antibodies' (H. Bazin, ed.), pp. 53–68. CRC Press, Boca Raton.

Blankenstein, T., Qin, Z, Li, W. and Diamantstein, T. (1990) DNA rearrangement and constitutive expression of the interleukin 6 gene in a mouse plasmacytoma. *J. Exp. Med.* **171**, 965–970.

Böttcher, I., Hämmerling, G. and Kapp, J.-F. (1978) Continuous production of monoclonal mouse IgE antibodies with known allergenic specificity by a hybrid cell line. *Nature* **275**, 761–762.

Boyse, E. A., Hubbard, L., Stockert E. and Lamm, M. E. (1970) Improved complementation in the cytotoxic test. *Transplantation* **10**, 446–449.

Bradley, L. M. and Shiigi, S. M. (1980) Secondary immunization to nitrophenyl haptens. *In* 'Selected Methods in Cellular Immunology' (B. B. Mishell and S. M. Shiigi, eds), pp. 45–47. Freeman, San Franciso.

Brown, J. P., Wright, P. W., Hart, C. E., Woodbury, R. G., Hellström, K. E. and Hellström, I. (1980) Protein antigens of normal and malignant human cells identified by immunoprecipitation with monoclonal antibodies. *J. Biol. Chem.* **255**, 4980–4983.

Burtonboy, G., Bazin, H., Deckers, C., Beckers, A., Lamy, M. and Heremans, J. F. (1973) Transplantable immunoglobulin-secreting tumors in rats. 3. Establishment of immunoglobulin-secreting cell lines from LOU-Ws1 strain rats. *Eur. J. Cancer* **9**, 259–262.

Butler, J. E., Ni, L., Joshi, K. S., Chang, J., Rosenberg, B. and Voss, E. W. (1993) The immunochemistry of sandwich ELISAs. VI. Greater than 90% of monoclonal and 75% of polyclonal anti-fluorescyl capture antibodies (CAbs) are denatured by passive adsorption. *Mol. Immunol.* **30**, 1165–1175.

Calabresi, P. and Chabner, B. A. (1990) Antineoplastic agents. *In* 'The Pharmacological Basis of Therapeutics' (A. G. Gilman, T. W. Rall, A. S. Nies and P. Taylor, eds) 8th edition, pp 1209–1227. McGraw-Hill, New York.

Catt, K. J. and Tregear, G. (1967) Solid-phase radioimmunoassay in antibody-coated tubes. *Science* **158**, 1570–1572.

Chapel, H. M. and August, P. J. (1976) Report of nine cases of accidental injury to man with Freund's complete adjuvant. *Clin. Exp. Immunol.* **24**, 538–541.

Coffino, P., Baumal, R., Laskov, R. and Scharff, M. (1972) Cloning of mouse myeloma cells and detection of rare variants. *J. Cell Physiol.* **79**, 429–440.

Cook, W. D. and Scharff, M. D. (1977) Antigen-binding mutants of mouse myeloma cells. *Proc. Natl Acad. Sci. USA* **74**, 5687–5691.

Cox, J. C. and Coulter, A. R. (1992) Advances in adjuvant technology and application. *In* 'Animal Parasite Control Utilizing Biotechnology' (W. K. Yong, ed.). CRC Press, Boca Raton.

Croce, C. M., Linnenbach, A., Hall, A., Steplewski, Z. and Koprowski, H. (1980) Production of human hybridomas secreting antibodies to measles virus. *Nature* **288**, 488–489.

Culvenor, J. G., Harris, A. W., Mandel, T. E., Whitelaw, A. and Ferber, E. (1981) Alkaline phosphatase in hematopoietic tumor cell lines of the mouse: high activity in cells of the B lymphoid lineage. *J. Immunol.* **126**, 1974–1977.

Darveau, A., Chevrier, M-C., Neron, S., Delage, R. and Lemieux, R. (1993) Efficient preparation of human monoclonal antibody-secreting heterohybridomas using peripheral B lymphocytes cultured in the CD40 system. *J. Immunol. Methods* **159**, 139–143.

Douillard, J. Y., Hoffman, T. and Herberman, R. B. (1980) Enzyme-linked immunosorbent assay for screening monoclonal antibody production: use of intact cells as antigen. *J. Immunol. Methods* **39**, 309–316.

Dresser, D. W. (1962) Specific inhibition of antibody production. II. Paralysis induced in adult mice by small quantities of antigen. *Immunology* **5**, 378–388.

Edwards, P. A. W., Smith, C. M., Neville, A. M. and O'Hare, M. J. (1982) A human-human hybridoma system based on a fast-growing mutant of the ARH-77 plasma cell leukemia-derived line. *Eur. J. Immunol.* **12**, 641–647.

Engvall, E. (1980) Enzyme immunoassay ELISA and EMIT. *Methods Enzymol.* **70**, 419–438.

Engvall, E. and Pesce, A. J. (1978) Quantitative Enzyme Immunoassay. *Scand. J. Immunol.* **8** (Suppl. 7).

Eshhar, Z., Ofarim, M. and Waks, T. (1980) Generation of hybridomas secreting murine reaginic antibodies at anti-DNP specificity. *J. Immunol.* **124**, 775–780.

Everson, L. K., Buell, B. N. and Rogentine, G. N. (1973) Separation of human lymphoid cells into G_1, S and G_2 cell cycle populations by use of a velocity sedimentation technique. *J. Exp. Med.* **137**, 343–358.

Ey, P. L., Prowse, S. J. and Jenkin, C. R. (1978) Isolation of pure IgG1, IgG2a and IgG2b immunoglobulins from mouse serum using protein A-Sepharose. *Immunochemistry* **15**, 429–436.

Falkenberg, F. W., Weichert, H., Krane, M., Bartels, I., Palme, M., Nagels, H.-O. and Fiebig, H. (1995) *In vitro* production of monoclonal antibodies in high concentration in a new and easy to handle modular fermenter. *J. Immunol. Methods* **179**, 13–29.

Farr, A. G. and Nakane, P. K. (1981) Immunohistochemistry with enzyme labeled antibodies: a brief review. *J. Immunol. Methods* **47**, 129–144.

Fazekas de St. Groth, S. and Scheidegger, D. (1980) Production of monoclonal antibodies: strategy and tactics. *J. Immunol. Methods* **35**, 1–21.

Frackelton, A. R. and Rotman, B. (1980) Functional diversity of antibodies elicited by bacterial β-D-galactosidase. *J. Biol. Chem.* **255**, 5286–5290.

Freshney, R. I. (1993) 'Culture of Animal Cells'. Wiley/Liss, New York.

Galfrè, G. and Milstein, C. (1981) Preparation of monoclonal antibodies: strategies and procedures. *Methods Enzymol.* **73**, 1–46.

Galfrè, G., G. Howe, S. C., Milstein, C., Butcher, G. W. and Howard, J. C. (1977) Antibodies to major histocompatibility antigens produced by hybrid cell lines. *Nature* **266**, 550–552.

Galfrè, G., Milstein, C. and Wright, B. (1979) Rat × rat hybrid myelomas and a monoclonal anti-Fd portion of mouse IgG. *Nature* **277**, 131–133.

Garcia-Rozas, C., Plaze, A., Diaz-Espada, F., Kreisler, M. and Martinez-Alonso, C. (1982) Alkaline phosphatase activity as a membrane marker for activated B cells. *J. Immunol.* **129**, 52–55.

Gatti, R. A., Östborn, A. and Fagraeus, A. (1974) Selective impairment of cell antigenicity by fixation. *J. Immunol.* **113**, 1361–1368.

Gefter, M. L., Margulies, D. H. and Scharff, M. D. (1977) A simple method for polyethylene glycol-promoted hybridization of mouse myeloma cells. *Somat. Cell. Genet.* **3**, 231–236.

Gignac, S. M., Brauer, S., Hane, B., Quentmeier, H. and Drexler, H. G. (1991) Elimination of mycoplasma from infected leukemia cell lines. *Leukemia* **5**, 162–165.

Gignac, S. M., Uphoff, C. C., MacLeod, R. A. F., Steube, K., Voges, M. and Drexler, H. G. (1992) Treatment of mycoplasma-contaminated continuous cell lines with mycoplasma removal agent (MRA) *Leukaemia Res.* **16**, 815–822.

Goding, J. W. (1978) Use of staphylococcal protein A as an immunological reagent. *J. Immunol. Methods* **20**, 241–253.

Goding, J. W. (1980) Antibody production by hybridomas. *J. Immunol. Methods* **39**, 285–308.

Goding, J. W. (1982) Asymmetrical surface IgG on MOPC-21 plasmacytoma cells contains one membrane heavy chain and one secretory heavy chain. *J. Immunol.* **128**, 2416–2421.

Gray, B. M. (1979) ELISA methodology for polysaccharide antigens: protein coupling of polysaccharides for adsorption to plastic tubes. *J. Immunol. Methods* **28**, 187–192.

Griffin, F. M., Ashland, G. and Capizzi, R. L. (1981) Kinetics of phototoxicity of Fischer's medium for L5178Y leukemic cells. *Cancer Res.* **41**, 2241–2248.

Hager, D. A. and Burgess, R. (1980) Elution of proteins from sodium dodecyl sulfate-polyacrylamide gels, removal of sodium dodecyl sulfate, and renaturation of enzymatic activity: results with sigma subunit of *Escherichia coli* RNA polymerase, wheat germ DNA topoisomerase, and other enzymes. *Anal. Biochem.* **109**, 76–86.

Hämmerling, G. J., Hämmerling, U. and Kearney, J. F. (1981) 'Monoclonal Antibodies and T-Cell Hybridomas', p. 569. Elsevier/North-Holland, Amsterdam.

Handman, E. and Jarvis, H. M. (1985) Nitrocellulose-based assays for the detection of glycolipids and other antigens: Mechanism of binding to nitrocellulose. *J. Immunol. Methods* **83**, 113–123.

Handman, E. and Remington, J. (1980) Serological and immunochemical characterisation of monoclonal antibodies to *Toxoplasma gondii*. *Immunology*, **40**, 579–588.

Harahap, A. R. and Goding, J. W. (1988) Distribution of the murine plasma cell antigen PC-1 in non-lymphoid cells. *J. Immunol.* **141**, 2317–2320.

Harrach, B. and Robenek, H. (1991) Immunocytochemical localization of apolipoprotein A-I using polyclonal antibodies raised against the formaldehyde-modified antigen. *J. Microscopy* **161**, 97–108.

Harris, J. F., Hawley, R. G. and Crawford-Sharpe, G. C. (1992) Increased frequency of both total and specific monoclonal antibody producing hybridomas using a fusion partner that constitutively expresses recombinant IL-6. *J. Immunol. Methods* **148**, 199–207.

Hay, R. J., Macy, M. L. and Chen, T. R. (1989) Mycoplasma infection of cultured cells. *Nature* **339**, 487–488.

Heeneman, S., Deutz, N. E. P. and Buurman, W. A. (1993) The concentrations of glutamine and ammonia in commercially available cell culture medium. *J. Immunol. Methods* **166**, 85–91.

Hengartner, H., Luzzati, A. L. and Schreier, M. (1978) Fusion of *in vitro* immunized lymphoid cells with X63Ag8. *Current Topics Microbiol. Immunol.* **81**, 92–99.

Hewish, D. R. and Werkmeister, J. A. (1989) The use of an electroporation apparatus for the production of murine hybridomas. *J. Immunol. Methods* **120**, 285–289.

Hoogenraad, N., Helman, T. and Hoogenraad, J. (1983) The effect of pre-injection of mice with pristane on ascites tumour formation and monoclonal antibody production. *J. Immunol. Methods* **61**, 317–320.

Hopert, A., Uphoff, C. C., Wirth, M., Hauser, H. and Drexler, H. G. (1993) Specificity and sensitivity of polymerase chain reaction (PCR) in comparison with other methods for the detection of mycoplasma contamination in cell lines. *J. Immunol. Methods* **164**, 91–100.

Howard, F. D., Ledbetter, J. A., Mehdi, S. Q. and Herzenberg, L. A. (1980) A rapid method for the detection of antibodies to cell surface antigens: a solid phase radioimunoassay using cell membranes. *J. Immunol. Methods* **38**, 75–84.

Hudson, L. and Hay, F. C. (1976) 'Practical Immunology.' Blackwell, Oxford.

Hurn, B. A. L. and Chantler, S. M. (1980) Production of reagent antibodies. *Methods Enzymol.* **70**, 104–142.

Jakoby, W. B. and Pastan, I. H. (1979) Cell culture. *Methods Enzymol.* LVIII.

Jaspert, R., Geske, T., Teichmann, A., Kabner, Y-M., Kretzschmar, K. and L'age-Stehr, J. (1995) Laboratory scale production of monoclonal antibodies in a tumbling chamber. *J. Immunol. Methods* **178**, 77–87.

Johnson, A. M., McNamara, P. J., Neoh, S. H., McDonald, P. J. and Zola, H. (1981) Hybridomas secreting monoclonal antibody to Toxoplasma gondii. *Aust. J. Exp. Biol. Med. Sci.* **59**, 303–306.

Johnson, D. A., Gautsch, J. W., Sportsman, J. R. and Elder, J. H. (1984) Improved technique utilizing nonfat dry milk for analysis of proteins and nucleic acids transferred to nitrocellulose. *Gene Anal. Techn.* **1**, 3–8.

Karpas, A., Fischer, P. and Swirsky, D. (1982a) Human myeloma cell line carrying a Philadelphia chromosome. *Science* **216**, 997–999.

Karpas, A., Fischer, P. and Swirsky, D. (1982b) Human plasmacytoma with an unusual karyotype growing in vitro and producing light-chain immunoglobulin. *Lancet* **i**, 931–933.

Kearney, J. F., Radbruch, A., Liesengang, B. and Rajewsky, K. (1979) A new mouse myeloma cell line that has lost immunoglobulin expression but permits the construction of antibody-secreting hybrid cell lines. *J. Immunol.* **123**, 1548–1550.

Kennett, R. H. (1980) Enzyme-linked antibody assay with cells attached to polyvinyl chloride plates. *In* 'Monoclonal Antibodies' (R. H. Kennett, T. J. McKearn and K. B. Bechtol, eds), pp. 376–377. Plenum Press, New York.

Kilmartin, J. V., Wright, B. and Milstein, C. (1982) Rat monoclonal antitubulin antibodies derived by using a new nonsecreting rat cell line. *J. Cell Biol.* **93**, 576–582.

Kincade, P. W., Lee, Sun, L. and Watanabe, T. (1981) Monoclonal rat antibodies to murine IgM determinants. *J. Immunol. Methods* **42**, 17–26.

Köhler, G. (1979) Soft agar cloning of lymphoid tumor lines: detection of hybrid clones with anti-SRBC activity. *In* 'Immunological Methods' (I. Lefkovits, and B. Pernis, eds), pp. 397–401. Academic Press, London and Orlando.

Köhler, G. and Milstein, C. (1975) Continuous cultures of fused cells secreting antibody of predefined specificity. *Nature* **256**, 495–497.

Kudo, T., Saeki, H. and Tachibana, T. (1991) A simple and improved method to generate human hybridomas. *J. Immunol. Methods* **145**, 119–125.

Ledbetter, J. A. and Herzenberg, L. A. (1979) Xenogeneic monoclonal antibodies to mouse lymphoid differentiation antigens. *Immunol. Rev.* **47**, 63–90.

Lee, S-K., Singh, J. and Taylor, R. B. (1975) Subclasses of T cells with different sensitivities to cytotoxic antibody in the presence of anesthetics. *Eur. J. Immunol.* **5**, 259–262.

Lernhardt, W., Anderson, J., Coutinho, A. and Melchers, F. (1978) Cloning of murine transformed cell lines in suspension culture with efficiencies near 100%. *Exp. Cell. Res.* **111**, 309–316.

Lo, M. S., Tsong, T. Y., Contad, M. K., Strittmatter, S. M., Hester, L. D. and Snyder, S. H. (1984) Monoclonal antibody production by receptor-mediated electrically induced cell fusion. *Nature* **310**, 792–794.

Loken, M. R. and Stall, A. M. (1982) Flow cytometry as an analytical and preparative tool in immunology. *J. Immunol. Methods* **50**, 85–112.

Luben, R. A. and Mohler, M. A. (1980) In vitro immunization as an adjunct to the production of hybridomas producing antibodies against the lymphokine osteoclast activating factor. *Mol. Immunol.* **17**, 635–639.

Luben, R. A., Brazeau, P., Böhlen, P. and Guillemin, R. (1982) Monoclonal antibodies to hypothalamic growth hormone-releasing factor with picomoles of antigen. *Science* **218**, 887–889.

Maggio, E. T. (1980) 'Enzyme-immunoassay'. CRC Press, Florida.

McKearn, T. J. (1980) ^{51}Cr-release cytotoxicity assay. *In* 'Monoclonal Antibodies' (R. H. Kennett, T. J. McKearn and K. B. Bechtol, eds), pp. 393–394. Plenum Press, New York.

Metcalf, D. (1977) 'Hemopoietic Colonies. *In vitro* cloning of normal and leukemic cells', *Recent Results in Cancer Research, Vol. 61*. Springer-Verlag, Berlin.

Mowles, J., Moran, S. and Doyle, A. (1989) Mycoplasma control. *Nature* **340**, 352.

Mowles, J. M. (1988) The use of ciprofloxacin for the elimination of mycoplasma from naturally infected cell lines. *Cytotechnology* **1**, 355–358.

Mueller, V. W., Hawes, C. S. and Jones, W. R. (1986) Monoclonal antibody production by hybridoma growth in Freund's adjuvant primed mice. *J. Immunol. Methods* **87**, 193–196.

Nelson-Rees, W. and Flandermeyer, N. R. (1976) HeLa cultures defined. *Science* **191**, 96–98.

Nilsson, K. (1971) Characteristics of established myeloma and lymphoblastoid cell lines derived from an E myeloma patient: a comparative study. *Int. J. Cancer* **7**, 380–396.

Nilsson, K., Bennich, H., Johansson, S. G. O. and Ponten, J. (1970) Established immunoglobulin producing myeloma (IgE). *Clin. Exp. Immunol.* **7**, 477–489.

Nordan, R. P. and Potter, M. (1986) A macrophage-derived factor required by plasmacytomas for survival and proliferation *in vitro*. *Science* **233**, 566–569.

Norwood, T. H., Zeigler, C. J. and Martin, G. M. (1976) Dimethyl sulfoxide enhances polyethylene glycol-mediated cell fusion. *Somat. Cell Genet.* **2**, 263–270.

Nowinski, R. C., Boglund, J., Lane, M., Lostrum, M., Bernstein, I., Young, W., Hakomori, S., Hill, L. and Cooney, M. (1980) Human monoclonal antibody against Forsman's antigens. *Science* **210**, 537–539.

Nowinski, R. C., Lostrom, M. E., Tam, M. R., Stone, M. R. and Burnette, W. N. (1979) The isolation of hybrid cell lines producing monoclonal antibodies against the p15(E) protein of ecotropic murine leukemia viruses. *Virology* **93**, 111–126.

O'Sullivan, M. J., Gnemmi, E., Simmonds, A. D., Chieregatti, G., Heyderman, E., Bridges, J. W. and Marks, V. (1979) A comparison of the ability of β-galactosidase and horseradish peroxidase enzyme-antibody conjugates to detect specific antibodies. *J. Immunol. Methods* **31**, 247–250.

Oi, V. T. and Herzenberg, L. A. (1980) Immunoglobulin-producing hybrid cell lines. *In* 'Selected methods in Cellular Immunology' (B. B. Mishell and S. M. Shiigi, eds), pp. 351–372. Freeman, San Francisco.

Oi, V. T., Jones, P. P., Goding, J. W., Herzenberg, L. A. and Herzenberg, L. A. (1978) Properties of monoclonal antibodies to mouse Ig allotypes, H-2 and Ia antigens. *Curr. Top. Microbiol. Immunol.* **81**, 115–129.

Olsson, L. and Kaplan, H. S. (1980) Human–human hybridomas producing monoclonal antibodies of predefined antigenic specificity. *Proc. Natl Acad. Sci. USA* **77**, 5429–5434.

Ormerod, M. G. (1990) 'Flow Cytometry: A Practical Approach'. IRL Press, Oxford.

Pannel, R. and Milstein, C. (1992) An oscillating bubble chamber for laboratory scale production of monoclonal antibodies as an alternative to ascitic tumours. *J. Immunol. Methods* 146, 43–48.

Parkhouse, R. M. E. and Guarnotta, G. (1978) Rapid binding test for detection of alloantibodies to lymphocyte surface antigens. *Curr. Top. Microbiol. Immunol.* **81**, 142.

Parks, D. R., Bryan, V. M., Oi, V. T. and Herzenberg, L. A. (1979) Antigen-specific identification and cloning of hybridomas with a fluorescence-activated cell sorter. *Proc. Natl Acad. Sci. USA* **76**, 1962–1966.

Phillips, D. R. and Morrison, M. (1970) The arrangement of proteins in the human erythrocyte membrane. *Biochem. Biophys. Res. Commun.* **40**, 284–289.

Pierce, S. K. and Klinman, N. (1976) Allogeneic carrier-specific enhancement of hapten-specific secondary B-cell responses. *J. Exp. Med.* **144**, 1254–1262.

Ponder, B. A. and Wilkinson, M. M. (1981) Inhibition of endogenous tissue alkaline phosphatase with the use of alkaline phosphatase conjugates in immunohistochemistry. *J. Histochem. Cytochem.* **29**, 981–984.

Posner, M. R., Antoniou, D., Griffin, J., Schlossman, S. F. and Lazarus, H. (1982) An enzyme-linked immunosorbent assay (ELISA) for the detection of monoclonal antibodies to cell surface antigens on viable cells. *J. Immunol. Methods* **48**, 23–31.

Ravoet, A. M. and Bazin, H. (1990) Fusion procedure. *In* 'Rat Hybridomas and Rat Monoclonal Antibodies' (H. Bazin, ed.), pp. 87–95. CRC Press, Florida.

Rawadi, G. and Dussurget, O. (1995) Advances in PCR-based detection of mycoplasmas contaminating cell cultures. *PCR Methods Applications* **4**, 199–208.

Reedman, B. M. and Klein, G. (1973) Cellular localization of an Epstein–Barr virus (EBV)-associated complement-fixing antigen in producer and non-producer lymphoblastoid cell lines. *Int. J. Cancer* **11**, 499–520.

Rottem, S. and Barile, M. F. (1993) Beware of mycoplasmas. *Trends Biotechnol.* **11**, 143–151.

Secher, D. S. and Burke, D. C. (1980) A monoclonal antibody for large-scale purification of human leucocyte interferon. *Nature* **285**, 446–450.

Shamma, S., Mantovani, T. L., Dalla-Favera, R. and Casali, P. (1993) Generation of human monoclonal antibodies by transformation of lymphoblastoid B cells with *ras* oncogene. *J. Immunol. Methods*. **160**, 19–25.

Shapiro, H. M. (1994) 'Practical Flow Cytometry', 3rd edn. Wiley-Liss, New York.

Sharon, J., Morrison, S. L. and Kabat, E. A. (1980) Formation of hybridoma clones in soft agarose: effect of pH and medium. *Somat. Cell. Genet.* **6**, 435–441.

Smith, A. D. and Wilson, J. E. (1986) A modified ELISA that selectively detects monoclonal antibodies recognizing native antigen. *J. Immunol. Methods*. 94, 31–35.

Smolarsky, M. (1980) A simple radioimmunoassay to determine binding of antibodies to lipid antigens. *J. Immunol. Methods* **38**, 85–93.

Springer, T. A. (1980) *Cell*-surface differentiation in the mouse. Characterization of 'jumping' and 'lineage' antigens using xenogeneic rat monoclonal antibodies. *In* 'Monoclonal Antibodies' (R. H. Kennett, T. J. McKearn and K. B. Bechtol, eds), pp. 185–217. Plenum Press, New York.

Staehelin, T., Durrer, B., Schmidt, J., Takacs, B., Stocker, J., Miggiano, V., Stahli, C., Rubinstein, M., Levy, W. P., Hershberg, R. and Pestka, S. (1981) Production at hybridomas secreting monoclonal antibodies to the human leukocyte interferons. *Proc. Natl Acad. Sci. USA* **78**, 1848–1852.

Stähli, C., Staehelin, T., Miggiano, V., Schmidt, J. and Haring, P. (1980) High frequencies of antigen-specific hybridomas: dependence on immunization parameters and prediction by spleen cell analysis. *J. Immunol. Methods* **32**, 297–304.

Stenger, D. A., Kubiniec, R. T., Purucker, W. J., Liang, H. and Hui, S. W. (1988) Optimization of electric parameters for efficient production of murine hybridomas. *Hybridoma* **7**, 505–518.

Stewart-Tull, D. E. S. (1995) 'The Theory and Practical Application of Adjuvants'. John Wiley and Sons, New York.

Stitz, L., Hengartener, H., Althage, A. and Zinkernagel, R. M. (1988) An easy and rapid method to screen large numbers of antibodies against internal cellular determinants. *J. Immunol. Methods* **106**, 211–216.

Stocker, J. W. and Heusser, C. H. (1979) Methods for binding cells to plastic: application to a solid-phase radioimmunoassay for cell-surface antigens. *J. Immunol. Methods* **26**, 87–95.

Stoien, J. D. and Wang, J. (1974) Effect of near-ultraviolet and visible light on mammalian cells in culture. II. Formation of toxic photoproducts in tissue culture medium by blacklight. *Proc. Natl Acad. Sci. USA* **71**, 3961–3965.

Taylor, R. B., Duffus, W. P. H., Raff, M. C. and de Petris, S. (1971) Redistribution and pinocytosis of lymphocyte surface immunoglobulin molecules by anti-immunoglobulin antibody. *Nature* (New Biol.) **233**, 225–229.

Tijssen, P. and Kurstak, E. (1984) Highly efficient and simple methods for the preparation of peroxidase and active peroxidase-antibody conjugates for enzyme immunoassays. *Anal. Biochem.* **136**, 451–457.

Trowbridge, I. S. (1978) Interspecies spleen-myeloma hybrid producing monoclonal antibodies against mouse lymphocyte surface glycoprotein T200. *J. Exp. Med.* **148**, 313–323.

Trucco, M. M., Stocker, J. W. and Ceppellini, R. (1978) Monoclonal antibodies against human lymphocyte antigens. *Nature* **273**, 666–668.

Tsu, T. T. and Herzenberg, L. A. (1980) Solid-phase radioimunoassays. *In* 'Selected Methods in Cellular Immunology' (B. B. Mishell and S. M. Shiigi, eds), pp. 373–397. Freeman, San Francisco.

Uphoff, C. C., Brauer, S., Grunicke, D., Gignac, S. M., MacLeod, R. A. F., Quentmeier, H., Steube, K., Tummier, M., Voges, M., Wagner, B. and Drexler, H. G. (1992) Sensitivity and specificity of five different mycoplasma detection assays. *Leukemia* **6**, 335–341.

Vennegoor, C., Polak-Vogelzang, A. A. and Hekman, A. (1982) Monoclonal antibodies against *Mycoplasma hyorhinis*. A secondary effect of immunization with cultured cells. *Exp. Cell. Res.* **137**, 89–94.

Voller, A., Bidwell, D. E. and Bartlett, A. (1979) 'The Enzyme Linked Immunosorbent Assay (ELISA). A Guide with Abstracts of Microplate Applications'. Dynatech Europe, Guernsey.

Wang, R. J. (1976) Effect of room fluorescent light on the deterioration of tissue culture medium. *In Vitro* **123**, 19–22.

Weisberg, W. G., Tully, J. G., Rose, D. L., Petzel, J. P., Oyaizu, H., Yang, D., Mandelco, L., Sechrest, J., Lawrence, T. G., van Etten., J., Maniloff, J. and Woese, C. R. (1989)

A phylogenetic analysis of the mycoplasmas: basis for their classification. *J. Bacteriol.* **171**, 6455–6467.

Werkmeister, J. A., Tebb, T. A., Kirkpatrick, A. and Shukla, D. D. (1991) The use of peptide-mediated electrofusion to select monoclonal antibodies directed against specific and homologous regions of the potyvirus coat protein. *J. Immunol. Methods* **143**, 151–157.

Wilson, M. B. and Nakane, P. K. (1978) Recent developments in the periodate method of conjugating horseradish peroxidase (HRPO) to antibodies. *In* 'Immunofluorescence and Related Staining Techniques' (W. Knapp, K. Holubar and G. Wick, eds.), pp. 215–224. Elsevier/North-Holland, Amsterdam and New York.

Wojchowksi, D. M. and Sytkowski, A. J. (1986) Hybridoma production by simplified avidin-mediated bridging. *J. Immunol. Methods* **90**, 173–177.

Wojciezsyn, J. W., Schlegel, R. A., Limley-Sapanski, K. and Jacobson, K. A. (1983) Studies on the mechanism of polyethylene glycol-mediated cell fusion using fluorescent membrane and cytoplasmic probes. *J. Cell. Biol.* **96**, 151–159.

Yelton, D. E., Diamond, B. A., Kwan, S-P. and Scharff, M. D. (1978) Fusion of mouse myeloma and spleen cells. *Curr. Top. Microbiol. Imunol.* **81**, 1–7.

Young, W. W., Regimbal, J. W. and Hakomori, S. (1979) Radioimmunoassay of glycosphingolipids: application for the detection of Forssman glycolipid in tissue extracts and cell membranes. *J. Immunol. Methods* **28**, 59–69.

Zola, H. and Brooks, D. (1982) Techniques for the production and characterization of monoclonal hybridoma antibodies *In* 'Monoclonal Hybridoma Antibodies: Techniques and Applications' (J. G. Hurrell, ed.), pp. 1–57. CRC Press, Florida.

9 Purification, Fragmentation and Isotopic Labelling of Monoclonal Antibodies

In many cases, purification of monoclonal antibodies is not necessary. For example, if fluorescein-conjugated anti-immunoglobulin is used as a second step in indirect immunofluorescence experiments (Chapter 12), there is little reason to purify the first antibody. Similarly, most analytical-scale experiments in which antigens are isolated by immunoprecipitation can be performed using unfractionated culture supernatants or serum from hybridoma-bearing animals. However, if monoclonal antibodies are to be coupled to fluorochromes, biotin or solid-phase affinity matrices, or if it is desired to make proteolytic fragments, at least partial purification is essential.

Typical antibody concentrations in serum or ascites of hybridoma-bearing mice range from 2 to 10 mg/ml, and thus represent a significant fraction of all protein present. In contrast, the antibody levels in culture supernatants of hybridomas are of the order of 5–50 μg/ml, unless the newer culture techniques for large-scale production are used (see Section 8.12.1). It is therefore obvious that purification of antibodies from serum or ascites will be much easier than from culture supernatants. If antibodies must be purified from culture supernatants, affinity chromatography is usually the method of choice.

The physical properties of antibodies vary somewhat between species, and procedures that have been developed for a particular species (e.g. human, rabbit) do not always work well when applied to other species. This is especially true when attempts are made to generate defined proteolytic fragments. Sometimes, the procedures for purification of antibodies were designed for polyclonal antisera and were optimized to make a reasonable compromise between yield and purity. The homogeneity of monoclonal antibodies makes them much easier to purify, but in order to obtain good results, it is advisable to individualize the conditions for each antibody. The methods described in this chapter have proved reliable and usually involve minimal effort.

The descriptions of antibody purification will concentrate on IgG and IgM, because these are the classes most frequently encountered. Fortunately, IgG is both the most common antibody and the easiest to purify. The purification and fragmentation of mouse IgA, IgD and IgE are specialized subjects, and the primary literature should be consulted (Grey *et al.*, 1970; Tomasi and Grey, 1972; Goding, 1980; Perez-Montgort and Metzger, 1982; Gorevic *et al.*, 1985; Hayzer and Jaton, 1985; Ishizaka, 1985; Mestecky and Kilian, 1985; Spiegelberg, 1985; Haba and Nisonoff, 1991). The purification and characterization of serum immunoglobulins has been reviewed in the volume edited by Di Sabato (1985). Much useful practical information on protein purification is provided in the monograph by Scopes (1993).

Most of the procedures described in this chapter can be used for either mouse or rat monoclonal antibodies. Where there are important differences, such as binding to staphylococcal protein A or streptococcal protein G, these are indicated where appropriate. For more details on the purification and fragmentation of rat monoclonal antibodies, the reader is referred to the excellent monograph by Bazin (1990).

The methods described in this chapter can be carried out with a minimum of expensive equipment. Excellent and rapid purifications are possible using more sophisticated and expensive equipment such as FPLC (Pharmacia), but this type of equipment is expensive. The principles described in this chapter can be applied to FPLC with very little change, and the manufacturer's instructions should be consulted for further details.

HPLC fractionation of antibodies is generally limited to ion exchange and gel filtration. Reverse-phase HPLC has not been commonly used with antibodies, perhaps because their large size and hydrophobic nature may give poor recoveries and excessively tight binding to the beads. Some authors have used hydrophobic interaction HPLC with a TSKgel Phenyl-5PW column, binding in 1 M ammonium sulfate and elution with a decreasing gradient of ammonium sulfate (Morimoto and Inouye, 1992; Inouye and Morimoto, 1993, 1994).

9.1 Determination of Antibody Class

Knowledge of antibody class and subclass is a great help in determining the strategy of purification. Unless special immunization and screening procedures have been used, the most frequent antibodies encountered will be IgM and IgG. Hybridomas secreting IgE are rare (Bottcher *et al.*, 1978; Eshhar *et al.*, 1980; Liu *et al.*, 1980), and hybridomas secreting IgA are usually only obtained when the lymphocytes for fusion are from gut-associated lymphoid tissue (Komisar *et al.*, 1982). If the screening procedure uses staphylococcal protein A or streptococcal protein G (see Sections 5.3.5, 8.10.3 and 8.10.4), it is unlikely that classes other than IgG will be detected. IgM antibodies are more unstable, have lower affinity, are more cross-reactive and more prone to nonspecific effects than other classes. They are generally best avoided unless there is no other choice.

There are now numerous commercially available kits for the isotyping of mouse monoclonal antibodies. Many are based on ELISA, either in 96-well trays or dipsticks. In recent years, competition has tended to bring the price of these kits down to a more reasonable level, and it is no longer cost-effective to produce one's own typing sera. However, it may occasionally be useful to perform Ouchterlony analysis, and for this reason the method is outlined in the following section.

9.1.1 Ouchterlony (Double Diffusion) Analysis

When a polyclonal antiserum reacts against a large polymeric antigen such as a protein or a polysaccharide, the resulting cross-linking often results in the formation of a precipitate. If the antigen and antibody are allowed to diffuse towards each other through agar, the precipitate will be in the form of a visible line. This technique was developed independently by Oudin and Ouchterlony, but the Ouchterlony technique is preferred because it is more convenient (Silverstein, 1989).

A typical analysis is shown in Fig. 9.1. Glass slides (25 × 75 mm) are coated with 3 ml molten agar (1% in phosphate-buffered saline, pH 7.4) and the agar is allowed to set. Alternatively, the gel may be poured on the hydrophilic side of 'Gelbond' flexible plastic sheets (Marine Colloids division of FMC Corporation). The use of Gelbond allows a permanent unbreakable record to be mounted in a notebook. Holes of 1–2 mm diameter are punched, usually in a circular pattern around a central hole. The distance between holes is typically 3–5 mm.

The best source of test immunoglobulin is probably culture supernatant from densely grown hybridoma cells obtained after cloning. This ensures that the only mouse immunoglobulin present is the desired antibody. The testing of antibody class from serum or ascites may cause ambiguous results because of the concurrent presence of normal immunoglobulins, although these can often be diluted out.

The optimal concentration of immunoglobulin for detection in Ouchterlony analysis is 100–1000 μg/ml, when strong anti-immunoglobulin sera are used. This concentration may be achieved by diluting hybridoma antibody-containing serum or ascites by about 20-fold. Culture fluid will have to be concentrated by a similar factor. This is easily achieved by addition of an equal volume of saturated ammonium sulfate (see Section 9.2.1) followed by centrifugation and resuspension of the pellet in an appropriate volume of

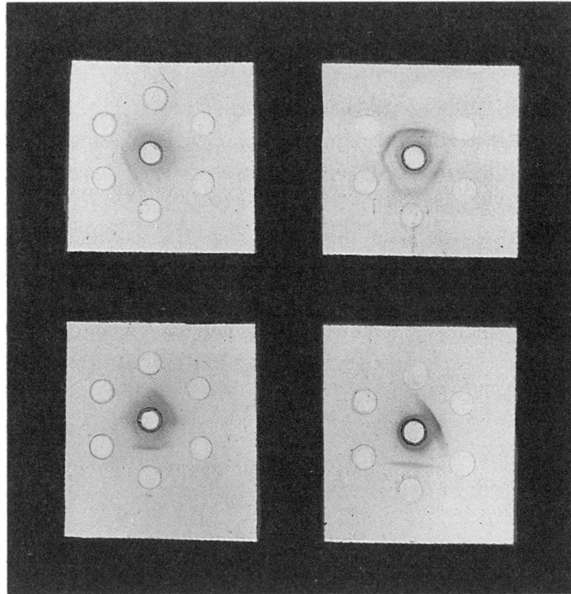

Fig. 9.1. Ouchterlony (double diffusion) analysis of monoclonal antibodies. The wells contain the following proteins (clockwise from top): IgG1κ, IgG2aλ, IgGbκ, IgG3λ, IgAκ and IgG3κ. The central wells contain anti-κ (upper two panels) and anti-λ (lower two panels). The two right-hand panels contain 3% polyethylene glycol, while the two left panels do not. It can be seen that the typing sera are specific, but the IgG3λ protein contains κ-bearing contaminants. Precipitation lines in the right-hand panels are strengthened by the presence of polyethylene glycol.

phosphate-buffered saline (PBS), and dialysing against the same buffer overnight to remove ammonium sulfate.

The central well is filled with the immunoglobulin-containing solution (typically 2–3 μl), and the peripheral wells are filled with undiluted class-specific anti-immunoglobulin antisera, which are available from many commercial suppliers. It is recommended that any laboratory engaged in the production of monoclonal antibodies should possess a set of these antisera. Diffusion is allowed to proceed at room temperature for several hours.

Strong 'home-made' class-specific anti-immunoglobulin antibodies will usually produce intense precipitin lines in 3–12 h, using antigen at 1.0 mg/ml. Commercial antibodies are usually much weaker, and appear to be 'watered down' to the maximum acceptable dilution, and often beyond. They should generally be used undiluted, and the antigen should be at about 100 μg/ml. Under these conditions, faint precipitin lines should appear after 12–18 h.

If, as is often the case with commercial antisera, the reaction is too weak to be visible, the lines may become visible after staining with Coomassie blue. Before staining is carried out, unprecipitated protein must be removed. This may be accomplished by soaking the wet slide in PBS with agitation for 1–2 days, but in my experience the agar often lifts off the slide. A better method is to wrap the slide in Whatman No. 1 filter paper that has been wetted with water and allow it to dry overnight at 37°C. Most of the uncomplexed protein and buffer salts will diffuse into the filter paper. Once the slide is dry, the paper is wetted again with water and removed. The slide is then soaked overnight in PBS to allow residual uncomplexed protein to diffuse out of the gel. The slide is then stained for 5 min in 0.1% Coomassie blue in 40% methanol: 10% acetic acid, and destained in 12% ethanol: 7% glacial acetic acid until the background is acceptable. It may then be dried at room temperature for a permanent record.

Sometimes the precipitin lines are poor, even after staining. Addition of 3% (w/v) polyethylene glycol (molecular weight 4000–6000) to the agar before pouring the plate will often enhance precipitation (Fig. 9.1).

9.2 Methods Used in Antibody Purification

9.2.1 Precipitation with Ammonium Sulfate

One of the oldest and most useful methods of purification of immunoglobulins is based on the observation that they are precipitated by lower concentrations of ammonium sulfate than most other serum proteins. Precipitation of immunoglobulins by ammonium sulfate is gentle, effective and simple. While it is not possible to purify immunoglobulins to homogeneity by this method, it provides a substantial enrichment, and reduces the protein load on subsequent purification steps.

Ammonium sulfate precipitation of immunoglobulins is simple to perform, but there are a few technical points which can greatly influence the degree of purification obtained. It is preferable to add ammonium sulfate slowly in the form of a saturated aqueous solution, rather than as solid crystals. High local concentrations of ammonium sulfate cause unwanted precipitation of proteins such as albumin, and thus degrade the degree of purification.

Saturated ammonium sulfate may be prepared by adding 1 kg of the crystals to 1 l of distilled water, and stirring for 1–2 days. To be sure of saturation, there must be undissolved crystals in the bottom of the vessel. The pH should be approximately 6–7. The solution may be stored at 4°C. It is customary to perform ammonium sulfate precipitations at 4°C, although most antibodies can be fractionated at room temperature without adverse effects. The percentage of saturation changes slightly with temperature, but for practical purposes the effect is negligible.

The serum to be fractionated should be placed in a beaker on a magnetic stirrer. Stirring should be slow, as frothing denatures protein. Saturated ammonium sulfate is added dropwise, allowing each drop to disperse before the next is added. When the concentration of ammonium sulfate reaches about 20% of saturation, the serum will begin to turn milky. Most immunoglobulins will be precipitated by 35–40% of saturation, although it may be necessary occasionally to use up to 50%. Higher concentrations do not increase the yield of immunoglobulins, but will cause increasing contamination by other proteins, especially transferrin and albumin.

The suspension should be stirred for 15–30 min, and then centrifuged at 2000 ***g*** or preferably 10 000 ***g***. The supernate should be clear, but may contain a layer of lipoprotein on the top. The pellet should be washed 2–3 times in 50% saturated ammonium sulfate. Washing the pellet considerably reduces contamination with nonimmunoglobulin proteins and does not lower the yield significantly. The volume of the pellet should not change greatly after washing. (In a previous edition of this book, washing with 40% saturated ammonium sulfate was recommended, but I have since encountered occasional monoclonal antibodies that were soluble in 40% ammonium sulfate.)

It should be noted that ammonium sulfate solutions are dense and resist mixing. Care must be taken that the stock solutions are adequately mixed. A few end-over-end mixings are desirable, as the mixing by magnetic flea may fail to disperse a dense lower layer.

Solutions of saturated ammonium sulfate should not be stored in containers that were previously used for laboratory wash-up detergents. I have found that traces of these detergents can inhibit precipitation of immunoglobulins by ammonium sulfate.

Finally, the precipitate is dissolved in PBS. It will virtually always redissolve easily. Ammonium ions will interfere with many subsequent procedures, especially conjugation of antibodies with fluorochromes, biotin or agarose. Ammonium sulfate may be removed by overnight dialysis against 500–1000 volumes of PBS, or any buffer compatible with subsequent manipulations. The dialysis fluid should be changed several times at intervals of a few hours.

9.2.2 Precipitation at Low Ionic Strength

In the early days of serum fractionation, it was noted that some serum proteins were soluble in distilled water, while others were not. The soluble fraction was known as *albumin* and the insoluble fraction as *globulin*. Subsequently, it was found that many globulin-like proteins were soluble in distilled water. These proteins were termed *pseudoglobulins*, to distinguish them from the water-insoluble *euglobulins*. Subsequently, the definition of globulins was broadened to include proteins precipitated by 50% saturated ammonium sulfate. Later still, the globulins were redefined to include any protein whose anodic mobility on electrophoresis was less than that of albumin (see Section 5.5, Fig. 5.3). The term pseudoglobulin is now obsolete, but the term euglobulin persists.

The euglobulin fraction of serum is composed mainly of a subpopulation of immunoglobulins, notably the majority of IgM molecules and a significant fraction of IgG and other classes. Precipitation of immunoglobulins at low ionic strength (euglobulin precipitation) is still occasionally used as a preliminary purification method. Some monoclonal antibodies are insoluble at low ionic strength, while others are not. Euglobulin properties are common among mouse IgM, IgG2b and IgG3 proteins, but rare in IgG1.

Euglobulin precipitation is usually carried out by dialysing the protein against distilled water. The pH should be about 6–7 because, like all proteins, immunoglobulins are least soluble near their isoelectric point. If necessary, low concentrations (*c.*5 mM) of citrate–phosphate buffer, pH 5.8, may be used. Dialysis should proceed at 4°C for 24–96 h, with several changes. The resulting precipitate is harvested by centrifugation, washed twice in the dialysis fluid, and resuspended in isotonic PBS. One disadvantage of euglobulin precipitation is that it is sometimes difficult to redissolve the precipitate. Addition of NaCl to a final concentration of 0.3–0.5 M may help.

According to Garvey *et al.* (1977), euglobulins precipitate more readily when dialysed against 2% boric acid (pH ≈6), and the boric acid precipitate is more readily solubilized than other precipitates. The effectiveness of the boric acid is said to derive from its low ionic strength (it ionizes only weakly) and the fact that its pH is close to the isoelectric point of many immunoglobulins, especially IgM. In addition, the complexing of borate with the carbohydrate moieties on glycoproteins may aid precipitation. The boric acid–glycoprotein complexes that are insoluble at low ionic strength are dissociable by higher salt concentrations.

9.2.3 Ion Exchange Chromatography

Ion exchange chromatography is one of the simplest and most useful ways of purifying immunoglobulins, especially IgG. It is particularly suitable for the purification of monoclonal antibodies, because simple gradient elution allows the correct compromise between yield and purity to be made without guesswork. The technique is gentle, and is capable of very high capacity. A 10 ml column will easily handle 100–200 mg protein, and the recovery should be virtually 100%. Modern ion exchange resins are inexpensive, easy to use and require a minimum of preparation. Ion exchange chromatography can be done without special equipment, although work is greatly facilitated by the use of a gradient mixer, an ultraviolet absorbance monitor and a fraction collector (Fig. 9.2). These should be standard equipment in most biological laboratories.

The most commonly used matrix consists of cellulose or agarose, to which is attached ionizable diethylaminoethyl (DEAE) groups. At pH 8, the matrix has a strong positive charge which must be balanced by negative ions (anions). The bound anions may be inorganic (e.g. chloride ions) or organic, such as the charged side chains of proteins, or a mixture of both. Hence, DEAE-cellulose is known as an anion exchanger.

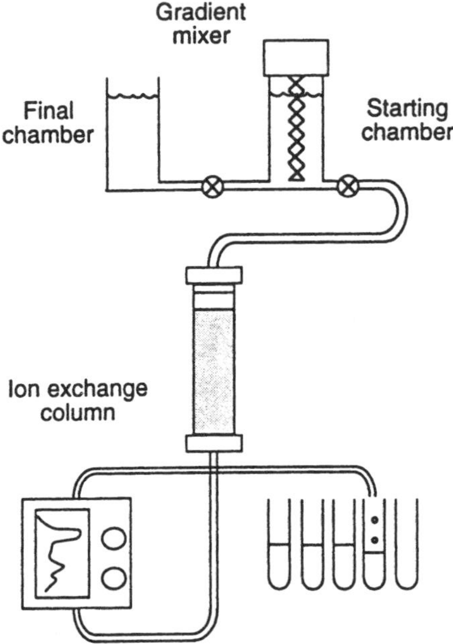

Fig. 9.2. Purification of monoclonal IgG antibodies by ion exchange chromatography. The antibody-containing solution is bound to the column at low ionic strength and pH 8.0, and eluted by a linear gradient of NaCl.

The immunoglobulins are among the most basic of serum proteins, with isoelectric points in the range 6–8. At pH 8, almost all serum proteins will be negatively charged, and thus bind to DEAE-cellulose at low ionic strength. With increasing concentration of competing anions (e.g. Cl^-) or lowering of pH, the proteins will be eluted, approximately in order of their isoelectric points. Because the immunoglobulins are the most basic of the major serum proteins, they will be the first to be eluted.

Ion exchange chromatography has the distinct advantage that the volume of the purified protein is independent of the volume of the sample applied to the beads. Indeed, the binding of antibody to a small ion exchange column at low ionic strength (10 mM Tris-HCl, pH 8.0) and subsequent elution in 1 M NaCl in the same buffer is an excellent way of concentrating antibodies from dilute solutions.

Preparation of the ion exchanger

The most convenient matrices are those which are supplied pre-swollen. DEAE-Sepharose (agarose) or DEAE-Sephacel (beaded cellulose) are suitable and there is little to choose between them for immunoglobulin purification. They are supplied in aqueous suspension, with Cl^- counterions.

The ion exchanger beads must be thoroughly equilibrated with the starting buffer prior to use. Unlike the older matrices, DEAE-Sephacel and DEAE-Sepharose do not require 'precycling' prior to use. The slurry (50–100 ml) should be poured into a large conical flask, and 1 l of starting buffer (10 mM Tris-HCl pH 8.0) added. After the beads have settled, the supernatant may be decanted, and another 1 l of buffer added. The process should be repeated several times. The beads are then ready for use and may be transferred to a smaller container. Alternatively, the beads may be equilibrated by washing with a few litres of buffer on a sintered glass funnel or a Büchner funnel, or simply poured into a small column and equilibrated *in situ*.

The size of the gel bed will be determined by the mass of protein to be bound. The column shape is not very important; the Bio-Rad 'Econocolumns' (cat. no. 731–1550) are ideal. For optimal resolution, no more than 10% of the available capacity should be used, but this rule can often be broken. As a rough approximation, the bed volume should be at least 10 ml for each 100 mg of protein. Immunoglobulins are, in general, fairly resistant to the action of proteases, and ion exchange chromatography can usually be performed at room temperature.

Air is less soluble in water at room temperature than at 4°C. If it is desired to work at room temperature and the gel slurry and the buffers have been stored in the cold, they should be allowed to stay at room temperature for a few hours or degassed by application of vacuum for a few seconds before pouring the column. Otherwise, the column will become filled with air bubbles, degrading its performance.

The pouring of the column is straightforward, and no special precautions are necessary. Gravity feed is adequate; a head of pressure of 100–200 mm of water will be sufficient. The pressure head is measured from the free end of the outlet tube to the free surface of the buffer reservoir.

Preparation of the sample

To ensure that all the protein in the sample will bind to the resin, the sample should be equilibrated with the starting buffer before application. If the salt concentration in the sample is physiological (i.e. *c.* 150 mM for serum), simple dilution in 20–50 volumes of starting buffer may be adequate. Alternatively, the sample may be dialysed overnight against 500–1000 volumes of starting buffer.

Immunoglobulins will occasionally precipitate in the low ionic strength starting buffer (see Section 9.2.2). For this reason, transfer into the starting buffer by desalting on small gel filtration columns (e.g. Sephadex) is only recommended if it is known that the protein will remain in solution and not precipitate in the column. However, even immunoglobulins which precipitate in the starting buffer may be fractioned by ion exchange chromatography. The precipitate is simply loaded onto the column, where it will bind. As the salt concentration is raised during elution, the protein will usually go back into solution, and the recovery and resolution are often indistinguishable from normal.

After the sample has been loaded, the column should be washed with 2–3 times its volume of starting buffer. Only occasional exceptionally basic immunoglobulins will emerge. Once the ultraviolet absorbance monitor tracing is steady, elution may be performed.

Step gradient elution

Bound proteins may be eluted by applying steps of increasing salt concentration (10 mM Tris-HCl, pH 8, plus 50 mM, 100 mM, 150 mM, 200 mM NaCl). Most IgG molecules will emerge by 200 mM salt.

Step gradient elution has the attraction of extreme simplicity, but it also has many disadvantages. It is necessary to try a number of conditions before the appropriate ones are found, and the process must be repeated for each new monoclonal antibody. Step gradient elution results in sharp leading edges of peaks, but the trailing edges are often broad, leading to excessive dilution. Step gradients are also susceptible to artefacts. The successive application of two steps may result in the elution of two successive peaks of protein, yet when analysed these fractions may be found to consist of the same protein, even if the sample is perfectly homogeneous. These cases are probably manifestations of heterogeneity in the strength of binding sites of the beads.

Linear gradient elution

Elution of monoclonal antibodies by a linear gradient of NaCl in 10 mM Tris-HCl, pH 8, is simple to perform, and gives much better results than step gradients. There is no need for individualization of conditions for each monoclonal antibody, because the gradient can be constructed to cover all elution conditions. Gradient elution subjects the trailing edge of each protein to a higher salt concentration than the leading edge. The leading edge is therefore relatively retarded, while the trailing edge is advanced, and the eluted peaks are sharp and symmetrical with minimal dilution.

Simple gradient mixers are available commercially (e.g. Pharmacia). Alternatively, they may be made in any workshop, or improvised by connecting two beakers via a short length of capillary tubing. Mixing is achieved using a magnetic 'flea' (see Fig. 9.2).

If the two chambers have the same shape and dimensions, it can be calculated that the gradient will be linear, with the starting conditions being those in the mixing chamber and the final conditions being those in the reservoir chamber. For purification of monoclonal IgG, one might typically start with 10 mM Tris-HCl, pH 8.0, and end with the same buffer containing 300–500 mM NaCl. The volume of the gradient is arbitrary; good results are obtained using 10–20 times the bed volume of the ion exchanger. Larger volumes will result in somewhat better separation at the expense of increased dilution. The gradient should be run slowly (over a period of 4–12 h) for best resolution. The speed may be adjusted by altering the head of pressure.

A typical fractionation might proceed as follows. Ten millilitres of mouse serum containing a monoclonal IgG antibody is made up to 50% saturated ammonium sulfate by slow addition of 10 ml saturated ammonium sulfate (Section 9.2.1). After two washes in 50% saturated ammonium sulfate, the precipitate is redissolved in PBS and the optical density at 280 nm measured.

Assuming that all the protein is IgG, 1 mg/ml will have an absorbance of 1.4 in a cell with 10 mm path length.

Let us assume that a yield of protein of 250 mg is obtained. The protein is dialysed against two changes of 1 l of 10 mM Tris-HCl, pH 8, and applied to a 20 ml column of DEAE-Sephacel equilibrated with the same buffer. After washing with 50 ml starting buffer, no protein emerges. (Occasionally, a monoclonal IgG antibody with an unusually basic isoelectric point may 'drop through' the column under these conditions, in which case no further purification is needed.) A gradient mixer is then set up with 100 ml of starting buffer in the mixing chamber and 100 ml of the same buffer plus 300 mM NaCl in the reservoir chamber (Fig. 9.2). The gradient is allowed to run over 4 h, and 40 × 5 ml fractions collected. The monoclonal antibody is the first sharp peak to emerge (at approximately 50 ml), and is recovered in 3–4 tubes (yield 70 mg). The purity of individual fractions may be assessed by agarose gel electrophoresis (see Section 5.5; Fig. 5.3).

Isolation of ribonuclease and protease-free monoclonal antibodies on DEAE Affi-Gel Blue

The use of ion exchange chromatography for purification of mouse monoclonal antibodies has been extended by Bruck *et al.* (1982), who reported the use of DEAE Affi-gel blue (Bio-Rad) for the one-step purification of IgG. Ascitic fluid was centrifuged at 1000 ***g*** to remove cells, then 100 000 ***g*** for 30 min to remove debris and fibrin clots. The fluid was then dialysed overnight against 100 volumes of 20 mM Tris-HCl, pH 7.2, and centrifuged again at 100 000 ***g***. The protein was then passed over a column of DEAE Affi-gel blue equilibrated with the same buffer. The column was washed with the same buffer containing added NaCl (25 mM) to elute transferrin. Finally, monoclonal antibodies (IgG1 and IgG2) were eluted with three bed volumes of 20 mM Tris-HCl pH 7.2 plus 50 mM NaCl. Antibodies were found to be free of contamination by protease, nuclease or albumin. Recovery was said to be 77–80%.

9.2.4 Gel Filtration

Gel filtration separates proteins according to their size (see Scopes, 1993). The procedure is simple to perform and is capable of good recoveries, but is slow and cumbersome and always results in dilution of the sample. If an antibody is to be purified by gel filtration and ion exchange chromatography, it is strongly advisable to perform the gel filtration first because the ion exchanger can be used to reconcentrate the antibody with minimal losses. There is usually a much smaller factor of purification than ion exchange or affinity chromatography. The main role of gel filtration in purification of IgG is an adjunct to other methods, if a higher degree of purification is needed. Gel filtration has a very important role in the purification of IgM.

Gel filtration is usually carried out in long, thin columns (typically 0.5–1 m long and 15–30 mm diameter). The gel itself occupies about 70% of the volume of the column, and the space between the beads the remaining 30%. Very small molecules are freely permeable into the beads, and hence emerge at a volume equal to the *total volume* (V_T) of the column ($\pi r^2 h$). Very large molecules are totally excluded from the beads, and therefore emerge after a volume equal to about 30% of V_T. This is known as the void volume (V_o) of the column. Thus, all fractionation must occur between V_o and V_T. The emergence of proteins after V_T is indicative of adsorption to the beads, which is an undesirable property.

The elution position of a protein is best described by the term K_{av}, where

$$K_{av} = (V_E - V_0) \div (V_T - V_0)$$

and V_E is the elution volume.

As a general rule, the best separations will result when the gel type is selected to give a K_{av} of about 0.3–0.6 for the protein of interest. It should be obvious that the selection of a gel that totally excludes the desired protein (e.g. Sephadex G-200 for IgM) will only result in chromatographic separation of the undesired impurities; the desired protein will not be chromatographed. It is surprising how often this combination has been described in the literature.

Selection of a suitable gel

The range of gels available has increased enormously in recent years. The main types of gel are granulated dextran (Sephadex), acrylamide (Bio-gel P), agarose (Sepharose, Bio-gel A) and various combinations of these. Sephacryl is a mixture of dextran and acrylamide.

The main requirements are resolution of the desired molecular weight range, lack of nonspecific adsorption, and ease of use. The newer matrices (e.g. Sephacryl) are much more resistant to compression and easier to pack. They may be run at higher pressures and faster flow rates than the classical gels

Table 9.1 Suitable gels for separation of immunoglobulins and their fragments

Immunoglobulin	Molecular Weight	Suitable Gel
IgG	150 000	Sephacryl S-300 (Pharmacia)
IgM	900 000	Sepharose 6B or Sephacryl S-500 (Pharmacia)
Fab	50 000	Sephadex G-100 (Pharmacia)
$F(ab')_2$	100 000	Sephacryl S-200 or S-300

such as Sephadex. Their resolution is at least as good, and often better. A list of suitable gels is given in Table 9.1.

Pouring the column

Sephadex is supplied as dry powder, and must be swollen in water for 1–2 days prior to use. The more loosely cross-linked forms of Sephadex (e.g. G-100, G-200) swell by up to 40 ml/g of dry powder, and must be swollen in a large excess of water. Sepharose and Sephacryl are supplied pre-swollen.

Gel filtration of immunoglobulins can usually be carried out at room temperature. It is essential to degas the buffers and gel if they have been stored in the cold. It may be possible to omit degassing if buffers and gel are allowed to reach room temperature before pouring the column. In general it is wise to degas, because bubbles in the column will seriously degrade its performance.

Resolution in gel filtration is critically dependent on a well-poured column. The column should be poured as follows:

(1) Close the outlet valve. Pour in a small quantity of buffer (10% of bed volume). Arrange the outlet tubing such that the vertical distance between the free end of the tube and the top of the gel is equal to the desired packing pressure. It is a good idea to include a 'safety loop' of inlet tubing which dips below the free end of the outlet tubing. This will prevent the column from running dry and cracking if the buffer runs out.

(2) Attach a reservoir cylinder to the top of the column.

(3) Mix a thick slurry of gel (*c*.50% suspension) by inverting the bottle. Do not use a magnetic 'flea', as it will fragment the beads.

(4) Gently pour the gel into the column, preferably down the side of the column along a glass rod. It is important that enough gel is poured to fill the column in one operation.

(5) Immediately open the outlet valve and start flow. The column should be packed at a slightly greater pressure than the desired running pressure (typically 150–400 mm water). Pharmacia recommends that Sephacryl gels be poured with the aid of a peristaltic pump. Run through at least 2–3 bed volumes of buffer before using the column.

(6) It is a good idea to run a small amount (1–3% of bed volume) of 10 mg/ml blue dextran (Pharmacia) to check that the column is packed evenly. The leading edge of the dye should run as a horizontal straight line.

Running the column

The buffer should have a pH of 7–8 and a total salt concentration of at least 100 mM to prevent adsorption effects. It is also desirable to include a preservative such as 10 mM sodium azide to inhibit microbial growth. The sample should have a volume of 2–5% of the column volume, and should not be too viscous (maximum 50 mg/ml). It must be free of any particulate or insoluble material.

The best resolution will be obtained if the flow rate is slow. The column should take 1–2 days to run, although the newer matrices with very fine particle sizes are capable of good resolution in shorter times. Band broadening caused by diffusion is not significant at realistic flow rates. Gel filtration using FPLC is capable of excellent separations in far shorter times.

9.2.5 Purification of Antibodies by Affinity Chromatography: General Aspects

Affinity chromatography is a method of separation which is based on the irreversible immobilization of one component of a system on a solid-phase matrix (usually agarose), and the subsequent binding and elution of a complementary *ligand* in free solution (Fig. 9.3). The question of whether the solid or liquid phase component is designated the ligand is arbitrary.

Fractionation of monoclonal antibodies by affinity chromatography offers

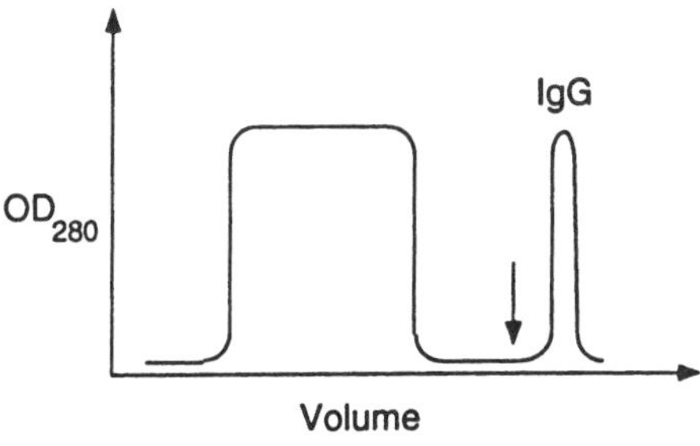

Fig. 9.3. Purification of IgG by affinity chromatography on protein A-Sepharose. The IgG-containing mixture is passed over a column of protein A-Sepharose, and the column washed until the ultraviolet absorption tracing is at baseline. Elution buffer is then applied (arrow; see text), releasing the bound IgG.

excellent purification in a single step, even if the sample is very dilute. The volume of recovered protein is independent of the sample volume, and the technique therefore has the major advantage of concentrating the antibody. Affinity chromatography is therefore ideally suited to purifying monoclonal antibodies from culture fluid. The main disadvantages of affinity chromatography are the relatively low capacity and high cost of immobilized ligands, and the need to subject the antibody to denaturing conditions for elution. Provided that they are looked after well, and not allowed to become clogged with particulate matter or contaminated by microorganisms or proteases, affinity columns may be used for a virtually unlimited number of times, so the cost per separation decreases with increasing use.

The principles involved in purification of antigen or antibodies by affinity chromatography are identical. The use of monoclonal antibodies in affinity chromatography will be discussed in detail in Chapter 11.

Cross-contamination of antibodies by repeated use of one affinity column

A great deal of care should be taken if a single affinity column is to be used to purify a number of different monoclonal antibodies. If elution is not complete, it is possible for a batch of antibody to be contaminated by traces of antibody from a previous purification. The counsel of perfection would, of course, be to use a given column for only one antibody. This is seldom economically possible. Alternatively, it is wise to 'pre-elute' each column immediately before it is used because some material from a previous run may remain on the column. 'Stripping' a column under very harsh conditions (e.g. 0.1 M glycine-hydrochloride pH 2.0 or 6 M guanidine-hydrochloride, pH 7.0) just before each use might also be considered. Finally, appropriate controls should be built into each experiment, so that if contamination has occurred, it can be detected.

Cross-contamination caused by inadequate stripping of affinity columns is a real phenomenon (Bill *et al.*, 1995). The US Food and Drug Administration and the European guidelines are strict concerning cross-contamination. All cleaning operations, including regeneration of chromatographic matrices, must be validated to show their effectiveness. The Guidelines of the European Community state that columns intended for reuse must be validated to ensure that the matrix retains the appropriate properties over the intended regeneration (CPMP Note for Guidance, 1992; US FDA, 1993; Bill *et al.*, 1995; see also Section 9.3.3).

Affinity chromatography on protein A or protein G

A more generally useful way to purify monoclonal antibodies, particularly from tissue culture supernatants, is to bind to and elute from immobilized

Table 9.2 Conditions for binding to and elution from staphylococcal protein A

IgG Subclass	Species	Binds	Elutes	Reference	Comments
IgG1*	Mouse	pH >8.0 in 1 M NaCl	pH <6.0	1–3; 7	Binding is strong at alkaline pH if the salt is increased to 1 M
IgG2a	Mouse	pH > 7.0	pH <4.5	1–3	Allotype influences binding [2]
IgG2b		pH > 7.0	pH <3.5	1–3	May aggregate after elution.
IgG3		pH > 7.0	pH <4.5	1–3	
IgG1	Rat	Weak and variable binding at pH 8.0		4,5	Occasional monoclonal antibodies bind [4]
IgG2a	Rat	No significant binding at pH 8.0		4,5	
IgG2b	Rat	Very weak binding at pH 8.0		4,5	
IgG2c	Rat	pH > 7.0	pH 4–5	4,5	Rat homologue of mouse IgG3 [6] Monoclonals which bind most strongly are most basic [5]

* Monoclonal mouse IgG1 may be purified on Affi-Gel Protein A (Bio-Rad) using a special buffer (Monoclonal Antibody Purification Scheme; MAPS).
References: 1, Ey *et al.* (1978); 2, Sepällä *et al.* (1981); 3, Watanabe *et al.* (1981); 4, Ledbetter and Herzenberg (1979); 5. Rousseaux *et al.* (1981); 6, Der Balian *et al.* (1980); 7, Malm (1987).

anti-immunoglobulin antibodies, staphylococcal protein A or streptococcal protein G (see Sections 5.3.5, 9.3.2 and 9.3.3). If the antibody is of a class and species that binds to protein A (Table 9.2), affinity chromatography on protein A-Sepharose is extremely attractive because of its commercial availability, high capacity and relatively gentle elution conditions.

However, there are many situations in which monoclonal antibodies do not bind to protein A, notably the majority of rat immunoglobulins, mouse IgG1, and classes other than IgG. If the antibody is IgG, protein G-agarose (Pharmacia or Pierce) may be used. Alternatively, anti-immunoglobulin may be substituted for protein A. Purification of antibodies by affinity chromatography on protein A-Sepharose and protein G-Sepharose are discussed in more detail in Section 9.3.2.

Affinity chromatography on anti-immunoglobulin antibodies coupled to agarose beads

Consider the purification of rat antibodies from culture fluid by affinity chromatography on goat antirat immunoglobulin antibodies coupled to agarose beads. A typical hyperimmune goat antirat immunoglobulin antiserum might contain 10 mg/ml total IgG and 1.0 mg/ml of specific antibody. In order to prepare a column of binding capacity of 100 mg (adequate for 1–2 l of hybridoma supernatant), it would be necessary to use at least 100 ml of goat serum. If the IgG fraction were partially purified by precipitation with 40% ammonium sulfate and coupled to cyanogen bromide-activated Sepharose (protein binding capacity ≈10 mg/ml), a column of 100 ml of gel would be needed. (All the above assume that no antibody activity is lost during the coupling).

Far better results will be achieved if the affinity matrix is coupled with affinity-purified antibodies rather than an ammonium sulfate fraction. This will allow a much smaller column (one-tenth the size), and has the very significant advantage that the antibodies have been selected to release their ligand under defined conditions.

Perhaps the ideal affinity matrix for purification of monoclonal antibodies would be a column of immobilized monoclonal anti-immunoglobulin antibodies. These might be selected for appropriate specificity and affinity, such that they release their ligand under defined gentle conditions. It seems likely that such matrices will eventually become commercially available.

Once made, anti-immunoglobulin-agarose should last virtually indefinitely, provided that care is taken to prevent drying or contamination with micro-organisms. Prior to first use, the column should be 'precycled' with normal rat serum followed by elution with glycine-hydrochloride, pH 2.5 or 0.05 M diethylamine, pH 11.5, to saturate any irreversibly binding antibodies. The column should then be returned to neutral pH and stored in PBS containing 0.005% merthiolate.

The column should also be precycled prior to each use with glycine-hydrochloride pH 2.5, or 0.05 M diethylamine pH 11.5, to remove any loosely bound material remaining from previous uses. Cross-contamination is a real problem. Culture fluid should then be passed over the column and washed through with PBS until no further protein emerges. (An ultra-violet absorbance monitor is useful). The bound material is then eluted with glycine-hydrochloride, pH 2.5 or 0.05 M diethylamine, pH 11.5, and dialysed to restore neutral pH. Finally, the column is regenerated as in the previous paragraph.

Affinity chromatography on immobilized antigen

If the antigen recognized by a monoclonal antibody is available in large (milligram) quantities, it may be coupled to agarose beads and used to isolate the antibody (see Chapter 11). The antibody should be bound at neutral pH and physiological salt concentration. Depending on the individual antibody, elution may be accomplished by minor changes in pH or ionic strength (Herrmann and Mescher, 1979), or by more drastic denaturing conditions such as 0.1 M glycine-hydrochloride, pH 2.5 or 0.05 M diethylamine, pH 11.5.

Other eluants include potassium thiocyanate (up to 3.5 M), urea (up to 8 M), guanidine-hydrochloride (up to 6 M), or propionic acid (1 M; see Chapter 11). The denaturing agent should be removed immediately by neutralization, dilution or dialysis.

If the antigen is available in milligram quantities, there may be little reason for making monoclonal antibodies. In addition, purification of monoclonal antibodies by more gentle techniques such as ion exchange chromatography is less likely to cause denaturation and aggregation, and should result in a product of adequate purity.

It is also possible to affinity-purify minute quantities of antibodies by elution from blots of protein antigens adsorbed to nitrocellulose or other membranes from Western blots or blots of bacteria or bacteriophages (see Section 15.5 for references). Although the antigen is not attached covalently to the membrane, binding is sufficiently tight that it is not disrupted by elution of antibodies under extremes of pH, perhaps because the mechanism of binding of the protein to the membrane is predominantly hydrophobic.

The membrane should be blocked thoroughly in bovine serum albumin, casein or nonfat powdered milk (typically 3–5% v/v) and then incubated with the antibody, and then washed thoroughly. The bound antibody may be eluted by glycine-hydrochloride pH 2.2–2.5 or diethylamine pH 11.5, or by 3.5 M potassium thiocyanate. Because of the extremely small amounts of antibody that can be isolated from any reasonable size membrane (typically much less than 1 μg), it is a good idea to add some 'carrier' protein such as bovine serum albumin (1 mg/ml) to the eluting buffer to reduce losses from adsorption to the tubes or dishes. The eluted protein should then be neutralized immediately. This procedure can be very useful for obtaining enough antibody to probe a Western blot or a colony immunoassay, but because of the very limited capacity and surface area of the membrane it would be impossible to purify enough antibody for large-scale purposes, such as to prepare an affinity column.

9.3 Purification of Monoclonal IgG

9.3.1 Purification of Monoclonal IgG by Ammonium Sulfate Precipitation and Ion Exchange Chromatography

This procedure is the method of choice for monoclonal IgG-containing ascites fluid and serum. It is suitable for any species and IgG subclass. The main purification occurs during ion exchange chromatography; the ammonium sulfate fractionation serves mainly to reduce the protein load on the column.

(1) Slowly add saturated ammonium sulfate to the serum, with constant stirring at room temperature, to a final concentration of 50%, and stir for 30 min.

(2) Centrifuge at 2000–1000 **g** for 5 min. Discard supernatant, which should be clear.
(3) Resuspend pellet in original volume of 50% saturated ammonium sulfate. Repeat centrifugation. The volume of the pellet should not diminish significantly. Discard supernatant.
(4) Repeat step 3.
(5) Resuspend pellet in PBS.
(6) Dialyse against 2–3 changes of 1 l 10 mM Tris-HCl, pH 8 (starting buffer), at 4°C overnight.
(7) Take a small aliquot (*c.* 100 μl) of the dialysed protein; dilute 1:10 in PBS. Measure optical density at 280 nm. A 1 mg/ml solution will have an optical density of ≈1.4 (1 cm cell). Calculate the total amount of protein.
(8) Set up a DEAE-Sephacel column, with a bed volume of at least 10 ml for each 100 mg protein (see Section 9.2.3; Fig. 9.2). The DEAE-Sephacel must be thoroughly equilibrated with starting buffer.
(9) Load sample. Wash column with starting buffer until chart recorder is stable.
(10) Elute with linear gradient of 0–300 mM NaCl in starting buffer (see Section 9.2.3). Collect fractions.
(11) Pool the first peak. As a general rule, IgG2 antibodies will elute at a lower salt concentration than IgG1. The latter class may be slightly contaminated with transferrin (molecular weight 80 000). Most monoclonal antibodies will emerge in the first half of the gradient. The yield should be 3–10 mg IgG per ml serum or ascites, and the monoclonal antibody will often be more than 95% pure. If necessary, further purification may be achieved by gel filtration on Sephacryl S-300 (Section 9.2.4).

If the fusion was performed using a total nonproducer myeloma such as Sp2, NS/0 or X63-Ag8.653, the monoclonal antibody should emerge as a single symmetrical peak. If mixed molecules are present (e.g. NS-1 hybrids), there may be more than one peak, reflecting the separation of the different species. If this occurs, each peak should be tested for antibody activity separately.

9.3.2 Purification of Monoclonal IgG by Affinity Chromatography on Protein A-Sepharose

The general properties of staphylococcal protein A have been discussed in Section 5.3.5. The conditions required for binding and elution are given in Table 9.2. There is considerable variation in the strength of binding among the different IgG subclasses in each species. In certain species, some IgM and IgA also binds. The binding of the various IgG subclasses of mouse and rat has been intensively studied (Ey *et al.*, 1978; Ledbetter and Herzenberg, 1979; Rousseaux *et al.*, 1981; Seppälä *et al.*, 1981; Watanabe *et al.*, 1981; Langone, 1982; Bazin, 1990). Some of this information is summarized in Tables 5.2 and 9.2.

Protein A, covalently attached to agarose beads, is available commercially from Pharmacia or many other companies. The following information applies to Protein A-Sepharose from Pharmacia. Five millilitres of wet gel (1.5 g dried beads) is capable of binding 100–125 mg IgG. The matrix will last virtually indefinitely, provided it is protected from drying and microbial contamination.

The conditions for elution of IgG from protein A are fairly mild, depending on the IgG subclass, species and allotype (Table 9.2). Affinity chromatography on protein A-Sepharose is the method of choice for the purification of appropriate IgG subclasses from hybridoma supernatants.

As a general rule, purification of mouse IgG on protein A-agarose is most satisfactory for IgG2a, IgG2b and IgG3. The binding of mouse IgG1 to protein A is rather weaker, but is strengthened by alkaline pH (greater than 8.0; Ey *et al.*, 1978) and at high salt concentration (greater than 1 M; see Malm, 1987). Combining high salt and high pH allows strong binding of mouse IgG1 to protein A, and makes the system reliable for purification of mouse IgG1.

Bio-Rad have produced a system that allows purification of mouse IgG1 on protein A-agarose, consisting of an affinity matrix (Affi-gel Protein A) and a special buffer system, the details of which have not been disclosed. Recovery of IgG1 is close to 100%. As many monoclonal antibodies are IgG1, the system may become very popular.

As in all affinity chromatography, it is good practice to 'precycle' the column with the eluting buffer just prior to each use. The tissue culture supernatant should be cleared of cells and debris by centrifugation, and sodium azide added (final concentration 10 mM) to discourage microbial growth. Unless special culture systems are used (see Section 8.12), the concentration of monoclonal antibody in culture supernatants is unlikely to exceed 50–100 μg/ml, and batches of 1 l may be processed on a 5 ml column. (The concentration of protein A-binding IgG in fetal calf serum is usually negligible.) The column is washed with PBS until the chart recorder reaches baseline (Fig. 9.3), and then the eluting buffer is applied. Although the information given in Table 9.2 may help in choosing the mildest conditions for elution, it is usually more convenient to elute all immunoglobulins using 0.1 M glycine-hydrochloride, pH 2.5. If it is found that the acid conditions inactivate the antibody, a trial of elution at higher pH is indicated.

In general, acid-eluted antibodies should be neutralized immediately with 1 M Tris-HCl, pH 8.0. Alternatively, the pH may be returned to neutrality by dialysis against a large volume of ice-cold PBS. It is fairly unusual for antibodies to be inactivated by a brief encounter of an acidic environment. There have been a few instances, however, in which affinity chromatography on protein A has caused aggregation of loss of activity of monoclonal antibodies. The reasons are obscure; the problem may be more common with mouse IgG3.

Table 9.3 Binding of mouse and rat IgG subclasses to streptococcal protein G-Sepharose

Species	Subclass	Binding to protein G-Sepharose
Mouse	G1	++
	G2a	++
	G2b	++
	G3	+
Rat	G1	++
	G2a	+
	G2b	+
	G2c	++

Binding to protein G-Sepharose 4 Fast Flow or protein G-Superose (Pharmacia) is considerably stronger than binding to protein G alone (see Akerström *et al.*, 1985). For many species, binding of IgG to protein G occurs equally strongly between pH 6 and 9, and at physiological ionic strength. However, it should be noted that Pharmacia recommends the use of low ionic strength buffers with protein G-Superose and protein G-Sepharose 4 Fast Flow for mouse and rat IgG. Binding of mouse IgG is in 20 mM sodium phosphate at pH 7.0, while binding of rat IgG is in 2 mM sodium phosphate, pH 7.0 (see Pharmacia literature). It may be presumed that the choice of low ionic strength buffers is not accidental and perhaps contributes to strengthening of binding by nonspecific electrostatic effects. Elution from protein G-Sepharose is in 0.1 M glycine-hydrochloride, pH 2.7.

Affinity chromatography on protein A-Sepharose may also be used to purify immunoglobulins from serum or ascites of hybridoma-bearing animals. However, the capacity of ion exchange chromatography is greater, and ion exchange also provides some separation of the monoclonal antibody from the normal immunoglobulins which are present in mouse serum. Occasionally, the columns may become blocked by fibrin deposition if ascites are applied directly. It is far cheaper to ruin a column of DEAE-cellulose than a column of protein A-Sepharose.

9.3.3 Purification of Monoclonal IgG by Affinity Chromatography on Protein G-Sepharose

When one has to purify IgG subclasses that do not bind strongly to protein A, such as many rat IgG subclasses, the use of protein G-Sepharose will often provide a solution. Protein G binds IgG from many species, including all mouse IgG subclasses. Although the binding of rat IgG is weaker than IgG from other species, it is still strong enough to be used successfully for affinity chromatography (Åckerström *et al.*, 1985; Björk and Åckerström, 1990; see also Section 5.3.5). The binding of various mouse and rat IgG subclasses to protein G is given in Tables 5.2 and 9.3.

Binding of IgG to protein G-Sepharose is said to be optimal at about pH 5 (Björck and Åckerström, 1990), although Pharmacia recommends loading the column in sodium phosphate buffer, pH 7.0 and elution in 0.1 M glycine-HCl, pH 2.7. As discussed previously (see Section 5.3.5 and Tables 5.2 and 9.3), binding of mouse and particularly rat IgG to protein G is relatively weak compared with other species, and in some brochures Pharmacia recommends the use of low ionic strength buffers for loading protein G-Superose (20 mM sodium phosphate pH 7.0 for mouse IgG and 2 mM sodium phosphate pH 7.0 for rat IgG). This may reflect the need for an additional nonspecific electrostatic component in addition to the specific component to achieve adequate binding (see Scopes, 1993 for discussion). However, good recoveries of mouse and rat antibodies are often obtained using protein G with binding at physiological salt concentration and slightly acidic pH (5.5–7). The use of low ionic strength buffers may cause problems for some monoclonal antibodies with euglobulin properties (some mouse IgG2b and IgG3; some rat IgG2c), which may precipitate out of solution under these conditions (see Section 9.2.2).

Except for the buffer conditions, the basic principles and practical details for affinity purification of IgG on protein G-Sepharose are very similar to those for protein A-Sepharose, as described in the previous section.

As with all affinity matrices, particular care should be taken in regenerating protein G-Sepharose, to avoid cross-contamination with antibodies from previous purifications. Ideally, the column would be used for only one sort of antibody, although this is seldom economically possible in research laboratories. It has recently been shown that the use of alkaline regeneration buffers (pH 11) for protein G-Sepharose is associated with increasing difficulty in elution and steadily diminishing effectiveness of the columns, suggesting that the affinity of the matrix for IgG is increased by the alkaline conditions (Bill *et al.*, 1995). It has been recommended that protein G columns be eluted at pH 3.0. Pharmacia recommends a pH of 2.7.

9.3.4 Purification of Monoclonal IgG by Affinity Chromatography on a Thiophilic Adsorbent ('T Gel')

Immunoglobulins, particularly IgG, have an unexplained affinity for a matrix made by coupling mercaptoethanol to Sepharose 6B that has been activated and cross-linked by divinylsulfone (Belew *et al.*, 1987). The matrix is known as 'T Gel'. The basis for this affinity is not clear, but it is probably related to 'salting out chromatography' because it is promoted by high salt concentrations. It does not involve thiol or disulfide interchange, as the matrix contains a sulfone and a thioether bond but does not contain free thiol groups. In addition to salting out, which is relatively nonspecific, there is almost certainly a semispecific affinity of immunogobulins for the sulfur atoms of the sulfone and thioether groups.

The overall structure of the side chain of T Gel is as follows:

$$P\text{-}O\text{-}CH_2\text{-}CH_2\text{-}SO_2\text{-}CH_2\text{-}CH_2\text{-}S\text{-}CH_2\text{-}CH_2OH$$

where P is the polysaccharide of the agarose beads.

Purification is as follows. The antibody-containing solution (serum or ascites) is adjusted to be 0.5 M in K_2SO_4, and the mixture stirred gently and clarified by centrifugation at 2000 ***g*** for 5 min. The clear supernatant is loaded onto a T gel column equilibrated with 50 mM sodium phosphate, pH 8.0, containing 0.5 M K_2SO_4 (adsorption buffer). The column is washed with adsorption buffer until all unbound material is eluted, and then the antibodies are eluted with the same buffer but without K_2SO_4. All purification can be carried out at 20°C.

It was shown that recovery of mouse IgG1, IgG2a, IgG2b and IgG3 was excellent but, for unknown reasons, recovery of mouse IgM was less satisfactory. Purity was very good, but not quite as good as can be achieved by affinity chromatography on protein A or protein G.

T Gel is available from Pierce and other suppliers. It has high capacity (more than 20 mg IgG per ml gel) and is relatively inexpensive. It is likely to be particularly attractive for large-scale purification. A more recent paper suggests that the use of modified thiophilic gels may be a useful extension of the technique (Knudsen *et al.*, 1992).

9.4 Purification of IgM

The purification of IgM is not as easy as that of IgG. IgM may be prepared in good yield and reasonable purity by precipitation with 40% saturated ammonium sulfate followed by gel filtration on Sepharose 6B or Sephacryl S-500. It is important to choose a gel filtration medium which includes IgM, so that it is chromatographed rather than simply dropped through the column (see Section 9.2.4). The choice of an appropriate gel filtration medium allows separation of IgM from high molecular weight lipoproteins and protein aggregates, which would emerge with IgM in Sephadex G-200. Even on an appropriate gel, there is likely to be some contamination of IgM with α_2-macroglobulin (M_r 725 000). Contamination with α_2-macroglobulin may be reduced by substitution of euglobulin precipitation (see Section 9.2.2) for ammonium sulfate in the first step. However, euglobulin precipitation causes some denaturation, as shown by the frequent difficulties in redissolving the precipitates. In addition, not all IgM proteins are euglobulins.

Jehanli and Hough (1981) have shown that monoclonal human IgM may be purified to homogeneity by filtration on Ultrogel AcA 34 followed by ion exchange chromatography on DEAE-Sepharose CL6B with salt gradient elution. The IgM was bound to the ion exchanger in 0.05 M phosphate/citrate buffer at pH 6.8, and eluted with a linear gradient to 0.1M phosphate/citrate, pH 5.0. Recovery was approximately 70%. This procedure may become the method of choice, and would probably be applicable to mouse IgM without modification. It would be preferable to use Ultrogel AcA 22, which chromatographs IgM, rather than AcA 34, which does not.

Inouye (1991) purified rat IgM from ascites by precipitation with 60% saturated ammonium sulfate, followed by gel filtration on TSKgel Toyopearl HW-55 (Tosoh, Tokyo, Japan) in 100 mM citrate buffer, pH 4.5.

There are numerous other published protocols for purification of IgM. Tatum (1993) published a large-scale method involving sequential precipitation with ammonium sulfate followed by resuspension and reprecipitation with polyethylene glycol. No chromatographic steps were needed. Other procedures for purification of IgM are given by Hayzer and Jaton (1985).

Knudsen *et al.* (1992) showed that thiophilic adsorption chromatography can be used to purify human and mouse IgM, although the purity was not assessed by polyacrylamide gel electrophoresis.

9.5 Fragmentation of Monoclonal Antibodies

In certain circumstances, it may be desirable to generate antigen-binding fragments of monoclonal antibodies. For example, when cells with receptors for the Fc portion of IgG are present (e.g. macrophages, monocytes), intact IgG molecules may bind nonspecifically via their Fc portions (see Section 5.3.4). Removal of the Fc will prevent this mode of binding. The production of small fragments of IgM (intact M_r 900 000) may aid penetration into the tissues for cytochemical studies.

The older immunological literature is replete with accounts of the fragmentation of mouse antibodies, often using conditions that were established for other species such as rabbit or human. In many cases, it was assumed that the procedures worked and no experimental verification was offered. In a few, it is now possible to state with hindsight that they could not possibly have worked. In others, the fragments were identified incorrectly. More recently, there has been an increasing awareness of the problem and the current literature is more solidly based.

It is important to examine the claims in the literature on fragmentation of mouse immunoglobulins with a critical eye. One should not be too surprised if one is unable to reproduce them. In the succeeding paragraphs, I will summarize what I consider to be the most reliable procedures and point out areas where uncertainty still exists.

9.5.1 Preparation of Fab Fragments of Mouse and Rat IgG

Cleavage of IgG at the hinge by the nonspecific protease papain is one other more straightforward way of producing defined fragments of mouse and rat IgG. It is easy to find conditions for complete cleavage. However, it should be pointed out that even in this case, evaluation of the literature is not as straightforward as it would appear.

Part of the problem arises from the fact that papain is a thiol protease (i.e. it has an SH group in the active site, which must be in the reduced form for activity). The presence of reducing agents necessary for the activation of papain (cysteine, mercaptoethanol or dithiothreitol, in order of increasing potency) may also facilitate digestion by their effects on the substrate. Cysteine was a popular reducing agent in the older literature. However, it is very easily oxidized to cystine by air, and results of such experiments may be poorly reproducible unless oxygen is excluded during digestion. All these reducing agents have the potential of cleaving the labile interchain disulfide bonds of immunoglobulins, but the extent to which these effects occur depends on the individual reducing agent and its concentration. In some published work, it is difficult to decide whether apparent heterogeneity in digestion is due to proteolytic cleavage, partial reduction, or a mixture of the two.

Papain is inactivated by heavy metals, which complex with its sulfhydryl group. It is often supplied as 'mercuripapain' to prevent autodigestion. Regardless of whether mercury is present or not, it is good practice to chelate any divalent cations with EDTA to ensure maximal activity of the enzyme. Papain should be 'activated' by a brief incubation in 0.1 M Tris-HCl, pH 8.0, containing 2 mM EDTA and 1 mM dithiothreitol, just prior to use.

Digestion may be terminated by the irreversible alkylation of the thiol group of papain with iodoacetamide. A twofold molar excess of iodoacetamide over all thiol groups present is recommended. Reduction of iodoacetamide by thiol groups causes the release of HI, so the reaction should be well buffered.

Iodoacetamide absorbs light significantly at 280 nm. In addition, iodoacetamide is light-sensitive, and must be stored and used in the dark. The presence of any brown colour is an indication of decomposition, and such preparations should not be used.

Practical procedure

The extent of digestion of mouse and rat IgG is highly dependent on the concentration of reducing agent. The procedure given below (Oi and Herzenberg, 1979) should be monitored by SDS–PAGE (reducing conditions). Digestion should result in complete disappearance of the γ heavy chain (M_r 55 000) and appearance of the Fd fragment of γ (M_r 27 000) and light chains (M_r 22 000–25 000). An example of a typical digestion is shown in Fig. 9.4.

(1) Activate papain (1–2 mg/ml) in 0.1 M Tris-HCl, pH 8.0, containing 2 mM EDTA and 1 mM dithiothreitol, for 15 min at 37°C.

(2) Digest IgG (1–10 mg/ml) in the same buffer at 37°C, using an enzyme:substrate ratio of 1:100, for 1 h.

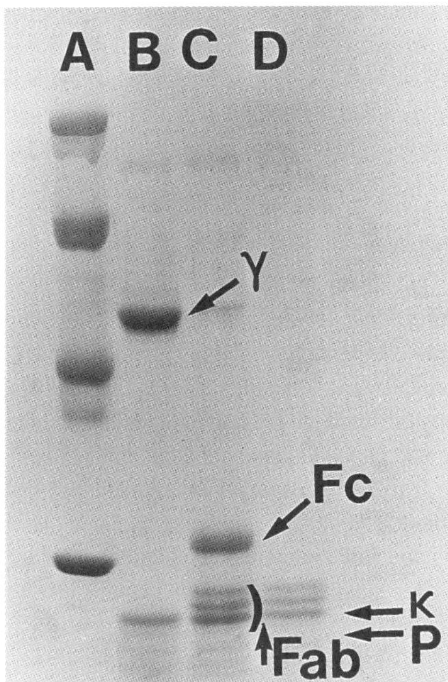

Fig. 9.4. SDS–PAGE of mouse IgG1 and its fragments after digestion with papain. Digestion was terminated with iodoacetamide. The concentration of acrylamide was 10%, and all samples were reduced. A, Molecular weight standards (top to bottom): 95 000; 68 000; 43 000; 30 000. B, Undigested IgG. C, IgG after digestion with papain. D: Fab fragment after purification. The lowest band is κ chain. The upper bands are Fd fragments, although it is not clear why two bands are present. The mobility of papain (M_r 20 000) is indicated (P) although there is not enough enzyme present to be visible. I thank Annette Blane for providing this gel.

(3) Terminate digestion by addition of iodoacetamide (final concentration 20 mM), and hold on ice for 1 h, protected from light.
(4) Dialyse overnight against phosphate-buffered saline, to remove iodoacetamide.

Comments It is absolutely essential to monitor the completeness of digestion by SDS–PAGE, as the individual IgG subclasses vary in their susceptibility to digestion. The rate of digestion is also influenced by the concentration of reducing agent. Mouse IgG2a and IgG2b are extremely susceptible to cleavage, even in 0.1 mM dithiothreitol, while IgG1 is more resistant, and requires at least 1 mM dithiothreitol, and sometimes more. At 1 mM, there will be considerable reduction of interchain disulfide bonds, but antibody activity will probably remain intact. If cleavage is incomplete, the problem will usually be solved by increasing the concentration of reducing agent.

All rat IgG subclasses are easily cleaved by papain in the presence of reducing agent (Rousseaux *et al.*, 1980, 1983; Rousseaux and Rousseaux-Prevost,

1990). It was found that Fab fragments could be produced from all rat IgG subclasses by treatment with 1% (w/w) papain at pH 7.0 in the presence of 10 mM cysteine for 2–4 h at 37°C. Completeness of digestion must be ascertained by SDS–PAGE in every case. Fragments could be separated by gel filtration, ion exchange chromatography or protein A-Sepharose (IgG2c only).

Separation of Fab and Fc

Separation of Fab fragments from Fc and residual intact IgG may usually be accomplished with ease. In the case of IgG subclasses with strong affinity for staphylococcal protein A (mouse IgG2a, IgG2b and IgG3; rat IgG2c), the digest is passed over a column of protein A-Sepharose (Goding, 1976, 1978). (Failure to inactivate the papain with iodoacetamide may result in destruction of the column.) The Fab fragments pass through the column, while Fc and undigested IgG are bound (Fig. 9.5, upper panel). Note that the Fc fragment of mouse IgG3 is very insoluble in water and may precipitate. The column may be regenerated and the bound material recovered by elution with 0.1 M glycine-hydrochloride, pH 2.5.

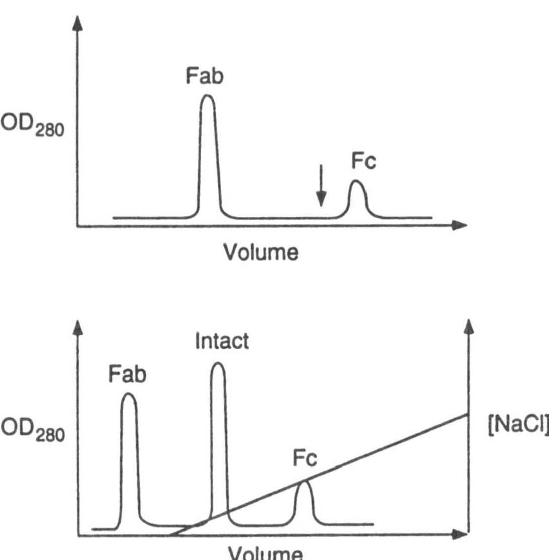

Fig. 9.5. Two methods for separation of Fab fragments of IgG from Fc fragments. Upper panel: separation on protein A-Sepharose. The IgG digest is passed over protein A-Sepharose; the Fab fragment drops through the column. When the pen recorder reaches baseline, the elution buffer (arrow; see text) is applied, and the Fc fragment and any undigested IgG eluted. Lower panel: separation by ion exchange chromatography. The digest is passed over a column of DEAE-cellulose. The Fab fragment drops through the column or elutes early in the gradient, followed by undigested IgG, and finally Fc.

Alternatively, the fragments may be separated by ion exchange chromatography (see Section 9.2.3), which is the method of choice for IgG subclasses that do not bind to protein A. The digest should be transferred to 0.01M Tris-HCl, pH 8.0, by gel filtration on a small column or by dialysis or dilution. (The Fc fragment of IgG3 will precipitate in this buffer, and should be removed by centrifugation.) The Fab fragments are generally more basic than the Fc or the intact molecule; they will emerge in the 'drop-through' or early in the salt gradient (Fig. 9.5, lower panel). The intact molecules and Fc fragments emerge at higher salt concentrations. (Occasionally, the Fab and Fc fragments may have similar charge, and may coelute from DEAE-cellulose. This is mainly a problem for classes other than IgG1, and is an indication for use of protein A rather than DEAE to separate them.)

A potential disadvantage of Fab fragments of monoclonal antibodies

If both antigen-combining sites of an intact IgG molecule have simultaneous and equal access to identical determinants on a multimeric antigen, the functional affinity (avidity) may approach the product of the individual affinities, because if one site temporarily unbinds, the antibody remains bound via the other site (see Fig. 2.4). The strength of binding of monovalent Fab fragments might therefore be weaker than that of the intact molecule. Thus, if a monoclonal antibody were to have a low affinity for its antigen (as if often the case) the problem would be aggravated by the use of Fab fragments. Normally stable antigen–antibody complexes might dissociate during washing or prolonged incubations.

The effect of valency on the strength of binding is vividly shown in a series of elegant experiments (Ways and Parham, 1983; Parham 1984). A monoclonal antibody to HLA antigens was shown to bind bivalently to cells. It was found that the binding constant for intact or $F(ab')_2$ fragments of this antibody was at least 68-fold greater than for monovalent fragments (Ways and Parham, 1983), and the true difference may be much greater. Bivalency was found to be essential for binding of two other monoclonal antibodies to HLA, as binding was undetectable if monovalent Fab fragments were used (Parham, 1984).

9.5.2 Preparation of $F(ab')_2$ Fragments of Mouse and Rat IgG

Pepsin is a nonspecific protease which is only active at acid pH, and is irreversibly denatured at neutral or alkaline pH as it moves from the stomach into the duodenum. *It is therefore essential to make up the pepsin solution from powder in an acidic buffer. If the pepsin is made up in PBS at neutral pH, it will be denatured and the digestion will not work.*

Treatment of human or rabbit immunoglobulin with pepsin (enzyme:substrate 1:100 at 37°C overnight at pH 4.5 in acetate buffer) usually results in cleavage at the *C*-terminal side of the inter-heavy chain disulfide bonds (Figs 2.2 and 5.1). The resulting large fragment is named $F(ab')_2$, because it contains two antigen-combining sites (Fig. 2.4). $F(ab')_2$ fragments do not bind to Fc receptors. The production of $F(ab')_2$ fragments of monoclonal antibodies would be very useful, because they would be expected to have considerably higher avidity than Fab (Fig. 2.4).

Unfortunately, the rote application of these procedures to mouse and rat IgG does not always produce satisfactory results. While there have been many claims for production of $F(ab')_2$ fragments of mouse IgG, experience using rigorous techniques to assess the digestion products indicates that the recommended procedures do not always work (see Oi and Herzenberg, 1979).

Mouse IgG

The problem of production of $F(ab')_2$ fragments of mouse IgG has been investigated by Lamoyi and Nisonoff (1983). Immunoglobulins were purified by conventional techniques, and digestion was carried out at pH 4.2 (IgG1 and IgG2a) or pH 4.5 (IgG3). The enzyme:substrate ratio was 1:33 (w/w) and the temperature was 37°C. No reducing agents were used. $F(ab')_2$ fragments were obtained in good yield from IgG1, IgG2a and IgG3 proteins, with the relative rates of digestion being IgG3 > IgG2a > IgG1. One IgG1 protein out of four tested was rapidly degraded, as were all of five different IgG2b proteins.

Optimal digestion times varied depending on the antibody class, and times which resulted in no undigested IgG generally resulted in somewhat lower yields, presumably owing to further digestion into smaller fragments. Suitable times were found to be 8 h for IgG1 (yield *c.* 70%), 4–8 h for IgG2a (yield 25–50%) and 15 min for IgG3 (yield *c.* 60%). It was not possible to find suitable conditions for production of $F(ab')_2$ fragments from mouse IgG2b. In general, it appears that the best yields are obtained by accepting a certain amount of undigested IgG, which may be removed by gel filtration (Table 9.1).

The production of $F(ab')_2$ fragments of mouse IgG has also been analysed in detail by Parham (1983). It was found that IgG1 was rather resistant to cleavage with pepsin, but stable $F(ab')_2$ fragments could be produced by cleavage in 0.1 M citrate, pH 3.5 at 37°C, with an IgG concentration of 1–2 mg/ml and a pepsin concentration of 25 μg/ml. Cleavage was generally complete by 8 h, and yields varied from 25–90% for seven different proteins. The Fc portion was completely degraded. An optimal pH of 3.5 for pepsin digestion of mouse IgG1 was also found by Morimoto and Inouye (1992).

Other procedures for production of $F(ab')_2$ fragments from mouse IgG1 have been published by Kurkela *et al.* (1988) who used papain digestion in the absence of thiols, and found that no Fab fragments were obtained, but $F(ab')_2$ fragments could be recovered in 50% yield. Mariani *et al.* (1991) found that digestion of mouse IgG1 with the thiol protease ficin in 50 mM Tris-HCl, 2 mM EDTA, pH 7.0 containing 1 mM cysteine at 37°C at an enzyme:substrate ratio of 1:30 for 2–8 h gave high yields of $F(ab')_2$ fragments. The reaction was stopped by the addition of 1:10 v/v of 100 mM *N*-ethylmaleimide. The procedure was less effective for mouse IgG2a and IgG2b, as these classes tended to progress rapidly to Fab fragments under the given conditions.

In the case of mouse IgG2a, Parham found that a pH of 4.1 was optimal for pepsin digestion for the three proteins examined, but that it was not possible to find conditions that were general and ideal. Pepsin was capable of cleaving on both sides of the three inter-heavy-chain disulfide bonds. At short times, $F(ab')_2$ predominated, but was contaminated by intact IgG. At longer times, a monomeric Fab′ fragment became the major product, and a compromise between yield and purity was unavoidable.

Digestion of mouse IgG2b with pepsin was unsatisfactory and resulted in asymmetrical cleavage. IgG3 was not examined.

Rea and Ultee (1993) made the novel observation that 0.5–0.8 M ammonium sulfate could be used to control the digestion of mouse monoclonal antibodies by pepsin. They found that pepsin sometimes bound to the antibodies and precipitated them, possibly owing to an electrostatic interaction that could be overcome with ammonium sulfate, which accelerated the digestion in some cases and slowed it in others. In some cases, ammonium sulfate could be used successfully to control the digestion of mouse IgG2a by pepsin, but not for IgG2b. The effect of ammonium sulfate appeared to depend to a great extent on the individual antibody.

Yamaguchi *et al.* (1995) have recently shown that mouse IgG2a and IgG2b can be cleaved into $F(ab')_2$ and Fc using lysyl endopeptidase (see also Kim *et al.*, 1994a, b). The reaction conditions were 50 mM Tris-HCl pH 8.5 at 37°C for 1–6 h, using an enzyme:substrate ratio of 1:50 to 1:1000, and the reaction was terminated by addition of TLCK (tosyl-L-phenylalanine-chloromethyl ketone) to 30 mM final concentration. The reaction products could be separated by affinity chromatography on protein A-Sepharose. This is the first report of reliable production of $F(ab')_2$ fragments of mouse IgG2b, and it will be interesting to see if the results are generally applicable to all monoclonal antibodies of this class.

As there are often very large differences in amino acid sequences between mouse IgG subclasses of different allotypes (see Section 6.1), it cannot be assumed that the results obtained with BALB/c immunoglobulins can be applied to IgG of other allotypes.

In summary, the production of $F(ab')_2$ fragments from mouse IgG1 is possible using pepsin or ficin, under carefully controlled conditions. Production of $F(ab')_2$ fragments of mouse IgG3 is also possible using pepsin.

The production of $F(ab')_2$ fragments from mouse IgG2a is sometimes possible, and cleavage into $F(ab')_2$ and Fc by lysyl endopeptidase is very promising. Production of $F(ab')_2$ fragments of mouse IgG2b has been very problematic, but recent work suggests that lysyl endopeptidase may provide a solution, but it will be necessary to determine whether this protocol works for all IgG2b monoclonal antibodies.

Rat IgG

Rousseaux and colleagues have studied the production of $F(ab')_2$ fragments of rat IgG in detail (Rousseaux *et al.*, 1983; Rousseaux and Rousseaux-Prévost, 1990). Their results may be summarized as follows. Rat IgG1 and to a lesser extent IgG2a were resistant to pepsin, but complete cleavage to $F(ab)_2$ fragments occurred if the IgG (15 mg/ml) was dialysed against 0.1 M formate buffer, pH 2.8 for 16 h at 4°C, followed by 0.1 M acetate buffer, pH 4.5 before peptic digestion. Cleavage into $F(ab')_2$ was complete after 4 h at 37°C with 1% (w/w) pepsin, and no other fragments were seen. It would appear that the pH 2.8 treatment caused an irreversible change in the conformation of the Fc, probably the $C_{\gamma2}$ domain.

Cleavage of rat IgG2b with pepsin was less satisfactory, and gave results reminiscent of mouse IgG2a (see Parham, 1983, and above). Optimal conditions for pepsin digestion of rat IgG2b consisted of an enzyme:substrate ratio of 5% and an incubation time of 18 h at 37°C at pH 4.5, without the preliminary treatment at pH 2.8.

A more satisfactory way to generate $F(ab')_2$-like fragments of rat IgG2b in good yield was found using staphylococcal V8 protease (3% w/w) in 0.1 M sodium phosphate pH 7.8, for 4 h at 37°C. Digestion was stopped by rapid freezing.

Rousseaux *et al.* (1983) found that IgG2c was the most susceptible of the rat IgG subclasses to peptic cleavage. Using an enzyme:substrate ratio of 1% (w/w) for 4 h at 37°C in 0.1 M citrate pH 4.5, rat IgG2c was completely cleaved into $F(ab')_2$, and the Fc subfragment pFc′ was also obtained. Longer times resulted in the appearance of monomeric Fab′ fragments.

The foregoing should serve to emphasize the importance of testing individual IgG proteins for peptic cleavage, and warn against the rote use of 'recipes'. Fragmentation must always be monitored by SDS–PAGE, and failure to do so invites trouble.

Rousseaux and colleagues have recently published a procedure for the fragmentation of rat IgE into $F(ab')_2$-ε and Cε4 fragments (Rousseaux and Rousseaux-Prévost, 1990). IgE (10 mg/ml in PBS) was incubated for 10 min at 37°C with 10 mM cysteine, and then digested with papain (1% w/w) at 37°C in PBS pH 7.4 for 10–30 min, and the reaction stopped with iodoacetamide. The fragments were purified by gel filtration.

9.5.3 Proteolytic Fragmentation of Mouse IgM

Information concerning the proteolytic fragmentation of mouse IgM has been limited, but recently several procedures have been published. Simply applying the procedures developed for human IgM is unlikely to result in success. Bourgois *et al.* (1977) published a method for the fragmentation of mouse IgM into Fab and Fc fragments using tryptic digestion of the native proteins at 37°C, but the results have not been reproduced by others (Matthew and Reichardt, 1982). Shimizu *et al.* (1974) were able to digest human IgM with trypsin in the presence of 5 M urea, resulting in Fab and $(Fc)_5$. The Fc began at residue 326, and the majority of the $C_{\mu}2$ domain was destroyed. Kehry *et al.* (1982) digested mouse IgM with trypsin in the presence of 5M urea (enzyme:substrate 1:100; 25°C for 18 h). They apparently obtained a cleavage at residue 220 (near the end of C_H1 domain), cleaving the IgM into Fab and $(Fc)_5$ fragments. The Fab and $(Fc)_5$ fragments were separated by gel filtration on Ultrogel AcA 22. However, the yield was poor and the procedure had to be modified for individual monoclonal IgM proteins (L. Hood, personal communication).

Another protocol was described by Matthew and Reichardt (1982). Mouse IgM (1 mg/ml in 50 mM Tris-HCl, 150 mM NaCl, 29 mM $CaCl_2$, pH 8.0) was digested with 0.01 mg/ml TPCK trypsin for 5 h at 37°C. Mercaptoethanol was added to 10 mM. After five additional minutes at 37°C, the solution was adjusted to 60 mM iodoacetamide and left at room temperature for 10 min. The samples were then dialysed against four changes of 1 l PBS over 48 h. The combination of tryptic digestion followed by reduction and alkylation produced active fragments of molecular weight 230 000 and 110 000. Both trypsinization and reduction were necessary. Analysis of the fragments indicated that the light chains were intact and the heavy chains intact or nearly intact. The M_r 230 000 fragments appeared to consist of two light chains and two heavy chains, while the M_r 110 000 fragments consisted of one light chain and one heavy chain.

The mechanism of this cleavage is not entirely clear. The authors postulated disulfide rearrangements and proteolytic attack on the J chain, reducing stability of the pentamer. It is interesting to note that the *C*-terminal 20 residues contain three potential tryptic cleavage sites (Kehry *et al.*, 1982). This region of the molecule is thought to be crucial to the polymerization process, and cleavage at these residues might result in depolymerization with virtually no change in apparent size of the heavy chain. Contamination of IgM with the trypsin inhibitor α_2-macroglobulin (Travis and Salvesen, 1983) may prevent digestion.

A new procedure for fragmentation of rat IgM was recently described by Inouye and Morimoto (1994). This procedure worked for four different rat IgM monoclonal antibodies and although it has not been tried, it may work for mouse IgM. Rat IgM was incubated at 1–2 mg/ml in 100 mM citrate buffer, pH 4.5. Use of lower pH resulted in random cleavage. Porcine pepsin was then added at a ratio of 1:200 w/w. Digestion proceeded with gentle stirring at 37°C for 2 h, and was stopped by raising the pH to denature the pepsin. The heavy chains (75 kDa intact) were digested into two fragments of 44 and 48 kDa, and the ratio of intensity of each did not change during the digestion. The light chain appeared to remain intact. The digest was purified by hydrophobic interaction chromatography on a TSKgel Phenyl-5PW column, eluted with a decreasing gradient of ammonium sulfate. Further analysis indicated that the fragments so produced were $F(ab')_{2\mu}$, with an intact molecular mass of about 147–153 kDa under non-reducing conditions. Recovery was about 50% of theoretical yield. The heterogeneity of the heavy chain fragments was ascribed to differences in carbohydrate content.

Yet another method for production of $F(ab')_2$ fragments from mouse IgM was published by Pascual and Clem (1992), who obtained 90% yield by digesting with pepsin in 0.02 M sodium acetate pH 4.0, with an enzyme:substrate ratio of 1:25 (w/w) at 4°C for 18–24 h, after which less than 2% of IgM was intact. The fragments were purified by gel filtration using Bio-gel A5M, and it was shown that they continued to bind antigen and were divalent.

9.6 Radiolabelling of Monoclonal Antibodies

Radiolabelling of antibodies involves the same principles and procedures that apply to proteins in general, and the reader is referred to Chapter 10 for a more detailed discussion.

9.6.1 Radioiodination of Antibodies

Antibodies may be radioiodinated to high specific activity with ^{125}I by the chloramine-T method (Section 10.4.1). Incorporation of one iodine per IgG molecule corresponds to a specific activity of approximately 12 μCi/μg. Experience with affinity-purified polyclonal antibodies from goat, sheep or rabbit indicates that it is realistic to expect at least 20–40% of TCA-precipitable counts to bind to antigen; presumably, the remaining molecules are damaged by the iodination procedure.

Damage to the antibodies could occur by at least two mechanisms. The first is by exposure to oxidizing agents, which are a feature of the chloramine-T method, the Iodogen method, Iodobeads, and the lactoperoxidase method, but may be avoided by the use of the iodine monochloride method (see below) or the Bolton–Hunter reagent. The second cause of damage by iodination could be due to the iodination of a crucial tyrosine residue in the antibody combining site. The majority of monoclonal antibodies can be iodinated on tyrosine residues without damaging the binding site, but if damage does occur, methods that avoid labelling tyrosine can be used, such as the Bolton–Hunter reagent.

Mouse antibodies in particular have a reputation for being easily damaged by radioiodination. However, a detailed study of the distribution of murine Ia antigens was carried out using mouse alloantibodies labelled with ^{125}I by the lactoperoxidase technique (Goding *et al.*, 1975). Tsu and Herzenberg (1980) were able to radiolabel mouse anti-allotype antibodies using the chloramine-T method. In this case, their combining sites were protected from damage by performing the iodination while the antibody was bound to antigen.

Ballou *et al.* (1979) used the Iodogen method (Section 10.4.5) to radiolabel antitumour antibodies. There are numerous other reports of the successful use of radioiodinated monoclonal antibodies. However, it would appear that many unpublished failures have also occurred, and that loss of antigen binding after iodination is a relatively common event (Nussenzweig *et al.*, 1982).

If a particular monoclonal antibody is damaged by radioiodination using the chloramine-T procedure, a trial of other procedures (Iodogen, lactoperoxidase, or Iodobeads) is worthwhile.

Yet another procedure for radioiodination of antibodies is the iodine monochloride method (Contreras *et al.*, 1983; Nicola and Metcalf, 1988), which avoids the use of oxidizing agents, and works by isotopic exchange. The only oxidizing agent that the antibody comes into contact with is iodine monochloride. The procedure is described in more detail in Chapter 10.

All the above methods label tyrosine residues, and failure to retain activity might indicate the presence of tyrosine in the antigen-binding site. The Bolton–Hunter reagent (see Section 10.4.4) labels lysine residues, and may be coupled under extremely mild conditions. In some cases, however, even the Bolton–Hunter reagent may cause damage (Nussenzweig *et al.*, 1982). If radioiodination cannot be carried out successfully using any of these methods, one might try a different antibody, a different isotope, or the use of ^{125}I-labelled anti-immunoglobulin or staphylococcal protein A. Finally, one might consider conjugation of the antibody with biotin, and use of ^{125}I-labelled avidin or streptavidin as a sandwich reagent (see Section 10.3).

9.6.2 Extrinsic Labelling of Antibodies with Tritium and ^{35}S

Although antibodies may be radiolabelled to the highest specific activity with iodine, there are a number of situations where tritium labelling may be useful. The very low energy of tritium (E_{max} = 18 keV) allows very precise localization of grains in autoradiographic experiments (Cuello *et al.*, 1980).

Tack *et al.* (1980) have shown that antibodies and other proteins may be tritiated to very high specific activities by reductive methylation. The specific activities obtained approach those routinely achieved by radioiodination. The method was applied by Wilder *et al.* (1979) to antibodies and staphylococcal protein A and was the only successful method of extrinsically labelling the monoclonal antidendritic cell antibody of Nussenzweig *et al.* (1982). The practical details are given in Section 10.5.

An alternative procedure involves the use of *N*-succinimidyl [2,3-^{3}H]-propionate. The compound is available from Amersham at a specific activity of 30–60 Ci/mmol, or about 1–3% of that of the ^{125}I-labelled analogue. It may be useful in situations where the low energy of tritium is an advantage, and where low specific activity is acceptable. A related ^{35}S-labelled compound is also available from Amersham at a specific activity of *c.* 1000 Ci/mmol. The compound is known as ^{35}SLR (*t*-butoxycarbonyl-L-[^{35}S]methionine-*N*-hydroxysuccinimide ester). Conditions for labelling are exactly as for the Bolton–Hunter reagent (see Section 10.4.4).

It should be noted that succinimide esters are rapidly hydrolysed in water. They must be stored dessicated, and allowed to warm to room temperature before opening the bottle, to avoid condensation. The reaction buffer must not contain any nucleophiles such as amines, Tris buffer or azide. They should be dissolved in dimethyl sulfoxide immediately before use (see Section 8.8.3).

9.6.3 Biosynthetic Labelling of Monoclonal Antibodies

The biosynthetic incorporation of radiolabelled amino acids into monoclonal antibodies is easily performed. The principles are discussed in detail in Section 10.6.2. If the highest possible specific activity is desired, the use of [^{35}S]-methionine (*c.* 1000 Ci/mmol) or [^{3}H]-lysine, arginine, leucine or phenylalanine (20–200 Ci/mmol) are recommended (Galfrè and Milstein, 1981).

It should be noted that [^{35}S]-methionine is commonly supplied in aqueous solution containing 0.1% (1.2×10^{-2} M) mercaptoethanol. Myeloma and hybridoma cells are very sensitive to mercaptoethanol, and cannot withstand concentrations of more than *c.* 5×10^{-5} M. If higher concentrations of isotope are needed, the methionine must be lyophilized immediately prior to use to remove the mercaptoethanol. Similar considerations apply to amino acids supplied in solutions containing ethanol (see Section 10.6.2).

Typically, 10^{7} hybridoma cells are labelled for 2–5 h in 1–2 ml medium containing 50 μCi isotope/ml. At the end of the labelling period, cells are removed by centrifugation. It is advisable to add a small amount of nonradioactive amino acid at the end of the labelling period. Immunoglobulin will be the major labelled protein in the supernatant (Ledbetter *et al.*, 1979), which may often be used without further purification. Specific activities of up to 1000 Ci/mmol may be achieved by biosynthetic incorporation (Cuello *et al.*, 1980).

9.7 Conditions for Stability and Storage of Monoclonal Antibodies

As mentioned repeatedly throughout this book, individual monoclonal antibodies may have highly individual characteristics. It is thus to be expected that they may differ greatly in their susceptibility to damage by environmental factors, and it is impossible to formulate conditions, under which every clonal product would be unconditionally stable. However, adherence to a few simple principles will minimize the chances of damage. If attention is paid to the avoidance of denaturation and proteolysis, the long-term storage of monoclonal antibodies presents few problems.

9.7.1 Denaturation of Antibodies

Denaturation may range from a slight and reversible conformational change to a drastic loss of solubility and massive irreversible aggregation. An appreciation of the forces that govern protein folding (Anfinsen, 1973) has led to the understanding that even fully denatured proteins may sometimes be renatured, with full restoration of biological activity. Provided that the disulfide bonds are allowed to reform correctly, totally denatured ribonuclease may be restored to full enzymic activity (Anfinsen, 1973). These principles have been applied successfully to many other enzymes (Hager and Burgess, 1980).

Recovery of antigen-binding activity by renaturation of extensively denatured antibody in the absence of antigen (Haber, 1964; Whitney and Tanford, 1965) was classical proof that their primary structure was sufficient to specify their final tertiary structure and biological activity. Antibody molecules can often, but not always, recover from extensive denaturation by urea, guanidine and extremes of pH (see also Chapter 11). Polyclonal antisera generally require quite harsh conditions (6–8 M urea, 5–6 M guanidine-hydrochloride, pH < 3 or > 10) before the antigen–antibody bond is disrupted. Once physiological conditions are restored, antibody activity usually returns.

It is to be expected that some monoclonal antibodies will be irreversibly denatured by such conditions. The subject of the reversibility of the antigen–antibody bond is discussed in more detail in Chapter 11.

Antibodies may be denatured by other physical processes. Excessive heat should be avoided. Most antibodies will survive heating at 56°C for 30 min, but some will not. Denaturation of antibodies by heat is likely to be irreversible. Similarly, frothing of protein solutions is a potent cause of irreversible denaturation. In the case of ovalbumin, denaturation by heat or frothing results in fried eggs and meringues, respectively (McGee, 1984).

Freeze–thaw cycles are also potentially damaging, particularly to IgM and murine IgG3. Some IgG antibodies will survive multiple freeze–thaw cycles, while others will become irreversibly aggregated. Most monoclonal antibodies

Table 9.4 Storage of monoclonal antibodies

Form of antibody	Recommended storage*
Serum or Ascites	Short-term storage at –20°C, long term at –70°C, in small aliquots. Long-term storage of serum at 4°C is usually satisfactory with addition of an equal volume of saturated ammonium sulfate. Ascites may contain protease activity, and should be frozen.
Culture Supernatant	Short-term storage at 4°C as a sterile solution containing 10 mM sodium azide or 0.005% merthiolate is usually safe, provided infection does not occur. Alternatively, freeze in aliquots, preferably at –70°C.
Purified IgG and IgM	Neutral pH; salt concentration 100–200 mM; protein concentration 1–10 mg/ml. Storage at 4°C with 10 mM sodium azide or 0.005% merthiolate is usually safe over long periods, but occasionally proteolytic breakdown occurs. Storage at –20°C in PBS in 50% glycerol is usually safe over long periods. (50% glycerol does not freeze at –20°C.) Alternatively, store at –70°C. IgM is particularly susceptible to denaturation by freeze–thaw cycles.
Antibodies conjugated with biotin, fluorochromes, enzymes etc.	Very prone to aggregation by freezing. Store at 4°C as for IgG and IgM. Alternatively, store in 50% glycerol at –20°C.

* Frozen samples should be stored in multiple small aliquots, and the number of freeze-thaw cycles kept to a minimum. Some domestic freezers have automatic defrost cycles, and some are barely capable of freezing reliably. Repeated freeze-thaw cycles are a potent cause of denaturation.

will survive one or two freeze–thaw cycles, but occasional ones may lose activity. The number of freeze–thaw cycles should be limited by freezing in multiple small aliquots. Samples stored at –20°C are not fully frozen, and storage at –70°C is much more reliable, particularly for the longer term.

A seldom-used but useful technique for storage of purified antibodies is as a slurry in 50% saturated ammonium sulfate at 4°C (see Table 9.4).

9.7.2 Degradation of Antibodies by Proteases

Provided that they are not exposed to proteases, purified monoclonal antibodies may be stored at 4°C for years without significant loss of activity. Proteolysis upon storage may arise from proteases in the antibody source, or perhaps more commonly from microbial contamination. Solutions may be protected from microbial growth by preservatives such as sodium azide (10 mM) or merthiolate (0.005%), although a few organisms are resistant to these agents. Sterilization by membrane filtration is safer, but more inconvenient.

Although serum contains a number of proteases, they are mostly present in an inactive form (Travis and Salvesen, 1983). Serum contains a number of protease inhibitors, such as α_2-macroglobulin and α_1-antitrypsin. Ascites is more prone to proteolytic breakdown than serum. Bruck *et al.* (1982) described a simple procedure for obtaining protease-free IgG from ascitic fluid.

In some respects, the requirements for minimization of proteolytic activity (freezing) and minimization of denaturation (not freezing) are contradictory. Proteolytic degradation may be eliminated by lyophilization (freeze-drying), but it would appear that many monoclonal antibodies are irreversibly denatured by the process. Dráber *et al.* (1995) recently showed that mouse IgM monoclonal antibodies can be lyophilized with preservation of biological activity if lyophilization is performed in the presence of 0.25 M trehalose (α-D-glucopyranosyl-α-D-glycopyranoside). Recommended storage conditions for monoclonal antibodies are given in Table 9.4.

References

Åkerström, B., Brodin, T., Reis, K. and Björck, L. (1985) Protein G: A powerful tool for binding and detection of monoclonal and polyclonal antibodies. *J. Immunol.* **135**, 2589–2592.

Anfinsen, C. B. (1973) Principles that govern the folding of protein chains. *Science* **181**, 223–230.

Ballou, B., Levine, G., Hakala, T. R. and Solter, D. (1979) Tumor location detected with radioactively labeled monoclonal antibody and external scintigraphy. *Science* **206**, 844–847.

Bazin, H. (1990) 'Rat Hybridomas and Rat Monoclonal Antibodies'. CRC Press, Boca Raton.

Belew, M., Juntti, N., Larsson, A. and Porath, J. (1987) A one-step purification method for monoclonal antibodies based on salt-promoted adsorption chromatography on a 'thiophilic' adsorbent. *J. Immunol. Methods* **102**, 173–182.

Bill, E., Lutz, U., Karlsson, B-M., Sparrman, M. and Allgaier, H. (1995) Optimization of protein G chromatography for biopharmaceutical monoclonal antibodies. *J. Molec. Recognition* **8**, 90–94.

Björck, L. and Akerström, B. (1990) Streptococcal protein G. *In* 'Bacterial Immunoglobulin-Binding Proteins' (M. D. P. Boyle, ed.), pp. 113–126. Academic Press, New York.

Bourgois, A., Abney, E. and Parkhouse, R. M. E. (1977) Mouse immunoglobulin receptors on lymphocytes: Identification of IgM and IgD molecules by tryptic cleavage and a postulated role for cell surface IgD. *Eur. J. Immunol.* **7**, 210–213.

Böttcher, I., Hammerling, G. and Kapp, J-F. (1978) Continuous production of monoclonal mouse IgE antibodies with known allergenic specificity by a hybrid cell line. *Nature* **275**, 761–762.

Bruck, C., Portetelle, D., Glineur, C. and Bollen, A. (1982) One-step purification of mouse monoclonal antibodies from ascitic fluid by DEAE Affi-gel blue chromatography. *J. Immunol. Methods* **53**, 313–319.

Commission of the European Communities, CEC (1992) 'Note for Guidance: Biotech Headings for Notice to Applicants, Part II'.

Contreras, M. A., Bale, W. and Spar, I. L. (1983) Iodine monochloride (ICl) iodination techniques. *Methods Enzymol.* **92**, 277–292.

Cuello, A. C., Milstein, C. and Priestly, J. V. (1980) Use of monoclonal antibodies in immunocytochemistry with special references to the central nervous system. *Brain Res. Bull.* **5**, 575–587.

Der Balian, G. P., Slack, J., Clevinger, B. L., Bazin, H. and Davie, J. M. (1980) Subclass restriction of murine antibodies. III. Antigens that stimulate IgG3 in mice stimulate IgG2c in rats. *J. Exp. Med.* **152**, 209–218.

Di Sabato, G., Langone, J. J. and Van Vunakis, H. (1985) Immunochemical Techniques. Section 1. Purification and Characterization of Serum Immunoglobulins. *Methods Enzymol.* **116**, 1–200.

Dráber, P., Dráberova, E. and Nováková, M. (1995) Stability of monoclonal IgM antibodies freeze-dried in the presence of trehalose. *J. Immunol. Methods* **181**, 37–43.

Eshhar, Z., Ofarim, M. and Waks, T. (1980) Generation of hybridomas secreting murine reaginic antibodies of anti-DNP specificity. *J. Immunol.* **124**, 775–780.

Ey, P. L., Prowse, S. J. and Jenkin, C. R. (1978) Isolation of pure IgG1, IgG2a and IgG2b imunoglobulins from mouse serum using protein A-Sepharose. *Immunochemistry* **15**, 429–436.

Galfrè, G. and Milstein, C. (1981) Preparation of monoclonal antibodies: strategies and procedures. *Methods Enzymol.* **73**, 3–46.

Garvey, J. S., Cremer, N. E. and Sussdorf, D. H. (1977) 'Methods in Immunology', 3rd edn. (W. A. Benjamin, ed.). Reading, MA.

Goding, J. W. (1976) Conjugation of antibodies with fluorochromes: modifications to the standard methods. *J. Immunol. Methods* **13**, 215–226.

Goding, J. W. (1978) Use of staphylococcal protein A as an immunological reagent. *J. Immunol. Methods* **20**, 241–253.

Goding, J. W. (1980) Structural studies of murine lymphocyte surface IgD. *J. Immunol* **124**, 2082–2088.

Goding, J. W., Nossal, G. J. V., Shreffler, D. C. and Marchalonis, J. J. (1975) Ia antigens on murine lymphoid cells: distribution, surface movement and partial characterization. *J. Immunogenet.* **2**, 9–25.

Gorevic, P. D., Prelli, F. C. and Frangione, B. (1985) Immunoglobulin G (IgG). *Methods Enzymol.* **116**, 3–26.

Grey, H. M., Sher, A. and Shalitin, N. (1970) The subunit structure of mouse IgA. *J. Immunol.* **105**, 75–84.

Haba, S. and Nisonoff, A. (1991) Proteolytic digestion of mouse IgE. *J. Immunol. Methods* **138**, 15–23.

Haber, E. (1964) Recovery of antigen specificity after denaturation and complete reduction of disulfides in a papain fragment of antibody. *Proc. Natl Acad. Sci. USA* **52**, 1099–1106.

Hager, D. A. and Burgess, R. R. (1980) Elution of proteins from sodium dodecyl sulfate-polyacrylamide gels, removed of sodium dodecyl sulfate, and renaturation of enzymatic activity: results with sigma subunit of *Escherichia coli* RNA polymerase, wheat germ DNA topoisomerase, and other enzymes. *Anal. Biochem.* **109**, 76–86.

Hayzer, D. J. and Jaton, J-C. (1985) Immunoglobulin M (IgM). *Methods Enzymol.* **116**, 26–37.

Herrmann, S. H. and Mescher, M. F. (1979) Purification of the H-2K^k molecule of the murine major histocompatibility complex. *J. Biol. Chem.* **254**, 8713–8716.

Inouye, K. (1991) Chromatographic behaviours of proteins and amino acids on a gel-filtration matrix, TSK-GEL Toyopearl. *Agric. Biol. Chem.* **55**, 2129–2139.

Inouye, K. and Morimoto, K. (1993) Single-step purification of F(ab')$_{2\mu}$ fragments of mouse monoclonal antibodies (immunoglobulins M) by hydrophobic interaction high-performance liquid chromatography using TSKgel Ether-5PW. *J. Biochem. Biophys. Methods* **26**, 27–39.

Inouye, K. and Morimoto, K. (1994) Preparation of F(ab')$_{2\mu}$ fragments from rat IgM and their application to the enzyme immunoassay of mouse interleukin-6. *J. Immunol. Methods* **171**, 239–244.

Ishizaka, K. (1985) Immunoglobulin E (IgE). *Methods Enzymol.* **116**, 76–95.

Jehanli, A. and Hough, D. (1981) A rapid procedure for the isolation of human IgM myeloma proteins. *J. Immunol. Methods* **44**, 199–204.

Kehry, M. R., Fuhrman, J. S., Schilling, J. W., Rogers, J., Sibley, C. H. and Hood, L. E. (1982). Complete amino acid sequence of a mouse μ chain: homology among heavy chain constant region domains. *Biochemistry* **21**, 5414–5424.

Kim, H., Matsunaga, C., Yoshino, A., Kato, K. and Arata, Y. (1994a) Dynamical structure of the hinge region of immunoglobulin G as studied by ^{13}C nuclear magnetic resonance spectroscopy. *J. Mol. Biol.* **236**, 300–309.

Kim, H., Yamaguchi, Y., Masuda, K., Matsunaga, C., Yamamoto, K., Irimura, T., Takahashi, N., Kato, K. and Arata, Y. (1994b) *O*-glycosylation in hinge region of mouse immunoglobulin G2b. *J. Biol. Chem.* **269**, 12345–12350.

Knudsen, K. L., Hansen, M. B., Henriksen, L. R., Andersen, B. K. and Lihme, A. (1992) Sulfone-aromatic ligands for thiophilic adsorption chromatography: purification of human and mouse immunoglobulins. *Anal. Biochem.* **201**, 170–177.

Komisar, J. L., Fuhrman, J. A. and Cebra, J. J. (1982) IgA-producing hybridomas are readily derived from gut-associated lymphoid tissue. *J. Immunol.* **128**, 2376–2378.

Kurkela, R., Vuolas, L, and Vihko, P. (1988) Preparation of $F(ab')_2$ fragments from mouse IgG1 suitable for use in radioimaging. *J. Immunol. Methods* **110**, 229–236.

Lamoyi, E. and Nisonoff, A. (1983) Preparation of $F(ab')_2$ fragments from mouse IgG of various subclasses. *J. Immunol. Methods* **56**, 234–243.

Langone, J. J. (1982) Protein A of *Staphylococcus aureus* and related immunoglobulin receptors produced by streptococci and pneumonococci. *Adv. Immunol.* **32**, 157–252.

Ledbetter, J. A. and Herzenberg, L. A. (1979) Xenogeneic monoclonal antibodies to mouse lymphoid differentiation antigens. *Immunol. Rev.* **47**, 63–90.

Ledbetter, J. A., Goding, J. W., Tsu, T. T. and Herzenberg, L. A. (1979) A new mouse lymphoid alloantigen (Lgp 100) recognized by a monoclonal rat antibody. *Immunogenetics* **8**, 347–360.

Liu, F-T., Bohn, J. W., Ferry, E. L., Yamamoto, H., Molinaro, C. A., Sherman, L. A. Klinman, N. R. and Katz, D. H. (1980) Monoclonal dinitrophenyl-specific IgE antibody: preparation, isolation and characterization. *J. Immunol.* **124**, 2728–2737.

Malm, B. (1987) A method suitable for the isolation of monoclonal antibodies from large volumes of serum-containing hybridoma cell culture supernatants. *J. Immunol. Methods* **104**, 103–109.

Mariani, M., Camagna, M., Tarditi, L. and Seccamani, E. (1991) A new enzymic method to obtain high-yield $F(ab')_2$ suitable for clinical use from mouse IgG1. *Mol. Immunol.* **28**, 69–77.

Matthew, W. D. and Reichardt, L. F. (1982) Development and application of an efficient procedure for converting mouse IgM into small, active fragments. *J. Immunol. Methods* **50**, 239–253.

McGee, H. (1984) 'On Food and Cooking. The Science and Lore of the Kitchen'. Charles Scribner's Sons, New York.

Mestecky, J. and Kilian, M. (1985) Immunoglobulin A (IgA). *Methods Enzymol.* **116**, 37–76.

Morimoto, K. and Inouye, K. (1992) Single-step purification of $F(ab')_2$ fragments of mouse monoclonal antibodies (immunoglobulins G1) by hydrophobic interaction high performance liquid chromatography using TSKgel Phenyl-5PW. *J. Biochem. Biophys. Methods* **24**, 107–117.

Nicola, N. A. and Metcalf, D.(1988) Binding, internalization, and degradation of ^{125}I-multipotential colony-stimulating factor (interleukin-3) by FDCP-1 cells. *Growth Factors* **1**, 29–39.

Nussenzweig, M. C., Steinman, R. M., Witmer, M. D. and Gutchinov, B. (1982) A monoclonal antibody specific for mouse dendritic cells. *Proc. Natl Acad. Sci. USA* **79**, 161–165.

Oi, V. T and Herzenberg, L. A. (1979) Localization of murine Ig-1b and Ig-1a (IgG2a) allotypic determinants detected with monoclonal antibodies. *Mol. Immunol.* **16**, 1005–1017.

Parham, P. (1983) On the fragmentation of monoclonal IgG1, IgG2a and IgG2b from BALB/c mice. *J. Immunol.* **131**, 2895–2902.

Parham, P. (1984) The binding of monoclonal antibodies to cell surface molecules. Quantitative analysis of the reactions and cross-reactions of an antibody (MB40.3) with four HLA-B molecules. *J. Biol. Chem.* **259**, 13077–13083.

Pascual, D. W. and Clem, L. W. (1992) Low temperature pepsin proteolysis. An effective procedure for mouse IgM F(ab')$_2$ fragment production. *J. Immunol. Methods.* **146**, 249–255.

Perez-Montfort, R. and Metzger, H. (1982) Proteolysis of soluble IgE-receptor complexes: localization of sites on IgE which interact with the Fc receptor. *Mol. Immunol.* **19**, 1113–1125.

Rea, D. W. and Ultee, M. E. (1993) A novel method for controlling the pepsin digestion of antibodies. *J. Immunol. Methods* **157**, 165–173.

Rousseaux, J. and Rousseaux-Prévost, R. (1990) Fragmentation of rat monoclonal antibodies. *In* 'Rat Hybridomas and Rat Monoclonal Antibodies' (H. Bazin, ed.), pp. 209–217. CRC Press, Boca Raton.

Rousseaux, J., Biserte, G. and Bazin, H. (1980) The differential enzyme sensitivity of rat immunoglobulin G subclasses to papain and pepsin. *Mol. Immunol.* **17**, 469–482.

Rousseaux, J., Picque, M. T., Bazin, H. and Biserte, G. (1981) Rat IgG subclasses: differences in affinity to protein A-Sepharose. *Molec. Immunol.* **18**, 639–645.

Rousseaux, J., Rousseaux-Prévost, R. and Bazin, H. (1983) Optimal conditions for the preparation of Fab and F(ab')$_2$ fragments from monoclonal IgG of different rat IgG subclasses. *J. Immunol. Methods* **64**, 141–146.

Scopes, R. K. (1993) 'Protein Purification', 3rd edn. Springer-Verlag, New York.

Seppälä, I., Sarvas, H., Péterfy, F. and Mäkelä, O. (1981) The four sub-classes of IgG can be isolated from mouse serum by using protein A-Sepharose. *Scan. J. Immunol.* **14**, 335–342.

Shimizu, A., Watanabe, S., Yamamura, Y. and Putnam, F. W. (1974) Tryptic digestion of immunoglobulin M in urea: conformational lability of the middle part of the molecule. *Immunochemistry* **11**, 719–727.

Silverstein, A. M. (1989) 'A History of Immunology'. Academic Press, San Diego, p. 316.

Spiegelberg, H. (1985) Immunoglobulin D (IgD). *Methods Enzymol.* **116**, 95–101.

Tack, B. F., Dean, J., Eilat, D., Lorenz, P. E. and Schechter, A. N. (1980) Tritium labeling of proteins to high specific activity by reductive methylation. *J. Biol. Chem.* **255**, 8842–8847.

Tatum, A. H. (1993) Large scale recovery of biologically active IgM (95% pure) from human plasma obtained by therapeutic plasmapheresis. *J. Immunol. Methods* **158**, 1–4.

Tomasi, T. B. and Grey, H. M. (1972) Structure and function of immunoglobulin A. *Prog. Allergy* **16**, 81–213.

Travis, J. and Salvesen, G. S. (1983) Human plasma proteinase inhibitors. *Annu. Rev. Biochem.* **52**, 655–709.

Tsu, T. T. and Herzenberg, L. A. (1980) Solid-phase radioimmunoassays. *In* 'Selected Methods in Cellular Immunology' (B. B. Mishell and S. M. Shiigi, eds), pp. 373–397. Freeman, San Francisco.

US Food and Drug Administration (1993) 'Guide to Inspections of Validation of Cleaning Processes'. FDA.

Watanabe, M., Ishii, T. and Nariuchi, H. (1981) Fractionation of IgG1, IgG2a, IgG2b and IgG3 immunoglobulins from mouse serum by protein A-Sepharose column chromatography. *Japan. J. Exp. Med.* **51**, 65–70.

Ways, J. P. and Parham, P. (1983) The binding of monoclonal antibodies to cell-surface molecules. A quantitative analysis with immunoglobulin G against two alloantigenic determinants of the human transplantation antigen HLA-A2. *Biochem J.* **216**, 423–432.

Whitney, P. L. and Tanford, C. (1965) Recovery of specific activity after complete unfolding and reduction of an antibody fragment. *Proc. Natl Acad. Sci. USA* **53**, 524–532.

Wilder, R. L., Yuen, C. C., Subbarao, B., Woods, V. L., Alexander, C. B. and Mage, R. G. (1979) Tritium (^{3}H) radiolabeling of protein A and antibody to high specific activity: application to cell surface antigen radioimmunoassays. *J. Immunol. Methods* **28**, 255–266.

Yamaguchi, Y., Kim, H., Kato, K., Masuda, K., Shimada, I. and Arata, Y. (1995) Proteolytic fragmentation with high specificity of mouse immunoglobulin G. *J. Immunol. Methods* **181**, 259–267.

10 Analysis of Antigens Recognized by Monoclonal Antibodies

Monoclonal antibodies are the most highly selective yet versatile of all biochemical isolation tools. Their usefulness in the identification and isolation of particular molecules contained in extremely complex mixtures is unsurpassed. Antibodies can provided the information that a particular structure is present, localize it with extreme accuracy, and allow it to be isolated on an analytical or preparative scale. Subsequent analysis of the structure of the antigen may be made by standard biochemical techniques.

This chapter will discuss the use of monoclonal antibodies in the isolation and structural analysis of antigens. The major emphasis will be on proteins, but carbohydrate antigens will be mentioned briefly.

10.1 Cellular Distribution of Antigens Detected by Monoclonal Antibodies

It is not necessary to know the structure of the antigen to exploit the usefulness of monoclonal antibodies. One may use monoclonal antibodies as a probe for identification and classification of cells or other structures on a purely empirical basis. This sort of approach is not without its hazards, however. In several cases, the initial impression that an antigen detected by a particular monoclonal antibody is confined to a functional lineage of cells may be disproven by a more extensive tissue survey.

For example, the monoclonal antibody OKT-9 was originally thought to be specific for a subpopulation of immature human thymocytes (Reinherz *et al.*, 1980). Subsequently, the antigen detected by OKT-9 was found to be present on all cells undergoing proliferation, and was shown to be the receptor for the iron transport protein transferrin (Goding and Burns, 1981; Goding and Harris, 1981; Sutherland *et al.*, 1981; Trowbridge and Omary, 1981). Similarly, the plasma cell membrane glycoprotein PC-1 was initially thought to be confined to plasma cells (Takahashi *et al.*, 1971; Goding and Shen, 1982), but later work showed that it is also present on chondrocytes, epididymis, capillary endothelium in brain, and the distal convoluted tubule of the kidney (Harahap and Goding, 1988).

In some cases, the cellular distribution of a particular antigen can be seen to make biological sense. The association of the OKT-9 antigen with cellular proliferation is easily understood when it is known that it is critically involved in the uptake of iron.

In other instances, the distribution of cellular antigens may be harder to understand. A particular antigen may be found in a wide variety of cell types with no obvious relationship in differentiation lineage or function. Milstein and Lennox (1980) named these molecules 'jumping' differentiation antigens. Springer (1980) noted that jumping antigens are usually glycolipids or glycoproteins with a high carbohydrate content, and that their distribution often changes when a cell moves to a different tissue. Jumping antigens were postulated to be involved in cell migration.

10.2 Determination of the Biochemical Nature of the Antigen

In general, the naturally occurring antigens are proteins or carbohydrates. Carbohydrates may be present as part of more complex structures such as glycoproteins or glycolipids. Lipids lacking carbohydrate are seldom recognized by antibodies, because they tend to form micelles and membranes, and thus remove themselves from the aqueous environment.

The preliminary steps to determine whether an antigen is a protein or carbohydrate may involve a number of simple tests. Protein antigens are likely to be sensitive to proteases such as pronase or trypsin, destroyed by heating to 100°C, and to survive treatment with periodate. The converse is true for carbohydrate antigens (Layton, 1980). None of these tests is absolutely diagnostic. For example, some proteins are extremely resistant to proteolysis (Handman *et al.*, 1981) or heat (Gullick *et al.*, 1981; Alterman *et al.*, 1990). Certain amino acids, notably tyrosine and tryptophan, may be destroyed by periodate (Geoghegan *et al.*, 1980). Methionine is also susceptible to oxidation by periodate (Yamasaki *et al.*, 1982). A particular monoclonal antibody may recognize a glycoprotein by either the protein moiety or the carbohydrate prosthetic group. The conformation of a glycoprotein may be influenced by modification or absence of its carbohydrate (Varki, 1993). Despite all these caveats, the overall pattern of the effects of treatment of the antigen with heat, proteases and periodate will often give a hint of its nature.

10.2.1 Glycolipid and Other Carbohydrate Antigens

Carbohydrates and glycolipids are highly immunogenic and antigenic (see Section 4.8). When mice are immunized with whole cells from rats of humans, antibodies against glycolipids are commonly produced (Hakomori and Kannagi, 1986; Stern *et al.*, 1978; Young and Hakomori, 1981). While the distribution of individual glycolipid antigens is often confined to certain cell types (Feizi, 1981; Hakomori, 1981; Hakomori, 1984; Wiegandt, 1985; Rademacher *et al.*, 1988), it is usually difficult to make any biological sense of their distribution (see Section 10.2).

In many cases, carbohydrate antigens have a highly repetitive polymeric structure, with large numbers of identical antigenic determinants. In these cases, the strength of the 'signal' obtained by binding of monoclonal antibodies may be much greater than in cases in which there is only a single determinant per molecule. The multivalent structure of the antigen may also allow the formation of very strong precipitin lines in Ouchterlony analysis (see Section 9.1.1).

In addition to generating extremely strong signals, the polymeric nature of many carbohydrate antigens facilitates their detection by allowing 'two-site' immunoradiometric assays (IRMA). In these assays, the antigen is immobilized by a nonradioactive monoclonal antibody attached to a solid surface such as nitrocellulose, and is then detected by enzyme- or ^{125}I-labelled monoclonal antibody. In other words, the antigen is the 'meat in the sandwich' between two monoclonal antibodies. The assay may be used with the two antibodies having the same or different specificity; the only requirement is that both antibodies must be able to bind simultaneously to the same molecule. Two-site assays of this type allow antigens to be detected and quantified in very complex mixtures without any need for purification.

The power of such an approach is exemplified by the analysis of the glycoconjugate of *Leishmania*, a protozoan parasite that grows in macrophages. The glycoconjugate acts as a receptor via which the parasite attaches to and penetrates the macrophage (Handman and Goding, 1985). Detection of the glycoconjugate by conventional biosynthetic labelling with tritiated sugars requires many days of autoradiographic exposure, but a simple nitrocellulose 'dot blot' assay based on binding of ^{125}I-labelled monoclonal antibodies allows detection in a few hours (Handman *et al.*, 1984; Handman and Jarvis, 1985; Fig. 10.1). Further improvements in sensitivity, particularly using enzyme-conjugated antibodies and readout by enhanced chemiluminescence could probably reduce the readout time to a matter of seconds.

If it is desired to select for hybridoma clones secreting monoclonal antibodies against glycolipid antigens, the assays of Smolarsky (1980) and Handman and Jarvis (1985) may be useful. The analysis of the structure of complex carbohydrates and glycolipids is a highly specialized subject, and will not be dealt with in this book. For further information, the review of

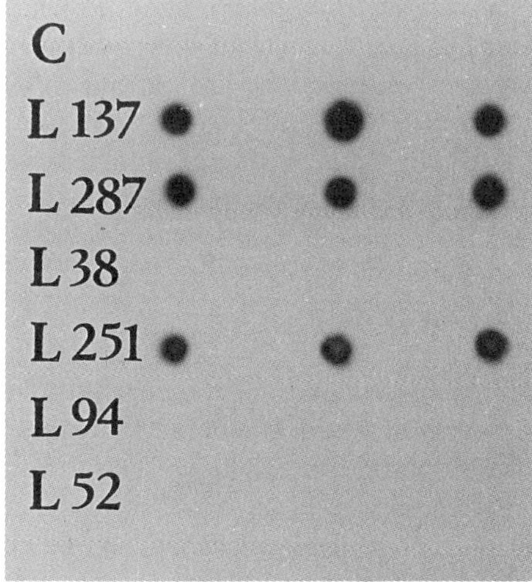

Fig. 10.1. Detection of antigen by a two-site dot blot immunoradiometric assay (IRMA). Nitrocellulose was coated with a monoclonal antibody that recognizes a polymeric carbohydrate antigen of Leishmania. *After saturation of any remaining nonspecific binding sites with 5% nonfat powdered milk in PBS (BLOTTO), the antigen-containing mixture (2 µl) was dotted onto the damp membrane. (The membrane must not be allowed to dry until the assay is completed.) Unbound antigen was washed off with BLOTTO, and the bound antigen detected by adding ^{125}I-monoclonal antibody (3 × 10^5 c.p.m./ml in BLOTTO). After further washing in BLOTTO, the membrane was autoradiographed at −70°C using an intensifying screen. Typical incubation times are 15–30 min and typical exposure times are 1–4 h. Reproduced with permission from Handman* et al. (1984), Oxford University Press, Oxford.

Weigandt (1985) is recommended. Assays for glycolipids are also discussed in Section 8.10.5.

10.2.2 Labelling of Antigen—Some Preliminary Remarks

The biochemical characterization of antigens often requires extremely high sensitivity. In addition, the analytical procedures that are in common use often require that additional proteins such as antibodies be added to the antigen mixture during analysis. These added proteins could cause confusion if they cannot be easily distinguished from the original antigen mixture. For both these reasons, it is common to use radioactivity or a chemical marker such as biotin to label the antigen-containing mixture. The final detection system is organized such that only the labelled molecules are detected (e.g. autoradiography for radioactive molecules and the use of enzyme-labelled avidin or streptavidin for biotin-labelled molecules). In this way, a combination of experimental flexibility and extremely high sensitivity can be achieved.

Until fairly recently, the only option with adequate sensitivity was radioactive labelling. However, in 1983 a way was found to increase the sensitivity of chemiluminescence by about 1000-fold (enhanced chemiluminescence or ECL). This is beginning to have revolutionary impact on antigen analysis, and is replacing many systems in which the use of radioactivity was previously essential. In many cases, ECL is combined with the use of enzyme-labelled antibodies and/or the biotin–avidin system. Production of peroxidase-conjugated antibodies is discussed in Section 8.10.7

10.3 The Biotin–Avidin System

When animals are fed on a diet containing raw egg-white as their sole source of protein, they develop a condition known as egg-white injury, characterized by neuromuscular disorders, dermatitis and loss of hair. Investigations by Györgi (1940) established that egg-white injury was due to deficiency of the water-soluble vitamin biotin. Subsequently it was shown that egg-white contains a protein which binds biotin with extremely high affinity.

The biotin-binding protein was named avidin (reviewed by Green, 1975). Avidin is a tetramer of identical subunits, each of M_r 15 000. Its isoelectric point is 10.5, and its affinity for biotin approximately 10^{-15} M. The biotin–avidin complex has been widely used in biology (reviewed by Bayer *et al.*, 1979; Wilchek and Bayer, 1988, 1990; Savage *et al.*, 1992; see also Section 12.9).

In recent years, the avidin–biotin system has started to replace the use of radioactive isotopes for the labelling of antigens and antibodies. When used in conjunction with peroxidase- or alkaline phosphatase-coupled avidin or streptavidin and ECL readout (Section 10.10.9), extremely high sensitivity can be achieved, with autoradiographic exposures often measured in seconds rather than hours or days. However, the extremely basic nature of avidin (pI = 10.5) may cause it to bind electrostatically to acidic structures such as DNA. The nonspecific binding of avidin to DNA may be diminished by raising the salt concentration (0.3 M KCl; Heggeness, 1977), and may also be diminished by competition with a large excess of the basic protein cytochrome *c*. A combination of KCl (0.3–0.5 M) and cytochrome *c* (1 mg/ml) during staining considerably lowers the background in histological sections. Avidin is a glycoprotein, and the binding of cellular proteins to its glycan moities is also a potential source of artefacts and nonspecific effects.

An alternative to avidin is *streptavidin*, a biotin-binding protein from *Streptomyces*. Streptavidin has four identical chains of similar molecular weight to those of avidin, but has a much less basic isoelectric point, and shows less nonspecific binding (see Section 12.9.3). It is available as a pure protein or conjugated to horseradish peroxidase or alkaline phosphatase from many different manufacturers. Streptavidin may also bind to cellular receptors through its RYD sequence (see Section 12.9.3).

As mentioned in Section 12.9.3, most tissue culture media contain biotin, which could cause inhibition of binding to avidin or streptavidin in some experiments.

10.3.1 Conjugation of Soluble Proteins with Biotin Succinimide Ester

Biotin must be derivatized before it can be coupled to protein; the most convenient method of coupling uses the *N*-hydroxysuccinimide ester of biotin, which is commercially available from numerous suppliers. Succinimide esters are extremely useful compounds for derivatization of antibodies because they allow formation of a stable amide bond under very mild conditions.

The structures of biotin and its *N*-hydroxysuccinimide ester are shown in Fig. 10.2. The reaction with proteins involves nucleophilic attack of the unprotonated ε-amino of lysine on the ester and displacement of *N*-hydroxysuccinimide (Fig. 10.2). The reaction proceeds most efficiently at slightly alkaline pH. Coupling is inhibited by extraneous amines, Tris or azide.

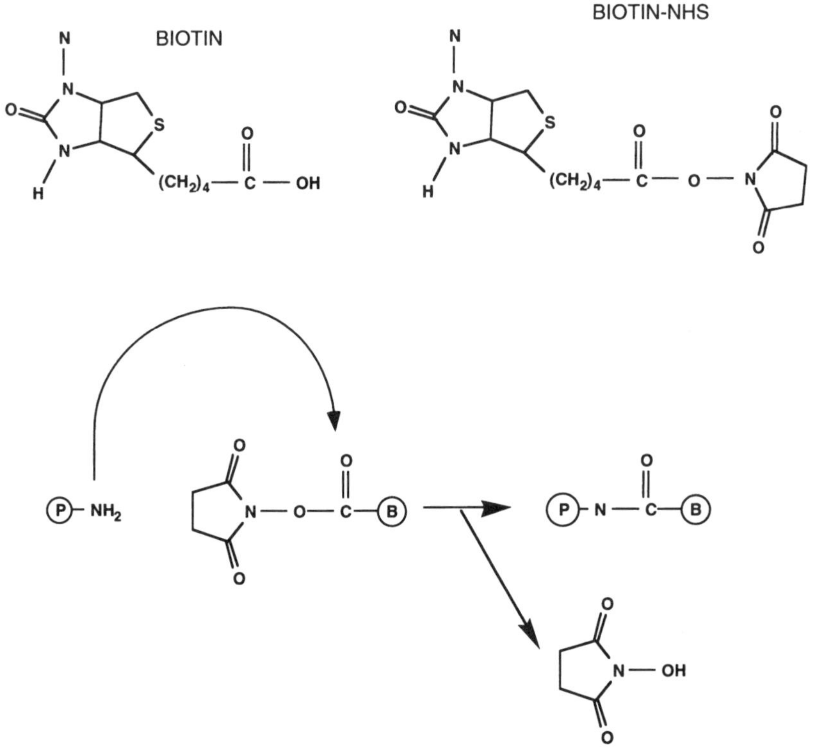

BIOTIN-NHS

BIOTIN

Fig. 10.2. Structure of biotin and biotin succinimide ester, and mechanism of coupling to proteins. The nucleophilic unprotonated ε-amino groups of lysine residues attack the activated ester, resulting in an amide bond and release of N*-hydroxysuccinimide.*

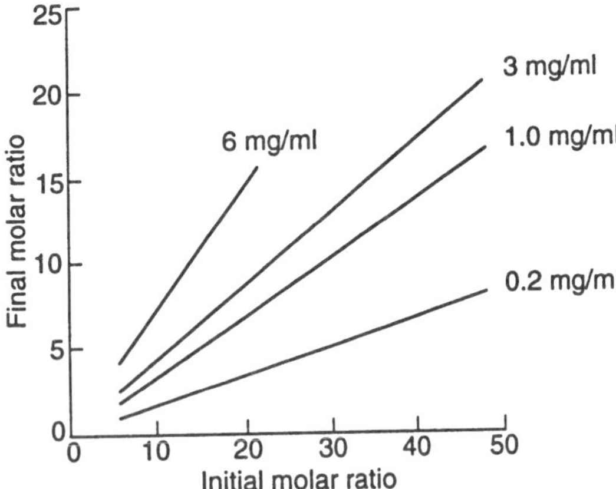

Fig. 10.3. Schematic illustration of the effect of protein concentration on the efficiency of coupling with biotin succinimide ester. At high protein concentrations, the reaction is quite efficient. At low protein concentrations, the competing hydrolysis reaction predominates.

Succinimide esters are very susceptible to hydrolysis, and during coupling there is a competing hydrolytic reaction. The extent to which coupling and hydrolysis compete depends on the protein concentration. Figure 10.3 illustrates this point. Bovine IgG was conjugated with the hapten 3-nitro-4-hydroxy-5-iodophenyl-(NIP), using the succinimide ester derivative. Unlike biotin, the NIP group absorbs in the visible range, allowing spectrometric determination of the conjugation ratio. Assuming that the succinimide ester of biotin has a similar degree of reactivity, one may use the curves shown in Fig. 10.3 to make a rough estimate of the likely efficiency of coupling.

If, as is often the case, the antigen to be biotinylated is part of a very complex and poorly defined mixture of proteins, these quantitative considerations may mean very little, and it may be necessary to adopt a rather empirical attitude to the amount of biotin succinimide ester to use. If it is possible to measure the total protein, the guidelines given above may be helpful. If not, it will probably be necessary to perform some pilot experiments to establish appropriate conditions.

Practical procedure

The protein should be dialysed overnight against 0.1 M $NaHCO_3$, pH 8.0–8.3. If precipitation with ammonium sulfate has been used, it is advisable to change the dialysis fluid several times, as ammonium ions will inhibit the reaction. Following dialysis, the protein concentration should be estimated and, if possible, the concentration adjusted to 1.0 mg/ml by dilution with 0.1 M $NaHCO_3$.

Biotin succinimide ester must be warmed to room temperature before the bottle is opened. The ester is then weighed out, and dissolved in dimethyl sulfoxide to 1.0 mg/ml immediately before use. The ester is freely soluble in dimethyl sulfoxide (DMSO), which is important because any delays in dissolving it in water will result in hydrolysis. The half-life of succinimide esters in 0.1 M $NaHCO_3$ is of the order of minutes.

The ester solution is then added to the protein, and mixed immediately. If the conditions of pH and protein concentration given above are followed, it will be found that 120 μl of the ester solution per ml of protein solution will usually give good conjugation. Sometimes it may help to vary the ratio of biotin:protein in the range 60 μl/ml to 240 μl/ml. Underconjugation results in low sensitivity; overconjugation may result in loss of antigenic determinants or protein aggregation.

The mixture should be held at room temperature for 1–2 h, by which time the reaction will have gone to completion. Free biotin may be removed by dialysis or gel filtration on a Sephadex PD-10 column (Pharmacia).

Meier *et al.* (1992) labelled ovalbumin (1.0 mg/ml in 0.1 M borate buffer, pH 8.8, 150 mM NaCl) at 25 and 250 μg of ester/mg protein for 15 min, 1 h and 4 h, followed by quenching with NH_4Cl and removal of unreacted ester and free biotin on Sephadex G-25. They found that labelling reached almost maximal levels after 15 min. Sensitivity of detection by horseradish peroxidase–streptavidin (Amersham or Vector Laboratories) was such that 2 ng of biotinylated ovalbumin could be detected after 1 min exposure of ECL, using 250 μg ester/mg ovalbumin. The sensitivity using 25 μg ester/mg ovalbumin was about fivefold lower. The higher concentration of ester caused a slight increase in apparent molecular weight of the protein on SDS–polyacrylamide gel electrophoresis, and would be expected to result in a significant lowering of the isoelectric point of the protein and the acquisition of considerable charge heterogeneity that would be visible on two-dimensional gels.

10.3.2 Biotinylation of Cell Surface Proteins

Biotin succinimide ester is moderately hydrophobic, and would probably penetrate the lipid bilayer of cell membranes, although Meier *et al.* (1992) found that it was possible to label lymphocyte membrane proteins on intact cells with very little evidence of cytoplasmic labelling. They felt that the slight degree of cytoplasmic labelling observed may have been attributable to the presence of small numbers of dead cells.

Sulfosuccinimidylbiotin is a derivative of biotin succinimide ester. Its negative charge would be expected to prevent it from crossing cell membranes. It is commercially available as sulfo-NHS-biotin from Pierce Chemical Company. Studies by Cole *et al.* (1987) and Altin and Pagler (1995) indicated that labelling was confined to the outside of the cell, and that intracellular proteins were not labelled.

Reaction conditions are not critical, but the cells should be washed several times in protein-free PBS prior to labelling, and should be highly viable to prevent labelling of cytoplasmic proteins. Labelling of cells is carried out in PBS, pH 8.0, as it is expected to be somewhat more efficient at slightly alkaline pH, using a final concentration of 2.5–5.0 mM sulfo-NHS biotin. As the ester is hydrophilic, it is not necessary to dissolve it in DMSO, but care must be taken to keep the powder dry and to dissolve it immediately prior to use. Cole *et al.* (1987) labelled for 30 min at 4°C, while Altin and Pagler (1995) labelled for the same time but at room temperature, which gave severalfold higher sensitivity. Labelling was more efficient at pH 8.0 than at pH 7.0, as expected from the reaction mechanism, in which only the unprotonated form of lysine is reactive.

It is interesting to note that the selectivity of labelling by biotin succinimide esters is different to that of radioiodination, and some lymphocyte membrane proteins are labelled with the former that were undetected using radioiodination (Terashima *et al.*, 1994; Kim *et al.*, 1994).

10.4 Radioiodination of Soluble Protein Antigens

Until quite recently, the extreme sensitivity of radioisotopic procedures was essential for the detection of the minute quantities of antigens present in complex biological mixtures. Radiolabelling also allows the addition of extraneous nonradioactive proteins (e.g. antibodies) without their detection in the final readout. The advent of enhanced chemiluminescence (ECL) has changed the situation dramatically (see the previous section and Section 10.10.9). However, there are still many situations where radioactive iodine is the method of choice, and these will now be explored.

Radioiodination with ^{125}I($t_{1/2}$ = 60 days) is a very useful method of labelling soluble proteins (Amersham, 1993). It is inexpensive and simple to perform, and results in labelled proteins of high specific activity. The main disadvantage of radioiodination is that it may occasionally damage the protein, resulting in loss of antigenic determinants or biological activity.

Many of the methods for protein iodination involve oxidizing agents, such as chloramine-T or hydrogen peroxide. These may cause oxidation of methionine to methionine sulfoxide, and tryptophan to oxindole (Salacinski *et al.*, 1981). Other methods, notably the iodine monochloride method (see Section 10.4.7) and Bolton–Hunter reagent (see Section 10.4.4), avoid these problems, but may introduce problems of their own. The relative merits of the various iodination procedures are listed in Table 10.1.

The susceptibility of individual proteins and antigenic determinants to damage by radioiodination is unpredictable, but may be minimized by attention to detail. Sometimes one method of iodination will damage a protein while another will not. In other cases, all methods of radioiodination may

Table 10.1 Procedures for radioiodination of soluble proteins

Method	Exposure to oxidizing agents	Specific activity obtainable	Efficiency	Alteration of native charge	Convenience
Chloramine-T	Yes	High	High	No	+++
Lactoperoxidase	Yes	Moderate	Moderate	No	++
Iodogen	Yes (solid state)	High	High	No	+++
Bolton-Hunter	No	Moderate	Low-moderate[a]	Yes[b]	±[c,d]
Iodobeads	Yes (solid state)	High	Moderate-high	No	+++
Iodine Monochloride[e]	Minimal	Moderate	Moderate	No	–

[a] Efficiency depends on protein concentration. Typical figures are 5% for 0.1 mg/ml, 10–20% at 1.0 mg/ml, and 50–90% at 10 mg/ml.

[b] Loss of one positive charge for each substitution.

[c] Reagent is extremely susceptible to hydrolysis and it is advisable to use one whole vial at a time.

[d] A reagent similar to the Bolton-Hunter reagent is *N*-succinimidyl [2,3–^{3}H] propionate which is available at 30–60 Ci/mmol from Amersham. The specific activity of labelled proteins would be much less than the corresponding iodinated proteins.

[e] The iodine monochloride method is by far the gentlest way of labelling proteins on tyrosine, and is much better suited to labelling small amounts of protein to high specific activity than the Bolton-Hunter reagent, which does not work well for small amounts of protein due to competing hydrolysis (see text).

damage the protein (e.g. Nussenzweig *et al.*, 1982), and alternative approaches such as tritium labelling by reductive methylation may be required (see Section 10.5).

When pure proteins are to be radioiodinated, it is customary to aim for an average substitution of approximately one iodine atom per molecule of protein (see below). If carrier-free ^{125}I is used, the substitution of a single ^{125}I atom results in a specific activity of *c.* 18 μCi/μg for a protein of molecular weight 100 000, or 180 μCi/μg for a protein of molecular weight 10 000. (One microcurie is equal to 2.2×10^6 disintegrations per minute (d.p.m.). The recommended unit is now the Becquerel, which is one disintegration per second. To date, most workers have preferred to use the older unit.)

Substitution of proteins with an average of less than one iodine per molecule will result in a proportion of unlabelled molecules. Unlabelled molecules merely degrade the sensitivity by competing with labelled molecules. Higher average substitution may result in increased sensitivity at the cost of increased radiation damage because the decay event may damage the antigen, and also because of irradiation of neighbouring molecules. The substitution of one iodine per molecule ensures that once decay has occurred, that molecule will no longer be detected.

If the antigen to be studied is contained within a complex mixture of proteins, it is impossible to predict its specific activity because the iodination efficiency of each protein will be different. In many practical situations, even the total protein concentration will be unknown.

10.4.1 The Chloramine-T Method

The chloramine-T method (Greenwood *et al.*, 1963) is the most widely used method of protein iodination. The main amino acid to be labelled is tyrosine, although histidine may be iodinated if the pH is greater than 8.0–8.5. The method has the advantages of simplicity and efficiency, and in most cases the degree of damage to the antigen is minimal.

Chloramine-T is an oxidizing agent and high concentrations may result in damage of the antigen. As a general rule, only 1–10 μg chloramine-T per reaction are necessary, rather than the 100 μg originally used. It is customary to terminate the iodination by addition of the reducing agent sodium metabisulfite, but this step may do more harm than good. Termination by addition of tyrosine (final concentration 1 mM) is equally effective, and has no potential for damage to protein, although iodotyrosine can bind noncovalently to carrier proteins such as bovine serum albumin. An alternative stopping reagent is sodium azide (final concentration 1 mM), which appears gentle and less likely to cause artefacts.

To prevent exposure to volatile iodine or aerosols, iodinations should always be performed in a fume hood. It is good practice to add a small amount of cold $Na^{127}I$ (final concentration 1 mM) at the end of all iodinations to help terminate the reaction and to minimize adsorption of radioiodide to the glassware. *This must not be done until after the oxidizing reagent such as chloramine-T has been removed by gel filtration, because the added iodine will be oxidized and cause massive additional iodination of the protein with nonradioactive iodine.*

A number of substances are capable of inhibiting the chloramine-T iodination of proteins. These include reducing agents and thiocyanate ions (George and Schenck, 1983). It is rare to find a protein which will not iodinate. (One such example appears to be ovalbumin.) Sodium azide also inhibits. If difficulty is experienced, it is recommended that the protein be dialysed against several changes of phosphate-buffered saline to remove inhibitors.

The procedure works best if protein concentration is fairly high (>1.0 mg/ml). The protein may be in Tris-HCl or phosphate buffer, pH 7–8.

Practical procedure

(1) Weigh out chloramine-T powder just before use, and make up to 1.0 mg/ml in distilled water.
(2) Pipette 10 μg protein into a small tube on ice.
(3) Add 10 μl 0.1 M sodium phosphate, pH 7.5, to buffer the NaOH which is often present in the radioactive iodide to prevent oxidation to iodine. (More recent formulations of ^{125}I are buffered close to pH 7 rather than being supplied in NaOH.)
(4) Add desired volume of ^{125}I (typically 5–10 μl).
(5) Add 1–10 μl chloramine-T Mix. Although the reaction occurs virtually instantaneously, it is customary to wait a few minutes.
(6) Add 2.5 ml phosphate-buffered saline; transfer to a disposable PD-10 Sephadex column (Pharmacia) equilibrated with the same buffer. Allow the sample run in; discard effluent. Add 3.5 ml of the same buffer; collect effluent, which contains the iodinated protein, free of unreacted iodide.

Aliquots (typically 1.0 μl) should be removed for direct counting or (preferably) counting after TCA precipitation. Typically, 50–90% of input counts should be incorporated into protein. If the protein concentration is very low (less than 50 μg/ml), the percentage incorporation of ^{125}I may be much less, although in some experiments this may still be acceptable.

10.4.2 TCA Precipitation

After any radiolabelling procedure, it is advisable to measure incorporation by TCA precipitation. If the percentage of total counts in protein is high, liquid-phase precipitation is adequate. However, if the fraction of total counts in protein is low (for example, after cell surface iodination or biosynthetic labelling), more accurate results will be obtained by collection and washing of the precipitate on a filter.

Liquid-phase procedure

(1) Take 1 μl aliquots of labelled protein (triplicate), pipette into an Eppendorf centrifuge tube containing carrier solution (500 μl phosphate buffered saline plus 0.5% fetal calf serum or 0.25 mg/ml bovine serum albumin, and 1 mM KI or 1 mM of the appropriate amino acid if radioactive amino acids are present).
(2) Add 500 μl 10% TCA. Mix.
(3) Let stand for 30 min at room temperature, or overnight at 4°C. The solution will go very slightly cloudy.
(4) Spin in Eppendorf centrifuge for 5 min (*c.* 10 000 ***g***).
(5) Remove and save 500 μl supernatant.
(6) Count the 500 μl of removed supernatant (S) and the remaining volume (R).
(7) Calculate as follows:

$$\% \text{ TCA precipitable counts} = \frac{R - S}{R + S} \times 100$$

Filter procedure

The procedure requires a side-arm flask connected to a vacuum (water pump) and a filtration apparatus for TCA precipitation. Also required are Millipore type HA 0.45 μm filters (25 mm diameter; cat. no. HAWP 2500).

(1) Perform steps 1–3 of liquid-phase procedure.
(2) Assemble filtration apparatus with filter in place. Wet filter with carrier solution; apply vacuum.
(3) Add sample to filter.
(4) Wash through with 10 ml 5% TCA.
(5) For ^{125}I, place into tube for counting.
(6) For ^{35}S, ^{14}C or ^{3}H, dry filter under a heat lamp; and place into Triton-based scintillation fluid for counting.

10.4.3 Lactoperoxidase Method

In some cases, the chloramine-T method may result in damage to the antigen owing to oxidation. The lactoperoxidase method (Marchalonis, 1969; Phillips and Morrison, 1970) may prove more gentle. The iodination is almost exclusively confined to tyrosine.

(1) Pipette 10 μl of protein (1.0 mg/ml) into a tube. Add 10 μl 0.1M Tris-HCl or sodium phosphate, pH 8, to provide additional buffering.
(2) Add 5 μl lactoperoxidase (0.25 mg/ml).
(3) Add desired volume of ^{125}I. Mix.
(4) Add 10 μl H_2O_2 (diluted 1:30 000 in PBS from 30% stock).
(5) Incubate at room temperature for 10 min.
(6) Separate free iodide from protein as for the chloramine-T method.

Uptake of 30–70% of input counts is typical. The most critical parameter is the concentration of H_2O_2. Too little peroxide provides insufficient substrate; too much poisons the enzyme. If difficulties are experienced, try varying the concentration of H_2O_2. A 1:1000 dilution of 30% H_2O_2 in phosphate buffered saline should have an optical density of *c.*0.7 at 230 nm (Phillips and Morrison, 1970).

Some proteins may be damaged by the H_2O_2. The peroxide concentration may be minimized by the use of glucose oxidase to generate continuously micromolar amounts of H_2O_2 in a coupled system (Hubbard and Cohn, 1975). Beads coated with a mixture of glucose oxidase and lactoperoxidase are available commercially (Enzymobeads, Bio-Rad).

10.4.4 Bolton–Hunter Method

A third method of radioiodination involves the use of *N*-succinimidyl 3-(4-hydroxy,–5-[^{125}I]iodophenyl) propionate, which reacts with the ε-amino group of lysine (Bolton and Hunter, 1973; Amersham, 1993). The reaction is extremely gentle, as no oxidizing agents are used. The main disadvantages are that the reagent is more expensive than sodium iodide, and it is not practicable to dispense multiple aliquots from a vial of Bolton–Hunter reagent on different days.

The net charge of the protein is decreased by one unit for each modified lysine. The introduction of artefactual charge heterogeneity may be of little importance in many situations, but it should be borne in mind if analysis by two-dimensional gels is contemplated.

The Bolton–Hunter reagent is available commercially at 2000 Ci/mmol or 4000 Ci/mmol (i.e., close to the theoretical maximum specific activities for mono- and di-substituted tyrosine). As is the case for all succinimide esters, the Bolton–Hunter reagent is extremely susceptible to hydrolysis (half-life of the order of minutes in alkaline aqueous solution). *Iodination of the protein and hydrolysis of the Bolton–Hunter reagent are therefore simultaneous competing reactions. The competition may be biased in favour of protein iodination by a high protein concentration.* A concentration of greater than 1.0 mg/ml is advisable, and 10 mg/ml is preferable (Table 10.1).

Extraneous nucleophiles such as free amines (e.g. Tris) or azide ions will inhibit the reaction.

Practical procedure

The Bolton–Hunter reagent is available from Amersham or New England Nuclear, supplied in dry benzene, in several different pack sizes. It should be stored refrigerated, not frozen. Aliquoting is not recommended because even traces of water result in hydrolysis.

(1) Evaporate the benzene by inserting a needle as an intake for a gentle stream of dry nitrogen or dry air, through the septum of the vial, just above the surface of the benzene. A charcoal trap may be used to contain any radioiodine released. Evaporation must be gentle, to concentrate the ester in the apex of the 'v' container.

(2) Add the protein in as small a volume and as concentrated form as possible (typically 10 μl of 10 mg/ml). Suitable buffers include borate or phosphate, pH 7.4–8.5. The efficiency is higher at the alkaline end.

(3) Incubate for 1 h at room temperature (New England Nuclear recommends 4°C for up to 18 h for maximal uptake).

(4) Add 10 μl 10 mM ethanolamine, Tris-HCl, glycine or sodium azide to terminate reaction.

(5) Pass the mixture over a PD-10 Sephadex column equilibrated with PBS plus 0.1% gelatin. (The Bolton–Hunter reagent binds to albumin via hydrophobic interactions.)

Typical uptake of ^{125}I is 5–30% (Table 10.1).

10.4.5 The 'Iodogen' Method

Fraker and Speck (1978) described a novel method of protein iodination which uses the water-insoluble compound 1,3,4,6-tetrachloro-3α,6α-diphenyl glycoluril (Iodogen, available from Pierce Chemicals). The method has achieved considerable acceptance (Markwell and Fox, 1978; Salacinksi *et al.*, 1981). Damage to proteins is claimed to be less than by using the chloramine-T method, although this has not always been rigorously substantiated. The main amino acid to be iodinated is tyrosine.

Iodogen is dissolved in chloroform or dichloromethane and pipetted into borosilicate glass tubes or iodination vials, and the solvent gently blown off with a stream of dry nitrogen. Iodogen forms a thin coating on the surface of the tube. Removal of the solvent by simple evaporation results in flaking of the coating (Markwell and Fox, 1978). Coated tubes may be stored for at least 6 months at room temperature. *Iodogen is light-sensitive, and must be stored in foil-wrapped tubes in a dessicator.*

The tube should be rinsed with buffer just prior to use, to remove any loose flakes. Iodination of proteins is performed by simply adding the protein to the tube, followed by the desired volume of ^{125}I. Reaction time is typically 5–10 min at room temperature, It is recommended that tubes be coated with not more than 10 μg Iodogen/100 μg of protein.

Unreacted iodide should be removed by gel filtration using a Sephadex G-25 column (PD-10; Pharmacia).

10.4.6 The 'Iodo-beads' Method

Chloramine-T, covalently attached to polystyrene beads, has been shown to be an efficient and gentle reagent for protein iodination (Markwell, 1982). The beads are available from Pierce Chemicals under the name 'Iodobeads'. The reaction is simple and efficient.

On the day of use, Iodo-beads are washed twice in PBS, and blotted dry on filter paper. Then, 100 μg protein in 500 μl PBS is added, followed by 1 mCi of $Na^{125}I$. The iodination reaction is initiated by addition of one or more Iodo-beads, and allowed to proceed for 15 min at room temperature. The reaction is terminated by removal of the liquid and gel filtration to remove unreacted iodide.

The reaction volume is not critical, and shaking is evidently not necessary. The reaction is not inhibited by SDS, 8M urea, or 2% Nonidet P-40, although it is abolished by reducing agents such as mercaptoethanol. The efficiency of incorporation was 35% with one bead, and virtually 100% with five beads.

10.4.7 The Iodine Monochloride Method

The iodine monochloride method is perhaps the most gentle method of iodination of proteins, and may allow retention of biological activity when all other methods fail. The protocol given below is based on that of Contreras *et al.* (1983), and was kindly supplied by Dr N. Nicola (see Nicola and Metcalf, 1988).

Preparation of stock solutions

Solution A: 0.15 g KI in 8 ml 6 M HCl
Solution B: 0.10 g K iodate (KIO_3) in 2 ml H_2O
Add B to A using a vortex mixer. (I_2 is precipitated but redissolved) and make up to 40 ml with pure water. Add 5 ml CCl_4 and bubble air through solution for 1 h. Remove CCl_4 phase in a separating funnel and add pure water to a final volume of 45 ml. This solution will be 0.02 M in KI, 0.011 M in IO_3 and 0.031 M in ICl.

Iodination

Note: All procedures must be carried out at room temperature (20°C) in a well-ventilated hood. Gaseous $^{125}I_2$ is generated, which could be inhaled and taken up by the thyroid.

(1) Add 40 μl 0.2 M sodium phosphate pH 7.2 containing 0.02% v/v Tween 20 to 1 mCi $Na^{125}I$.
(2) Add 1–2 μg protein in 2–40 μl of same buffer.
(3) In a separate tube, add 5 μl stock ICl solution to 0.5 ml 2 M NaCl in H_2O while vortex mixing.
(4) Add 3 μl of diluted ICl solution to protein/^{125}I mixture while vortex mixing.
(5) Add a further 6 μl of diluted ICl solution while vortex mixing.
(6) Separate ^{125}I protein free from free ^{125}I by gel filtration in a Pharmacia PD-10 column.

10.5 Radiolabelling of Soluble Proteins with Tritium by Reductive Methylation

Occasionally, the protein of interest may be inactivated by all of the above methods of radioiodination (Nussenzweig *et al.*, 1982). In one case, the successful alternative procedure involved labelling with tritium by reductive methylation (Tack *et al.*, 1980).

The respective disintegration rates of ^{125}I, ^3H and ^{14}C at 100% isotopic abundance are 1.7×10^4, 9.8×10^3 and 4.6 Ci/g. Until this procedure was developed, radiolabelling with ^{125}I was the only method capable of achieving the high specific activities needed for highly sensitive radioimmunoassay (RIA) or receptor studies. Tack *et al.* (1980) have shown that it is feasible to label haemoglobin, complement and immunoglobulins to very high specific activities by reductive methylation.

Advantages of reductive methylation include minimal disruption to the conformation of the protein, preservation of native charge, and long half-life (12 years). Disadvantages include the need to handle extremely large amounts of a chemically unstable radioactive compound and the additional work involved in use of scintillation methods of detection of tritium.

10.5.1 Practical Procedure (after Tack *et al.*, 1980)

NaB^3H_4 of specific activity 40–60 Ci/mmol was obtained from New England Nuclear. A quantity corresponding to 500 mCi was weighed out, dissolved in anhydrous methylamine and dispensed in 100 mCi aliquots in dry glass vials. The solvent was removed under vacuum, and nitrogen gas introduced.

Tritiated borohydride is available from New England Nuclear in a much more convenient form. It is supplied at a specific activity of 70 Ci/mmol, in 25 mCi batches frozen in 1 ml 10 mM NaOH, and may be stored for several months at –70°C. It should not be frozen and thawed repeatedly, but rather should be stored in small aliquots.

The protein (5–10 mg/ml) was dialysed against 0.2 M borate, pH 8.9, and 200 μl transferred to 1.0 ml 'Reacti-vials' (Kontes Glass, or Pierce). The vials were placed on ice for 1 h, and all subsequent operations were carried out in a well-ventilated fume hood.

The crystalline NaB^3H_4 was dissolved in 0.01 M NaOH to a final concentration of 1.0 Ci/ml just prior to use, and a stock solution of 12.4 M formaldehyde diluted with water to 0.1–0.2 M. The formaldehyde (10–20 μl) and NaB^3H_4 (30–50 μl) were added sequentially to each vial, which was then sealed with a rubber septum and held on ice for 10 min. Protein solutions were removed by needle aspiration through the septum and desalted by gel filtration on disposable Sephadex G-25 columns. (In view of the extremely large quantities of low molecular weight isotope present, it is questionable whether this procedure would be adequate to remove all the unreacted borohydride.)

10.6 Radiolabelling of Cellular Proteins

10.6.1 Cell Surface Radioiodination

The easiest way to obtain radiolabelled membrane proteins of high specific activity is via cell-surface radioiodination of intact, living cells. The two procedures that are most commonly used are lactoperoxidase-catalysed iodination (Phillips and Morrison, 1970; Marchalonis *et al.*, 1971) and an adaptation of the Iodogen method (Markwell and Fox, 1978). The efficiency of the two procedures is not greatly different. A typical experiment might start with 10^7 cells and 500 μCi ^{125}I (i.e. *c.* 10^9 c.p.m.). About 1% of the counts will be incorporated into protein and a somewhat larger percentage into lipid.

The mere presence of ^{125}I-labelled material in such experiments does not prove that such material was synthesized by the cell. This seemingly obvious point has led many people astray, and the common association of serum albumin has probably been confused with immunoglobulin heavy chains on more than one occasion (Sidman, 1981).

Similarly, the presence of ^{125}I-labelled protein does not prove that the protein was present on the cell surface. If the recommended conditions for iodination are changed appreciably (e.g. by increasing the iodide concentration), cytoplasmic labelling may occur. Similarly, if there are any dead cells present, extensive cytoplasmic labelling is unavoidable.

Proof that the label is confined to the cell surface may be obtained in several ways. If the labelled protein can be removed from intact cells by protease treatment, this constitutes good evidence for a surface location (Felsted and Gupta, 1982). Electron microscope autoradiography of labelled cells (Marchalonis *et al.*, 1971; Tartakoff and Vassalli, 1979) is another approach, but the method is cumbersome and technically demanding. A third way of assessing the degree of cytoplasmic labelling is to analyse the whole lysate by two-dimensional (charge-size) polyacrylamide gel electrophoresis. The autoradiograph should be laid over the dried, stained gel, and the spots on the film compared with the stained spots. Only cytoplasmic proteins are sufficiently abundant to be seen by staining. It therefore follows that there should be no correspondence between stained spots and spots on the film. The most prominent stained protein will probably be actin, which is an acidic protein of M_r 43 000. Presence of ^{125}I in actin should be taken as evidence of cytoplasmic labelling.

Assessment of cell viability by acridine orange/ethidium bromide staining

If it is crucial that the labelling be confined to the cell surface, the cell population should have very few dead cells (<3%). Viability may be assessed by the acridine orange/ethidium bromide method (Lee *et al.*, 1975), which is much more accurate and unambiguous than eosin or trypan blue exclusion. Both dyes are potentially carcinogenic, and should be handled with care to avoid ingestion or skin contact.

One drop of cells in PBS is placed on a glass slide, and one drop of PBS containing 1 p.p.m. acridine orange and 1 p.p.m. ethidium bromide is added, followed by a cover slip. Cells are examined by fluorescence microscopy, using filters for fluorescein (see Chapter 12). Living cells stain bright green, while dead cells stain bright orange.

Lactoperoxidase-catalysed cell-surface iodination

(1) Wash cells three times in PBS to remove serum and loosely adsorbed protein. *After washing, transfer cells to a fresh tube to prevent iodination of protein adsorbed to the sides of the tube.*
(2) Resuspend cells (typically 10^7) in 200 μl PBS (do not alter volume). All steps may be carried out at room temperature, unless otherwise directed.
(3) Just before iodination, dilute 10 μl of 30% H_2O_2 in 10 ml PBS (i.e. 1:1000). The optical density of the 1:1000 stock should be about 0.7 at 230 nm (Phillips and Morrison, 1970). Then prepare serial threefold dilutions in PBS, giving final dilutions of 1:3000; 1:9000 and 1:27 000.
(4) Add 50 μl of 0.2 mg/ml lactoperoxidase to the cells, then 500 μCi ^{125}I.
(5) At 1-min intervals, add successively 10 μl of 1:27 000; 1: 9000; 1:3000 H_2O_2.

(6) One minute after the last addition, wash the cells twice in 3–5 ml PBS.

(7) Solubilize the cells in 300 μl – 3.0 ml PBS containing 0.5% Triton X-100 or Nonidet P-40, at 4°C for 30–60 min (see Section 10.7). If desired, the solubilization buffer may contain the protease inhibitors EDTA, *N*-ethyl maleimide and phenylmethyl sulfonyl fluoride (1 mM of each). The last two must be made up freshly.

All subsequent steps should be at 4°C.

(8) Nuclei and insoluble debris may be removed by centrifugation at 2000–10 000 ***g*** for 5–10 min.

A good iodination should result in 10–40% of the input counts remaining with the cells after washing. Of these, only 1–10% will be TCA-precipitable, indicating incorporation into protein. The remainder will be in free iodide or lipid. TCA precipitations can be performed on membrane filters (see Section 10.4.2).

The level of incorporation of ^{125}I into protein (1–3% of input counts) is low, but this is the best that can be achieved. The concentration of enzyme is not very critical. The most important parameter is the concentration of peroxide; too low a concentration results in poor efficiency, while too high a concentration poisons the enzyme and may damage the cells. The increasing pulses of H_2O_2 guarantee that the optimal concentration will be reached, yet the total concentration is not excessive.

If the concentrations of H_2O_2 employed cause problems, continuous micromolar amounts of H_2O_2 may be generated by glucose and glucose oxidase (Hubbard and Cohn, 1975). If cells containing significant amounts of peroxidase are to be labelled, the lactoperoxidase method may result in cytoplasmic labelling. In these cases, the Iodogen method may be preferred (see below).

Use of Iodogen to label cell-surface membrane proteins

Markwell and Fox (1978) have shown that the Iodogen method for iodination of soluble proteins (see Section 10.4.5) may be adapted to the specific labelling of the externally exposed proteins of cell membranes. The pattern of labelling of proteins is virtually identical to that obtained using the lactoperoxidase technique (Felsted and Gupta, 1982). Advantages of the Iodogen method include the absence of any extraneous added protein or H_2O_2. Tubes may be coated with Iodogen as described in Section 10.4.5. Alternatively, if adherent cells such as fibroblasts are to be labelled, Iodogen may be coated onto a glass cover slip, which is gently floated above the cell monolayer (Markwell and Fox, 1978).

Cells should be washed three times in PBS to remove extraneous protein. As for the lactoperoxidase method, cell viability must be very high. The Iodogen-coated iodination vial should be rinsed with PBS just prior to use, to remove any loosely attached material. Cells (typically 10^7) are added in 1.0 ml PBS, followed by 500 μCi ^{125}I, and the reaction allowed to proceed for 10 min at room temperature, with occasional agitation.

The cells are then harvested by centrifugation in PBS, and washed once in the same buffer. Conditions for solubilization are identical to those for the lactoperoxidase technique (see also Section 10.7).

Typical amounts of Iodogen per vial are 5–10 μg. Higher amounts are likely to kill the cells and result in cytoplasmic labelling.

10.6.2 Biosynthetic Incorporation of Radioactive Amino Acids

The choice of amino acid and isotope warrants careful consideration. The aim will usually be to incorporate the greatest number of counts at the least expense. Useful information about the choice of amino acids has been published by Vitetta *et al.* (1976) and Coligan *et al.* (1983).

In many cases, [^{35}S]-methionine will be the amino acid of choice (E_{max} = 167 KeV; $t_{1/2}$ = 87 days). Its high specific activity (>1000 Ci/mmol) and good uptake by cells more than compensates for its relative rarity in proteins. The next highest specific activity amino acids are those labelled with tritium. The maximum specific activities available range from 20–200 Ci/mmol. Tritiated amino acids have a radioactive half-life of 12 years, although they usually deteriorate at 1–3% per month. The relatively low energy of tritium (E_{max} = 18 KeV) was once a serious problem, but the advent of simple fluorographic techniques (Bonner and Laskey, 1974; Laskey, 1980, 1993) have made autoradiography with tritium a simple matter. Tritium is detected by soaking the gel in a solution containing a scintillant. The scintillant is then precipitated in the gel by a change of solvent or by drying. The low-energy β-emission of tritium is converted with high efficiency into photons, and the film is thus exposed by light rather than direct ionizing radiation. Experimental details and recommendations are given in Section 10.9.1.

Labelling with [^{14}C]-amino acids ($t_{1/2}$ = 5760 years; E_{max} = 155 KeV) is seldom justified. Their specific activity is much lower than the tritiated amino acids (typically 300 mCi/mmol), and they are much more expensive.

The uptake and incorporation of amino acids does not exactly parallel their relative abundance in cellular proteins. Good results are obtained with methionine, arginine, leucine, lysine and tyrosine. Although cysteine can be obtained at as high specific activity as methionine and is usually more abundant, it generally gives poorer labelling. Cysteine receives its sulfur from methionine, and if [^{35}S]-cysteine is used, it is a good idea to use [^{35}S]-methionine as well. Mixtures of high specific activity [^{35}S]-cysteine and [^{35}S]-methionine are now commercially available from many suppliers.

The buffer that the isotopes are supplied in should be checked for its effect on cells. Methionine and cysteine may be supplied in aqueous solution containing 0.1% (1.2×10^{-2} M) mercaptoethanol. The maximum concentration of mercaptoethanol that most cells can tolerate is around 5×10^{-5} M (this corresponds to 50 μCi/ml). Tritiated or [^{14}C]-amino acids are usually supplied in water plus 2% ethanol to scavenge free radicals. A concentration of 0.2–0.3% ethanol should not be exceeded. If higher concentrations of isotope are required, the ethanol or mercaptoethanol may be removed by lyophilization. Because these reagents are added to stabilize the amino acid, lyophilization should be performed just prior to labelling, and only in the quantity needed.

[^{35}S]-methionine and -cysteine may be stored at –70°C or in liquid nitrogen. When a vial is first opened, the isotope should be aliquoted into small tubes (0.5 mCi/tube). Do not repeatedly freeze and thaw. Tritiated and [^{14}C]-amino acids and sugars should be stored at 4°C, protected from light. In general, aqueous solutions of ^{3}H-labelled compounds should not be frozen, because molecular clustering by freezing accelerates self radiolysis. While the β-emission of ^{35}S and ^{14}C will not penetrate a thin layer of plastic or glass, the resulting interaction may generate penetrating Brehmsstrahlung radiation and cause fogging of adjacent films unless lead shielding is used.

Practical procedure

(1) Prepare medium lacking appropriate amino acid. (Kits are available from Gibco or numerous other suppliers.)
(2) Add 10% fetal calf serum (preferably dialysed against PBS to remove any free amino acids).
(3) Set up cultures containing 10^{6}–10^{7} cells in 1–2 ml medium containing desired amount of radioactive amino acid (typically 50–100 μCi/ml).
(4) Culture at 37°C for desired period (generally 1–4 h). If the culture is for less than about 4 h, the cells need not be sterile, although gross bacterial contamination must be avoided.
(5) Add nonradioactive amino acid to *c.* 1 mM.
(6) Harvest cells by centrifugation. If desired, save supernatant.
(7) Wash cells twice in medium lacking isotope but containing non-radioactive amino acid.
(8) Solubilize cells as for surface iodination.
(9) Check incorporation of isotope by TCA precipitation using a membrane (see Section 10.4.2).

It is advisable to check that the kinetics of uptake of radioactivity are linear. Nonlinear kinetics indicate that something is changing during the culture period (e.g. the cells are dying). Linear kinetics are essential for 'pulse-chase' experiments. The achievement of sufficient incorporation of isotope during pulse-chase experiments may require depletion of nonradioactive cellular amino acid pools by pre-incubation for 1–2 h at 37°C in medium lacking the relevant amino acid. The chase may be carried out by addition of the non-radioactive amino acid to a final concentration of *c.* 1 mM, which is a vast excess over that of the radioactive species. In the case of long incubations of rapidly metabolizing cells, it may be necessary to add small (micromolar) amounts of nonradioactive species to prevent amino acid starvation.

10.6.3 Biosynthetic Labelling with ^{32}P

Cells should be cultured for 2–4 h in phosphate-free culture medium containing 5% fetal calf serum and 0.5–5.0 mCi carrier-free [^{32}P]-orthophosphate (Radke and Martin, 1979; Radke *et al.*, 1980; Cooper and Hunter, 1981; Garrison, 1993; Hardie, 1993). Labelling is terminated by washing in PBS. Radioactive phosphate will be incorporated into phosphoproteins, phospholipids, DNA and RNA. The presence of labelled DNA and RNA may cause severe background problems in electrophoretic analyses of proteins, and it may be necessary to treat the cell lysates with nucleases (Radke and Martin, 1979; Radke *et al.*, 1980). Cells may receive substantial amounts of irradiation during labelling.

10.6.4 Biosynthetic Labelling of the Carbohydrate Moieties of Glycoproteins and Glycolipids

Biosynthetic labelling of carbohydrate may be accomplished using ^{3}H-labelled mannose, fucose, galactose or glucosamine (Melchers, 1973; Tartakoff and Vassalli, 1979; Vasilov and Pleogh, 1982). Extensive metabolic conversion to other sugars or amino acids is not a severe problem, except in the case of [^{3}H]-glucosamine, which may be converted to sialic acid (Sharon and Lis, 1982). Glycoproteins cannot be labelled by addition of [^{3}H]-sialic acid to the cultures because it is not taken up by cells.

The practical details of biosynthetic labelling with radioactive sugars are little different from those for amino acids, except that glucose must be omitted from the medium during labelling. The specific activities that can be achieved are generally very low.

10.6.5 Radioactive Labelling of Carbohydrate by Oxidation Followed by Reduction with Tritiated Borohydride

The externally disposed sugars of glycoproteins and glycolipids of cells may be labelled by oxidation with galactose oxidase or periodate followed by reduction with tritiated borohydride (Gahmberg and Hakomori, 1973; Steck and Dawson, 1974; Gahmberg *et al.*, 1976; Gahmberg and Anderson, 1977).

The specific activities obtainable are fairly low, and certain glycoproteins and glycolipids may be labelled much more strongly than others.

Tritiated borohydride is available from New England Nuclear or Amersham at a specific activity of up to 70 Ci/mmol, frozen in NaOH, and may be stored for several months at –70°C. It should not be frozen and thawed repeatedly, but rather should be stored in small aliquots.

Galactose oxidase procedure

The procedure is from Gahmberg *et al.* (1976). Twenty million cells in 1.0 ml Dulbecco's PBS are treated with 5 U galactose oxidase (with or without 12.5 U neuraminidase) for 1 h at 37°C. They are then washed three times in PBS, and resuspended in 0.5 ml PBS. Tritiated borohydride (0.5 mCi) is then added, and the cells left at room temperature for 30 min. Finally, the cells are washed several times.

Periodate procedure

As an alternative to the use of enzymes, the cell-surface carbohydrate may be oxidized with sodium periodate. Provided that the periodate concentration is low, the incubation times are kept short and the cells kept cold, labelling may be restricted to the cell surface (Gahmberg and Anderson, 1977).

Approximately 3×10^7 cells are treated with 1 mM sodium periodate in 1.0 ml PBS in the dark on ice for 5 min. Glycerol is then added to quench the reaction (200 μl of a 0.1 M solution), and the cells are washed three times. They are then treated with tritiated borohydride (0.5 mCi in 0.5 ml PBS) for 30 min at room temperature, and finally washed three times in PBS.

Typical incorporation was $2–4 \times 10^3$ c.p.m. per 10^6 cells (Gahmberg and Anderson, 1977). By comparison, the labelling of a similar number of cells with 0.5 mCi ^{125}I by the lactoperoxidase technique results in typical incorporations of *c.* 10^6 c.p.m. per 10^6 cells (see Section 10.6.1).

10.7 Solubilization of Membrane Proteins

Before the classical techniques of biochemical purification and analysis can be applied to membrane proteins, they must be converted into a water-soluble form. In a few limited and special cases, solubilization of membrane proteins may be achieved by detachment from the membrane using proteolytic cleavage, or relatively minor changes in ionic conditions. In almost all other cases, however, the solubilization of membrane proteins in intact and native form can only be achieved by the use of detergents. Successful isolation of membrane antigens requires an understanding of the forces that hold membranes together, and of the mechanism of action of detergents.

10.7.1 Basic Principles

The cell membrane consists of an essentially fluid bilayer of lipids, arranged with their hydrophobic portions facing inwards and their polar head-groups interacting with the aqueous environment (Singer and Nicolson, 1972). In discussions of membranes, the lumen of intracellular organelles such as the endoplasmic reticulum and Golgi apparatus is considered to be topographically extracellular. Membrane lipids exhibit varying degrees of asymmetry in their disposition; most of the carbohydrate of glycolipids and glycoproteins lies on the extracellular face, while phosphatidyl ethanolamine lies mainly on the cytoplasmic face (Rothman and Lenard, 1977). Lateral diffusion of lipids is rapid, but 'flip-flop' movement across the membrane is less common. Recent evidence suggests that the asymmetry of membrane lipids is maintained by 'flippase' enzymes (reviewed by Higgins, 1994).

In marked contrast to lipids, the asymmetrical disposition of membrane proteins is absolute. This asymmetry is a consequence of the fact that most membrane proteins are inserted into the membrane during their synthesis, and once inserted, the energy required to move a polar region of protein across a nonpolar lipid bilayer is so great that the process rarely occurs.

Integral and peripheral membrane proteins

Membrane proteins are held in or on the membrane by two distinct mechanisms. Some are attached by electrostatic or other noncovalent interactions, and may be released by relatively small changes in pH or ionic strength. This class is known as *peripheral membrane proteins* (Singer and Nicolson, 1972). Once released, peripheral membrane proteins are usually soluble in water in the absence of detergents. Most, but not all, peripheral membrane proteins do not span or penetrate the lipid bilayer. A well-known example of a peripheral membrane protein is β_2-microglobulin, which is a subunit of the major histocompatibility antigens, and is noncovalently attached to the extracellular portion of their heavy chains. Isolated β_2-microglobulin is soluble in water in the absence of detergents.

The other major class of membrane proteins has been termed *integral* because their removal and solubilization requires disruption of the membrane with detergents. Integral membrane proteins generally possess at least one uninterrupted stretch of around 20–25 uncharged amino acids. This region spans or penetrates the lipid bilayer, and its extreme hydrophobicity ensures that the protein remains firmly embedded in the membrane.

Type I and Type II membrane proteins

Integral membrane proteins may be further subdivided. Many integral membrane proteins possess a single transmembrane sequence. These may be divided into *type I* membrane proteins, which have a cleavable *N*-terminal signal sequence and a transmembrane sequence that is usually situated close to the *C* terminus. *Type II* membrane proteins have a noncleavable hydrophobic transmembrane region close to the *N* terminus, which serves as a combined signal/anchor sequence. Examples of type I membrane proteins include the histocompatibility antigens, glycophorin and membrane immunoglobulin. Examples of type II membrane proteins include the transferrin receptor, the asialoglycoprotein receptor, and many ecto-enzymes and glycosyl transferases.

Many integral membrane proteins span the membrane more than once, and often many times. Either terminus may be inside or outside the cell. Proteins with multiple transmembrane domains include a large family of G-protein-coupled receptors such as rhodopsin, the coloured visual pigments, and receptors for many small molecules, as well as many pumps and channels.

Glycophosphatidylinositol (GPI)-linked membrane proteins and glycolipids

A further class of membrane proteins is attached to the membrane via a glycosyl phosphatidyl inosityl (GPI) anchor (Ferguson and Williams, 1988). Examples include the T cell antigen Thy-1, erythrocyte cholinesterase and the trypanosomal variant coat antigen. GPI-linked proteins often require special conditions for solubilization. They may not be adequately solubilized by nonionic detergents such as Triton X-100 (see below), but may be fully soluble in deoxycholate (Ferguson and Williams, 1988). Recent work has shown that GPI-linked proteins are associated with substantial quantities of glycolipids and cholesterol that are not soluble in nonionic detergents, and it has been speculated that these proteins may be localized to special glycolipid microdomains in the membrane (Brown, 1992; Fiedler, 1994; Kurzchalia *et al.*, 1995; Casey, 1995). The major lipophosphoglycan of *Leishmania* is also anchored in the membrane via a GPI anchor (McConville *et al.*, 1987).

10.7.2 Solubilization of Membrane Proteins by Detergents

Membrane solubilization by detergents has been reviewed in detail (Helenius and Simons, 1975; Helenius *et al.*, 1979). The essential points are as follows. Detergents are amphiphilic molecules which exist in aqueous solution as monomers and micelles. The micelles consist of aggregates of 2–100 monomers, with their hydrophobic portions buried in the centre and their hydrophilic portions at the surface. At low detergent concentrations, most detergent molecules exist as monomers. With increasing detergent concentration, the concentration of monomers rises until at a poorly defined concentration called the *critical micelle concentration*, the monomer concentration ceases to rise, and further increase in detergent concentration results from an increase in micelle concentration. The micelle size is largely independent of detergent concentration, but is influenced to a variable extent by the type of detergent, the salt concentration and the pH.

Solubilization of integral membrane proteins occurs by replacement of the planar lipid bilayer with a micelle of detergent (Helenius and Simons, 1975). The detergent micelle binds to the hydrophobic transmembrane sequence, with the hydrophilic part of the detergent facing outwards. Nonionic and weakly ionic detergents interact much more weakly or not at all with the hydrophilic cytoplasmic or extracellular portions of membrane proteins and the majority of water-soluble proteins.

It is the monomer concentration which determines the solubilizing power. Thus, the concentration of detergent should always exceed the critical micelle concentration, so that the monomer concentration is as high as possible (Helenius *et al.*, 1979). The use of lower concentrations is fraught with the risk of incomplete solubilization and a wide variety of resultant artefacts. Concentrations much higher than 10–20 times the critical micelle concentration are best avoided because there is no increase in solubilizing power, and the effects of impurities in the detergent may become significant (Ashani and Catravas, 1980). Highly purified grades of Triton X-100 suitable for membrane research are available from Calbiochem and Boehringer.

The other major requirement for solubilization concerns the total mass of detergent. Membrane lipid and detergent may be regarded as competing with each other for the hydrophobic regions of membrane proteins, so the total mass of detergent should be at least 10 times the total mass of cell lipid. This requirement is easily satisfied in most analytical experiments, in which the mass of membrane lipid may be considered negligible. However, the total mass of detergent should be carefully considered when large-scale purifications are planned.

Removal of detergent from membrane proteins usually results in uncontrolled aggregation and precipitation. It is therefore essential that buffers should contain an adequate concentration of detergent at all stages during their isolation and handling.

10.7.3 Choice of Detergent

The detergents that have been found useful for membrane solubilization may be divided into three broad groups: nonionic, weakly ionic and strongly ionic.

As a general rule, the order given correlates with increasing solubilizing power, but also with increasing disruption of protein–protein interactions and denaturation.

A great deal of knowledge about use of detergents has been obtained empirically, and there is a need for individualization of choice of detergent for solubilization of a particular membrane protein. Although hundreds of different detergents might be considered, the majority of membrane proteins are adequately solubilized by Triton X-100 or the closely related Nonidet P-40. In some cases, other detergents will need to be tried. Detergents such as the zwitterionic sulfobetaines (Gonenne and Ernst, 1978) and CHAPS (Hjelmeland, 1980) have been claimed to be especially effective, although the question of whether increased solubilization power can be achieved without a concomitant increase in denaturing ability has not yet been resolved.

Some detergents such as digitonin and octyl glucoside are said to be particularly gentle in that they cause minimal disruption to protein–protein interactions, and allow the isolation of macromolecular complexes in a form that is thought to mirror the form that exists in the cell (Oettgen *et al.*, 1986; Nakamura and Rodbell, 1990; Cambier *et al.*, 1994; Kim *et al.*, 1994; Terashima *et al.*, 1994).

Non-ionic detergents

The most widely used detergent in membrane solubilization is Triton X-100. Nonidet P-40 is virtually identical in structure. Triton X-100 has a very low critical micelle concentration (*c.* 3×10^{-4} M or 0.02%). It is a very effective membrane solubilizer, usually has minimal effect on protein–protein interactions, and leaves the nucleus intact. Sometimes, protein–protein interactions are affected, and the use of digitonin or octyl glucoside may be necessary to isolate macromolecular complexes in an intact form (see above). Antigen–antibody interactions are usually unaffected.

A typical protocol for solubilization is as follows. To $1–5 \times 10^7$ cells, add 0.5–2.0 ml 0.5% Triton X-100 in PBS, and mix gently. Leave on ice for 15–60 min, then remove nuclei and debris by centrifugation. (In many cases, solubilization is virtually instantaneous but it is customary to leave the mixture on ice for 30–60 min, just in case it is not). For many applications, a low-speed centrifugation (say 3000 ***g*** for 10 min) is sufficient. For rigorous demonstration that solubilization has been achieved, it is necessary to centrifuge at 100 000 ***g*** for 30–60 min.

Removal of Triton X-100 by dialysis is extremely slow owing to its low critical micelle concentration and alternative detergent removal methods using hydrophobic beads (Holloway, 1973) may result in drastic loss of membrane protein. Octyl glucoside may often be substituted for Triton X-100 (Baron and Thompson, 1975). It has a very high critical micelle concentration (*c.* 25 mM) and is easily removed by dialysis, because the dialysable monomer concentration is high (Helenius *et al.*, 1979).

Digitonin

Triton X-100 or Nonidet P-40 sometimes disrupt protein–protein interactions. This has been particularly well documented in the case of certain subunits of the T and B cell receptors for antigen (Oettgen *et al.*, 1986; Cambier *et al.*, 1994; Kim *et al.*, 1994; Terashima *et al.*, 1994). In these cases, associations between subunits that are not seen when the cells are solubilized in Triton X-100 may be apparent when the cells are solubilized in digitonin.

Digitonin is prepared as a 2% w/v stock solution by adding the solid detergent to boiling water, stirring for 2 min, allowed to stand at room temperature for a week and then filtered (Bridges, 1977; Oettgen *et al.*, 1986). It is used at a typical final concentration of 1% (w/v).

Bile salts

A second class of detergents in common use are the bile salts. The most widely used is sodium deoxycholate, which is more effective than Triton X-100 in solubilizing certain proteins, such as the Thy-1 antigen (Barclay *et al.*, 1975). It is possible that the reason is related to 'domains' of unusual lipids surrounding GPI anchors (Ferguson and Williams, 1988; Casey, 1995).

Deoxycholate has a greater ability than Triton X-100 or digitonin to disrupt protein–protein interactions, and may sometimes denature proteins. It will lyse the nucleus, causing release of DNA. The pK_a or deoxycholate is 6.2, and it forms an insoluble gel at a pH of 7.2 or lower. Deoxycholate is also precipitated by divalent cations. Unlike Triton X-100, deoxycholate does not absorb light appreciably at 280 nm.

Antigen–antibody interactions are usually, but not always preserved. Herrmann and Mescher (1979) found that the monoclonal anti-H-$2K^k$ antibody 11–4.1 failed to bind antigen in the presence of deoxycholate. It appears that deoxycholate causes a significant conformational change in the antigen (Herrmann *et al.*, 1982).

Sodium dodecyl sulfate

SDS is an example of a strong ionic detergent. It is highly denaturing, and very effective at disrupting protein–protein interactions, especially when combined with heat. Provided they are not heated, the subunits of proteins such as the class II histocompatibility antigens remain attached to each other in the presence of SDS (Springer *et al.*, 1977). Few proteins cannot be solubilized in sodium dodecyl sulfate; those that cannot include keratins and other proteins that have extensive covalent cross-linking between subunits.

Solubilization of cells in SDS or deoxycholate will result in lysis of the nucleus and release of DNA, which can make the sample very viscous and difficult to handle, particularly for loading onto SDS gels for Western blots; however, the viscosity can be reduced by shearing the DNA (Section 10.10.1).

Removal of detergent and concentration of dilute protein solutions for analysis by SDS polyacrylamide gel electrophoresis

Removal of sodium dodecyl sulfate from proteins is difficult, but can be achieved by adding nine volumes of acetone or methanol, holding at −20°C for 1 h, and centrifuging at 13 000 ***g*** for 10 min. The SDS will remain in solution.

Alternatively, the protein may be precipitated with methanol/chloroform (Wessel and Flügge, 1984). An aliquot (0.4 ml) of methanol is added to 0.1 ml of the protein solution and the samples are vortexed and centrifuged (10 s at 9000 ***g***). To the pellet is added 0.1 ml chloroform, and the samples are vortexed and centrifuged again. For samples containing a large amount of detergent or lipid, 0.2 ml chloroform may be used. Then, 0.3 ml water is added, and the sample vortexed and centrifuged for 1 min at 9000 ***g***. The upper phase is discarded, and a further 0.3 ml methanol added to the lower phase and the interphase with the precipitated protein. After mixing again, the sample is centrifuged at 9000 ***g*** for 2 min to pellet the protein. The supernatant is discarded, and the pellet is dried by a stream of air. The method gives high recoveries, even when only 2–3 μg of protein is used (Wessel and Flügge, 1984).

Another procedure, involving extraction of the protein into phenol followed by ether and drying under reduced pressure, allows quantitative recovery of extremely dilute proteins (10 ng/ml) from solutions containing detergents or salts (Sauvé *et al.*, 1995). Proteins can sometimes be renatured by transfer into 6 M guanidine-hydrochloride containing 1mM dithiothreitol, followed by dilution into physiological buffer (Hager and Burgess, 1980).

Most antigen–antibody bonds are disrupted by SDS, although occasionally it is possible to carry out immunoprecipitation procedures if a large excess of Triton X-100 is included. Under these conditions, the SDS is incorporated into Triton X-100 micelles, lowering its effective concentration. Antigens denatured by SDS are capable of eliciting surprisingly strong antibody responses when injected into animals (Stumph *et al.*, 1974; Tijan *et al.*, 1975; Carroll *et al.*, 1978; Lane and Robbins, 1978; Granger and Lazarides, 1979; see Section 15.1.4).

10.7.4 Enrichment of Membrane Proteins by Fractionation in Triton X-114

When heated, aqueous solutions of the Triton series of detergents become cloudy owing to the formation of large micellar aggregates. For Triton X-100, the 'cloud point' occurs at about 65°C. For Triton X-114, which is very similar in its solubilizing powers, the cloud point in 150 mM NaCl is at about 22°C. If cells are solubilized in Triton X-114 and the detergent is heated to 37°C, the detergent micelles may be centrifuged to the bottom of the tube as an oily droplet containing about 11% detergent, which will contain the great majority of the membrane proteins (Fig. 10.4). The hydrophilic water-soluble cytoplasmic proteins will stay in the detergent-depleted upper phase (Bordier, 1981, 1988; Brusca and Radolf, 1994; Fig. 10.5). This method of fractionation of membrane proteins is simple, inexpensive and effective. It provides a typical enrichment factor for membrane proteins of about 20-fold, which presumably reflects their abundance as a fraction of total cellular protein (Brusca and Radolf, 1994). Although the method is very reliable, an occasional membrane protein behaves anomalously, for reasons that are not well understood (Maher and Singer, 1985; Brusca and Radolf, 1994). Fractionation in Triton X-114 is also highly effective in removal of endotoxin (LPS) from water-soluble proteins (Aida and Pabst, 1990).

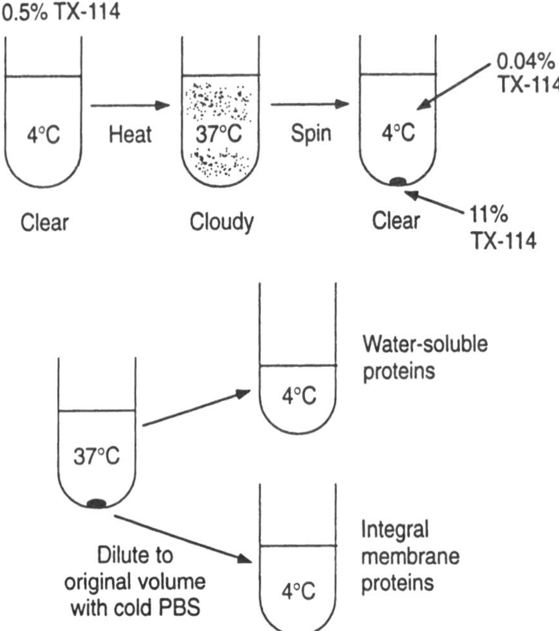

Fig. 10.4. Phase separation in Triton X-114.

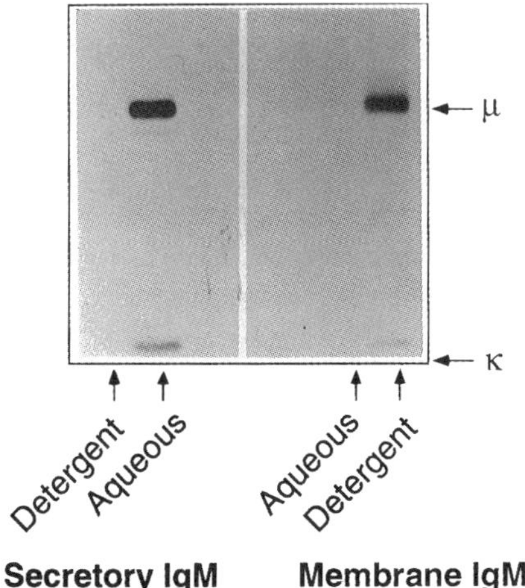

Fig. 10.5. Separation of membrane IgM and secretory IgM in Triton X-114.

The preparation of stock solutions for Triton X-114 fractionation is given in Table 10.2.

Practical procedure

(1) Solubilize cells (10^7) in 0.5 ml 0.5% Triton X-114 in PBS, on ice, for 30–60 min. As a rough guide, for 100 μl packed cells, use at least 1–2 ml of 0.5% Triton X-114. It is vital that the total mass of detergent (i.e. the concentration × volume) is more than 10 times the mass of membrane lipid in the cells, or there will not be adequate solubilization. If the protein concentration in the lysate is too high, there may be nonspecific aggregation and contamination of the detergent phase by insoluble cytoskeletal proteins. The addition of a small amount of bromophenol blue to the cell lysate will colour the detergent phase blue, making it easier to see.

Because the procedure requires fairly lengthy incubations at 37°C, it is a good idea to add protease inhibitors to the lysis buffer. One possibility is to add a cocktail including 2 mM PMSF (freshly made from the powder), 1 mM iodoacetamide, 1 mM EDTA, and 5 μg/ml leupeptin.

(2) Spin out nuclei in Eppendorf centrifuge at 4°C for 10 min, or use the Sorvall centrifuge (SS34 rotor at 14 000 rpm for 20 min). Add 10 μl of bromophenol blue (stock solution 1 mg/ml in water). The bromophenol

Table 10.2 Preparation of stock solutions for fractionation in Triton X-114

	Solution	Recipe
A.	For sucrose cushion; 6% sucrose, 0.06% TX-114 in PBS	10 ml PBS 0.6 g sucrose 50 μl 11.4% TX-114 (see below)
B.	0.5% TX-114 in PBS	10 ml PBS 500 μl 11.4% TX-114 (see below)
C.	Pre-condensed 11.4% TX-114 in PBS	See text

Triton X-114 may be purchased from FLUKA AG, CH-9470 Buchs, Switzerland (Catalogue No. 93421; 250 ml bottles). It is recommended that the Triton X-114 be pre-condensed to remove any small amounts of detergent with chain lengths different from that of the pure detergent (Bordier, 1981). Add 2 ml TX-114 to 100 ml PBS in a measuring cylinder. Add 1.6 mg butylated hydroxytoluene as an anti-oxidant. Mix well by inversion until homogeneous. Put cylinder in 37°C room overnight. Next day, two layers will be visible. The lower layer (10–20 ml) is enriched in detergent, and the upper layer is depleted in detergent. Remove and discard the upper layer by suction using water pump. Make lower layer up to 100 ml with PBS. Mix well, and repeat pre-condensation twice, as above. Save lower phase, which consists of 11.4% Triton X-114. The 11% solution of Triton X-114 is very viscous, but will be easier to pipette using a Pipetman tip which has had the end cut off with scissors before use, to create a larger orifice. The addition of a trace of bromophenol blue to the stock TX-114 solution allows easy visibility during pre-condensation and use.

blue will partition into the detergent phase, making the oily droplet easier to see.

(3) In an ice bucket, have a tube ready containing solution A (sucrose cushion; volume equal to volume of lysate).

(4) Take cleared lysate from step 2, and carefully load over the sucrose cushion.

(5) Place tube in 37°C water bath 3 min. Contents will go cloudy, especially upper part.
(If the volume is larger than about 1 ml, longer times may be needed for the solution to reach the desired temperatures.)

(6) Spin in an Eppendorf centrifuge at room temperature for 2–3 min. (If you are quick, the solution will not cool in the short times used).

(7) Transfer supernatant above sucrose cushion into fresh tube on ice.

(8) Remove the sucrose cushion by gentle suction, taking care to leave the blue oily droplet at the bottom. (The detergent-rich oily drop in the bottom is very small, estimated at 50–100 μl).

(9) Resuspend the oily drop containing integral membrane proteins in ice-cold PBS, to volume of original lysate. The oily droplet is difficult to resuspend and may require vigorous mixing. It may be resuspended in a smaller volume, giving a higher detergent concentration, and concentrating the membrane proteins.

(10) (Optional). Repeat steps 4–7 and pool second detergent-depleted supernatant with first, on ice.

(11) (Optional). To the pooled depleted supernates, add 50 μl of 11.4% TX-114. Mix, and place in 37°C bath for 3 min. Spin as above. Discard oily pellet. This step is to deplete the aqueous phase of any contaminating membrane proteins.

The supernatant will contain the water-soluble proteins.

Membrane proteins isolated by Triton X-114 fractionation contain large amounts of the detergent, which will prevent proper running of SDS–polyacrylamide gels. The detergent can be removed by the procedures given in Section 10.7.3.

10.8 Isolation of Radiolabelled Antigens by Immunoprecipitation

In addition to providing an exquisitely sensitive basis for its detection, radiolabelling or biotinylation of the antigen has the major attraction that nonlabelled proteins such as antibodies may be added without detection in the final readout (scintillation counting, autoradiography or Western blots using labelled avidin or streptavidin). In other words, labelled antigen may be specifically precipitated by nonlabelled antibody. The antigen–antibody immunoprecipitates are collected by centrifugation, washed free of material which is not bound to antibody, and the antigen recovered and analysed (Fig. 10.6). This approach, which was pioneered by Schwartz and Nathenson (1971), has been widely adopted to study hundreds of different antigens.

The attractions of immunoprecipitation include simplicity, flexibility and extreme sensitivity. It is feasible to isolate and analyse 10–30 different antigens in one experiment. The minimum quantity of antigen detectable depends mainly on the specific activity of the radiolabelled antigen. If ^{125}I-labelled antigen is used, detection in the picogram–nanogram range may be achieved without difficulty.

The main disadvantage of immunoprecipitation is that it may occasionally create ambiguity. If a particular molecular entity is precipitated by a monoclonal antibody, the possibility may exist that the antigenic determinant recognized by the antibody is not on that molecule, but rather on another molecule that is physically attached to it. The second molecule could escape detection by being nonlabelled, running off the end of a polyacrylamide gel, or being of a class of molecule that escaped detection in the analytical system used. For example, if one used an anti-β_2-microglobulin antibody to isolate HLA antigens, and analysed the proteins on a 7.5% acrylamide gel, a single band of molecular weight 45 000 would be seen. The 'real' antigen would have escaped detection by running off the end of the gel. Provided that one is aware of the problem and approaches immunoprecipitation intelligently, the method will be found to be very useful. The problem of identification of the polypeptide which actually contains the relevant antigenic determinant may sometimes be solved by Western blots (see Section 10.10).

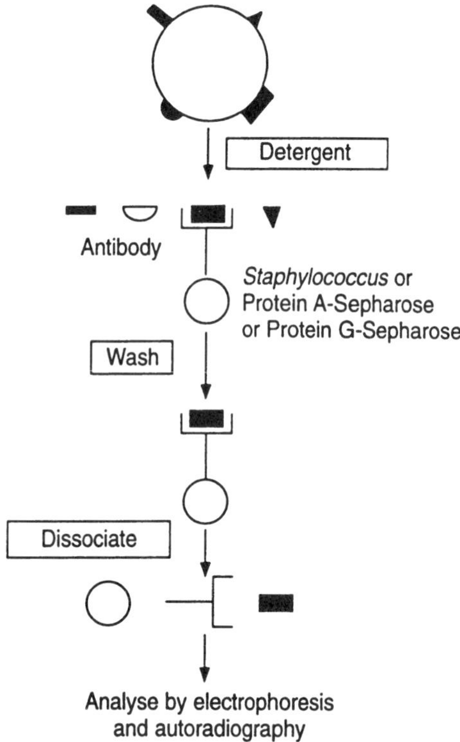

Fig. 10.6. Isolation of antigens by immunoprecipitation.

The use of biosynthetic labelling rather than surface iodination or biotinylation of cells places much greater demands on the specificity of the immunoprecipitation system, because the proportion of TCA-precipitable counts incorporated into any one protein will be a much smaller fraction of the total counts. In addition, biosynthetic labelling will incorporate a substantial proportion of radioactivity into cytoskeletal proteins, which are often on the borderline of solubility, and tend to come out of solution or bind non-specifically to the beads during immunoprecipitation, degrading the specificity and increasing the background bands on gels. Preliminary enrichment of membrane proteins by fractionation in Triton X-114 (see Section 10.7.4) will be found very useful, although occasional membrane proteins may behave anomalously (Maher and Singer, 1985).

If the protein of interest is known to be glycosylated (as are the majority of membrane and secretory proteins), preliminary fractionation on lectin columns may sometimes give useful enrichment. Lentil lectin is popular because its specificity for mannose allows most (but not all) glycoproteins to bind, and its affinity is low enough to allow efficient elution by competing sugars (Hayman and Crumpton, 1972). The volume of the matrix need only be 5–50 μl beads per 10^7–10^8 cells. It is important to appreciate that any one lectin will not bind all the various modified forms of carbohydrates of glycoproteins, and the recovery is generally low.

Antibodies are not the only proteins in serum that can bind to other proteins and care should be taken to avoid artefacts. In particular, the iron transport protein transferrin binds to cell-surface receptors. The possibility of artefacts due to transferrin has been suggested (Goding and Burns, 1981), and a case has been documented (Harford, 1984). There are many other examples of serum proteins that bind to cells, such as low density lipoprotein. Similarly, streptavidin binds to certain cell surface receptors via its RYD sequence (see Section 12.9.3).

In summary, success in immunoprecipitation depends very largely on having strong, specific antisera or monoclonal antibodies, together with biotinylated or highly radioactive soluble antigens. If these requirements are met, it is unlikely that the remaining steps will present difficulties.

Immunoprecipitation analysis of membrane proteins may be considered as occurring in four distinct phases: binding of the antibody, precipitation, washing and elution.

10.8.1 Binding of Antibody

For most practical purposes, antibodies bind almost equally well at 37°C, room temperature and 4°C. It is advisable to perform all steps on ice, however, to minimize the chance of proteolytic degradation. For the same reason, work should be performed as rapidly as possible.

It is preferable to perform the precipitation step on the same day as the labelling step because freezing and thawing of the antigen-containing mixture may result in protein denaturation and aggregation, leading to high background on gels or disruption of protein–protein interactions (Moosic *et al.*, 1980). If it is impossible to avoid freezing, the antigen should be snap-frozen in a dry ice-ethanol bath and stored at −70°C. The extract should always be centrifuged after thawing to remove any insoluble aggregated proteins.

The amount of antiserum required will depend on a number of factors. If the antibody is added in large excess, incubation times may be kept short (30–60 min) and precipitation will often be quantitative. However, it is often not clear how much antibody constitutes an excess, especially in early stages of experimentation. As a rough guide, typical amounts of strong polyclonal antisera for quantitative precipitation of individual membrane antigens from 10^7 cells range from 5–50 μl. Corresponding figures for monoclonal antibodies are in the range 0.1–1.0 μl ascites or serum, and 20–200 μl culture supernatant.

The length of time of incubation with the antiserum is not very critical. For strong polyclonal antisera used in excess, 30–60 min are usually adequate. Occasionally, weak or low-affinity antisera may require longer to reach equilibrium. Some authors recommend overnight incubations. There is little evidence that this is necessary if the antibody is in large excess.

Usually, the incubation will be carried out at physiological salt concentration and pH (150 mM salt, pH 7.2–7.4). Occasionally, individual monoclonal antibodies may precipitate spontaneously in the cold (cryoglobulins) or at low ionic strength (euglobulins). Some monoclonal antibodies may fail to bind when seemingly minor changes in pH or salt are made (Herrmann and Mescher, 1979).

10.8.2 Precipitation

Precipitation is usually carried out in 1.5 ml Eppendorf tubes or equivalent. The extremely small mass of antigen is usually insufficient to allow a precipitate to form. Even if the mass of antigen were sufficient, monoclonal antibodies may not always allow the formation of the three-dimensional lattice which is necessary for precipitation. It is therefore virtually universal practice to increase the mass of the precipitate by addition of a solid phase, consisting of agarose beads to which are coupled staphylococcal protein A, streptococcal protein G, or anti-immunoglobulin antibodies (Fig. 10.6). Heat-killed and formalin-fixed protein A-containing staphylococci can also be used (see Kessler, 1981), although most workers now use protein A-Sepharose.

If the antibody to be used binds staphyococcal protein A, protein A-Sepharose is usually the reagent of choice to collect the immune complexes for precipitation. If the first antibody does not bind protein A strongly (see Tables 5.2 and 9.2), it may sometimes be possible to strengthen binding by altering the buffer conditions. For example, mouse IgG1 will bind strongly to protein A-Sepharose in a buffer that combines high pH and high salt concentration (typically Tris-HCl, pH 8.6 plus 1.0 M NaCl). Alternatively, one may add an anti-immunoglobulin 'second antibody' that binds protein A, such as rabbit antimouse immunoglobulin. Rabbits have only one IgG subclass, and all rabbit IgG binds protein A. Somewhat surprisingly, it is found that even goat or sheep second antibodies give excellent recovery when used with protein A-Sepharose in immunoprecipitation, even though their major IgG1 subclass does not bind protein A. The reason may be because multivalency of the complexes increases the strength of binding of IgG1, or because there is sufficient amount of the minor class IgG2 which does bind protein A (Kessler, 1981).

In the interests of quantitative precipitation, the amount of second antibody should be enough to bind all the first antibody. It will usually be necessary to add 5–15 μl of second antibody for each microlitre or microgram of first antibody. A somewhat higher ratio of second:first antibody may be needed if the first antibody is serum or ascites from hybridoma-bearing mice. If the first antibody is hybridoma supernatant, one might need roughly 20 μl second antibody per millilitre culture fluid. These figures should be regarded as guidelines only, and will vary depending on immunoglobulin concentrations in the first antibody, and the antibody concentration in the second antiserum. Although it has been customary to leave the mixture for an hour or so to allow the first antibody to bind before adding the second antibody, this may not be necessary, and good results are often obtained even if the first and second antibody are added simultaneously.

The use of solid-phase precipitation rather than the older practice of allowing the formation of antigen–antibody precipitates in the liquid phase has many advantages. The total amount of IgG in the final pellet is much less, resulting in less overloading of polyacrylamide gels and less nonspecific precipitation. Much shorter incubation times are possible, as the binding of immune complexes to staphylococci or protein A-Sepharose is virtually instantaneous. After adding the staphylococci or beads and mixing for a few seconds, washing may be commenced.

In the interests of quantitative precipitation, the amount of solid-phase adsorbent should be sufficient to bind all the immune complexes in the tube. The quantity of solid-phase adsorbent required may be calculated from the following information: 100 μl of a 10% (v/v) suspension of staphylococci or 50 μl of 10% v/v protein A-Sepharose (Pharmacia) will bind about 100 μg of IgG (Johnsson and Kronvall, 1974). The IgG concentration in immune serum or serum from hybridoma-bearing mice is typically 2–10 mg/ml. Antibody levels in hybridoma supernatants are typically 10–100 μg/ml. It is convenient to use a slight excess of bacteria or protein A-Sepharose so that one may be sure that all the IgG in solution will bind. The use of an excess of solid phase avoids the need for individualizing the conditions for each new antibody. Typically, one might use 1.0 μl hybridoma serum or ascites (containing 1–10 μg IgG) followed by 50 μl of a 10% suspension of staphylococci or 25 μl of a 10% suspension of protein A-Sepharose. In this case, the solid phase has an IgG-binding capacity of 50 μg, which is a comfortable excess.

It is often necessary to 'preclear' the cell lysate by adding staphylococci or protein A-Sepharose, and then removing them by centrifugation prior to adding antibodies. This will deplete the antigen-containing solution of molecules that bind nonspecifically to the matrix.

10.8.3 Washing

After immune complexes have been bound to the solid phase, unbound material must be washed away. The two main points that require consideration are the amount of washing and the composition of the wash buffer.

Washing is accomplished by addition of a convenient volume of wash buffer (see below), followed by centrifugation. Staphylococci require centrifugation at 3000–5000 ***g*** for 10 min to be pelleted, although lower forces and shorter times may be satisfactory if the bacteria are highly aggregated. Protein A-Sepharose requires less than 400 ***g*** for 2–3 min. If one uses the popular Eppendorf bench-top microcentrifuge or equivalent (typically 13 000 r.p.m.), centrifugation times of 1 min are usually sufficient for staphylococci, and a few seconds for Protein A-Sepharose.

Each wash should dilute the soluble phase by a factor of approximately 100-fold. The degree of purification of the desired antigen might be expected to be approximately equal to the dilution factor, assuming no loss of antigen or nonspecific binding of contaminants to the solid phase. After two to three washes, the main limiting factor is likely to be nonspecific binding of contaminants, and little is gained by further washing.

It is important to realize that some monoclonal antibodies may have a fast 'off' rate or be of low affinity. In such cases, excessive washing may result in severe loss of antigen. If no antigen is precipitated, the number of washes should be decreased, and the washes should be carried out as rapidly as possible. Sometimes it will be possible to reduce the centrifugation to one brief spin by layering the mixture containing the beads over a small cushion of 10% (w/v) sucrose in PBS. If the supernatant is very carefully removed (sucking from the surface and the sides of the tube) a single centrifugation may be all that is required.

Composition of the wash buffer

The earliest studies used isotonic buffers at neutral pH, containing 0.5% Triton X-100 to maintain solubility of membrane proteins. Modified wash buffers may lower the nonspecific binding to protein A-Sepharose and staphylococci.

Nonspecific binding of staphylococci is highly dependent on pH and salt concentration, suggesting a predominantly electrostatic mechanism. The background is greatest at acidic pH and low salt, suggesting that the bacteria act as a cation exchanger. This would be consistent with loss of positive charges during formaldehyde fixation. The mechanisms of nonspecific binding to Sepharose beads are less clear, but may relate to precipitation on the beads of cytoskeletal proteins that are on the borderline of solubility.

Fortunately, alkaline pH tends to stabilize the binding of immunoglobulins to protein A (Ey *et al.*, 1978), as does a high salt concentration. The combination of high pH and high salt is particularly effective at increasing the strength of binding of mouse IgG1 to protein A. Good results have been found using a wash buffer consisting of 50 mM Tris-HCl, pH 8.3, with 0.6 M NaCl and 0.5% Triton X-100 (Goding and Herzenberg, 1980). Addition of bovine serum albumin (BSA) or ovalbumin to the wash buffer may also lower the background by saturating nonspecific protein adsorption sites, but care should be taken to choose materials without protease contamination. Other variations of the wash buffer have been listed by Kessler (1981). It must be emphasized, however, that certain monoclonal antibodies may fail to bind after apparently minor changes such as high salt buffers (Herrmann and Mescher, 1979).

10.8.4 Elution of Antigen from Immune Complexes

Elution of antigen from antibody will generally require denaturing conditions. Many different agents are suitable (MacSween and Eastwood, 1981). The major constraint will usually be the subsequent analytical step.

If analysis is to consist of SDS–polyacrylamide gel electrophoresis, the dry pellet should be resuspended in 30–150 μl SDS sample buffer, placed in a boiling water bath for 1–5 min, and the beads or staphylococci removed by centrifugation. More prolonged heating may cleave peptide bonds (Kubo, 1995). It is usual to load 30–50 μl per gel. The remainder may be kept frozen for subsequent analysis.

If the analytical step is isoelectric focusing or nonequilibrium pH gradient electrophoresis (NEPHGE), elution is performed by the addition of 200 μl of a buffer containing 9.5 M urea, 2% pH 3–10 Ampholines, 0.5% Triton X-100 and 50 mM dithiothreitol. Heating is both unnecessary and undesirable, as it will result in carbamylation of proteins and artefactual charge heterogeneity (Anderson and Hickman, 1979). The beads or bacteria are removed by centrifugation.

In some cases, it may be desired to maintain as much of the native state of the antigen as possible. If the antigen has a single proteolytic cleavage site, it is sometimes possible to liberate a fragment by gentle proteolysis of the precipitate (Goding, 1980). When different monoclonal antibodies of the same antigen are used, this method may allow the release of individual portions of the antigen and allow topographical mapping (see Section 10.11).

10.9 Electrophoretic Methods for Analysis of Protein Antigens

10.9.1 SDS-polyacrylamide Gel Electrophoresis

Polyacrylamide gel electrophoresis in the presence of sodium dodecyl sulfate (SDS–PAGE) is perhaps the most widely used single method of protein analysis. It was introduced by Shapiro *et al.* in 1967. When proteins are heated to 100°C for 2–5 min in the presence of SDS and reducing agents, they unfold and bind approximately 1.4 g SDS/g of protein. This corresponds to approximately one SDS molecule for each two amino acids. The exact mechanism of binding is still poorly understood, but probably involves the binding of SDS monomers and micelles to the protein. Many proteins retain a considerable degree of secondary structure in SDS, and models in which the SDS–protein complexes are shown as elongated rods with negative charges buttered along them are oversimplified. SDS binding imparts a very strong negative charge to the protein, dominating its native charge. Thus, the charge:mass ratio becomes constant for virtually all proteins. Under these conditions, the electrophoretic mobility in acrylamide is inversely proportional to the logarithm of the molecular weight (Weber and Osborne, 1969).

In 1970, Laemmli combined SDS-polyacrylamide gel electrophoresis with the discontinuous (disc) buffer system of Ornstein (1964) and Davis (1964). The Ornstein–Davis system exploits the differing mobilities of chloride and glycine ions at pH 6.8 to generate a moving boundary that gathers the sample into an extremely narrow band in a short upper 'stacking' gel with negligible sieving, prior to separation or 'unstacking' of the proteins in the lower resolving gel. The width of the sample entering the resolving gel is thus virtually independent of the initial sample volume. The Laemmli system is thus capable of extremely high resolution. In view of its importance, some of its features will now be described in more detail.

Theory of the discontinuous buffer system

The Ornstein–Davis buffer system works as follows. The sample buffer is Tris-HCl pH 6.8, and is followed by Tris-glycine pH 8.3 in the running buffer. When the current is applied, the negatively charged glycinate ions start to move into the Tris-HCl sample buffer. At pH 6.8, however, glycinate ions have virtually no net charge, because their amino and carboxyl groups are both fully ionized. Glycinate ions therefore trail behind the fully ionized chloride ions, but their tendency to fall behind is counteracted by the fact that this would generate a zone of higher resistance, and thus higher voltage drop (field strength), which would cause them to move faster. The result is that a sharp moving boundary is formed, with chloride ions at its leading edge and glycinate at its trailing edge.

The large protein–SDS complexes (and the marker dye bromophenol blue) have a lower mobility than that of the small fully ionized chloride ions, but higher than that of the virtually uncharged glycinate ions. Thus, as the moving boundary passes through the sample, proteins and bromophenol blue are gathered into the extremely narrow gap between the chloride and glycinate ions and enter the resolving gel as a very narrow band, regardless of the initial sample volume. Once the boundary enters the resolving gel, the pH rises to 8.8. The glycinate ions become negatively charged again, and overtake the protein. The proteins then migrate according to their size.

Effect of reduction

If proteins are heated in SDS without reducing agents, they usually bind less than 1.4 g SDS/g of protein (Pitt-Rivers and Impiombato, 1968). This might be expected to cause the proteins to move more slowly than when reduced, owing to decreased charge. However, unreduced polypeptides that contain intrachain disulfide bonds generally migrate somewhat more rapidly when reduced (Peterson *et al.*, 1972; Kaufman and Strominger, 1982; Allore and Barber, 1984). This result may be due to a more extended shape when the protein is allowed to unfold after reduction, and to the fact that the mass of the detergent itself causes significant sieving effects (Kubo *et al.*, 1979). Molecular weight estimates of unreduced proteins are therefore generally less reliable than those of reduced proteins. The bands are also often more diffuse in unreduced gels.

Effect of glycosylation

The presence of carbohydrate on glycoproteins causes them to migrate more slowly than the corresponding nonglycosylated protein. If the amount of carbohydrate is small (less than 5% of the protein molecular weight), the mobility of the glycoprotein corresponds approximately to its total molecular weight (protein plus carbohydrate). However, the retardation due to carbohydrate is disproportionately great for low percentage gels, and molecular weight estimations of glycoproteins on SDS gels are not very accurate.

Effect of modifications of the charge of the protein on mobility in SDS–PAGE

To a first approximation, the binding of SDS completely dominates the charge of proteins and also causes extensive unfolding. However, it is apparent that when the same protein is compared before and after modifications that affect charge but have little or no effect on size, there is often a detectable change in mobility on SDS gels.

For example, when proteins are reduced and then *S*-carboxymethylated with iodoacetic acid, they acquire one additional negatively charged carboxyl group for each cysteine residue. Bovine serum albumin (M_r 68 000) has 17 cysteine residues. When reduced, it runs at 68 000, but after reduction and *S*-carboxymethylation, it runs at *c.* 80 000.

Similarly, mouse β_2-microglobulin has two allelic forms (Michaelson *et al.*, 1980; Goding and Walker, 1980). The *a* allele has aspartic acid at position 85, while the *b* allele has alanine. The *a* allele moves a little more slowly on SDS gels. Extensively carbamylated proteins also move more slowly (Anderson and Hickman, 1979), as do phosphorylated proteins (Wegener and Jones, 1984).

Thus the general trend is that, all things being equal, acquisition of increased negative charge causes proteins to migrate a little more slowly in SDS–PAGE.

Practical procedure for SDS-PAGE

Recipes for buffers and gels are given in Tables 10.3–10.5.

(1) Assemble apparatus. Add 100 μl 0.1% bromophenol blue to upper tank.
(2) Select desired acrylamide concentration (Table 10.4); prepare gel mixtures according to Table 10.4. Mix well; add ammonium persulfate and TEMED (*N*,*N*,*N*′,*N*′-tetramethylene-ethylenediamine) just before pouring.
(3) Pour lower gel; overlay by spraying with a fine mist of 0.1% SDS in water; allow to polymerize (this should take 45 min). Leave for a further hour or (preferably) overnight.
(4) Remove liquid by suction.
(5) Pour stacking gel just before use (Table 10.5).
(6) Insert comb; wait for polymerization (20–30 min).
(7) Remove comb; suck out excess liquid.
(8) Heat sample in sample buffer in boiling water bath for 2–5 min.
(9) Load sample. Length of sample must be less than length of stacking gel, as measured from bottom of well to main gel.
(10) Overlay sample with running buffer.
(11) Run gels at 20 mA, constant current, with anode (positive) at bottom, until bromophenol blue line reaches bottom.
(12) Turn off current; remove gel.
(13) The procedure for subsequent handling at the gel will depend on whether it is desired to stain for total protein, to process for autoradiography when radioactivity is present, or possibly both. The type of treatment will also depend on the nature of the isotope present (see below).

Staining for total protein

Place the gel into stain (0.1% Coomassie brilliant blue in 40% methanol, 10% acetic acid) for at least 1 h. Destain in 12% ethanol, 7% acetic acid. Sensitivity of total protein staining may be increased approximately 100-fold by silver staining. The procedure of Morrisey (1981) is recommended.

Use of intensifying screens for ^{125}I and ^{32}P

The use of intensifying screens greatly increases the sensitivity of detection of ^{125}I and ^{32}P (Laskey, 1980, 1993). If the correct combination of film and screen

Table 10.3 Solutions for SDS-polyacrylamide gel electrophoresis

Solution	Composition	Recipe
Sample buffer	62 mM Tris-HCl, pH 6.8 0.2% SDS, 50 mM dithiothreitol, 10% glycerol	To 400 ml distilled water, add 3.8 g Trizma base, 50 ml glycerol and 11.5 g SDS. Titrate carefully to pH 6.8. Add 3.9 g dithiothreitol, and make to 500 ml. Store at 4°C.
Lower gel buffer	1.5 M Tris-HCl, pH 8.8, 0.4% SDS	To 900 ml deionized distilled water, add 182 g Trizma base and 4.0 g SDS. Titrate to pH 8.8 with conc. HCl, then make up 1 l. Store at 4°C.
Upper gel buffer	0.5 M Tris-HCl, pH 6.8, 0.4% SDS	To 400 ml deionized water, add 30 g Trizma base and 2.0 g SDS. Titrate to pH 6.8 with conc. HCl, then make up to 500 ml. Store at 4°C.
Running buffer (10X stock)	25 mM Tris, 190 mM glycine pH 8.3, 0.1% SDS when diluted 1:10	To 10 l distilled water, add 303 g Trizma base, 144 g glycine and 100 g SDS. Do not titrate. Store at room temperature.
30% Acrylamide stock	bis:acrylamide = 1:36	Available pre-mixed. To make from ingredients, weigh out 300 g acrylamide, 8.2 g N,N′-methylene-bis-acrylamide. Add deionized distilled water to 1027 ml. Filter through Whatman No. 1 paper, store at 4°C, protected from light. Acrylamide is a neurotoxin; do not mouth pipette.
10% Ammonium persulfate		To 5 ml deionized distilled water, add 0.5 g ammonium persulfate. Store at 4°C, and make fresh solution weekly. Store ammonium persulfate crystals in desiccator. Ammonium persulfate concentration is important, and compound is unstable.
N,N,N′,N′,-tetra-methylethylene-diamine (TEMED)		Store at 4°C.

Table 10.4 Recipes for SDS-polyacrylamide gel electrophoresis (1 gel = 16 ml)

M_r range	% acrylamide	Deionized distilled water (ml)	Lower gel buffer (ml)	30% acrylamide (ml)	10% ammonium persulfate[a] (μl)	TEMED (μl)
70 000–200 000	5.0	9.3	4.0	2.7	40	20.0
40 000–150 000	7.5	8.0	4.0	4.0	30	15.0
20 000–100 000	10.0	6.7	4.0	5.3	25	12.5
10 000–70 000	12.5	5.3	4.0	6.7	20	10.0
8 000–50 000	15.0	4.0	4.0	8.0	15	10.0

[a]Adjust volume to allow polymerization in 30–45 minutes at room temperature.

Table 10.5 Recipes for stacking gels (1 gel = 5.0 ml)

	Distilled water (ml)	Upper gel buffer (ml)	30% acrylamide (ml)	urea (g)	10% ammonium persulfate (ml)	TEMED (ml)
Without urea	2.95	1.25	0.8	–	15	5
With urea (~ 4M)	2.3	1.25	0.8	1.35	10	5

is used, a protein containing 1000 c.p.m. will give a dark band with overnight exposure. (This can be easily verified by cutting out a band and counting it.) High-energy β or γ radiation from the gel passes through the X-ray film and strikes the intensifying screen, causing the release of photons which expose the film. Each disintegration causes the release of multiple photons, resulting in an improvement of sensitivity of 10–30-fold. The increase is obtained only at –70°C, because of the instability of activated silver halide crystals at higher temperatures (Laskey 1980, 1993). The screens are only effective for ^{32}P and ^{125}I, and do not work with ^{35}S, ^{14}C or ^{3}H, because their disintegration energy is not sufficient to penetrate the film and the protective coating of the screen (Fig. 10.7).

It is important to appreciate that the maximum sensitivity is achieved only with certain types of screens and films, such as Dupont Cronex Lightning Plus screens (calcium tungstate), which emit blue light (Laskey, 1980, 1993). It is essential to use a blue-sensitive film, such as Agfa Curix RP-2 or Fuji RX. The highest sensitivity is obtained with Kodak X-Omat AR (XAR) film, but this film is considerably more expensive.

If strict linear quantitation is needed, the films should be ‘preflashed’ with a brief pulse of light from a domestic photoflash and appropriate filters (Laskey, 1980, 1993). Preflashing increases the sensitivity of detection of the very weakest signals, but has no effect on stronger signals. The nonlinearity of

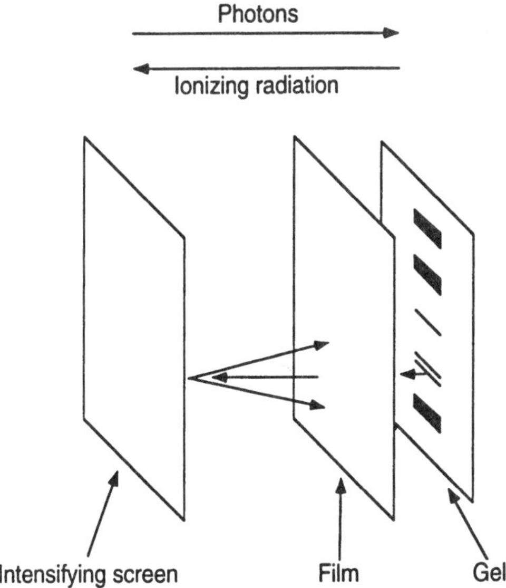

Fig. 10.7. Use of intensifying screens for autoradiography. High-energy β or γ radiation from the gel passes through the X-ray film and strikes the intensifying screen, causing the release of multiple photons which expose the film. The gel is usually dried before autoradiography. Drying the gel gives a permanent record for long-term storage and also gives a sharper image.

the response without preflashing will tend to suppress the weaker background bands, which may lead to prettier but perhaps misleading gels. However, strict quantification is not always needed, and preflashing can often be omitted. Intensifying screens are compatible with Coomassie blue stained gels.

Recently Dupont NEN have discontinued the Cronex Lightning Plus screens and replaced them by a new range of 'NEN Reflection' intensifying screens and film which are said to be more sensitive. Although the new screens give some enhancement at room temperature, exposure at −70°C is recommended, especially for exposures of more than a few hours. Intensifying screens do not enhance the sensitivity of enhanced chemiluminescence.

Detection of ^3H, ^{14}C and ^{35}S by impregnation of the gel with organic scintillants

^{35}S and ^{14}C may be detected by direct autoradiography at room temperature. There is no increase in sensitivity for direct autoradiography at −70°C. The gels may be stained with Coomassie blue without affecting sensitivity. Direct autoradiography cannot be used for ^3H, because its energy is too low to emerge from the gel or penetrate the gelatin coating of the X-ray film.

The efficiency of detection of weak β emitters such as ^3H, ^{14}C and ^{35}S is increased by impregnating the gel with organic scintillants which capture the radiation with high efficiency and convert it into photons, which are capable of leaving the gel (Laskey, 1980, 1993). The increase in sensitivity of detection of ^{3}H is increased enormously. The increase in sensitivity for ^{14}C and ^{35}S depends on the thickness of the gel. For 0.8 mm gels, the improvement is approximately three-fold. Thicker gels will give a bigger factor of improvement, but are seldom used because the bands will be more diffuse. For typical thin gels, the gain in sensitivity from the use of scintillants for ^{14}C and ^{35}S is not usually worthwhile.

Several different scintillants are available. In my opinion, the most satisfactory is 'Amplify' (Amersham). The gel is soaked in Amplify for 30 min, placed on Whatman 3MM paper and immediately dried on a gel drier (80°C). It is not necessary to fix the proteins in the gel prior to use of either product, provided they are processed as soon as electrophoresis is finished.

Fluorography is effective only at –70°C or below (Laskey 1980, 1993). As is the case for intensifying screens, linear quantification requires preflashing, although the consequences of not preflashing are only apparent for extremely faint bands. Staining of gels with Coomassie blue lowers on the sensitivity when the gels are impregnated with fluor (Higgins and Dahmus, 1982). This is apparently due to quenching of the fluors rather than direct absorption of the blue light, because Coomassie blue does not absorb light significantly at the relevant emission wavelengths of the fluors.

Drying acrylamide gels

To dry the gel, place it on a piece of kitchen plastic (Glad-Wrap, Saran-Wrap or equivalent). Cover with a piece of Whatman 3MM paper. Trim edges with scissors. Dry on a gel dryer (Bio-Rad model 224 or equivalent) attached to a good water pump or an oil vacuum pump. If the latter is used, a cold trap using dry-ice and ethanol is essential to prevent water from entering the pump and destroying it. Teflon-based pumps are now available and are said to be tolerant of organic solvents and acetic acid. They may be more suitable for drying gels. Once drying has started, if the vacuum is lost, even momentarily, the gel will shatter into hundreds of pieces. Drying takes 20–60 min, depending on the vacuum and gel thickness.

Alternatively, the gel may be dried with virtually no equipment, using the following procedure. Take two sheets of cellophane, which may be obtained at any stationery store, and wet them in water. Lay one sheet onto a glass plate. Lay the gel on top of it, and smooth out any bubbles. Lay the second sheet over the gel. Make a frame around the edge using thin strips of plastic or wood, held in place with foldback paper clips (Fig. 10.8). This frame prevents the cellophane from moving as it dries. Leave to dry at room temperature for 24–48 h. Remove clips and strips, and trim the cellophane. This procedure is suitable for

Fig. 10.8. Method of drying acrylamide gels between cellophane sheets. The resulting dried gel is ideal for photography of Coomassie blue or silver stained proteins, or can be autoradiographed. Similar methods have been described by Wallevik and Jensenius (1982) and Juang et al. *(1984).*

all forms of fluorography, or for photography. The gels should be stored in a book to prevent curling. A very similar procedure has been described by Wallevik and Jensenius (1982).

Setting up for autoradiography

Identification of the gel and alignment of the autoradiograph with stained bands is facilitated by marking the gel with adhesive labels and writing on them with radioactive ink. Use an ordinary fountain pen (cartridge pens are ideal), and add 1.0 μCi/ml of any nonvolatile ^{14}C- or ^{35}S-containing compound to the ink. Alternatively, nonradioactive light-emitting 'TrackerTape' (Amersham cat. no. RPN 2050) is very convenient, and may be written on with a normal pen or pencil.

The cassettes used in clinical radiology are convenient but expensive. Equally satisfactory results are obtained by using lightproof plastic bags (or discarded lightproof envelopes of films). Plastic bags may be made by folding and taping black plastic sheeting. It is important that the film be in intimate contact with the gel or the image will be blurred. Fold over the top of the envelope, and clamp between two boards with paper clips. The use of 1 mm stainless steel sheets is sufficient to shield the radiation of ^{125}I, and provides a flat surface that ensures that the gel and the film are in intimate contact.

Acrylamide poisoning

Acrylamide is a neurotoxin (Kuperman, 1958). It must never be pipetted by mouth, or allowed to contact the skin. Always wear a mask and gloves when weighing the powder. Premixed solutions of acrylamide and bis-acrylamide are now available commercially, and are highly recommended, as they eliminate the need to weigh out the powder. Use gloves when handling the poured gels. While it is true that the polymerized gel is harmless, 'polymerized' gels virtually always contain some unreacted monomer. The gels should therefore be handled with care.

Acrylamide poisoning causes peripheral neuropathy, and in severe cases, brain damage. Symptoms include numbness and weakness in hands, feet, arms or legs, tingling sensation in arms or legs, increased sweating and peeling of skin of hands. In more severe cases, there may be clumsiness, unsteadiness on the feet and difficulties with micturition. Recovery generally occurs after cessation of exposure, but may be very slow (months to years) and may never be complete (McCollister *et al.*, 1964; Garland and Patterson, 1967; Kesson *et al.*, 1977).

10.9.2 Two-dimensional (Charge, Size) Electrophoresis

The combination of isoelectric focusing (IEF) and SDS–PAGE results in an extremely high-resolution system which is capable of separating almost all cellular proteins from each other (O'Farrell, 1975). As originally described, the method was only suitable for acidic proteins, but these comprise the great majority of cellular proteins.

A later modification of the technique allows analysis of basic proteins. The technique is termed 'non-equilibrium pH gradient electrophoresis' (NEPHGE). The separation process is fundamentally very similar to IEF, but the duration is shorter. Acidic proteins become virtually stationary at their isoelectric points. Basic proteins are well resolved from each other, but do not stop moving (O'Farrell *et al.*, 1977). NEPHGE analysis is capable of resolving all but the most acidic and basic cellular proteins. Garrels (1979) has described a number of modifications and refinements to the original system.

Most secretory and externally disposed membrane proteins are glycosylated. When glycoproteins are analysed on two-dimensional gels, they do not run as discrete spots, but as fuzzy smears or 'families' of spots of similar M_r but differing in unitary steps of charge (e.g. Ledbetter *et al.*, 1979; see also Fig. 10.9). This heterogeneity is at least partly a reflection of their variable carbohydrate content (Ledbetter *et al.*, 1979).

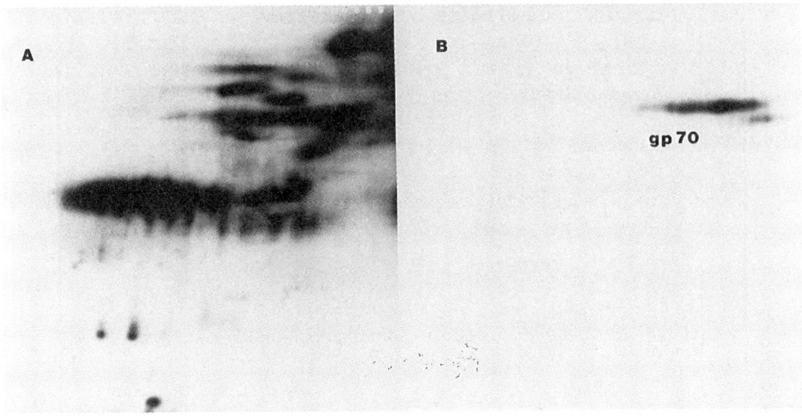

Fig. 10.9. Two-dimensional polyacrylamide gel electrophoresis of membrane proteins of the T-cell lymphoma EL-4. Membrane proteins were radiolabelled with ^{125}I by the lactoperoxidase technique, solubilized in Triton X-100, and analysed by nonequilibrium pH gradient electrophoresis (right to left) followed by SDS–polyacrylamide gel electrophoresis (top to bottom). Acidic proteins lie to the right. (A) Whole extract. (B) Murine leukaemia virus envelope protein gp 70, isolated from EL-4 cells by immunoprecipitation. Note that gp 70 is visible in the whole extract.

Isoelectric focusing (IEF)

Gels are run in 150 mm long, 2 mm inside diameter tubes. Recipes for gels are given in Table 10.6.

(1) Cap bottom of tubes with Parafilm. Place on corrugated cardboard around a bottle; hold in place with a rubber band.
(2) Make up gel mixture, and pour gels, leaving space for sample (20–200 μl) at top. Gently overlayer with distilled water. After polymerization (*c.*45 min), leave for a further 45 min, then flick out water and overlayer gel with IEF lysis buffer (20 μl) followed by distilled water. Gels may be run immediately, or left for 24 h.
(3) When ready to start, remove overlayer by flicking, and load gels into standard tube gel apparatus, with dilute phosphoric acid in bottom tank (anode).
(4) Overlayer with 20 μl IEF lysis buffer, followed by NaOH cathode solution. Prerun for 15 min at 200V, 30 min at 300V, and 30 min at 400V (anode at bottom).
(5) Turn off current, remove cathode buffer and overlayer, and apply samples in IEF lysis buffer. Resolution is essentially independent of sample volume, which can be up to 200 μl. Overlayer with a small volume of IEF lysis buffer slightly diluted with water, followed by NaOH cathode solution.
(6) Run at 300–400V (constant voltage) for a total of 4800 volt-hours, then increase voltage to 800V for 1 h.

Table 10.6 Recipes for isoelectric focusing (IEF) and non-equilibrium pH gradient electrophoresis

Solution	Composition	Recipe
IEF lysis buffer	9.5 M urea, 2% Triton X-100, 2% Ampholines, 50 mM dithiothreitol	28.5 g urea, 1.0 ml Triton X-100, 2.0 ml pH 5–7 Ampholines, 0.5 ml pH 3.5–10 Ampholines, 390 mg dithiothreitol, deionized distilled water to 50 ml. Store in 200 μl aliquots at –70°C.
NEPHGE gels (10)		2.75 g urea, 0.67 ml 30% acrylamide stock, 1.0 ml 10% Triton X-100, 1.0 ml deionized distilled water, 250 μl pH 3.5–10 Ampholines. Mix until all urea is dissolved (do not heat). Add 7.0 μl 10% ammonium persulfate and 4.5 μl TEMED. Mix. Pour gels. Overlayer with water until polymerized, then remove water and overlayer with IEF lysis buffer followed by water.
IEF gels (10)		Same as NEPHGE gels, except use 200 μl pH 5–7 Ampholines and 50 μl pH 3.5–10 Ampholines.
30% Acrylamide stock	bis:acrylamide = 1:17	Available pre-mixed. To make from ingredients, take 28.4 g acrylamide, 1.6 g N,N′-methylene-bis-acrylamide; add deionized distilled water to 100 ml. Filter through Whatman No. 1 paper; store at 4°C protected from light.
10% Triton X-100		10 ml Triton X-100 plus 90 ml deionized distilled water. Stir until homogeneous.
Anode solution	10 mM orthophosphoric acid	0.7 ml 85% orthophosphoric acid in 1 l deionized distilled water
Cathode solution	20 mM NaOH	1.6 g NaOH pellets in 2 l deionized distilled water. (Always make up freshly. For IEF, boil water for 10 minutes; add NaOH in 5 ml water, boil again for 5 min. Allow to cool.)

(7) Turn off power; remove gels from tubes with a 1.0 ml tuberculin syringe with plastic tubing attached. Apply pressure to bottom of gels. If gels are hard to extrude, try using SDS sample buffer instead of air in syringe.
(8) Equilibrate each gel with 5 ml SDS sample buffer (Table 10.3) for 1–2 h, with gentle rocking at room temperature.
(9) Gels may be snap-frozen in SDS sample buffer using a dry ice–ethanol bath any time after 30 min, and equilibration finished just prior to running second dimension.
(10) Pour second-dimension SDS gel, preferably on the day prior to use.
(11) Pour stacking gel for second dimension, 30 min before equilibration is due to finish. Use a straight Teflon sheet instead of a comb, or use combs upside-down. While gel is polymerizing, set up a boiling water bath, and tanks for second dimension. Place 0.2 g agarose in 10 ml SDS sample buffer and heat until dissolved.
(12) When stacking gel is polymerized, suck out excess liquid. Insert a small piece of Teflon between the plates at one end, to provide a well for molecular weight standards.
(13) Add 1–2 ml hot agarose, and quickly lie first dimension gel onto stacking gel. Leave agarose to set for 5 min. Remove Teflon piece and load molecular weight standards.
(14) Fill reservoirs with SDS running buffer (Table 10.3). Remove bubbles from under plate. Add 100 μl 0.1% bromophenol blue to upper reservoir.
(15) Run at 20 mA per gel, with anode at bottom, exactly as for one-dimensional gels.

Nonequilibrium pH gradient electrophoresis (NEPHGE)

Despite their long name, NEPHGE gels are actually much simpler to run than IEF gels. For most purposes, all that is required is separation—determination of the isolelectric point is of little biological importance. Unless there is good reason to use IEF gels, NEPHGE are to be preferred.

The gels are set up exactly as for IEF, except that all Ampholines are pH 3.5–10 (Table 10.6), the prerun is omitted, and the cathode (NaOH) is at the bottom. Gels are run at 500 V (constant voltage) for 5 h, and extruded by applying pressure to the lower end. It is not necessary to boil the cathode buffer.

Calibration and alignment of two-dimensional gels

Molecular weight standards for SDS gels are available commercially as kits of purified proteins. Suitable proteins include myosin (200 000); β-galactosidase (116 000), phosphorylase (95 000), bovine serum albumin (68 000), ovalbumin

(43 000), beef heart lactate dehydrogenase (30 000), soybean trypsin inhibitor (20 000) and cytochrome *c* (13 000). Inconsistencies of *c.*10% in molecular weight estimations between laboratories are very common, especially for glycoproteins.

The measurement of isoelectric points is complicated by the fact that pH measurements in the presence of high concentrations of urea may have large errors (O'Farrell, 1975). The published values for isoelectric points of proteins are often in poor agreement. Two published values for the isoelectric point of the human transferrin receptor differ by almost one pH unit (Bleil and Bretscher, 1982; Felsted and Gupta, 1982).

A more useful way to compare results between individual gels is to incorporate well-characterized nonradioactive proteins into the sample as internal standards. Suitable standards include ovalbumin (M_r 43 000; pI = 4.6) and bovine serum albumin (M_r 68 000; pI = 4.9). Commercial isoelectric point standard kits have proven disappointing. The use of carbamylated protein standards (Anderson and Hickman, 1979) holds some promise, and would be especially useful in conjunction with proteins of well-characterized isoelectric points.

Alignment of the autoradiograph with the gel is facilitated by marking the gel with TrackerTape (Amersham) or a fountain pen containing radioactive ink (see Section 10.9.1). This also allows identification of each film with the gel and experiment number.

10.9.3 Trouble-shooting Acrylamide Gels

The complexity of acrylamide gel electrophoresis is not very great, but there are a number of things that can go wrong. The following is a list of common problems, and suggestions for their remedy.

Poor polymerization

Polymerization is very temperature-sensitive, and is much slower on cold days. In the winter, the gels may fail to polymerize if the laboratory is not adequately heated. Polymerization is also inhibited by heavy metal ions, especially copper, other impurities in the water or reagents, and contaminants from the glass, combs, spacers and tubing. New combs, spacers and tubing may need 'conditioning' by several dummy runs until they are ready. Try to avoid contact of acrylamide and buffers with the inserts in the tops of bottles. Gel tubes must be rinsed very thoroughly if ethanol–KOH or chromic acid is used for cleaning.

Ammonium persulfate is unstable; it is hygroscopic and breaks down in aqueous solution. Store the crystals in a desiccator at room temperature. They should be free-running and dry. Although it is often recommended that ammonium persulfate be made up freshly for each gel or stored frozen in small aliquots, a 10% stock solution of ammonium persulfate is usable for at least a week if stored at 4°C.

High background on autoradiographs

Include an excess of nonradioactive iodine or amino acid after labelling to dilute out its specific activity and 'compete out' any tendency to stick non-specifically. If labelling proteins with ^{32}P, there may be background due to DNA or RNA. This may be eliminated by use of appropriate nucleases.

If the whole-cell extract or the whole of an *in vitro* translation mix is loaded onto a gel, there will be a large amount of unincorporated isotope. This may diffuse out into the stain, destain or fluor solution. It can then diffuse back into the gel, causing the whole of the outline of the gel to be black. The solution is simple. Wash the gel thoroughly with several changes of destain or water prior to adding the fluor solution or drying.

Diffuse uneven background is often caused by stray radiation. Check freezer or laboratory for sources of radiation. Use 1 mm steel sheets outside gels to shield them during autoradiography.

Coomassie blue staining faint

The dye solution may be used repeatedly but eventually becomes less effective. This is probably due to contamination with SDS from gels rather than the dye being 'used up'. The SDS competes with the dye for the protein. The solution is to make up fresh dye. To prevent contamination of the dye with SDS, and to achieve more reproducible staining, it may be a good idea to rinse the gel for 10–30 min in destain prior to staining.

Dye precipitates on surface of gel

Filter dye solution through filter paper before use. The dye may precipitate if alcohol or acetic acid is lost by evaporation; cover with kitchen plastic while staining. Keep gel fully immersed at all times as drying causes precipitation. Precipitates can be removed by gentle wiping with paper tissues; the gel is strengthened and tearing prevented by soaking it in 100% methanol for a few minutes prior to wiping. It is then transferred back into destain.

Poor autoradiographs

There must be intimate contact between gel and film. Are the recommendations concerning film, screens and fluors being followed exactly?

Replace developer monthly. The correct safelight for Kodak X-omat AR film is Kodak 6B. Check for light leaks. Fluorography and intensifying screens only work at –70°C.

Sometimes, a coarse granular image occurs when the fluor 'Enhance' is used. The cause appears to be excessive heating and patchy loss of fluor. This problem is not seen with 'Amplify'. Gels treated with 'Enhance' must be dried at 60°C. Gels treated with 'Amplify' can be dried at 80°C.

Wavy bands

Excessive heating during electrophoresis is the most likely culprit. Use a lower current. Remember that heating is proportional to the square of the current or voltage. Other causes include uneven polymerization, which may be caused by inadequate mixing, air bubbles, chemical contaminants, movement or vibration during polymerization, or uneven room heating. Do not use too much agarose to hold the first dimension gel in place (no more than 2 ml).

Gels brittle, milky or soft

This is generally caused by inadequate polymerization or incorrect concentration of acrylamide or bis. Milky gels arise from too much bis; soft gels from not enough. Make up new ammonium persulfate.

Double or broad bands in SDS gels; failure of stacking

The ionic composition of the sample must be approximately that of the loading buffer. Excessive salt or a pH that is more than 0.5 unit from 6.8 will cause the gels to run badly. The length of the sample must be less than the distance between the bottom of the sample well and the top of the resolving gel. Pour a longer stacking gel. Note that the dye marker is always fuzzy in gels of more than 11–12% acrylamide.

If the sharp blue dye marker line does not form, it probably means that one of the buffers is incorrectly made up. Check all buffers, by substitution with new ones if necessary. A conductivity meter is a great help in locating incorrectly made buffers. The pH of the buffers must be correct. Many pH electrodes give large errors with Tris. Is your electrode suitable for Tris? Is the pH meter working properly? The great majority of faults in pH meters are in the electrode rather than the electronics. Remember that Tris has a large temperature coefficient.

Individual lanes in SDS polyacrylamide gels vary in width

This is usually due to the ionic composition of the sample being too far removed from that of SDS sample buffer. It is important that the pH of the sample be not far from 6.8, and it should not contain excessive amounts of salt. The best solution is to dilute the sample in a larger volume of SDS sample buffer and only load a small percentage.

Gels run too fast or too slowly

Wrong composition or pH of buffer, or (less likely) faulty current or voltage measurement. Check pH of lower gel buffer.

Cracking of gel during drying

If the gels are no more than 0.8 mm in thickness, cracking during drying is seldom a problem. If it is absolutely essential to use thicker gels, the problem may be eased somewhat by soaking the gel in 55% methanol/2% glycerol overnight after destaining. The gel will shrink considerably, but is then more resistant to cracking. Other possible causes include excessive heat, poor quality acrylamide and loss of vacuum. Cracking is virtually inevitable if the vacuum is lost during drying.

Incomplete reduction

Dithiothreitol (DTT) is a better reducing agent than mercaptoethanol, and less susceptible to oxidation by air. None the less, DTT can deteriorate over a period of time and is best stored at 0.5 M in 500 μl aliquots at −70°C. Addition of the charged reducing agent sodium thioglycollate (1 mM) to the upper tank buffer of SDS gels provides a reducing environment within the gel during electrophoresis.

Poor isoelectric focusing or NEPHGE

Excessive salt in sample. Acrylic acid contamination in acrylamide (deionize with mixed bed resin such as Bio-Rad AG501-x8). Urea concentration too low. Gel tubes not adequately rinsed before use.

Crumbling of basic end of NEPHGE gels

A common problem. The collapse seems to be caused by the combination of high electric field strength (owing to depletion of ampholytes) and high pH.

See Tanaka (1981) and Tanaka *et al.* (1982) for an interesting discussion. Try deionizing the acrylamide. Wait 5–10 min after turning off current before extruding gel (broadening of bands will be negligible). Extrude gently.

First dimension gel falls off during run

This is only a problem if it occurs during the first 30 min of electrophoresis. Try a higher percentage agarose.

Artefactual charge heterogeneity in IEF and NEPHGE

Heating proteins in urea will result in artefactual charge heterogeneity, owing to carbamylation of lysines (Anderson and Hickman, 1979). Urea solutions should be made up freshly or stored at −70°C. They must not be heated.

Protein overload

This will result in distorted bands, smearing caused by incomplete solubility, and may result in physical weakening of the first dimension gel of two-dimensional systems.

Current rises during IEF or NEPHGE; no focusing

Electrodes are reversed! The anode (positive) must be at the acidic end.

10.9.4 Peptide Mapping by Limited Proteolysis

Proteins usually consist of a series of compact and protease-resistant domains joined by short amino acid sequences which are more susceptible to proteolysis. Partial proteolytic cleavage of native proteins may therefore yield large fragments which can provide useful information about the structure of the intact molecule.

Many proteases are active in the presence of SDS and reducing agents, and even when nonspecific proteases are used to digest partially denatured substrates, the resulting breakdown pattern is highly reproducible and characteristic of each individual protein (Cleveland *et al.*, 1977). SDS–PAGE of the proteolytic fragments can therefore be used as a simple and powerful way of assessing relatedness of proteins. Analysis of fragments is usually carried out on one-dimensional gels, although two-dimensional analysis has also been used (Hoh *et al.*, 1979; Whalen *et al.*, 1979).

The enzymes most commonly used include staphylococcal V8 protease (which cleaves at glutamic and aspartic acid), and the more nonspecific proteases papain, chymotrypsin, pronase, subtilisin and thermolysin. Trypsin has been used only rarely, and at very high concentrations, presumably because of inactivation by SDS and reducing agents. In a few cases, even SDS-treated proteins are highly resistant to proteolysis. In these cases, digestion in 4M urea will usually result in adequate cleavage (Handman *et al.*, 1981).

Frequently, high-resolution electrophoretic analysis of antigens immunoprecipitated by monoclonal antibodies reveals multiple spots. Peptide mapping may help decide whether they represent different modified forms of the same basic sequence (glycosylated, phosphorylated or degraded by protease), or distinct sequences of a multi-subunit structure.

Peptide mapping is often useful in conjunction with monoclonal antibodies as a way of testing relatedness between two similar proteins. For example, a protein consisting of two apparently identical disulfide-bonded M_r 95 000 polypeptides is precipitated from proliferating human cells by the monoclonal antibody OKT-9. The receptor for transferrin has a similar structure. Peptide maps of the protein bound by OKT-9 were identical to maps of a transferrin-binding protein of the same molecular weight and subunit structure, proving that OKT-9 recognizes the transferrin receptor (Goding and Burns, 1981).

Gullick *et al.* (1981) used monoclonal antibodies to immunoprecipitate proteolytic fragments of the acetylcholine receptor. Interestingly, many fragments were still recognizable by monoclonal antibodies after heating to 100°C in 2% SDS followed by dilution into 0.1% SDS and 0.5% Triton X-100.

Practical procedure

If the polypeptide is available in solution, digestion may be carried out in the liquid phase. If it has been isolated by immunoprecipitation, it may be solubilized by heating in SDS sample buffer at 100°C for 2–5 min. The beads or staphylococci should be removed by centrifugation and the solution allowed to cool prior to adding the protease.

Typical conditions are 0.1–50 μg/ml protease at room temperature for 30–90 min, but should be adjusted to give a good range of breakdown products. Digestion may be terminated by heating to 100°C, but this is not essential, as the sample may be loaded immediately onto the analytical gel. If two-dimensional gels are used, it is probably wise to remove the SDS by precipitation with acetone or chloroform/methanol (Hager and Burgess, 1980; Wessel and Flügge, 1984) prior to resuspension in IEF lysis buffer. NEPHGE and IEF gels will tolerate small amounts of SDS, especially if a large excess of Triton X-100 is present.

A very useful variant involves cutting out protein bands from polyacrylamide gels and eluting the protein electrophoretically, together with added protease, into the stacking gel of a 15% SDS gel (Handman *et al.*, 1981; Goding and Burns, 1981; Wiegers and Dernick, 1981; Goding, 1982). The stacking process allows the protease and its substrate to be brought together. The proteins may be stained or radioactive, and the gel spot may be wet (in which case it should be equilibrated with SDS sample buffer) or dried. The presence of 'Enhance' scintillant does not appear to influence digestion. If the spot is from a dried gel, it should be placed directly into the sample well of a 15% Laemmli gel and rehydrated for 1 h with sample buffer containing protease (1–5 μg/ml), then overlaid with running buffer in the usual way. As the bromophenol blue dye approaches the main gel, the current is turned off for 30–90 min, to allow digestion to take place in the stacking gel. Protease resistant polypeptides may be digested more readily if 4 M urea is incorporated into the stacking gel (Handman *et al.*, 1981; see Table 10.5).

10.9.5 Peptide Mapping by Chemical Cleavage

The Cleveland peptide mapping system has been adapted to use chemical cleavage rather than proteases. The advantages of chemical cleavage include less need to individualize conditions for each protein, and a wider choice of points of cutting. Perhaps the most useful of these procedures involves *N*-chlorosuccinimide, which cuts mainly at tryptophan (Lischwe and Ochs, 1982). Tryptophan is uncommon, so fragments are large. In addition, it is one of the most highly conserved amino acids, so the fragmentation tends to emphasize similarities rather than differences.

Other chemical cleavage methods include: hydroxylamine, which cuts at asparagine–glycine bonds (Saris *et al.*, 1983); dilute acid, which cuts at aspartyl–proline bonds (Rittenhouse and Marcus, 1984); and cyanogen bromide, which cuts at methionine (Nikodem and Fresco, 1979; Lam and Kasper, 1980; Lonsdale-Eccles *et al.*, 1981).

Although the acid cleavage method is very simple, it generates a very small number of large fragments. Acid cleavage mostly occurs at aspartyl–prolyl bonds, which might be expected to occur roughly every $20 \times 20 = 400$ amino acids. Accordingly, many smaller proteins will not have aspartyl–prolyl bonds, and will not be cleaved by acid. The cyanogen bromide method involves the use of a very toxic cleavage reagent and should only be used when all other methods fail.

In the next few sections, I will describe adaptations of these methods that have been used in my laboratory. The method using *N*-chlorosuccinimide is preferred because it gives excellent results virtually every time, and the reagent is inexpensive and has low toxicity.

Peptide mapping with N-chlorosuccinimide (NCS)

(1) On the day of the experiment, make up a solution of acetic acid/urea/water (AUW) by mixing 10 ml glacial acetic acid, 10 g urea and 10 ml H_2O.
(2) Identify and cut out desired spot from gel. (Keep spot as small as possible, preferably less than 2 mm in width.)
(3) Rehydrate spot in 3 ml AUW for 10 min at room temperature, with occasional agitation.
(4) Suck out liquid, and remove paper backing.
(5) Transfer slice to fresh 3 ml AUW.
(6) Rotate on vertical wheel for 20 min.
(7) Prepare 10 ml AUW containing 15 mM *N*-chlorosuccinimide (20 mg NCS/10 ml).
(8) Transfer gel slice to 3 ml NCS solution.
(9) Rotate on wheel for 20 min at room temperature.
(10) Suck out liquid.
(11) Add 10 ml 1.0 M Tris-HCl, pH 8.0 to neutralize acid.
(12) Rotate on wheel for 10–20 min.
(13) Suck out liquid. Replace with 10 ml reducing SDS sample buffer. Rotate on wheel for 20 min.
(14) Pour stacking gel onto a 15% SDS gel, with teeth of comb upwards to generate a flat surface. (Better results are obtained with gradient gels, such as 10–15% acrylamide.)
(15) Suck out liquid from gel slice. Load slice onto stacking gel, taking care to avoid bubbles under it.
(16) Overlay the gel slice with sample buffer, then running buffer.
(17) Run the gel in the usual way (no interruption of current).

Peptide mapping by cleavage of aspartyl–proline bonds in dilute acid

The method is based on that of Rittenhouse and Marcus (1984) with slight modifications. These authors used imidazole buffers in the stacking gel and sample buffer, but in my hands Tris-HCl is satisfactory provided the width of the gel slice is kept small.

(1) Cut out slice from dried gel. Rehydrate in 1.5 ml 15 mM HCl in water.
(2) Remove paper backing.
(3) Place sample in a boiling water bath for 10–20 min.
(4) Suck out acid. Replace with 1 ml SDS sample buffer. Change the sample buffer 2–3 times over 30 min until pH rises to *c.* 6.8 (pH paper). (It may be possible to simplify this step.)
(5) Go to step 13 of previous procedure.

It is also possible to cleave polypeptides in wet gels after Coomassie blue staining. Simply add 1 ml destain, place in a boiling water bath for 10–20 min, then go to step 11 of the previous procedure.

Peptide mapping by CNBr cleavage

(1) Rehydrate dried gel slice in 70% formic acid (FA) for 1–5 min.
(2) Remove paper (discard).
(3) Equilibrate gel slice in 5 ml FA 20 min on wheel at room temperature.
(4) Make up CNBr in dimethyl formamide (DMF) in a screw-cap glass vial (DMF dissolves plastic). Weigh vial, then add estimated 500 mg CNBr, then tightly cap tube, then weigh again. Then add same number of μl DMF as mg CNBr. This will make a solution of approximately 588 mg/ml.
(5) Take 340 μl CNBr and add to 5 ml of FA containing the gel slice. (Final concentration of CNBr = 40 mg/ml.)
(6) Rotate on wheel 1.5 h at room temperature.
(7) Remove liquid (see below for disposal of CNBr waste).
(8) Wash twice (20 min/wash) in 10 ml 10% acetic acid or water.
(9) Go to step 11 of the *N*-chlorosuccinimide procedure.

CNBr is extremely toxic, and must be handled with great care. It should only be handled in a fume hood, and should be stored in a tightly capped bottle within a second bottle, at 4°C or below. Waste containing CNBr should be placed into a large excess of sodium hypochlorite solution for 24 h, prior to disposal.

10.10 Western Blots

Immunoprecipitation analysis, as described in earlier portions of this chapter, has some fundamental limitations. It is seldom possible to know with certainty whether a radioactive band on a gel represents the polypeptide recognized by the antibody, or whether it is precipitated because it is physically attached to another (possibly nonradioactive) molecule which bears the antigenic determinant. A second problem with immunoprecipitation is that it may be difficult to radiolabel the antigen to sufficiently high specific activity.

These problems can sometimes be overcome by probing the proteins with antibodies *after* electrophoretic separation (Burnette, 1981). The procedure is often known as Western blotting (reviewed by Gershoni and Palade, 1983). The heterogeneous mixture of nonradioactive proteins is separated by electrophoresis in a polyacrylamide gel, and the proteins eluted electrophoretically by a transverse electric field onto a membrane which binds protein tightly. (Nitrocellulose is the most popular membrane.) The membrane is then probed with a radioactive antibody, washed and autoradiographed.

10.10.1 Sample Preparation for Western Blots

Some thought must be given to the preparation of the sample for Western blotting. If whole cells are lysed in SDS sample buffer, the resulting DNA release may cause the solution to be very viscous and difficult to handle. Viscosity can be reduced to a manageable level by rapidly passing the lysate though a 21-gauge needle several times. The sensitivity of the method can often be greatly improved by preliminary enrichment procedures, such as immunoprecipitation or fractionation in Triton X-114. Dilute protein solutions may be concentrated by precipitation with organic solvents, which will also serve to remove nonionic detergents that will interfere with the running of SDS gels (see Section 10.7.3).

10.10.2 Enrichment of Antigen by Immunoprecipitation

It is possible to combine Western blotting with immunoprecipitation. The antigen is immunoprecipitated as above (see Section 10.8), and processed as usual for SDS–PAGE. The advantages of preliminary immunoprecipitation are greatly increased sensitivity and improved signal:noise ratio (Wiser and Schweiger, 1986). The disadvantage is that the gel will contain the immunoprecipitating IgG, which will give a signal if anti-immunoglobulin reagents are used. This could be a major problem if the antigen of interest happened to co-migrate with IgG, but the problem can usually be overcome by judicious choice of reducing or nonreducing conditions to move the IgG to a part of the gel where it does not interfere with the desired signal.

10.10.3 Preliminary Enrichment for Membrane Antigens by Fractionation in Triton X-114

If the antigen is a membrane protein, fractionation of the antigen mixture in Triton X-114 provides an enrichment of about 20-fold (see Bordier, 1981, 1988; Brusca and Radolf, 1994). This simple procedure selectively partitions membrane proteins into a detergent-rich phase. It is simple, economical and effective. Details are given in Section 10.7.4. The presence of large amounts of Triton X-114 in the sample will cause SDS gels to run badly, and it is strongly recommended that the protein be precipitated with nine volumes of ice-cold acetone followed by centrifugation (5 min in Eppendorf centrifuge at 13 000 ***g***) prior to resuspending in SDS sample buffer (see section 10.7.3).

10.10.4 Combination of Biotin Labelling, Western Blotting, and Enhanced Chemiluminescence

The enzymic generation of light has been used as an analytical tool for many years (reviewed by Whitehead *et al.*, 1979). In 1983, it was discovered that the addition of certain substances, including phenylphenol, iodophenol or other aromatic compounds increase the sensitivity of luminescent assays by more than 1000-fold. The substrates are not particularly expensive, and the reaction is simple and convenient.

The advent of enhanced chemiluminescence (ECL) has had a major impact on the analysis of antigens, because it has made it possible to avoid the use of radioactive isotopes and at the same time to get results with shorter exposures. Instead of exposing autoradiographs for several days, exposure times can often be measured in minutes or seconds. The power of ECL is often exploited in conjunction with Western blots (see section 10.10.9).

The antigen mixture to be analysed is biotinylated (see Section 10.3), and then the desired antigen is isolated by immunoprecipitation. The precipitate is then dissociated and analysed by SDS–PAGE and electrophoretically transferred to a nitrocellulose membrane, which is probed with peroxidase–streptavidin, followed by detection with ECL. Exposure times in the order of seconds to minutes are often possible. This approach is likely to become very popular (see Terashima *et al.*, 1994, and Kim *et al.*, 1994, for some elegant examples).

10.10.5 Western Blotting: Practical procedure

After electrophoresis, the gel is removed from the apparatus and placed in a dish containing 20% methanol in 20 mM Tris base, 150 mM glycine, pH 8. The methanol prevents the gel from shrinking or expanding. A sandwich is then built up, consisting of a perforated Plexiglass support, a porous polyethylene sheet (Bel-Art) or Scotch-Brite scouring pads, three sheets of heavy filter paper (Whatman 3M), the gel, the membrane, more filter paper, a porous pad and a second perforated support sheet (Fig. 10.10). All components should be wetted in buffer, and special care must be taken to avoid trapping air bubbles between the gel and the nitrocellulose membrane. The membrane must not be touched with ungloved fingers at any stage.

The whole assembly is held together with rubber bands, and placed in a tank containing buffer and platinum electrodes. Suitable tanks are available commercially, or can be improvized using plastic lunch boxes.

The current must be applied with the nitrocellulose membrane on the anode side of the gel. The field strength may be determined empirically, but typical values are 6–8 V/cm for 16–24 h (Burnette, 1981). The popular 'mini-gel' systems often require only 1 h for transfer.

Excessive current may result in overheating and distorted bands. If possible, transfer should take place in a cold room, with stirring using a magnetic 'flea' to prevent the build-up of pH gradients.

Some variability in the efficiency of transfer has been noted. In general, large proteins (M_r > 100 000) may require higher field strengths and longer transfer times. Erickson *et al.* (1982) have shown that the addition of 0.1% SDS to the transfer buffer improves the transfer of high molecular weight

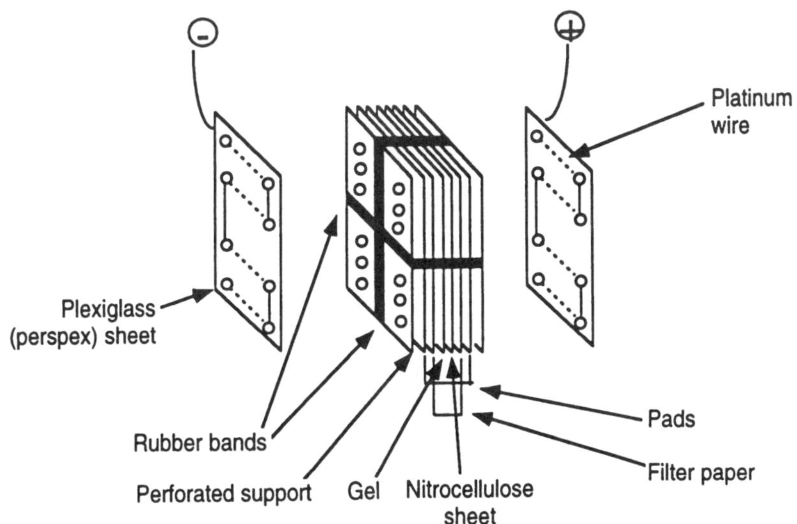

Fig. 10.10. Western blotting. The polyacrylamide gel containing the sample is overlaid with a sheet of nitrocellulose, and sandwiched between filter papers, kitchen scouring pads and perforated Perspex (Plexiglass) supports. The whole sandwich is held together by rubber bands, and submerged in a buffer consisting of 20 mM Tris base, 150 mM glycine and 20% methanol in water. The methanol prevents the gel from shrinking or expanding. An electric field is applied such that the nitrocellulose is on the anode side of the gel. Proteins in the gel are electrophoretically transferred to the nitrocellulose, to which they bind.

proteins. An additional source of variability is the brand and batch of the nitrocellulose. Some brands are treated with a nonionic detergent, which inhibits protein binding. Washing the membrane in distilled water or transfer buffer prior to use often improves the transfer.

10.10.6 Blocking the Western Membrane

After the transfer of proteins from the SDS gel, the membrane must be blocked so that further nonspecific binding of protein is prevented. Nonfat powdered milk (Bovine Lacto Transfer Technique Optimizer or BLOTTO) is a very effective substitute for BSA, which was previously the reagent of choice (Johnson *et al.*, 1984). BLOTTO consists of 5% (w/v) nonfat powdered milk in PBS plus 0.01% Antifoam A (Sigma) and 0.001% merthiolate as preservative. If the BLOTTO is used on the day that it is made, the merthiolate is unnecessary. The Antifoam is also optional.

Results using BLOTTO are generally much better than those using other blockers, and the cost is much less than BSA. BLOTTO is now the reagent of choice for blocking. BLOTTO may also substitute for BSA in other immunological procedures such as radioimmunoassays.

Many other blocking agents have been used, including high concentrations of BSA or casein. Batteiger *et al.* (1982) have shown that the detergent Tween 20 may be used as a blocker, but the use of nonionic detergents is risky as it may cause serious losses of protein from the membrane. Polyvinyl alcohol is said to be an effective blocker, and to act virtually instantaneously (Miranda *et al.*, 1993).

10.10.7 Probing the Membrane

After blocking, the membrane is placed in a dish containing the antibodies (typical concentration 1–50 μg/ml) in BLOTTO. After 30–60 min, the membrane is removed, washed in BLOTTO, and placed in the same buffer containing an appropriate dilution of peroxidase-conjugated anti-immunoglobulin or protein A, or 10^5 c.p.m./ml of ^{125}I-labelled staphylococcal protein A or affinity-purified anti-immunoglobulin. After a further 30–60 min, the membrane is removed, washed extensively, and processed for autoradiography.

10.10.8 Readout for Western Gels

Readout by ECL (see Section 10.10.9) is now the preferred method, as it is extremely sensitive and often requires only a few seconds to obtain adequate sensitivity. If the background staining of the membrane is excessive owing to the extreme sensitivity of ECL, dilution of the ECL substrate may be helpful (see Pampori *et al.*, 1995).

Alternatively, ^{125}I-labelled anti-immunoglobulin or protein A may be detected by autoradiography using intensifying screens (Fig. 10.7). The use of colour readout by ELISA is no longer recommended as it lacks sensitivity and often fades upon storage.

It has recently been reported that the sensitivity of detection of unreduced biotinylated proteins in Western blots may be greatly increased by soaking the gel in β-mercaptoethanol prior to transfer (Weston *et al.*, 1995). This is probably due to improved access of avidin to the biotin rather than improved efficiency of transfer.

The potential sensitivity of Western blotting is extremely high. Burnette (1981) has shown that is easily capable of detecting murine leukemia virus proteins from as few as 1000 cells. When used with ECL, even greater sensitivity would be possible.

10.10.9 Enhanced Chemiluminescence—Practical Procedure

Enhanced chemiluminescence systems are available using horseradish peroxidase- or alkaline phosphatase-conjugated antibodies or streptavidin.

Appropriate substrate mixtures are available commercially. Autoradiography is performed using blue-sensitive film without intensifying screens. Kits are available from Amersham, Dupont, Boehringer, Bio-Rad and other suppliers.

The following information is for the Amersham system. The membrane is probed with peroxidase-conjugated antibody or peroxidase-conjugated streptavidin, washed in PBS/BLOTTO, and washed twice more in PBS without BLOTTO. From this point on, one should work quickly, preferably in a dark room close to the X-ray developer. The membrane is placed in a plastic bag or Petri dish containing a freshly prepared mixture of each of the two ECL solutions (typically 125 $\mu l/cm^2$ of membrane), and incubated for precisely 1 min at room temperature. The reagents are then drained off. The membrane, while still damp but not wet, is placed between two plastic sheets and exposed to X-ray film with the protein-containing side facing the film.

If horseradish peroxidase is used, the light emission peaks at about 20 min, and slowly declines over a period of hours. The duration of light emission is somewhat longer if alkaline phosphatase system is used, but the relative sensitivity of the two systems has not been rigorously compared. As the turnover number for horseradish peroxidase is greater than alkaline phosphatase, the sensitivity for the former may be greater.

It will usually be necessary to take several exposures. The first exposure could be typically about 1 min, and subsequent exposures should be guided by results.

10.10.10 Use of Remazol Prestained Molecular Weight Standards with Western Blots

It will often be desirable to identify the molecular weight standards on Western blots, but it is not possible to use Coomassie blue to stain nitrocellulose blots, because the membrane binds the dye. Bound proteins can be stained with Ponceau red, but this stain is not very sensitive. A much better approach is to use Remazol blue stained standards, in which the dye is covalently attached to the proteins (Griffith, 1972). In this way, the prestained standards can be seen while the gel is running. The prestained standards can also act as a very useful guide for alignment of the gels and cutting up the membrane. The mobility of the Remazol stained standards is not quite the same as for unconjugated standards but the difference is minor. The sensitivity of Remazol is not as high as Coomassie blue, and slightly more protein will have to be added to the gel.

Remazol brilliant blue R (Sigma cat. no. R-8001, 25 or 100 g bottles) is a textile dye, and is the potassium salt of a sulfatoethyl sulfonic acid. It reacts with primary and secondary amines, alcoholic OH groups and SH groups. Binding is covalent. Coupling is inhibited by amines (e.g. Tris), thiols and hydroxyl-containing components (e.g. glycerol). These must be absent from the reaction buffer.

The dye solution is made up by dissolving 10 mg Remazol brilliant blue in 1.0 ml 10% SDS in water. The coupling is carried out in 0.2 M carbonate/bicarbonate buffer, pH 9.5 (0.86 g Na_2CO_3 plus 1.72 g $NaHCO_3$ to final volume of 100 ml distilled water; the pH should be about 9.3–9.5 without titration). Both the dye solution and the buffer can be stored frozen for subsequent use.

For low molecular weight standards, dilute two vials of Pharmacia low molecular weight standards (cat. no. 17-0446-01) into a total of 180 µl pH 9.5 coupling buffer. For high molecular weight standards, take 100 µl of the Bio-Rad HMW stock (cat no. 161-0303; supplied in 50% glycerol) and dilute to a final volume of 200 µl with water. Remove the glycerol from the Bio-Rad standards by overnight dialysis against two changes of 1 l of coupling buffer. Failure to dialyse out all the glycerol will result in failed coupling.

Practical procedure

(1) Add 50 µl Remazol mixture to the standard solution prepared as above.
(2) Heat at 70°C for 30–60 min (not critical). Dilute to 1 ml with reducing SDS sample buffer. Nonreduced standards are slightly less accurate and give more diffuse bands.
(3) Store in 50 µl aliquots at –20°C.
(4) Load 25–50 µl per track.

10.10.11 Western blots with Nonspecific Polyclonal Antisera

If the relevant antigen is available in pure form, or if it has detectable enzymic activity, it is possible to adapt the Western blot so that the antigen may be identified even when the antiserum contains numerous irrelevant antibodies (Muilerman *et al.*, 1982; van der Meer *et al.*, 1983). The gel is run and blotted as usual, and probed with the nonspecific antiserum. Instead of anti-immunoglobulin or staphylococcal protein A, the membrane is probed with ^{125}I-labelled pure antigen or the enzyme-containing mixture. After washing, the antigen is detected by autoradiography or by enzyme assay. The fact that antibodies are symmetrical (with two identical binding sites) means that any antibodies that have only one site bound to the nitrocellulose have one free site which allows the labelled antigen or enzyme to bind. Only the subpopulation of specific antibodies are thus detected.

10.10.12 Use of Western Blots with Monoclonal Antibodies

It is usually considered that proteins which have been heated to 100°C in SDS plus reducing agents are totally denatured. This view may not be strictly correct. Immunoglobulin γ-chains isolated from SDS gels and subjected to limited proteolysis in SDS-containing buffers were cleaved almost exclusively at regions between domains (Goding, 1982). The existence of many additional potential cleavage sites was revealed by the substantially larger number of breakdown products when digestion was performed in 4 M urea.

The conformations that proteins adopt after they have bound to nitrocellulose are not well understood, nor is the precise mechanism of binding to the membrane. It would seem likely that proteins bound to the membrane would be electrophoretically 'stripped' of bound SDS, which would continue to the anode. Accordingly, bound proteins might thus regain more of their native configuration.

It has been known for many years that the reactivity of immune serum made against native proteins is usually much weaker when tested on denatured proteins (see Chapter 4 for discussion). However, antisera against the native protein will almost always contain at least some clonal products that recognize the denatured protein, and vice versa. It is this fact that allows the Western technique to function.

In marked contrast to polyclonal antibodies, monoclonal antibodies against the native protein may or may not recognize the denatured product, in an all-or-none fashion. It may be expected that many clones (perhaps the majority) will fail to recognize the denatured antigen. It would be expected that the Western technique would work much more reliably with polyclonal antibodies, and experience is consistent with this view. None the less, Western blots sometimes work with monoclonal antibodies, and are well worth trying.

Goldstein *et al.* (1982) produced a monoclonal antibody against cytomegalovirus, which detected an M_r 80 000 protein in Western blots. Interestingly, the antibody did not precipitate any protein in standard immunoprecipitation system, perhaps because it only recognized the denatured form of the antigen. Barbour *et al.* (1982) were able to use the Western technique with a considerable proportion of monoclonal antibodies against variable proteins of *Borrelia hermsii*. Parham *et al.* (1982) found three out of 15 monoclonal anti-HLA antibodies worked in Western blots. O'Connor and Ashman (1982) found four out of five monoclonal anti-*Salmonella* antibodies detected protein in Western blots.

The chances of a monoclonal antibody working on Western blots are improved if the sample is analysed without reduction, presumably because intact disulfide bonds prevent complete unfolding and may help guide refolding. It may also help to avoid boiling the sample.

10.10.13 Recovery of Antibody, Antigen and Re-use of Western Blots

Antibody may be eluted off the nitrocellulose-immobilized protein by treatment with pH 2.2–2.5 glycine-HCl, preferably in the presence of 0.1–1.0%

BSA (Legochi and Verma, 1981; Erickson *et al.*, 1982). The nitrocellulose-immobilized protein remains firmly attached to the nitrocellulose. Indeed, it cannot be removed by SDS or deoxycholate, although it is efficiently eluted by Triton X-100, Nonidet P-40 or octyl glycoside (Lin and Kasamatsu, 1983). The general rule appears to be that nonionic detergents are good blockers and eluters, while anionic detergents neither block nor elute, perhaps because of repulsion from the negatively charged paper. The bound protein can also sometimes be recovered by dissolving the nitrocellulose in dimethyl sulfoxide, and may be used as antigen for immunization (Knudsen, 1985).

10.11 Topographical Analysis of Proteins by Monoclonal and Polyclonal Antibodies

Monoclonal antibodies usually recognize discrete sites on the surface of protein antigens. They may therefore be used as very precise probes for the topography of proteins (Schneider *et al.*, 1982).

Some of the principles involved are illustrated by studies on lymphocyte membrane IgD. Mouse IgD exhibits genetic polymorphism, first detected by the presence of anti-IgD antibodies in antisera made in C57BL/6 mice against CBA lymphocytes (Goding *et al.*, 1976). Subsequently, monoclonal antibodies were produced against IgD allotypic determinants (Pearson *et al.*, 1977; Oi *et al.*, 1978).

Treatment of intact B cells with trypsin abolished the binding of monoclonal antibody 11–6.3, but did not affect the binding of antibody 10–4.22 (Kessler *et al.*, 1979). Antibody 11–6.3 was found to bind to the Fab portion of IgD (Kessler *et al.*, 1979) while antibody 10–4.22 bound to the Fc portion (Goding, 1980). These results proved that IgD was attached to the membrane via its Fc portion.

When the anti-Fc monoclonal antibody was used to immunoprecipitate IgD from detergent lysates of cells, a Fab′ portion of IgD could be released from the precipitates by gentle proteolysis with staphylococcal V8 protease. The Fab′ fragment consisted of an M_r 40 000 heavy chain fragment (Fd′) disulfide-bonded to an intact light chain. Conversely, when antibodies against the Fab portion of IgD were used for immunoprecipitation, a fragment consisting of two disulfide-bonded M_r 20 000 heavy chain fragments was released from the precipitate by V8 protease. These results proved that the only inter-heavy-chain disulfide bonds in IgD must reside in the *C*-terminal 20 000 dalton portion (Goding, 1980). This result was confirmed by subsequent sequence studies (Tucker *et al.*, 1980; Cheng *et al.*, 1982).

The use of monoclonal antibodies to isolate well-defined large proteolytic fragments of IgD under mild conditions allowed an analysis of their detergent-binding properties. It was found that the Fc portion of membrane IgD bound large amounts of nonionic detergent, while the Fab did not (Goding,

1980). This result, which was consistent with anchorage of IgD by its Fc portion, strongly suggested that membrane IgD was an intrinsic membrane protein. Subsequent work revealed the presence of a typical hydrophobic transmembrane sequence close to the *C* terminus (Cheng *et al.*, 1982).

A second example of topographical analysis using antibodies concerns the plasma cell membrane glycoprotein PC-1 (phosphodiesterase I). PC-1 is a type II transmembrane protein with a large extracellular catalytic *C*-terminal domain and a short cytoplasmic *N*-terminal domain (van Driel and Goding, 1987; Buckley *et al.*, 1990; Belli and Goding, 1994). It was not clear which of two potential initiator methionines was the true initiator. Initiation at the first methionine would give a cytoplasmic tail of 76 amino acids for human PC-1, while initiation at the second methionine would give a cytoplasmic tail of only 24 amino acids. The issue was resolved by subcloning a short stretch of sequence between the two methionines into the bacterial expression vector pGEX-1 (Smith and Johnson, 1988), raising antisera to the resulting protein, and testing the antibodies on the native protein by Western blotting. The antibodies showed that the intact protein contained antigenic determinants encoded by the region between the two methionines, proving that the first methionine must have been the initiator (Belli and Goding, 1994).

This type of approach may be used to build up a picture of protein antigens, providing information about their orientation, ligand-binding sites and other topographical features. It may be particularly useful for membrane proteins, because they may possess topographical 'landmarks' such as carbohydrate (always extracellular), covalent lipid (presumably close to the lipid bilayer), and phosphorylated amino acids (always intracellular) (Omary and Trowbridge, 1980, 1981; Schneider *et al.*, 1982).

10.12 Use of Monoclonal Antibodies as Probes for Protein Conformation

Antibodies are very sensitive to protein conformation, and monoclonal antibodies are therefore well-suited to probe conformational changes in proteins. This is illustrated in the case of monoclonal antibody 4H4, which was raised against a putative 'EF Hand' calcium-binding region of the PC-1 antigen (Belli *et al.*, 1994). The EF Hand region from human PC-1 was subcloned into the bacterial expression vector pGEX-KT (Hakes and Dixon, 1992), and used to immunize mice. Monoclonal antibodies were screened on methanol-fixed mouse L cells that had been transfected with human PC-1. A monoclonal antibody (4H4) which reacted with methanol-fixed cells did not stain intact unfixed cells, suggesting that it recognized only an unfolded conformation of PC-1. However, the 4H4 antibody did stain PC-1 on intact living cells in the presence of EDTA. This result suggests that the release of calcium from the EF hand caused a conformational change that allowed the antibody to bind

(Belli *et al.*, 1994). In contrast, polyclonal antibodies to human PC-1, raised by immunizing mice against syngeneic mouse fibroblasts transfected with human PC-1, showed no difference in binding to PC-1 in the presence of calcium or EDTA. Taken together, these results suggest that the conformational change associated with calcium binding was restricted to the *C*-terminal portion of the protein (Belli *et al.*, 1994).

Calmodulin is also known to undergo large conformational changes upon binding of calcium. Wolf *et al.* (1995) have shown that monoclonal antibodies can be generated which detect these changes.

10.13 Use of Antibodies in the Production and Identification of Cloned DNA Sequences

Antibodies are now widely used in gene cloning. Some examples of cloning strategies involving antibodies are given in the following sections. It must be emphasized that the examples given are by no means exhaustive, and that the most appropriate strategy for any individual gene is best approached on a case by case basis. Three different strategies for cloning using antibodies will be presented.

The first, involving immunological screening of bacterial expression libraries, is potentially the simplest, but has a number of potential difficulties, especially when monoclonal antibodies are used.

The second strategy involves expression cloning in mammalian cells. This powerful method is now becoming extremely popular, and is ideally suited to the use of monoclonal antibodies.

The third approach consists of purifying the relevant protein by affinity chromatography, obtaining partial amino acid sequence, and using this information to design short degenerate oligonucleotides to screen libraries or as primers for the polymerase chain reaction (PCR). This approach is probably the most reliable method for cloning, but it is also the most laborious.

Before we explore these three approaches, let us consider the important issue of the source of DNA for cloning. The two main possibilities are genomic DNA and cDNA transcribed from mRNA by reverse transcriptase.

10.13.1 Genomic and cDNA Cloning

The average protein has a molecular weight of about 50 000, and consists of about 500 amino acids. If these were encoded without interruption, a typical gene would be $500 \times 3 = 1500$ base pairs in length. However, the great majority of genes in higher organisms are much larger, because they are made up of discontinuous segments. These segments are termed *exons* and *introns*. Exons are those sequences that remain in mature messenger RNA (mRNA). Introns, or intervening sequences, are removed during RNA maturation. The result is that typical genes are 3–50 kilobases in length, and occasionally as large as 200 kilobases or more.

The first decision in any cloning experiment is therefore whether to attempt to clone genomic DNA or cDNA. The factors that influence this decision are beyond the scope of this book (Sambrook *et al.*, 1989). In brief, genomic cloning requires vectors capable of accepting large DNA inserts (λ phage or cosmids). The great complexity of genomic DNA means that genomic cloning usually requires screening of larger numbers of clones than may be the case for cDNA cloning. However, cDNA cloning requires isolation of mRNA and a larger number of manipulations. In contrast to genomic cloning, the nucleotide sequence of cDNA clones usually allows unambiguous prediction of the full protein sequence, because the introns have been removed.

The cloning strategy for any individual gene should be individualized to exploit whatever special characteristics are available. In the case of cDNA cloning, these might include regulatory properties such as differential mRNA expression in various tissues or cells, or response to stimuli.

10.13.2 Direct Identification of Proteins Expressed in Bacteria

In many cases, expression of cloned protein antigens in bacteria will involve clones that are only partial length, and may be incorrectly folded. Accordingly, the antigenic determinants recognized by a particular monoclonal antibody may or may not be expressed. However, there are now many examples of successful identification of cloned genes detected by antibody screening of bacterial expression libraries. Polyclonal antibodies would be more likely than monoclonal antibodies to detect denatured, non-glycosylated or fragmented proteins (see Chapter 4), and experience has tended to bear this out. Although there have been some successes using monoclonal antibodies, their use for screening bacterial expression libraries is risky.

Choice of vector

Antibody screening can be used with both plasmid and bacteriophage vectors, and each type of vector can be used with both cDNA and genomic DNA. The main disadvantage of plasmid vectors is their low cloning efficiency. If the desired mRNA is of low abundance, it may be difficult to construct a library of adequate size to ensure detection of the corresponding cDNA clone. The problem of efficiency is exacerbated by the additional requirement of the correct frame and orientation for antibody recognition.

In order to overcome the limited efficiency of plasmid cloning, λ phage vectors suitable for cDNA have been developed. The *in vitro* packaging of DNA into phage particles allows a high transformation efficiency (typically 10^6–10^7 clones/μg cDNA).

The most popular vector has undoubtedly been λgt11 (see Young *et al.*, 1985), although commercial vectors are now available with many refinements including polylinkers with numerous restriction sites and automatic plasmid excision (e.g. λ Zap from Stratagene). A detailed account of the use of λgt11 is given by Huynh *et al.* (1985). In some cases, expression of the desired polypeptide has been detected with monoclonal antibodies (Goridis *et al.*, 1985; Young *et al.*, 1985). However, the majority of successful reports of gene isolation with λgt11 have used polyclonal antibodies, and it is to be expected that monoclonal antibodies would work only in a minority of cases (see Chapter 4).

Two recent technological advances have made the use of plasmid libraries much more attractive. The combined use of PCR to amplify cDNA (Belyavsky *et al.*, 1989), together with the extremely high efficiency of transformation of *Escherichia coli* by electroporation (up to 10^9 colonies/μg DNA) make this approach very powerful and convenient.

Antibody screening of expression libraries in Escherichia coli

A library is made by standard procedures. Both plasmid and phage vectors may be used, and antibody screening can be used for both genomic DNA and cDNA, although the latter has been more common.

The library is plated out on agar, and replicas made of the bacterial colonies or phage plaques by taking *duplicate* 'lifts' of nitrocellulose disks from the plates. Alignment of the filter with the plate is achieved by stabbing through the membrane with a hypodermic needle dipped in India ink prior to taking the lift.

After blocking the filter for nonspecific protein binding sites with BLOTTO (see Section 10.10.6), the filter is probed with antibodies in a manner that is virtually identical to the probing of Western blots, using either ^{125}I-labelled or enzyme-labelled antibodies, followed by autoradiography or visual inspection (see Huynh *et al.*, 1985; and Sambrook *et al.*, 1989, for further details). The use of ECL for readout is an attractive option (see Section 10.10.9), although its extreme sensitivity may cause excessive background. If so, the background may be able to be reduced to manageable levels by diluting the ECL substrate (Pampori *et al.*, 1995).

The big attraction of antibody screening is simplicity. The disadvantages are uncertain reliability and low efficiency. However, the approach may fail because the antibodies do not recognize the protein in the form synthesized in bacteria, or because the desired protein is toxic for *E. coli*, or because foreign

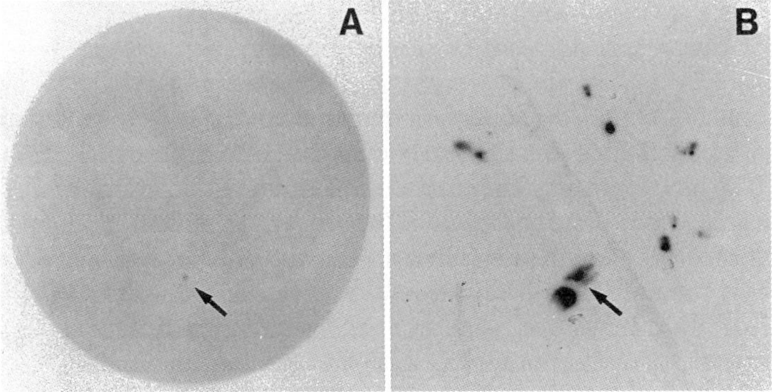

Fig. 10.11. Detection of cloned DNA sequences by antibody screening of proteins expressed in Escherichia coli. *Duplicate nitrocellulose filter 'lifts' from a mouse lymphocyte cDNA library were screened with ^{125}I-labelled polyclonal antibodies to the γ chain of mouse IgG (panel A) or ^{32}P-labelled cloned γ chain cDNA (panel B). Approximately 20 times more clones were detected with the ^{32}P-labelled probe than with the antibody. This discrepancy reflects the need for the clone to be in the correct reading frame and also in the correct orientation (one in six chance), and for the clone to direct the synthesis of a protein bearing epitopes that can be recognised by the antibodies. In this case, screening with polyclonal antibodies only detected one clone in 20. Screening with monoclonal antibodies would be expected to be far less efficient.*

proteins are sometimes degraded in *E. coli*. These variables are almost impossible to predict, and depend both on the gene itself and the properties of the host and vector. The use of an inducible promoter system may help the problem of toxicity, and the use of protease-deficient *lon* mutant *E. coli* strains may reduce the likelihood of proteolytic degradation.

A further source of inefficiency of antibody screening results from the fact that it depends on the cloned sequence being in the *correct reading frame* (one in three chance) and the *correct orientation* (one in two chance). Theoretically, only about one in six clones containing the desired sequence will be detected. In practice, the result is often a great deal worse, because partial-length clones may not be detected (see Fig. 10.11). When a cDNA expression library was screened with polyclonal anti-γ chain antibodies, only one clone in 20 that contained the relevant sequences was detected (Fig. 10.11). Results using monoclonal antibodies would be expected to be much worse, as they only detect single epitopes. If the clone that is being sought is very rare, the inefficiency of antibody screening may make this approach unworkable.

(On a slightly more optimistic note, it can be pointed out that occasionally one finds correct expression of a polypeptide despite its being in the 'wrong' reading frame. These instances are probably due to internal initiation of protein synthesis.)

10.13.3 Use of Monoclonal Antibodies to Screen cDNA Libraries Expressed in Mammalian Cells

The cloning of cDNA by expression in mammalian cells is an extremely powerful and general approach (Aruffo and Seed, 1987a, b; Simonsen and Lodish, 1994). The procedure may be summarized as follows. A cDNA library is made in a plasmid vector that is based on SV40, using techniques that encourage the production of full-length clones that are directionally oriented with respect to the vector promoter. Pools of plasmids are transfected into COS cells using DEAE-dextran, and the cells cultured for 3–4 days. The cells are then assayed for the desired activity, for example by ligand binding, immunofluorescence or secretion of a molecule with the desired activity.

The assay is usually chosen in such a way that the plasmid can be recovered from the cells. If the method of screening involves destruction of the cells, such as by fixation for cytoplasmic immunofluorescence, it is possible to take pools of cells and split them into two groups, one of which is assayed for the relevant antigen or activity while the other is kept for further use.

In some cases it may be possible to enrich for cells expressing the antigen, such as by 'panning' with antibody-coated Petri dishes (Aruffo and Seed, 1987a, b). The plasmids are recovered from the COS cells by a process called Hirt extraction and used to retransform *E. coli.* The DNA is then purified and used to start a new cycle of transfection. By successive rounds of screening smaller and smaller pools, one eventually gets to the point where the desired antigen is carried by a single clone.

While the method has been undoubtedly very successful, it has a reputation for difficulty and unreliability. It is not uncommon to find that despite obtaining unequivocal expression in COS cells, all the pools of clones recovered are negative for the desired antigen when rescreened, probably because the plasmids often undergo rearrangements and deletions in COS cells (Brakenhoff *et al.*, 1994; Simonsen and Lodish, 1994). This problem can be avoided by adopting a slightly different strategy. Instead of recovering plasmids from the COS cells, one simply uses the COS cells as an assay to tell that a given pool contains the desired clone. Positive pools are then subdivided into smaller and smaller pools until a single clone is shown to encode the desired antigen, as confirmed by the ability to express it when transfected into COS cells. The method has many attractions, including its generality. If an antibody is available, in principle it should be possible to clone the corresponding cDNA, and the resulting clones can be expected to be full length.

The cDNA expression vectors such as CDM8 developed by Seed and colleagues have a complex antibiotic selection system based on a mutant ampicillin resistance gene which is rescued by plating on an *E. coli* strain containing an amber suppressor tRNA gene (MC1061/P3). The logic behind this system appears to have been that expression of a functional β-lactamase gene may have been toxic to mammalian cells. There is little evidence for such a proposal, and vectors with simple ampicillin resistance gene selection such as pcDNA3 (Invitrogen) are now preferable. The newer vectors have much more flexible polylinkers containing many more restriction sites, as well as numerous other advantages such as inbuilt selectable markers for long-term expression in stable cell lines.

10.13.4 Cloning Strategies Based on Amino Acid Sequence

Protein purification for amino acid sequencing

Affinity chromatography using monoclonal antibodies may greatly facilitate protein purification for amino acid sequencing, which in turn can be used to clone the relevant gene or cDNA. While it is somewhat laborious, this is probably the most reliable way to clone a new gene. The special requirements for protein sequencing are described in Section 11.7. If further purification is needed, preparative SDS–PAGE provides high resolution and recovery. Details of elution of proteins from preparative SDS gels are given in Section 15.2. Alternatively, the protein may be transferred onto a PVDF membrane by Western blotting and subjected to direct *N*-terminal sequencing or sequencing after cleavage and repurification of fragments (see below).

The sequence can be obtained from any part of the protein. The options are usually the amino terminus or randomly chosen peptides. This choice may be influenced by at least two factors. First, the amino terminus is often 'blocked' to sequencing by acetylation, cyclization or other chemical modification. A second problem exists concerning the use of amino-terminal sequences for the design of oligonucleotide probes. This region is encoded by the 5′ end of the mRNA, which is often underrepresented in cDNA libraries primed with oligo-dT (van Driel *et al.*, 1985). This 3′ bias may be severe for large mRNA species, but may be overcome by constructing 'randomly primed' cDNA libraries (Ebina *et al.*, 1985; van Driel *et al.*, 1985).

It is therefore common practice to cleave the purified protein, to separate the fragments, and to obtain the amino acid sequence of the amino termini of each fragment. In the older literature, it was common practice to cleave with agents such as cyanogen bromide, which cleaves at the rare amino acid methionine, and to isolate the resulting large fragments. However, these large fragments are often insoluble in water and difficult to purify.

More recently, a much more popular strategy has been to cleave the protein with trypsin, which cleaves at the common amino acids lysine and arginine, resulting in peptides of an average length of about 10 amino acids. An average-sized protein of 500 amino acids would yield an average of 50 peptides, many of which would be sequenceable (Stearne *et al.*, 1985). Provided that the protein is reduced and carboxymethylated, cleavage with trypsin can usually be made to proceed to completion, which is essential for good recoveries and ease of separation. Even quite complex mixtures of tryptic peptides are usually well resolved by reverse-phase HPLC, and recovery is usually excellent. If a given peak is heterogeneous (i.e. consists of more than one tryptic peptide), it is relatively easy to rechromatograph the peak using slightly different conditions, such as altered pH or the presence or absence of ion-pairing agents such as trifluoroacetic acid. Under these conditions, it is often found that co-migrating peptides can be made to move slightly differently and be resolved from each other (van Driel *et al.*, 1985; Grego *et al.*, 1985, 1986; Stearne *et al.*, 1985a, b).

Provided certain precautions are taken, tryptic cleavage can be carried out in solution (van Driel *et al.*, 1984; Grego *et al.*, 1985, 1986; Stearne *et al.*, 1985a, b). More recently, methods have been developed for digesting proteins in gel slices or on nitrocellulose or PVDF membranes, which minimizes peptide losses due to handling (Kurzchalia *et al.*, 1992; Patterson, 1994; Hellman *et al.*, 1995; Jenö *et al.*, 1995; Moritz *et al.*, 1995).

The use of tryptic peptides also has the advantage that for each mole of protein, several different sequences can be obtained, greatly increasing the chances of finding a region with sufficiently low degeneracy for making oligonucleotide probes (Table 10.7).

Design of oligonucleotides for library screening

Under conditions of high salt concentration and low temperature, stable hybrids can be formed between complementary DNA strands that are as short as 14 bases. In '6x SSC' (i.e. 0.9 M NaCl, 90 mM citrate, pH 7.0), the minimum melting temperature of short hybrids is given by the formula:

$$t_m = (G + C) \times 4°C + (A + T) \times 2°C$$

where (G + C) is the total number of G and C residues, and (A + T) is the total number of A + T residues. The greater contribution of G or C residues to the stability of the hybrid is because a G-C pair involves three hydrogen bonds, while an A-T pair involves only two.

To give a concrete example, consider the sequence:

ACGTTAGCGAATGA

If such a sequence were used as a probe for isolating a given gene, it would remain hybridized to the desired clone after washing in 6x SSC, provided the temperature was below about 40°C.

How long should an oligonucleotide probe be? Most successful probes are at least 14 base pairs in length. At this length, a single mismatch will usually cause complete loss of hybridization, especially if the mismatch is towards the middle. With longer probes (≥ 25–35 bases), a greater degree of mismatch may be tolerated. Longer probes also have greater specificity (see below). Very long probes (say 60 bases) can also be made, and one may take a guess at the codon

Table 10.7 Reverse genetic code

		1	2	3	1	2	3
ALA	A	G	C	X			
ARG	R	C	G	X	A	G	A G
ASN	N	A	A	T C			
ASP	D	G	A	T C			
CYS	C	T	G	T C			
GLN	Q	C	A	A G			
GLU	E	G	A	A G			
GLY	G	G	G	X			
HIS	H	C	A	T C			
ILE	I	A	T	CT A			
LEU	L	C	T	X	T	T	A G
LYS	K	A	A	A G			
MET	M	A	T	G			
PHE	F	T	T	T C			
PRO	P	C	C	X			
SER	S	T	C	X	A	G	T C
THR	T	A	C	X			
TRP	W	T	G	G			
TYR	Y	T	A	T C			

Table 10.7 cont.

VAL	V	G	T	X
STOP		T	A	G (AMBER)
		T	G	A (OPAL)
		T	A	A (OCHRE)

This table displays the genetic code as a function of individual amino acids. The left-hand columns indicate the standard three-letter and one-letter abbreviations for each amino acid. The columns marked 1, 2 and 3 indicate the first, second and third base positions.

It may be seen that only two amino acids (methionine and tryptophan) are encoded by unique codons. Arginine, leucine and serine each have six possible codons. The remaining amino acids have 2–4 possible codons.

X = any base.

usage to minimize degeneracy. These 'guessmers' can be used to screen at higher stringency than shorter probes, and can sometimes work very well, depending at least partly on luck.

Let us now consider the use of such a probe to screen a large cDNA library. How selective will a short oligonucleotide probe be? The probability that an individual probe of length n bases will be exactly complementary to a randomly chosen sequence of the same length is 1 in 4^n. In other words, a particular probe of length 14 bases would be exactly complementary to one clone within a hypothetical library consisting of $4^{14} = 2.7 \times 10^8$ random clones, each of 14 bases in length.

More realistically, if we wished to screen a library of 10^6 clones of average length 1000 bases, the library would contain the equivalent of $10^6 \times 10^3 \div 14 = 7.1 \times 10^8$ 'random' 14-mer sequences, and few if any false positives would be expected. Conversely, a library of 10^7 clones of average length 10 000 bases would contain the equivalent of $10^7 \times 10^4 \div 14 = 7.1 \times 10^9$ random 14-mer sequences, and we would expect $7.1 \times 10^9 \div 2.7 \times 10^8 = 26$ perfectly matched clones to occur by chance alone.

The hypothetical examples given above assumed that we knew the exact sequence needed for probing the library. Unfortunately, the real-life situation is more complicated. The degeneracy of the genetic code means that for most amino acids there will be several possible codons, and although some codons are used more often than others, we cannot predict with certainty which will be used in a particular case.

Table 10.7 shows the genetic code in reverse. Only two amino acids (methionine and tryptophan) are encoded by single codons. Arginine, leucine and serine are each encoded by six possible codons. The remaining amino acids have 2–4 codons.

If we can determine a short stretch of protein sequence, preferably one in which there are five or more amino acids encoded by four or fewer codons, we can design mixed oligonucleotide probes that encompass all the possibilities. Such probes may consist of as many as 256 or more different sequences. The larger the number of sequences in the mixture, the smaller the fraction of radioactivity in the 'correct' sequence, but the resultant dilution of radioactivity is seldom a problem as the hybridization signals are usually very strong. A more serious problem is that such highly degenerate probes increase the risk of detecting false positives. Consider the example given earlier (a library of 10^6 clones of average length 1000 bases), and assume that instead of a single probe sequence, 256 different sequences were present. The chance of a perfect match with a randomly occurring 14-mer is increased 256-fold, to 1 in 10^6. Our library still contains 0.71×10^8 'equivalent 14-mer sequences', and thus we can expect 71 false positives. The desired clone would be only one of these.

The probability of obtaining the correct clone may be greatly increased if two nonoverlapping or only slightly overlapping probes are used to screen duplicate lifts (see Stearne *et al.*, 1985b, for an example). Clones that are positive with both probes are strong candidates. The chances of obtaining the correct clone are also greatly increased by the use of longer probes. Indeed, recent advances in oligonucleotide chemistry have made the synthesis of 30–40-mers routine. Such long probes are quite tolerant of mismatches, and it is feasible to choose arbitrarily the most commonly used codon for all positions. The 'long probe' or 'guessmer' approach has been very successful and has largely rendered short mixed probes obsolete, although the ability to design long probes depends on obtaining a suitable stretch of amino acid sequence (see Ebina *et al.*, 1985). Currently, it seems likely that this approach may, in turn, be superseded by the use of the PCR using degenerate primers (see below).

Screening libraries with oligonucleotide probes is probably the most reliable method for gene cloning. If the correct protein has been purified, and if sufficient reliable sequence can be obtained, success in cloning is virtually guaranteed. On the negative side, the procedure is laborious, and requires access to both protein sequencing and oligonucleotide synthesis facilities. However, the bulk of the work in sequencing of protein involves the initial purification, and only limited sequence information is needed; the minimum amount of sequence is five contiguous amino acids, each of which is encoded by four or fewer codons.

Isolation of genes by PCR using degenerate oligonucleotide primers

Another strategy that can be used to exploit partial amino acid sequences of a purified protein is to design degenerate oligonucleotides for the polymerase chain reaction (Gould *et al.*, 1989; McPherson *et al.*, 1991). PCR has been used very successfully to isolate clones with extremely degenerate primers because false priming from one primer is unlikely to give rise to a sequence that is false-primed by the primer from the other end. False products will therefore tend to be amplified linearly, while the desired product will be amplified exponentially.

A potential problem for the use of PCR using primers based on tryptic peptides is that one does not usually know the order of the tryptic peptides in the intact protein, and tryptic peptides are usually too short to allow the design of two primers facing each other from the same peptide. In addition, two random tryptic peptides may be separated in the intact protein by such a large distance that the PCR yield is very low. This problem can be tackled by guesswork. If one chooses two random peptides for design of primers, there will be a 50% chance that the primers will face each other in the cDNA. If one tries several different combinations of primers from different peptides, there is a good chance that one combination will work, as revealed by a PCR product that contains the correct amino acid sequence of each of the peptides from which the primers were designed.

10.13.5 Authentication of Clones

In all cloning experiments, it is essential to verify that the correct clone has been obtained. The verification procedure is necessary to eliminate false positives arising as a result of ambiguities in the screening procedure, such as those due to unexpected antibody reactivities or highly degenerate oligonucleotide probes.

It is not uncommon to obtain false positives by antibody screening of bacterial expression libraries. Authentication should be performed using procedures that are independent of those used to obtain the clones in the first place.

Verification could include checks on the presence of hybridizing mRNA of the appropriate size and abundance in expressing cells, compared with non-expressing cells, or hybridization of the cloned gene with an independent oligonucleotide probe from a different peptide. Correct prediction of amino acid sequences lying outside the region used for design of the probe would be good evidence that the correct clone has been obtained, as would the ability of the cloned gene to transfer appropriate biological or antigenic activity in an expression system such as COS cells.

Occasionally, papers are published in which the wrong gene was cloned. In order to avoid embarrassment, it is wise to use several independent methods to check the authenticity of clones, and to treat the exercise of authentication just as seriously as the effort to isolate the original clone.

References

Aida, Y. and Pabst, M. J. (1990) Removal of endotoxin from protein solutions by phase separation using Triton X-114. *J. Immunol. Methods* **132**, 191–195.

Allore, R. J. and Barber, B. H. (1984) A recommendation for visualizing disulfide bonding by one-dimensional sodium dodecyl sulfate–polyacrylamide gel electrophoresis. *Anal. Biochem.* **137**, 523–527.

Alterman, L. A., Crispe, I. N. and Kinnon, C. (1990) Characterization of the murine heat-stable antigen: an hematolymphoid differentiation antigen defined by the J11d, M1/69 and B2A2 antibodies. *Eur. J. Immunol.* **20**, 1597–1602.

Altin, J. G. and Pagler, E. B. (1995) A one-step procedure for biotinylation and chemical cross-linking of lymphocyte surface and intracellular membrane-associated proteins. *Anal. Biochem.* **224**, 382–389.

Amersham. (1993) 'Guide to Radioiodination Techniques'. Amersham International plc, Bucks, England.

Anderson, N. L. and Hickman, B. J. (1979) Analytical techniques for cell fractionation. XXIV. Isoelectric point standards for two-dimensional electrophoresis. *Anal. Biochem.* **93**, 312–320.

Aruffo, A. and Seed, B. (1987a) Molecular cloning of two CD7 (T-cell leukaemia antigen) cDNAs by a COS cell expression system. *EMBO J.* **6**, 3313–3316.

Aruffo, A. and Seed, B. (1987b) Molecular cloning of a CD28 cDNA by a high-efficiency COS cell expression system. *Proc. Natl Acad. Sci. USA* **84**, 8573–8577.

Ashani, Y. and Catravas, G. N. (1980) Highly reactive impurities in Triton X-100 and Brij 35: Partial characterization and removal. *Anal. Biochem.* **109**, 55–62.

Barbour, A. G., Tessier, S. L. and Stoenner, H. G. (182) Variable major proteins of *Borrelia hermsii. J. Exp. Med.* **156**, 1312–1324.

Barclay, A. N., Letarte-Muirhead, M. and Williams, A. F. (1975) Purification of the Thy-1 molecule from rat brain. *Biochem. J.* **151**, 699–706.

Baron, C. and Thompson, T. E. (1975) Solubilisation of bacterial membrane proteins using alkyl glycosides and dioctanoyl phosphatidylcholine. *Biochem. Biophys. Acta* **382**, 276–285.

Batteiger, B., Newhall, W. J. and Jones, R. B. (1982) The use of Tween 20 as a blocking agent in the immunological detection of proteins transferred to nitrocellulose membranes. *J. Immunol. Methods* **55**, 297–307.

Bayer, E. A., Shutelsky, E. and Wilchek, M. (1979) The avidin–biotin complex in affinity cytochemistry. *Methods Enzymol.* **62**, 308–315.

Belli, S., Sali, A. and Goding, J. W. (1994) Divalent cations stabilize the conformation of plasma cell membrane glycoprotein PC-1 (alkaline phosphodiesterase I). *Biochem. J.* **304**, 75–80.

Belli, S. I. and Goding, J. W. (1994) Biochemical characterization of human PC-1, an enzyme possessing alkaline phosphodiesterase I and nucleotide pyrophosphatase activities. *Eur. J. Biochem.* **226**, 433–443.

Belyavsky, A., Vinogradova, T. and Rajewsky, K. (1989) PCR-based cDNA library construction: general cDNA libraries at the level of a few cells. *Nucl. Acids Res.* **17**, 2919–2932.

Bleil, J. D. and Bretscher, M. S. (1982) Transferrin receptor and its recycling in HeLa cells. *EMBO J.* **1**, 351–355.

Bolton, A. E. and Hunter, W. M. (1973) The labelling of proteins to high specific activity by conjugation to a ^{125}I-containing acylating agent. *Biochem. J.* **133**, 529–538.

Bonner, W. M. and Laskey, R. A. (1974) A film detection method for tritium-labelled proteins and nucleic acids in acrylamide gels. *Eur. J. Biochem.* **46**, 83–88.

Bordier, C. (1981) Phase separation of integral membrane proteins in Triton X-114 solution. *J. Biol. Chem* **256**, 1604–1607.

Bordier, C. (1988) Analytical and preparative phase separation of glycolipid-anchored membrane proteins in Triton X-114 solution. *In* 'Post-translational modification of Proteins by Lipids. A laboratory Manual' (U. Brodeck and C. Bordier, eds), pp. 29–33. Springer Verlag, Berlin.

Brakenhoff, R. H., Knippels, E. M. C. and van Dongen, G. A. M. S. (1994) Optimization and simplification of expression cloning in eukaryotic vector/host systems. *Anal. Biochem.* **218**, 460–463.

Bridges, C. D. B. (1977) A method for preparing stable digitonin solutions for visual pigment extraction. *Vision Res.* **17**, 301–302.

Brown, D. A. (1992) Interactions between GPI-anchored proteins and membrane lipids. *Trends Cell. Biol.* **2**, 338–343.

Brusca, J. S. and Radolf, J. D. (1994) Isolation of integral membrane proteins by phase partitioning in Triton X-114. *Methods Enzymol.* **228**, 182–193.

Buckley, M. F., Loveland, K. A., McKinstry, W. J., Garson, O. M. and Goding, J. W. (1990) Plasma cell membrane glycoprotein PC-1: cDNA cloning of the human molecule, amino acid sequence and chromosomal location. *J. Biol. Chem.* **265**, 17506–17511.

Burnette, W. N. (1981) 'Western Blotting': electrophoretic transfer of proteins from sodium dodecyl sulfate–polyacrylamide gels to unmodified nitrocellulose and radiographic detection with antibody and radioiodinated protein A. *Anal. Biochem.* **112**, 195–203.

Cambier, J. C., Pleiman, C. M. and Clark, M. R. (1994) Signal transduction by the B cell receptor and its coreceptors. *Annu. Rev. Immunol.* **12**, 457–486.

Carrol, R. B., Goldfine, S. M. and Mehero, J. A. (1978) Antiserum to polyacrylamide gel-purified simian virus 40 T antigen. *Virology* **87**, 194–198.

Casey, P. J. (1995) Protein lipidation in cell signaling. *Science* **268**, 221–225.

Cheng, H.-L., Blattner, F. R., Fitzmaurice, L., Mushinkski, J. F. and Tucker, P. W. (1982) Structure of genes for membrane and secreted murine IgD heavy chains. *Nature* **296**, 410–415.

Cleveland, D. W., Fischer, S. G., Kirschner, M. W. and Laemmli, U. K. (1977) Peptide mapping by limited proteolysis in sodium dodecyl sulfate and analysis by gel electrophoresis. *J. Biol. Chem.* **252**, 1102–1106.

Cole, S. R., Ashman, L. K. and Ey, P. L. (1987) Biotinylation: an alternative to radioiodination for the identification of cell surface antigens in immunoprecipitates. *Mol. Immunol.* **24**, 699–705.

Coligan, J. E., Gates, F. T., Kimball, E. S. and Maloy, W. L. (1983) Radiochemical sequence analysis of biosynthetically labeled proteins. *Methods Enzymol.* **91**, 413–434.

Contreras, M. A., Bale, W. and Spar, I. L. (1983) Iodine monochloride (ICl) iodination techniques. *Methods Enzymol.* **92**, 277–292.

Cooper, J. A. and Hunter, T. (1981) Changes in protein phosphorylation in Rous sarcoma virus-transformed chicken embryo cells. *Mol. Cell. Biol.* **1**, 165–178.

Davis, B. J. (1964) Disc electrophoresis. II. Method and application to human serum proteins. *Ann. NY Acad. Sci.* **121**, 404–427.

Ebina, Y., Ellis, L., Jarnagin, K., Edery, M., Graf, L., Clausner, E., Ou, J., Masiarz, F., Kan, Y. W., Goldfine, I. D., Roth, R. A. and Rutter, W. J. (1985) The human insulin receptor cDNA: The structural basis for hormone-activated transmembrane signalling. *Cell* **40**, 747–758.

Erickson, P. F., Minier, L. N. and Lasher, R. S. (1982) Quantitative electrophoretic transfer of polypeptides from SDS polyacrylamide gels to nitrocellulose sheets: a method for their re-use in immunoautoradiographic detection of antigens. *J. Immunol. Methods* **51**, 241–249.

Ey, P. L., Prowse, S. J. and Jenkin, C. R. (1978) Isolation of pure IgG1, IgG2a and IgG2b immunoglobulins from mouse serum using protein A-Sepharose. *Immunochemistry* **15**, 241–249.

Feizi, T. (1981) Carbohydrate differentiation antigens. *Trends Biochem. Sci.* **6**. 333–335.

Felsted, R. L. and Gupta, S. K. (1982) A comparison of K-562 and HL-60 human leukemic cell surface membrane proteins by two-dimensional electrophoresis. *J. Biol. Chem.* **257**, 13211–13217.

Ferguson, M. A. J. and Williams, A. F. (1988) Cell-surface anchoring of proteins via glycosyl-phosphatidylinositol structures. *Annu. Rev. Biochem.* **57**, 285–320.

Fiedler, K., Parton, R. G., Kellner, R., Etzold, T. and Simons, K. (1994) VIP36, a novel component of glycolipid rafts and exocytic carrier vesicles in epithelial cells. *EMBO J.* **13**, 1729–1740.

Fraker, P. J. and Speck, J. C. (1978) Protein and cell membrane iodinations with a sparingly soluble chloramide, 1,3,4,6-tetrachloro-3α, 6α-diphenylglycoluril. *Biochem. Biophys. Res. Commun.* **80**, 849–857.

Gahmberg, C. G. and Andersson, L. C. (1977) Selective radioactive labeling of cell surface sialoglycoproteins by periodate-tritiated borohydride. *J. Biol. Chem.* **252**, 5888–5894.

Gahmberg, C. G. and Hakomori, S. (1973) External labeling of cell surface galactose and galactosamine in glycolipid and glycoprotein of human erythrocytes. *J. Biol. Chem.* **248**, 4311–4317.

Gahmberg, C. G., Hayry, P. and Andersson, L. C. (1976) Characterization of surface glycoproteins of mouse lymphoid cells. *J. Cell. Biol.* **68**, 642–653.

Garland, T. O. and Patterson, M. W. H. (1967) Six cases of acrylamide poisoning. *Brit. Med. J.* **4**, 134–138.

Garrels, J. I. (1979) Two-dimensional gel electrophoresis and computer analysis of proteins synthesized by clonal cell lines. *J. Biol. Chem.* **254**, 7961–7977.

Garrison, J. G. (1993) Study of protein phosphorylation in intact cells. *In* 'Protein Phosphorylation. A Practical Approach' (D. G. Hardie, ed.) pp. 1–30. IRL Press, Oxford.

Geoghegan, K. F., Dallas, J. L. and Feeney, R. E. (1980) Periodate inactivation of ovotransferrin and human serum transferrin. *J. Biol. Chem.* **255**, 11429–11434.

George, S. and Schenck, J. R. (1983) Thiocyanate inhibition of protein iodination by the chloramine-T method and a rapid method for measurement of low levels of thiocyanate. *Anal. Biochem.* **131**, 1–15.

Gershoni, J. M. and Palade, G. E. (1983) Protein blotting: principles and applications. *Anal. Biochem.* **131**, 1–15.

Goding, J. W. (1980) Structural studies of murine lymphocyte surface IgD. *J. Immunol.* **124**, 2082–2088.

Goding, J. W. (1982) Asymmetrical surface IgG on MOPC-21 plasmacytoma cells contains one membrane heavy chain and one secretory heavy chain. *J. Immunol.* **128**, 2416–2421.

Goding, J. W. and Burns, G. F. (1981) Monoclonal antibody OKT-9 recognizes the receptor for transferrin on human acute lymphocytic leukemia cells. *J. Immunol.* **127**, 1256–1258.

Goding, J. W. and Harris, A. W. (1981) Subunit structure of cell surface proteins: Disulfide bonding in antigen receptors, Ly-2/3 antigens and transferrin receptors of murine T and B lymphocytes. *Proc. Natl Acad. Sci. USA* **78**, 4530–4534.

Goding, J. W. and Herzenberg, L. A. (1980) Biosynthesis of lymphocyte surface IgD in the mouse. *J. Immunol.* **124**, 2540–2547.

Goding, J. W. and Shen, F.-W. (1982) Structure of the murine plasma cell alloantigen PC-1: Comparison with the receptor for transferrin. *J. Immunol.* **129**, 2636–2640.

Goding, J. W. and Walker, I. D. (1980) Allelic forms of β_2-microglobulin in the mouse. *Proc. Natl Acad. Sci. USA* **77**, 7395–7399.

Goding, J. W., Warr, J. W. and Warner, N. L. (1976) Genetic polymorphism of IgD-like cell surface immunoglobulin in the mouse. *Proc. Natl Acad. Sci. USA* **73**, 1305–1309.

Goldstein, L. C., McDougall, J., Hackman, R., Meyers, J. D., Thomas, E. D. and Nowinski, R. C. (1982) Monoclonal antibodies to cytomegalovirus: rapid identification of clinical isolates and preliminary use in diagnosis of cytomegalovirus pneumonia. *Infect. Immun.* **38**, 273–281.

Gonenne, A. and Ernst, R. (1978) Solubilization of membrane proteins by sulfobetaines, novel zwitterionic detergents. *Anal. Biochem.* **87**, 28–38.

Goridis, C., Hirn, M., Santoni, M. J., Gennarini, G., Deagostini-Baxin, H., Jordan, B. R., Kiefer, M. and Steinmetz, M. (1985) Isolation of mouse *N*–CAM-related cDNA: detection and cloning using monoclonal antibodies. *EMBO. J.* **4**, 631–635.

Gould, S. J., Subramani, S. and Scheffler, I. E. (1989) Use of the DNA polymerase chain reaction for homology probing: Isolation of partial cDNA or genomic clones encoding the iron-sulfur protein of succinate dehydrogenase from several species. *Proc. Natl Acad. Sci. USA* **86**, 1934–1938.

Granger, B. L. and Lazarides, E. (1979) Desmin and vimentin coexist at the periphery of the myofibril Z disc. *Cell* **18**, 1053–1063.

Green, N. M. (1975) Avidin. *Adv. Protein Chem.* **29**, 85–133.

Greenwood, F. C., Hunter, W. M. and Glover, J. S. (1963) The preparation of ^{131}I-labelled human growth hormone of high specific radioactivity. *Biochem. J.* **89**, 114–123.

Grego, B., van Driel, I. R., Goding, J. W., Nice, E. C. and Simpson, R. J. (1986) Use of microbore high-performance liquid chromatography for purifying subnanomole levels of polypeptides for microsequencing. Structural studies on the murine plasma cell antigen PC-1. *Int. J. Peptide Protein Res.* **27**, 201–207.

Grego, B., van Driel, I. R., Stearne, P. A., Goding, J. W., Nice, E. C. and Simpson, R. J. (1985) A microbore high-performance liquid chromatography strategy for the purification of polypeptides for gas-phase sequence analysis. Structural studies on the murine transferrin receptor. *Eur. J. Biochem.* **148**, 485–491.

Griffith, I. P. (1972) Immediate visualisation of proteins in dodecyl sulfate-polyacrylamide gels by pre-staining with Remazol dyes. *Anal. Biochem* **46**: 402–412.

Gullick, W. J., Tzartos, S. and Lindstrom, J. (1981) Monoclonal antibodies as probes of acetylcholine receptor structure. I. Peptide mapping. *Biochemistry* **20**, 2173–2180.

Györgi, P. (1940) A further note on the identity of vitamin H with biotin. *Science* **92**, 609–610.

Hager, D. A. and Burgess, R. R. (1980) Elution of proteins from sodium dodecyl sulfate-polyacrylamide gels, removal of sodium dodecyl sulfate, and renaturation of enzymatic activity: results with sigma subunit of *Escherichia coli* RNA polymerase, wheat germ topoisomerase and other enzymes. *Anal. Biochem.* **109**, 76–86.

Hakes, D. J. and Dixon, J. E. (1992) New vectors for high level expression of recombinant proteins in bacteria. *Anal. Biochem.* **202**, 293–298.

Hakomori, S. (1984) Glycosphingolipids as differentiation-dependent, tumor-associated markers and as regulators of cell proliferation. *Trends Biochem. Sci.* **9**, 453–458.

Hakomori, S. and Kannagi, R. (1986) Carbohydrate antigens in glycolipids and glycoproteins. *In* 'Handbook of Experimental Immunology' (D. M. Weir, L. A. Herzenberg, C. C. Blackwell and L. A. Herzenberg, eds) 4th ed. Blackwell, Edinburgh (5, 329–336).

Hakomori, S. I. (1981) Glycosphingolipids in cellular interaction, differentiation and oncogenesis. *A. Rev. Biochem.* **50**, 733–764.

Handman, E. and Goding, J. W. (1985) The *Leishmania* receptor for macrophages is a lipid-containing glycoconjugate. *EMBO J.* **4**, 329–336.

Handman, E. and Jarvis, H. M. (1985) Nitrocellulose-based assays for the detection of glycolipids and other antigens: Mechanism of binding to nitrocellulose. *J. Immunol. Methods* **83**, 113–123.

Handman, E., Mitchell, G. F. and Goding, J. W. (1981) Identification and characterization of protein antigens of *Leishmania tropica* isolates. *J. Immunol.* **126**, 508–512.

Handman, E., Greenblatt, C. L. and Goding, J. W. (1984) An amphipathic sulfated glycoconjugate of *Leishmania*: Characterization with monoclonal antibodies. *EMBO J.* **3**, 2301–2306.

Harahap, A. R. and Goding, J. W. (1988) Distribution of the murine plasma cell antigen PC-1 in non-lymphoid cells. *J. Immunol.* **141**, 2317–2320.

Hardie, D. G. (1993) Protein Phosphorylation. A Practical Approach. IRL Press, Oxford.

Harford, J. (1984) An artefact explains the apparent association of the transferrin receptor with a *ras* gene product. *Nature* **311**, 673–675.

Hayman, M. J. and Crumpton, M. J. (1972) Isolation of glycoproteins from pig lymphocyte plasma membrane using *Lens culinaris* phytohaemagglutinin. *Biochem. Biophys. Res. Commun.* **47**, 923–930.

Heggeness, M. H. (1977) Avidin binds to condensed chromatin. *Stain Technol.* **52**, 165–169.

Helenius, A. and Simons, K. (1975) Solubilization of membranes by detergents. *Biochem. Biophys. Acta* **415**, 29–79.

Helenius, A., McCaslin, D. R., Fries E. and Tanford C. (1979) Properties of detergents. *Methods Enzymol.* **56**:734–749.

Hellman, U., Wernstedt, C., Gonez, J. and Heldin, C-H. (1995) Improvement of an 'in gel' digestion procedure for the micropreparation of internal protein fragments for amino acid sequencing. *Anal. Biochem.* **224**, 451–455.

Herrmann, S. H. and Mescher, M. F. (1979) Purification of the H-2K^k molecule of the murine major histocompatibility complex. *J. Biol. Chem.* **254**, 8713–8716.

Herrmann, S. H., Chow, C. M. and Mescher, M. F. (1982) Proteolytic modifications of the carboxyl-terminal region of H-2K^k. *J. Biol. Chem.* **257**, 14181–14186.

Higgins, C. F. (1994) Flip-flop: the transmembrane translocation of lipids. *Cell* **79**, 393–395.

Higgins, R. C. and Dahmus, M. E. (1982) Tritium/PPO gel fluorographic efficiency is reduced by Coomassie blue staining. *Electrophoresis* **3**, 214–216.

Hjelmeland, L. M. (1980) A nondenaturing zwitterionic detergent for membrane biochemistry: design and synthesis. *Proc. Natl Acad. Sci. USA* **77**, 6368–6370.

Hoh, J. F. Y., Yeoh, G. P. S., Thomas, M. A. W. and Higginbotham, L. (1979) Structural differences in the heavy chains of rat ventricular myosin isoenzymes. *FEBS Lett.* **97**, 330–334.

Holloway, P. W. (1973) A simple procedure for removal of Triton X-100 from protein samples. *Anal. Biochem.* **53**, 304–308.

Hubbard, A. L. and Cohn, Z. A. (1975) Externally disposed plasma membrane proteins. I. Enzymatic iodination of mouse L Cells. *J. Cell. Biol.* **64**, 438–460.

Huynh, T. V., Young, R. A. and Davis, R. W. (1985) Construction and screening cDNA libraries in λgt10 and λgt11. *In* 'DNA Cloning: A Practical Approach' Vol. 1 (D. M. Glover, ed.), pp. 49–78. IRL Press, Oxford.

Jenö, P., Mini, T., Moes, S., Hintermann, E. and Horst, M. (1995) Internal sequences from proteins digested in polyacrylamide gels. *Anal. Biochem.* **224**, 75–82.

Johnson, D. A., Gautsch, J. W., Sportsman, J. R. and Elder, J. H. (1984) Improved technique utilizing nonfat dry milk for analysis of proteins and nucleic acids transferred to nitrocellulose. *Gene Anal. Techn.* **1**, 3–8.

Johnsson, S. and Kronvall, G. (1974) The use of protein A-containing Staphylococcus aureus as a solid phase anti-IgG reagent in radioimmunoassays as exemplified in the quantitation of α-fetoprotein in normal human adult serum. *Eur. J. Immunol.* **4**, 29–33.

Juang, R-H., Chang, Y-D., Sung, H-Y. and Su, J-C. (1984) Oven-drying method for polyacrylamide gel slab packed in cellophane sandwich. *Anal. Biochem.* **141**, 348–350.

Kaufman, J. F. and Strominger, J. L. (1982) HLA-DR light chain has a polymorphic *N*-terminal region and a conserved immunoglobulin-like *C*-terminal region. *Nature* **297**, 694–697.

Kessler, S. W. (1975) Rapid isolation of antigens from cells with a staphylococcal protein A-antibody adsorbent: Parameters of the interaction of antibody–antigen complexes with protein A. *J. Immunol.* **115**, 1617–1624.

Kessler, S. W. (1976) Cell membrane antigen isolation with the staphylococcal protein A-antibody adsorbent. *J. Immunol.* **117**, 1482–1490.

Kessler, S. W. (1981) Use of protein A-bearing staphylococci for immunoprecipitation and isolation of antigens from cells. *Methods Enzymol.* **73**, 442–459.

Kessler, S. W., Woods, V. L., Finkelman, F. D. and Scher, I. (1979) Membrane orientation and location of multiple and distinct allotypic determinants of mouse lymphocyte IgD. *J. Immunol.* **123**, 2772–2778.

Kesson, C. M., Baird, A. W. and Lawson, D. H. (1977) Acrylamide poisoning. *Postgrad. Med. J.* **53**, 16–17.

Kim, K-M., Adachi, T.k, Nielsen, P. J., Terashima, M., Lamers, M. C., Köhler, G. and Reth, M. (1994) Two new proteins preferentially associated with membrane immunoglobulin D. *EMBO J.* **13**, 3793–3800.

Knudsen, K. (1985) Proteins transferred to nitrocellulose for use as immunogens. *Anal. Biochem.* **147**, 285–288.

Korman, A. J., Knudsen, P. J., Kaufman, J. F. and Strominger, J. L. (1982) cDNA clones for the heavy chain of HLA-DR antigens obtained after immunopurification of polysomes by monoclonal antibody. *Proc. Natl Acad. Sci. USA* **79**, 1844–1848.

Kubo, K. (1995) Effect of incubation of solutions of proteins containing dodecyl sulfate on the cleavage of peptides bonds by boiling. *Anal. Biochem.* **225**, 351–353.

Kubo, K., Isemura, T. and Takagi, T. (1979) Electrophoretic behaviour of micellar and monomeric sodium dodecyl sulfate in polyacrylamide gel electrophoresis with reference to those of SDS-protein complexes. *Anal. Biochem.* **92**, 243–247.

Kuperman, A. S. (1958) Effects of acrylamide on the central nervous system of the cat. *J. Pharmacol. Exp. Therap.* **123**, 180–192.

Kurzchalia, T. V., Gorvel, J-P., Dupree, P., Parton, R., Kellner, R., Houthaeve, T., Gruenberg, J. and Simons, K. (1992) Interaction of *rab5* with cytosolic proteins. *J. Biol. Chem.* **267**, 18419–18423.

Kurzchalia, T. V., Hartman, E. and Dupree, P. (1995) Guilt by insolubility—does a protein's detergent insolubility reflect a caveolar location? *Trends Cell Biol.* **5**, 187–189.

Laemmli, U. K. (1970) Cleavage of structural proteins during the assembly of the head of the bacteriophage T4. *Nature* **227**, 680–685.

Lam, K. S. and Kasper, C. B. (1980) Sequence homology analysis of a heterogeneous protein population by chemical and enzymic digestion using a two-dimensional sodium dodecyl sulfate-polyacrylamide gel system. *Anal. Biochem.* **108**, 220–226.

Lane, D. P. and Robbins, A. K. (1978) An immunochemical investigation of SV40 T antigens. I. Production, properties and specificity of a rabbit antibody to purified simian virus 40 large-T antigen. *Virology* **87**, 182–193.

Laskey, R. A. (1980) The use of intensifying screens or organic scintillators for visualizing radioactive molecules resolved by gel electrophoresis. *Methods Enzymol.* **65**, 363–371.

Laskey, R. A. (1993) 'Efficient Detection of Biomolecules by Autoradiography, Fluorography or Chemiluminescence' (Review 23). Amersham International Ltd. Buckinghamshire, England.

Layton, J. E. (1980) Anti-carbohydrate activity of T cell-reactive chicken anti-mouse immunoglobulin antibodies. *J. Immunol.* **125**, 1993–1997.

Ledbetter, J. A. and Herzenberg, L. A. (1979) Xenogeneic monoclonal antibodies to mouse lymphoid differentiation antigens. *Immunol. Rev.* **47**, 63–90.

Ledbetter, J. A., Goding, J. W., Tsu, T. T. and Herzenberg, L. A. (1979) A new mouse lymphoid alloantigen (Lgp100) recognized by a monoclonal rat antibody. *Immunogenetics* **8**, 347–360.

Lee, S.-K., Singh, J. and Taylor, R. B. (1975) Subclasses of T cells with different sensitivities to cytotoxic antibody in the presence of anesthetics. *Eur. J. Immunol.* **5**, 259–262.

Legochi, R. P. and Verma, D. P. S. (1981) Multiple immunoreplica technique: screening for specific proteins with a series of different antibodies using one polyacrylamide gel. *Anal. Biochem.* **111**, 385–392.

Lin, W. and Kasamatsu, H. (1983) On the electrotransfer of polypeptides from gels to nitrocellulose membranes. *Anal. Biochem.* **128**, 302–311.

Lischwe, M. A. and Ochs, D. (1982) A new method for partial peptide mapping using *N*-chlorosuccinimide/urea and peptide silver staining in sodium dodecyl sulfate-polyacrylamide gels. *Anal. Biochem.* **127**, 453–457.

Lonsdale-Eccles, J. D., Lyndley, A. M. and Dale, B. A. (1981) Cyanogen bromide cleavage of proteins in sodium dodecyl sulfate/polyacrylamide gels. *Biochem. J.* **197**, 591–597.

MacSween, J. M. and Eastwood, S. L. (1981) Recovery of antigen from staphylococcal protein A-antibody adsorbents. *Methods Enzymol.* **73**, 459–471.

Maher, P. A. and Singer, S. J. (1985) Anomalous interaction of the actetylcholine receptor protein with the nonionic detergent Triton X-114. *Proc. Natl Acad. Sci. USA* **82**, 958–962.

Marchalonis, J. J. (1969) An enzymic method for the trace iodination of immunoglobulins and other proteins. *Biochem. J.* **113**, 299–305.

Marchalonis, J. J., Cone, R. E. and Santer, V. (1971) Enzymic iodination: A probe for cell surface proteins of normal and neoplastic lymphocytes. *Biochem. J.* **124**, 921–927.

Markwell, M. A. K. (1982) A new solid-state reagent to iodinate proteins. I. Conditions for the efficient labelling of antiserum. *Anal. Biochem.* **125**, 427–432.

Markwell, M. A. K. and Fox, C. F. (1978) Surface-specific iodination of membrane proteins of viruses and eucaryotic cells using 1,3,4,6-tetrachloro-3α,6α-diphenylglycoluril. *Biochemistry* **17**, 4807–4817.

McCollister, D. D., Oyen, F. and Rowe, V. K. (1964) Toxicology of acrylamide. *Toxicol. Appl. Pharmacol.* **6**, 172–181.

McConville, M. J., Bacic, A., Mitchell, G. F. M. and Handman, E. (1987) Lipophosphoglycan of *Leishmania major* that vaccinates mice against cutaneous leishmaniasis contains an alkylglycerophosphoinositol lipid anchor. *Proc. Natl Acad. Sci. USA* **84**, 8941–8945.

McPherson, M. J., Jones, K. M. and Gurr, S. J. (1991) PCR with highly degenerate primers. *In* PCR: A Practical Approach (M. J. McPherson, P. Quirke, and G. R. Taylor, eds). IRL Press, pp. 171–186, Oxford.

Meier, T., Arni, S., Malarkannan, S., Poincelet, M. and Hoessli, D. (1992) Immunodetection of biotinylated lymphocyte-surface proteins by enhanced chemiluminescence: a nonradioactive method for cell-surface protein analysis. *Anal. Biochem.* **204**, 220–226.

Melchers, F. (1973) Synthesis, surface deposition and secretion of immunoglobulin M in bone marrow-derived lymphocytes before and after mitogenic stimulation. *Transplant. Rev.* **14**, 76–130.

Michaelson. J., Rothenberg, E. and Boyse, E. A. (1980) Genetic polymorphism of murine β_2-microglobulin detected biochemically. *Immunogenetics* **11**, 93–95.

Milstein, C. and Lennox, E. (1980) The use of monoclonal antibody techniques in the study of developing cell surfaces. *Curr. Top. Dev. Biol.* **14**, 1–32.

Miranda, P. V., Brandelli, A. and Tezon, J. G. (1993) Instantaneous blocking for immunoblots. *Anal. Biochem.* **209**, 376–377.

Moosic, J. P., Nilson, A., Hämmerling, G. J. and McKean, D. J. (1980) Biochemical characterization of Ia antigens. I. Characterization of the 31K polypeptide associated with I-A subregion Ia antigens. *J. Immunol.* **125**, 1463–1469.

Moritz, R. L., Eddes, J., Ji, H., Ried, G. and Simpson, R. J. (1995). Rapid separation of proteins and peptides using conventional silica-based supports: identification of 2-D gel proteins following 'in gel' proteolysis. *In* 'Techniques in Protein Chemistry' VI (Crabb, J. W., ed), pp. 311–319. Academic Press, San Diego.

Morrissey, J. H. (1981) Silver stain for proteins in polyacrylamide gels: A modified procedure with enhanced uniform sensitivity. *Anal. Biochem.* **117**, 307–310.

Muilerman, H. G., ter Hart, H. G. J. and Van Dijk, W. V. (1982) Specific detection of inactive enzyme protein after polyacrylamide gel electrophoresis by a new enzyme-immunoassay method using unspecific antiserum and partially purified active enzyme: application to rat liver phosphodiesterase I. *Anal. Biochem.* **120**, 46–51.

Nakamura, S-I. and Rodbell, M. (1990) Octyl glucoside extracts GTP-binding regulatory proteins from rat brain 'synaptoneurosomes' as large, polydisperse structures devoid of βγ complexes and sensitive to disaggregation by guanine nucleotides. *Proc. Natl Acad. Sci. USA* **87**, 6413–6417.

Nicola, N. A. and Metcalf, D.(1988) Binding, internalization, and degradation of ^{125}I-multipotential colony-stimulating factor (interleukin-3) by FDCP-1 cells. *Growth Factors* **1**, 29–39.

Nikodem, V. and Fresco, J. R. (1979) Protein fingerprinting by SDS-gel electrophoresis after partial fragmentation with CNBr. *Anal. Biochem.* **97**, 382–386.

Nussenzweig, M. C., Steinman, R. M., Witmer, M. D. and Gutchinov, B. (1982) A monoclonal antibody specific for mouse dendritic cells. *Proc. Natl Acad. Sci. USA* **79**, 161–165.

O'Connor, C. G. and Ashman, L. K. (1982) Application of the nitrocellulose transfer technique and alkaline phosphatase conjugated anti-immunoglobulin for determination of the specificity of monoclonal antibodies to protein mixtures. *J. Immunol. Methods* **54**, 267–271.

O'Farrell, P. H. (1975) High resolution two-dimensional electrophoresis of proteins. *J. Biol. Chem.* **250**, 4007–4021.

O'Farrell, P. Z., Goodman, H. M. and O'Farrell, P. H. (1977) High resolution two-dimensional electrophoresis of basic as well as acidic proteins. *Cell* **12**, 1133–1142.

Oettgen, H. C., Pettey, C. L., Maloy, W. L. and Terhorst, C. (1986) A T3-like protein complex associated with the antigen receptor on murine T cells. *Nature* **320**, 272–275.

Oi, V. T., Jones, P. P., Goding, J. W., Herzenberg, L. A. and Herzenberg, L. A. (1978) Properties of monoclonal antibodies to mouse Ig allotypes, H-2, and Ia antigens. *Curr. Topics Microbiol. Immunol.* **81**, 115–129.

Omary, M. B. and Trowbridge, I. S. (1981) Biosynthesis of the human transferrin receptor in cultured cells. *J. Biol. Chem.* **256**, 12888–12892.

Ornstein, L. (1964) Disc electrophoresis I. Background and theory. *Ann. NY Acad. Sci.* **121**, 321–349.

Pampori, N. A., Pampori, M. K. and Shapiro, B. H. (1995) Dilution of the chemiluminescence reagents reduces the background in Western Blots. *BioTechniques* **18**, 588–590.

Patterson, S. D. (1994) From electrophoretically separated protein to identification: strategies for sequence and mass analysis. *Anal. Biochem.* **221**, 1–15.

Pearson, T., Galfé, G., Ziegler, A. and Milstein, C. (1977) A myeloma hybrid producing antibody specific for an allotypic determinant on 'IgD-like' molecules of the mouse. *Eur. J. Immunol.* **7**, 684–690.

Peterson, P. A., Cunningham, B. A., Berggard, I. and Edelman, G. M. (1972) β_2-microglobulin—a free immunoglobulin domain. *Proc. Natl Acad. Sci. USA* **69**, 1697–1702.

Phillips, D. R. and Morrison, M. (1970) The arrangement of proteins in the human erythrocyte membrane. *Biochem. Biophys. Res. Commun.* **40**, 284–289.

Pitt-Rivers, R. and Impiombato, F. S. A. (1968) The binding of sodium dodecyl sulfate to various proteins. *Biochem. J.* **109**, 825–830.

Rademacher, T. W., Parekh, R. B. and Dwek, R. A. (1988) Glycobiology. *Annu. Rev. Biochem.* **57**, 785–838.

Radke, K. and Martin, G. S. (1979) Transformation by Rous sarcoma virus: effects of src expression on the synthesis and phosphorylation of cellular polypeptides. *Proc. Natl Acad. Sci. USA* **76**, 5212–5216.

Radke, K., Gilmore, T. and Martin, G. S. (1980) Transformation of Rous sarcoma virus: a cellular substrate for transformation-specific protein phosphorylation contains phosphotyrosine. *Cell* **21**, 821–828.

Reinherz, E. L., Kung, P. C., Goldstein, G., Levey, R. H. and Schlossman, S. F. (1980) Discrete stages of human intrathymic differentiation: analysis of normal thymocytes and leukemic lymphoblasts of T lineage. *Proc. Natl Acad. Sci. USA* **77**, 1588–1592.

Rittenhouse, J. and Marcus, F. (1984) Peptide mapping in polyacrylamide gel electrophoresis after cleavage at aspartyl–prolyl bonds in sodium dodecyl sulfate-containing buffers. *Anal. Biochem.* **138**, 442–448.

Rothman, J. E. and Lenard, J. (1977) Membrane asymmetry. *Science* 195, 743–753.

Sabban, E., Marchesi, V., Adesnik, M. and Sabatini, D. D. (1981) Erythrocyte membrane protein band 3: biosynthesis and incorporation into membranes. *J. Cell Biol.* **91**, 637–646.

Salacinski, P. R. P., McLean, C., Sykes, J. E. C., Clement-Jones, V. V. and Lowry, P. J. (1981) Iodination of proteins, glycoproteins, and peptides using a solid-phase oxidizing agent, 1,3,4,6-tetrachloro-3α,6α-diphenyl glycoluril (Iodogen). *Anal. Biochem.* **117**, 136–146.

Sambrook, J., Fritsch, E. F. and Maniatis, T. (1989) 'Molecular Cloning. A Laboratory Manual' 2nd edn. Cold Spring Harbor Laboratory Press, New York.

Saris, C. J. M., van Eeenbergen, J., Jenks, B. G. and Bloemers, H. P. J. (1983) Hydroxylamine cleavage of proteins in polyacrylamide gels. *Anal. Biochem.* **132**, 54–67.

Sauvé, D. M., Ho, D. T. and Roberge, M. (1995) Concentration of dilute protein for gel electrophoresis. *Anal. Biochem.* **226**, 382–383.

Savage, D., Mattson, G., Desai, S., Nielander, G. W., Morgensen, S. and Conklin, E. J. (1992) Avidin-Biotin Chemistry: A Handbook. Pierce Chemical Company, Rockford, Illinois.

Schneider, C., Kurkinen, M. and Greaves, M. (1983) Isolation of cDNA clones for the human transferrin receptor. *EMBO J.* **2**, 2259–2263.

Schneider, C., Sutherland, R., Newman, R. and Greaves, M. (1982) Structural features of the cell surface receptor for transferrin that is recognized by the monoclonal antibody OKT9. *J. Biol. Chem.* **257**, 8516–8522.

Schwartz, B. D. and Nathenson, S. G. (1971) Isolation of H-2 alloantigens solubilized by the detergent NP-40. *J. Immunol.* **107**, 1363–1367.

Shapiro, A. L., Vinuela, E. and Maizel, J. B. (1967) Molecular weight estimation of polypeptide chains by electrophoresis in SDS-polyacrylamide gels. *Biochem. Biophys. Res. Commun.* **28**, 815–820.

Sharon, N. and Lis, H. (1982) Glycoproteins. *In* 'The Proteins' (H. Neurath and R. L. Hill, eds), pp. 1–144. Academic Press, New York.

Sidman, C. L. (1981) Lymphocyte surface receptors and albumin. *J. Immunol.* **127**, 1454–1458.

Simonsen, H. and Lodish, H. F. (1994) Cloning by function: expression cloning in mammalian cells. *Trends Pharmacol. Sci.* **15**, 437–441.

Singer, S. J. and Nicolson, G. L. (1972) The fluid mosaic model of the structure of cell membranes. *Science* **175**, 720–731.

Smith, D. B. and Johnson, K. S. (1988) Single-step purification of polypeptides expressed in *Escherichia coli* as fusion proteins with glutathione S-transferase. *Gene (Amst.)* **67**, 31–40.

Smolarsky, M. (1980) A simple radioimmunoassay to determine the binding of antibodies to lipid antigens. *J. Immunol. Methods* **38**, 85–93.

Springer, T. A., Kaufman, J. F., Siddoway, L. A., Mann, D. L. and Strominger, J. L. (1977) Purification of HLA-linked B lymphocyte alloantigens in immunologically active form by preparative sodium dodecyl sulfate-gel electrophoresis and studies on their subunit association. *J. Biol. Chem.* **252**, 6201–6207.

Springer, T. A. (1980) Cell-surface differentiation in the mouse. Characterization of 'jumping' and 'lineage' antigens using xenogeneic rat monoclonal antibodies. In 'Monoclonal Antibodies' (R. H. Kennett, T. J. McKearn and K. B. Bechtol, eds), pp. 185–217. Plenum Press, New York.

Stearne, P. A., van Driel, I. R., Grego, B., Simpson, R. J. and Goding, J. W. (1985a) The murine plasma cell antigen PC-1: purification and partial amino acid sequence. *J. Immunol.* **134**, 443–448.

Stearne. P. A., Pietersz, G. A. and Goding, J. W. (1985b) cDNA cloning of the murine transferrin receptor: Sequence of trans-membrane and adjacent regions. *J. Immunol.* **134**, 3474–3479.

Steck, T. L. and Dawson, G. (1974) Topographical distribution of complex carbohydrates in the erythrocyte membrane. *J. Biol. Chem.* **249**, 2135–2142.

Stern, P., Willison, K., Lennox, E., Galfré, G., Milstein, C., Secher, D., Zeigler, A. and Springer, T. (1978) Monoclonal antibodies as probes for differentiation and tumor associated antigens: A Forssman specificity on teratocarcinoma stem cells. *Cell* **14**, 775–783.

Stumph, W. E., Elgin, S. C. R. and Hood, L. E. (1974) Antibodies to protein dissolved in sodium dodecyl sulfate. *J. Immunol.* **113**, 1752–1756.

Sutherland, R., Delia, D., Schneider, C., Newman, R., Kemshead, J. and Greaves M. (1981) Ubiquitious cell-surface glycoprotein on tumor cells is proliferation-associated receptor for transferrin. *Proc. Natl Acad. Sci. USA* **78**, 4515–4519.

Tack, B. F., Dean, J., Eilat, D., Lorenz, P. E. and Schechter, A. (1980) Tritium labelling of proteins to high specific activity by reductive methylation. *J. Biol. Chem.* **255**, 8842–8847.

Takahashi, T., Old, L. J. and Boyse, E. A. (1971) Surface alloantigens of plasma cells. *J. Exp. Med.* **131**, 1325–1341.

Tanaka, T. (1981) Gels. *Scientific American*. **244**, 124–133.

Tanaka, T., Nishio, I., Sun, S-T. and Veno-Nishio, S. (1982) Collapse of gels in an electric field. *Science* **218**, 467–469.

Tartakoff, A. and Vassalli, P. (1979) Plasma cell immunoglobulin M molecules. Their biosynthesis, assembly, and intracellular transport. *J. Cell Biol.* **83**, 284–299.

Terashima, M., Kim, K-M., Adachi, T., Nielsen, P. J., Reth, M., Köhler, G. and

Lamers, M. C. (1994) The IgM antigen receptor of B lymphocytes is associated with prohibitin and a prohibitin-related protein. *EMBO J.* **13**, 3782–3792.

Tijian, R., Stinchcomb, D. and Losick, R. (1975) Antibody directed against bacillus subtilis and σ factor purified by sodium dodecyl sulfate slab gel electrophoresis. *J. Biol. Chem.* **250**, 8824–8828.

Trowbridge, I. S. and Omary, M. B. (1981) Human cell surface glycoprotein related to cell proliferation is the receptor for transferrin. *Proc. Natl Acad. Sci. USA* **78**, 3039–3043.

Tse, A. G. D., Barclay, A. N., Watts, A. and Williams, A. F. (1985) A glycophospholipid tail at the carboxyl terminus of the Thy-1 glycoprotein of neurons and thymocytes. *Science* **230**, 1003–1008.

Tucker, P. W., Liu, C.-P., Muskinski, J. F. and Blattner, F. R. (1980) Mouse immunoglobulin D: Messenger RNA and genomic DNA sequences. *Science* **209**, 1353–1360.

van der Meer, J., Dorssers, L. and Zabel, P. (1983) Antibody-linked polymerase assay on protein blots: a novel method for identifying polymerases following SDS–polyacrylamide gel electrophoresis. *EMBO J.* **2**, 233–237.

van Driel, I. R. and Goding, J. W. (1987) Plasma cell membrane glycoprotein PC-1: primary structure deduced from cDNA clones. *J. Biol. Chem.* **262**, 4882–4887.

van Driel, I. R., Stearne, P. A., Grego, B., Simpson, R. J. and Goding, J. W. (1984) The receptor for transferrin on murine myeloma cells: one-step purification based on its physiology, and partial amino acid sequence. *J. Immunol.* **133**, 3220–3224.

van Driel, I. R., Wilks, A. F., Pietersz, G. A. and Goding, J. W. (1985) Murine plasma cell membrane antigen PC-1: Molecular cloning of cDNA and analysis of expression. *Proc. Natl Acad. Sci. USA* **82**, 8619–8623.

Varki, A. (1993) Biological roles of oligosaccharides: all of the theories are correct. *Glycobiology* **3**, 97–130.

Vasilov, R. G. and Pleogh, H. L. (1982) Biosynthesis of murine IgD: heterogeneity of glycosylation. *Eur. J. Immunol.* **12**, 804–813.

Vitetta, E. S., Capra, J. D., Klapper, D. G., Klein, J. and Uhr, J. W. (1976) The partial amino sequence of an H-2K molecule. *Proc. Natl Acad. Sci. USA* **73**, 905–909.

Wallevik, K. and Jensenius, J. C. (1982) A simple and reliable method for the drying of polyacrylamide slab gels. *J. Biochem. Biophys. Methods.* **6**, 17–21.

Weber, K. and Osborne, M. (1969) The reliability of molecular weight determinations by dodecyl sulfate-polyacrylamide gel electrophoresis. *J. Biol. Chem.* **244**, 4406–4412.

Wegener, A. D. and Jones, L. R. (1984) Phosphorylation-induced mobility shift in phospholamban in sodium dodecyl sulfate-acrylamide gel electrophoresis. *J. Biol. Chem.* **259**, 1834–1841.

Wessel, D. and Flügge, U. I. (1984) A method for the quantitative recovery of protein in dilute solution in the presence of detergents and lipids. *Anal. Biochem.* **138**, 141–143.

Weston, S. A., Crossett, B., Tuckwell, D. S. and Humphries, M. J. (1995) Effect of β-mercaptoethanol on the detection of biotinylated proteins. *Anal. Biochem.* **225**, 28–33.

Whalen, R. G., Schwartz, K., Bouveret, P., Sell, S. M. and Gros, F. (1979) Contractile protein isozymes in muscle development: Identification of an embryonic form of myosin heavy chain. *Proc. Natl Acad. Sci. USA* **76**, 5197–5201.

Whitehead, T. P., Kricka, L. J., Carter, T. J. N. and Thorpe, G. H. G. (1979) Analytical luminescence: Its potential in the clinical laboratory. *Clin. Chem.* **25**, 1531–1546.

Wiegandt, H. (1985) Glycolipids. Elsevier, Amsterdam.

Wiegers, K. I. and Dernick, R. (1981) Peptide maps of labelled poliovirus proteins after two-dimensional analysis by limited proteolysis in sodium dodecyl sulfate. *Electrophoresis* **2**, 98–104.

Wilchek, M. and Bayer, E. A. (1988) The avidin–biotin complex in bioanalytical applications. *Anal. Biochem.* **171**, 1–31.

Wilchek, M. and Bayer, E. A. (eds) (1990). Avidin–biotin technology. *Methods Enzymol.* **184**.

Wiser, M. F. and Schweiger, H-G. (1986) Increased sensitivity in antigen detection during immunoblot analysis resulting from antigen enrichment via immunoprecipitation. *Anal. Biochem.* **155**, 71–77.

Wolf, T., Fleminger, G. and Solomon, B. (1995) Functional conformations of calmodulin: 1. Preparation and characterization of a conformational specific anti-bovine calmodulin monoclonal antibody. *J. Mol. Recognition* **8**, 67–71.

Yamasaki, R. B., Osuga, D. T. and Feeney, R. E. (1982) Periodate oxidation of methionine in proteins. *Anal. Biochem.* **126**, 183–189.

Young, R. A., Mehra, V., Sweetser, D., Buchanan, T., Clark-Curtiss, J., Davis, R. W. and Bloom, B. R. (1985) Genes for the major protein antigens of the leprosy parasite *Mycobacterium leprae. Nature* **316**, 450–452.

Young, W. W. and Hakomori, S. (1981) Therapy of mouse lymphoma with monoclonal antibodies. *Science* **211**, 487–489.

11 Affinity Chromatography

Affinity chromatography is a method of fractionation which exploits the bio-specific binding of a particular molecule to a second molecule, often termed the 'ligand' (reviewed by Jakoby, 1984; Dean *et al.*, 1985; Hermanson *et al.*, 1992). The technique is extremely powerful. Purification factors of 2000–20 000-fold are often possible, and it is sometimes possible to achieve purification to homogeneity in a single step.

Immobilized antibodies and antigens have been used as affinity reagents for many years (Campbell *et al.*, 1951; Wofsy and Burr, 1969; Ruoslahti, 1976), but the method suffered from problems of low binding capacity, harsh elution conditions, and specificity limited by the quality of the antibodies. As a result, antibody immunoadsorbents did not achieve the widespread usage that might have been expected.

The advent of monoclonal antibodies has revolutionized affinity chromatography. In theory at least, it is now possible to produce affinity columns of any desired specificity, with high capacity and very mild elution conditions (Herrmann and Mescher, 1979; Parham, 1979, 1983; Secher and Burke, 1980; Dalchau and Fabre, 1982; Stallcup *et al.*, 1981; Mescher *et al.*, 1983; Turkewitz *et al.*, 1983; Dean *et al.*, 1985).

Affinity chromatography should not be considered in isolation, but rather as part of an overall strategy for protein purification. Much useful background information may be obtained from the excellent books by Harris and Angal (1989), Hermanson *et al.* (1992) and Scopes (1993).

11.1 Choice of Affinity Matrix and Coupling Reaction

The ideal affinity matrix would have negligible nonspecific adsorption, lack charged groups, and be capable of binding antibody in a leakproof manner with full preservation of activity. To date, beaded agarose has come closest to meeting this ideal. In practice, the great majority of immunoadsorbents are based on this material.

Many different coupling procedures are available (Table 11.1). The most popular is the cyanogen bromide method (Axèn *et al.*, 1967; March *et al.*, 1974). Activation of agarose by cyanogen bromide is simple and inexpensive, but involves working with a volatile and toxic chemical. It is now possible to obtain preactivated agarose as a stable lyophilized powder (CNBr-activated Sepharose 4B; Pharmacia). The binding capacity of the reconstituted gel will remain high for a period of months to years, provided that the powder is kept dry. The cost of the commercial product is not excessive when the usual small columns are constructed, but if columns of more than 50 ml are contemplated, activation with CNBr in the laboratory should be considered.

The chemistry of activation of agarose by CNBr has undergone some re-evaluation since its first description (Jennissen, 1995). The major stable active intermediate is now known to be a cyanate ester (Ernst-Cabrera and Wilchek, 1986; Jennissen, 1995). If the activated gel is extensively washed at alkaline pH, the cyanate esters tend to hydrolyse to inert carbamate esters and coupling to proteins occurs via cyclic imidocarbonate to form uncharged carbamate ester derivatives (Kohn and Wilchek, 1981). Conversely, if the gel is washed in dilute HCl, as is the case for lyophilized and reconstituted CNBr-activated Sepharose from Pharmacia, the cyclic imidocarbonate groups are hydrolysed and the coupling to the amino groups of proteins occurs via cyanate esters to form positively charged isourea linkages. The extent to which these competing reactions occur outside experimental model situations is not completely clear.

If it is desired to generate stable uncharged linkages, extensive washing of the gel at alkaline pH may be carried out (Kohn and Wilchek, 1981), although totally uncharged linkages may be more easily and reproducibly obtained using activation with carbonyldiimidazole (see below) or chloroformate (Ernst-Cabrera and Wilchek, 1986).

The presence of a charged linkage will tend to cause nonspecific ion-exchange effects which could degrade the specificity of the gel when used for affinity chromatography. This ion-exchange effect will be most pronounced at very low salt concentrations, and may be overcome by performing the chromatography in buffers of high salt concentration (e.g. 0.5 M NaCl). However, if the affinity of the immobilized group for the desired ligand is low, the added nonspecific binding energy provided by the charged groups could actually be beneficial, and could make the difference between the column functioning or not functioning (Scopes, 1993).

The isourea linkage between the gel and the amino groups of lysine is not completely stable, and may result in a very slow leakage of the protein from the column over a period of months to years (Tesser *et al.*, 1974; Lowe, 1979). Leakage is accelerated by raised temperature and nucleophiles such as

Table 11.1 Properties of coupling methods for affinity chromatography

	CNBr[a]	*N*-hydroxysuccinimide[b]	Carbonyldiimidazole[c]	Toluene sulphonyl chloride
Ease of preparation of affinity matrix	+	–	+++	+++
Ease of binding of protein	+++	+++	+++	+++
Charge-free linkage[d]	No	Yes	Yes	No
Stability of linkage[e]	+	+++	+++	+++
Spacer arm[f]	No	Yes	No	No

[a] Available commercially as CNBr-activated Sepharose 4B (Pharmacia).

[b] Available commercially as Affigel 10 (Bio-Rad) or Activated CH-Sepharose 4B (Pharmacia).

[c] Available commercially as Reacti-Gel (6X) (Pierce).

[d] Charged linkages may cause nonspecific binding at low salt concentrations.

[e] The rate of leakage of protein from CNBr-activated agarose is very low (typically $< 0.02\%$ per day). Leakage is accelerated by alkaline pH, higher temperatures and presence of nucleophiles (e.g. amines such as Tris).

The spacer arm of Affigel 10 is attached to the gel by a stable ether linkage, while the spacer arm of Activated CH Sepharose is attached via a labile isourea bond. It is good practice to precycle all affinity columns with the eluting buffer just before use.

[f] Spacer arms may give rise to nonspecific adsorption. If the spacer is hydrophobic, nonspecific binding will be increased at high salt concentrations.

primary amines or proteins (Wilchek *et al.*, 1975). Kowal and Parsons (1980) have shown that treatment of the affinity matrix with the protein cross-linker glutaraldehyde (≈0.05%) greatly decreases the rate of leakage. It would appear prudent to store the columns at 4°C, at neutral or slightly acidic pH. There may be theoretical grounds for avoiding the nucleophiles Tris and azide, although most workers have not found them to be a problem. A suitable buffer would be phosphate-buffered saline (PBS) containing 0.005% merthiolate.

For most small-scale laboratory work, pre-activated and dried CNBr-activated agarose (Pharmacia or other brands) is still the most popular gel and may be considered the reagent of choice.

Proteins may also be coupled to agarose which has been derivatized by spacer arms with *N*-hydroxysuccinimide ester at their distal ends (Cuatrecasas and Parikh, 1972). Succinimide esters are very susceptible to nucleophilic attack by the ε-amino groups of lysine, resulting in the displacement of *N*-hydroxysuccinimide and the formation of a stable amide bond between the protein and the spacer arm. A slight practical disadvantage of these gels is that the progress and efficiency of coupling of protein cannot be followed by ultraviolet absorption, because *N*-hydroxysuccinimide also absorbs strongly. The chemistry of this very useful reaction is discussed in Section 10.3.

Activated succinimide esters coupled to agarose beads are available commercially as Affigel 10 (Bio-Rad) and Activated CH Sepharose (Pharmacia). Affigel 10 is preferable in most situations, because its hydrophilic spacer arm is attached to the gel via a stable ether bond. The spacer arm of Activated CH Sepharose is hydrophobic, and is coupled to the gel via the less stable isourea bond produced by CNBr activation. Hydrophobic spacer arms may give rise to nonspecific binding (Er-El *et al.*, 1972; O'Carra *et al.*, 1973; Lowe, 1979).

Several other activation procedures have been described. Bethell *et al.* (1979) have shown that Sepharose CL-6B may be activated with 1,1′-carbonyldiimidazole (Pierce), producing an imidazolyl carbamate derivative. The activation procedure is simple, and the product reacts smoothly with *N*-nucleophiles to form a stable bond without additional charge. The activated beads are available commercially as a stable suspension in dry acetone (Reacti-Gel, Pierce). The stability, lack of spacer arms, and high capacity of Reacti-Gel overcome many of the problems of other matrices.

Nilsson and Mosbach (1980) showed that agarose can be activated with *p*-toluene-sulfonyl chloride (tosyl chloride), to form corresponding esters (tosylates) that have excellent leaving properties in reactions with nucleophiles. The linkage to protein is very stable, but is charged at neutral pH. A closely related procedure uses tresyl chloride (Mosbach and Nilsson, 1981; see also Jennissen, 1995).

Several newer activated affinity matricies are now available. 3M Emphaze has been produced by Pierce Chemicals. It couples to *N*-nucleophiles by reaction with azlactone groups on the beads, resulting in a very stable amide bond.

The linkage is hydrophilic, and the beads are claimed to have very low leakage and nonspecific binding. ECH Sepharose (Pharmacia) has a long hydrophilic spacer arm, and is claimed to permit extremely stable coupling. EAH Sepharose (Pharmacia) is a similar product designed to provide stable coupling via carboxyl groups. The manufacturers' catalogues should be consulted for further information.

11.1.1 Stability of Agarose

Native agarose can tolerate undiluted ethanol, methanol, butanol, acetone and dioxane, 80% (v/v) aqueous pyridine and 50% (v/v) dimethyl formamide (Lowe and Dean, 1974), but is dissolved by dimethyl sulfoxide (DMSO) (Lowe, 1979). Freezing of native agarose results in irreversible changes in structure. However, the activation of agarose by CNBr introduces covalent cross-links between the polysaccharide chains, greatly improving its stability. After activation and coupling, the final product is able to withstand high concentrations of salt, urea, guanidine-hydrochloride, sodium dodecyl sulfate (SDS), deoxycholate and Triton X-100.

Agarose in which the polysaccharide chains have been cross-linked by 2,3 dibromopropanol is available commercially from Pharmacia as Sepharose CL®. It is highly resistant to disruption by heat, extremes of pH, and organic solvents (Lowe, 1979).

11.1.2 Preparation of CNBr-activated Agarose

Agarose beads (Sepharose 4B or Sepharose 4B-CL) must be thoroughly washed with distilled water prior to activation. Washing may be carried out on a large sintered glass funnel. Just prior to removal from the funnel, the gel should be washed with two volumes 2 M $NaHCO_3$–Na_2CO_3 buffer, pH ≈11. The gel is then removed from the funnel, resuspended in an equal volume buffer, and cooled to 4–5°C in an ice bath.

The reaction is initiated by adding CNBr (100 mg/g gel, dissolved in acetonitrile), with constant stirring, but not using a magnetic 'flea' which will tend to shatter the beads (March *et al.*, 1974; Nishikawa and Bailon, 1975). Activation takes 10–15 min, after which the gel is washed with ice-cold distilled water on a sintered glass funnel. The liquid should be collected into a side-arm flask containing ferrous sulfate to inactivate residual CNBr and cyanides. Washing after activation must be performed quickly, as the activated groups are susceptible to hydrolysis, especially at alkaline pH. After the alkaline buffer has been washed out, stability of the gel is enhanced by washing with 1 mM HCl in water. The gel should be used within 10–20 min.

CNBr is extremely toxic. All manipulations should be carried out in a fume hood. CNBr should not be weighed directly on an open balance. An estimated quantity should be transferred into preweighed glass screw-top bottles and the exact amount determined by subtraction.

11.1.3 Activation of Agarose by 1,1′-Carbonyldiimidazole

The procedure described is that given by Bethell *et al.* (1979). Sepharose CL-6B (3 g of moist cake) is washed sequentially with water, dioxane:water, 3:7; dioxane:water, 7:3; and dioxane (20 ml of each), and suspended in 5 ml dioxane. The activating agent, 1,1′-carbonyldiimidazole (120 mg), is added and the suspension shaken at room temperature for 15 min. The activated gel is then washed with dioxane (100 ml), and used immediately, although it is stable in anhydrous dioxane. Just prior to coupling with protein, it is advisable to reverse the wash procedure and transfer the gel back to water.

The commercial product (Reacti-Gel, Pierce) is supplied as a preswollen gel in anhydrous acetone, and should be washed in acetone:water (7:3), acetone:water (3:7), and then water, prior to use. Unlike *N*-hydroxysuccinimide ester-activated supports, which have hydrolysis half-lives measured in minutes (see Section 10.3), gels activated with 1,1′-carbonyldiimidazole require 30 h for complete hydrolysis at pH 8.5.

Coupling of proteins to Reacti-Gel is most efficient at a pH of about 9.5–10, although many proteins can be coupled at pH 8.5 in 0.1 M borate buffer. Imidazole is released during coupling, and because it absorbs light at 280 nm, one cannot use the optical density of the supernatant to follow the coupling of protein to the gel. Coupling can be monitored using the Pierce Coomassie blue protein assay reagent (cat. no. 23200).

11.1.4 Activation of Agarose by *p*-Toluene Sulfonyl Chloride

The procedure is that of Nilsson and Mosbach (1980, 1981). Wet Sepharose CL-6B is transferred into dioxane by washing with 3 × 10 gel volumes of water, water:dioxane (3:1, v/v), water:dioxane (1:3), dioxane, and finally dried dioxane, containing less than 0.01% water. Dry acetone may be used instead of dioxane.

Seven grams of Sepharose are then transferred to a round-bottomed flask containing 1 g *p*-toluene sulfonyl chloride (tosyl chloride) dissolved in 2 ml dried dioxane. Pyridine (1.0 ml) is added dropwise, with stirring. After 1 h reaction at room temperature, the gel is washed twice with 10 volumes dioxane and then gradually transferred back to water by reversing the above procedure.

It is stated that the gels were stored at 4°C in distilled water until used, but no information was given concerning the rate of hydrolysis of the activated gel under these conditions. It would seem prudent to use the gel within an hour or so of activation.

A closely related but possibly more attractive chemistry involves the use of tresyl chloride (Mosbach and Nilsson, 1981).

11.2 Coupling of Antibodies and Other Proteins to Activated Gels

The antibody preparation to be coupled to agarose beads need not be purified to homogeneity. For most purposes, ammonium sulfate precipitation, carried out as described in Section 9.2.1, will be sufficient. However, it is vital to dialyse extensively after ammonium sulfate precipitation, because ammonium ions will inhibit the coupling reaction (see below).

Regardless of the mechanism of activation, the subsequent coupling of proteins occurs by nucleophilic attack on the gel by the ε-amino groups of lysine. The amino groups must be unprotonated; the reaction therefore proceeds most rapidly at slightly alkaline pH. There is also a minor competing hydrolysis of the activated groups on the gel. It is essential to avoid the presence of extraneous nucleophiles. Tris, ammonium and azide ions will inhibit coupling.

The efficiency of coupling is greatest at slightly alkaline pH (7.5–8.5), but conditions which lead to maximal protein binding may lead to inactivation of antibodies (see Section 11.3). Suitable buffers include PBS, 0.1 M sodium borate or 0.1 M sodium bicarbonate. The reaction is usually completed in 1–2 h at room temperature, or overnight at 4°C, although somewhat longer times are recommended for gels activated by 1,1′-carbonyldiimidazole or tosyl chloride.

The protein-binding capacity of the gels varies depending on the degree of activation, the pH, and the individual protein. CNBr-activated Sepharose-4B will bind 10–15 mg protein/ml of wet gel; the total mass of antibody should be chosen to exceed this by a small amount. Typical concentrations of protein in the coupling buffer are 2–20 mg/ml, but the important parameter is the ratio of the total protein mass to the mass of activated gel. The kinetics of binding may be followed by removal of small aliquots from the reaction mixture and measurement of the ultraviolet absorbance at 280 nm after removal of beads by centrifugation.

As soon as the reaction is judged to be complete, it is important to inactivate any remaining activated groups. It is customary to inactivate with 1 M ethanolamine, titrated to pH 8 with HCl, for 1–2 h at room temperature. In some cases, inactivation with ethanolamine may lead to increased nonspecific binding (Heinzel *et al.*, 1976), and glycine may be preferred. Alternatively, the reactive groups may be left to hydrolyse for a few days. It is also advisable to wash the gel with several cycles of 0.5 M NaCl, 0.1 M acetate pH *c.*4, followed by 0.5 M NaCl, 0.1 M $NaHCO_3$, pH *c.*8.3, to remove any loosely adsorbed protein.

11.2.1 Coupling of Proteins to CNBr-activated Sepharose 4B

Lyophilized cyanogen bromide-activated Sepharose 4B is available from Pharmacia. Provided it is kept dry, it will last for years. If it is stored refrigerated be sure to allow the vial to warm to room temperature for at least 30 min before opening it, to avoid condensation and consequent deterioration of the gel. The following is a practical procedure for its use.

(1) Calculate the amount of gel needed. Assume that 0.3 g of dry gel will result in 1 ml wet gel, which will in turn bind 10 mg protein.
(2) Weigh out desired amount of dry gel.
(3) Rehydrate gel in 1 mM HCl in a beaker at room temperature for 15 min. Do not use a magnetic stirrer, as it will fragment the beads. Occasionally swirl to mix.
(4) Pour gel slurry into a coarse sintered glass funnel (Schott) attached to a side-arm flask and a water vacuum pump. (It is convenient to have a 'vacuum bypass' hole in the rubber stopper of the flask, to control the vacuum by finger pressure.)
(5) Wash the gel with *c.* 50 volumes 1mM HCl, without allowing it to dry out.
(6) Scoop the gel into a tube containing the protein in an appropriate buffer (typically of 0.1 M $NaHCO_3$).
(7) Immediately cap the tube, and place on a vertical rotating wheel, to give end-over-end-mixing.
(8) Allow coupling to proceed for 1–2 h at room temperature. The progress of coupling may be followed by removing aliquots of the supernatant and measuring the optical density at 280 nm.
(9) As soon as coupling is about 90% complete, quench the remaining active sites on the beads by adding one-tenth volume 0.1 M glycine hydrochloride or 0.1 M Tris-HCl, pH 8, and hold at room temperature for a further 1–2 h.
(10) Spin out the beads (400g, 5 min).
(11) Wash the beads, alternating three times between 0.1 M sodium acetate pH 4.0 plus 0.5 M NaCl, and 0.1 M Tris-HCl, pH 8.0, plus 0.5 M NaCl.
(12) Store the beads in PBS plus 0.005% merthiolate.

11.2.2 Coupling of Antibodies to Affigel 10

The succinimide ester active group of Affigel 10 (Bio-Rad) is very rapidly hydrolysed in water, with a half-life of minutes. It is therefore essential to add the protein to the gel without delay. A borate or bicarbonate buffer should be used (pH about 8), and nucleophiles such as Tris, azide, thiols or amines will inhibit coupling. The progress of coupling cannot be followed by measuring the absorbance of the supernatant, because *N*-hydroxy succinimide is released during coupling, and absorbs strongly at 280 nm. Coupling efficiency can be assessed by use of a dye-binding protein assay or by measuring the optical density at 280 nm after the supernatant has been dialysed to remove the *N*-hydroxysuccinimide.

11.3 Optimization of Antibody Activity of Immunoadsorbents

Provided that no extraneous inhibitors are present, the coupling of protein to the gels almost always proceeds without difficulty. Coupling efficiencies of 100% are easily achieved, but are undesirable. A distinction must be made between the efficient coupling of protein, which is easy, and the preservation of maximal antibody-binding activity after coupling, which may present problems.

The commercially available gels are extremely highly activated, and a frequent problem is that the bound protein is attached by so many sites that it is no longer capable of biological activity (Cuatrecasas and Anfinsen, 1971). It is often preferable to minimize multipoint coupling by the following strategies:

(1) The gel may be 'offered' a little more protein than it is capable of binding. Any unbound protein can easily be recovered and reused. It is unwise to use a large excess of gel over protein.

(2) The pH of coupling may be as low as 6.5, increasing the extent of protonation of the ε-amino groups of lysine residues. Coupling of many proteins is still efficient at this pH (Cuatrecasas and Anfinsen, 1971), although some monoclonal antibodies with very acidic isoelectric points may not couple. If difficulty is experienced, raise the pH.

(3) Coupling may be terminated by ethanolamine or glycine as soon as the majority of protein is coupled.

(4) The gel may be 'prehydrolysed' to reduce the density of activated groups. While this option is theoretically attractive, it may be difficult to guess the extent of hydrolysis achieved. In the final analysis, each monoclonal antibody may behave differently, and the choice of conditions favouring retention of activity is empirical. None the less, an appreciation of the variables will generally lead to success.

11.3.1 Covalent Binding of Antibodies to Protein A-Sepharose

As discussed in the previous section, the random attachment of antibodies to solid-phase matrices may result in significant loss of binding activity. In contrast, the binding of antibodies via their Fc portions to staphylococcal protein A (see Sections 5.3.5 and 9.3.2) leaves the antigen-combining site in the correct orientation for the binding of antigen. Gersten and Marchalonis (1978) showed that antibodies may be covalently cross-linked to protein A-Sepharose with dimethyl suberimidate, with preservation of antigen-binding activity. Schneider *et al.* (1982) have exploited this approach to produce highly active immunoadsorbents; they used dimethyl pimelimidate, which has a spacer with one extra carbon, and spans 9.2 Å compared with 8.6 Å for dimethyl suberimidate.

Practical procedure

The following procedure is based on that of Schneider *et al.* (1982). Protein A-Sepharose CL-4B (Pharmacia) is mixed with antibody in 0.1 M borate buffer, pH 8.2, for 30 min at room temperature, and excess antibody removed by washing with the same buffer. The gel is then washed with 0.2 M triethanolamine, pH 8.2, and resuspended in 20 volumes of 10–20 mM dimethyl pimelimidate dihydrochloride (Pierce) freshly made up in the same buffer. The mixture is agitated at room temperature for 45 min, and the reaction terminated by centrifugation and resuspension in an equal volume of ethanolamine, pH 8.2, of the same molarity as the dimethyl pimelimidate. After 5 min, the cross-linked beads are washed three times in borate buffer, pH 8.2, containing 0.02% sodium azide.

The concentration of cross-linking agent was not critical, and antibody activity was preserved over a range of 10–100 mM. Maximal antigen-binding capacity of the columns occurred when the protein A column was 50% saturated with antibody (Schneider *et al.*, 1982).

11.4 Use of Antibody Affinity Columns

Individual monoclonal antibodies may have quite different properties, and it is unlikely that the same conditions of binding and elution will apply in all cases. It is important to examine the conditions for binding and elution for each antibody individually. If possible, it would be preferable to test a number of monoclonal antibodies of the same specificity, and choose the antibody with the most desirable characteristics.

Affinity chromatography on antibody columns may be divided into five equally important phases; sample preparation, precycling, binding, washing and elution.

11.4.1 Sample Preparation

The first requirement is that the antigen be soluble and free of aggregated material or debris which would obstruct the column. The solubilization and handling of membrane antigens is discussed in Section 10.7. If there is any doubt concerning solubility or debris, or if the antigen mixture has been frozen, centrifugation at 10 000 ***g*** or even 100 000 ***g*** for 10–15 min is advisable before loading the gel. In general, if cell lysates are to be passed over affinity columns, it is not a good idea to use material that has been frozen and thawed. Freezing sometimes results in severe aggregation of the desired or undesired proteins. The aggregates will tend to get trapped in the column, slowing the flow rate and possibly dissolving during elution, seriously degrading the purification (Turkewitz *et al.*, 1983).

It is often worth considering the possibility of preliminary enrichment of the antigen prior to affinity chromatography. This might be achieved by fractionation on a lectin column in the case of glycoproteins. If the antigen to be purified is a membrane protein, two-phase separation in Triton X-114 will give perhaps 20–50-fold enrichment and will greatly improve the purity of the final product (see Section 10.7.4; see also Symons *et al.*, 1994).

11.4.2 Precycling the Column

Regardless of the type of linkage of antibody to the column, it should be assumed that slight leakage of antibody may occur during long-term storage. If the antibody is of very high affinity, and the antigen present in very small amounts, leakage might result in total failure of binding to the column. If the antigen emerged bound to antibody, it might not be detected; failure to recover the antigen from the column could lead to the erroneous conclusion that the antigen had been irreversibly bound to the column. In addition, traces of antibody in the antigen solution will degrade the overall degree of purification that may be achieved.

It is therefore vital that antibody columns be washed thoroughly immediately prior to use. Precycling of the column before each use with the eluting buffer, followed by extensive washing in the binding buffer, is strongly advisable. This procedure will remove any loosely bound nonspecifically adsorbed material, and any residual bound antigen from previous uses.

11.4.3 Binding of Antigen

It is customary to perform affinity chromatography in small columns. An excellent and inexpensive column is the Bio-Rad Econo-Column (cat. no. 731–1550), which can accommodate bed volumes from 0.2 ml to 10 ml. The use of columns rather than batch procedures allows the antigen-containing mixture to percolate slowly through the matrix, and gives optimal chances for interaction. Flow rates of 1–2 ml/min are typical.

Batch procedures are also feasible, but care must be taken to mix in such a way that all the liquid has a chance to interact with the beads. Mixing with a magnetic 'flea' is likely to fragment the beads, and end-over-end mixing on a rotating wheel for several hours is preferable. On the whole, columns are preferred for preparative affinity chromatography and there are sound reasons why recovery from columns is likely to be higher than from batch procedures (Scopes, 1993).

It is strongly advisable to pass the antigen-containing mixture over a 'pre-column' of agarose (underivatized or coupled with an irrelevant antibody) to remove any molecules that bind nonspecifically. The pre-column will also protect the main column from clogging.

The requirements for the buffer in which binding takes place are simple. The pH and salt concentration must be such that binding is strong, and if membrane proteins are present, the buffer must contain appropriate amounts of a suitable detergent. Typical loading buffers are PBS, pH 7.4, or 0.1 M Tris-HCl, pH 8.0.

The volume of antibody-coupled gel should be appropriate to the anticipated mass of antigen. A common mistake is to make the column much too big. Large columns are seldom necessary. Excessively large columns may produce worse results than columns of appropriate size, because any tendency for the matrix to bind nonspecifically will increase linearly with bed volume, while specific binding will increase only until all the antigen is bound (Fig. 11.1).

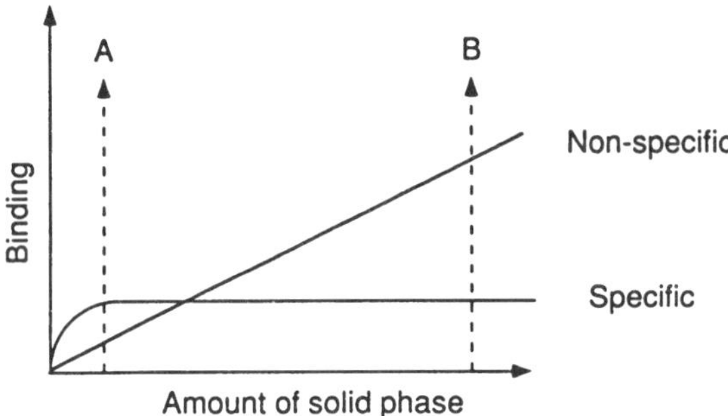

Fig. 11.1. Effect of bed volume of affinity columns on specific and nonspecific binding. If the column is offered a fixed mass of antigen, specific binding will increase with increasing bed volume until all antigen is bound. In contrast, nonspecific binding may increase in proportion to column size. It follows that the best purification factors will be achieved when the column is close to saturation (line A), rather than when the column is very large (line B).

11.4.4 Washing the Immunoadsorbent: The Problem of Nonspecific Binding

After the antigen-containing mixture has been loaded, unbound material must be removed by washing the gel. In many cases, the gel is simply washed with 10–20 column volumes of the same buffer used for loading. An ultraviolet absorbance monitor will be found extremely useful in deciding when adequate washing has been achieved. In a few cases involving low-affinity antibodies, excessive washing may remove some or all of the bound antigen.

It is important to appreciate that some examples of 'nonspecific' binding to affinity columns can have quite well-defined causes. For example, some lectin-like molecules may bind directly to agarose. Many serum proteins (e.g. hormones, low-density lipoprotein and transferrin) bind to cells via specific receptors. Transferrin is a very frequent contaminant in immunoglobulin preparations, and is therefore likely to be present on many affinity columns. The possibility of artefacts due to transferrin receptors has been suggested (Goding and Burns, 1981) and a case has been documented by Harford (1984).

An interesting and unexpected cause of 'nonspecific' binding was recently reported by Osborn *et al.* (1994), who demonstrated the presence of antibodies to galactosyl determinants in a rabbit antiserum against the protozoan parasite *Leishmania major*. These antibodies bound to agarose beads, causing an experimental artefact that was initially puzzling, but was quite easy to understand once the cause was understood. It is not known how often such antibodies are encountered, but the possibility should be borne in mind.

A certain degree of nonspecific binding is inevitable; some is due to properties of the matrix and some to properties of individual molecules in the antigen-containing mixture. Regardless of the location of the problem, nonspecific adsorption may often be reduced by use of modified wash buffers (see also Section 10.8.3).

Nonspecific binding due to electrostatic effects (i.e. interaction of charged groups on proteins with charged groups on the matrix) are most marked when the total salt concentration is low. Under low salt conditions, the gel may act as an ion exchange resin, and nonspecific binding of proteins may occur. These considerations suggest that the salt concentration should be kept high (0.3–0.5 M) to compete out electrostatic binding. Because the net charge of proteins varies with pH, nonspecific binding due to electrostatic effects is often pH sensitive. Accordingly, a trial of buffers of different pH may be worthwhile (Houwen *et al.*, 1975; Zoller and Matzku, 1976; Smith *et al.*, 1978; Kessler, 1981; Schneider *et al.*, 1982).

Another cause of nonspecific binding is the presence of hydrophobic interactions. These are usually fairly minimal when the very hydrophilic agarose beads are used, but may be greatly aggravated by the presence of hydrophobic spacer arms (Er-El *et al.*, 1972). Hydrophobic effects are increased at high salt concentrations (>1 M NaCl). In some cases, non-specific binding caused by hydrophobic interactions may be reduced by the presence of detergents (0.5% Triton X-100, 0.5% deoxycholate or 0.1% Tween 20; Smith *et al.*, 1978).

The extent to which increasingly harsh wash buffers may be used will depend on the effect that these conditions have on the conformation of individual antigens and antibodies. In many cases, buffers containing 0.5 M NaCl and/or 0.5% deoxycholate will have no effect on antibody–antigen interactions (Parham, 1979; Goding and Herzenberg, 1980; Schneider *et al.*, 1982).

In other cases, these relatively mild conditions may cause complete disruption of binding. Monoclonal anti-H-2K^k antibody 11–4.1 (Oi *et al.*, 1978) binds tightly to its antigen in 15 mM phosphate buffer, pH 7, containing 0.5% Nonidet P40, but the interaction is weakened by addition of 0.15 M NaCl, and almost totally disrupted by 0.5 M NaCl or by 0.5% deoxycholate (Herrmann and Mescher, 1979). Similar results were obtained with a second anti-H-2 antibody (Stallcup *et al.*, 1981). The disruption of binding by deoxycholate appears to be caused by a reversible conformational change in the antigen, because deoxycholate-treated H-2K^k molecules are more susceptible to digestion by trypsin and chymotrypsin (Herrmann *et al.*, 1982).

11.5 Elution of Antigen from Immunoadsorbents

The behaviour of polyclonal antibody affinity columns is quite different to that of monoclonal antibody columns. In the case of polyclonal antibodies, the behaviour of the column is likely to be dominated by the highest affinity subset of antibodies, because this subset must be disrupted before antigen can be released. Elution curves from polyclonal immunoadsorbents often have sharp leading edges and long trailing edges, indicating that some high-affinity interactions cannot be fully disrupted. The situation with monoclonal antibodies is completely different. The homogeneous nature of the interactions means that once appropriate elution conditions are found, peaks should emerge with minimal trailing. Elution conditions for monoclonal antibodies may sometimes be much milder, and recoveries greater.

In the case of monoclonal antibody 11–4.1, a very minor change in conditions was sufficient for elution of antigen (see Section 11.4.4). In the majority of cases, however, it may be expected that much harsher conditions will be required. None the less, it is strongly advisable to try a variety of elution conditions for each new monoclonal antibody. The most gentle conditions which are effective for elution should be chosen; this will maximize the chances of recovering biological activity and may also improve purity.

In a few special cases, it may be possible to elute biospecifically. Antibodies to haptens or peptides may often be eluted with a high concentration of hapten or peptide, provided the affinity is not excessive. Some enzymes undergo conformational changes in response to the binding of cofactors and it may occasionally be possible to exploit this property to disrupt antibody binding.

It is often assumed that the highest affinity interactions are the ones that require the strongest denaturing conditions to disrupt. There is no compelling theoretical reason why this should always be the case, and there is evidence that affinity and ease of disruption are not necessarily related (Parham, 1983).

In the great majority of cases, biospecific elution will not be feasible, and elution will have to depend on the induction of a conformational change in antigen, antibody or both by alteration of pH, dissociating agents such as urea or guanidine, or chaotropic ions. Antibodies will usually renature after removal of deforming agents (see Section 9.7.1). Table 11.2 lists some common ways in which elution may be performed.

Traditionally, the most popular method of elution of antibodies or antigens from immune complexes has involved the use of glycine-hydrochloride buffers at pH 2.2–2.8. Most antibodies will release their antigen under these conditions. In more recent years, elution at pH 11.5 has become popular and often gives excellent results. A variety of other elution condtions may be considered (Table 11.2).

If elution at extremes of pH is not successful in releasing the antigen, or if the antigen or antibody is damaged by these conditions, it is worthwhile trying chaotropic ions. The potency of these ions in disrupting antigen–antibody complexes approximately parallels the Hofmeister series:

$$SCN^- > I^- > ClO_4^- > NO_3^- > Br^- > Cl^- > CH_3COO^- > SO_4^{2-} > PO_4^{3-}$$
$$Ba^{2+} > Ca^{2+} > Mg^{2+} > Li^+ > Na^+ > K^+ > Cs^+ > NH_4^+$$

A trial of chaotropic ions should probably commence with NaSCN (3.5 M). Note, however, that thiocyanate is an extremely potent inhibitor of radio-iodination (George and Schenck, 1983). Samples must be dialysed with many changes of buffer for several days before iodination is attempted. $MgCl_2$ (2–4 M) has been used effectively and is sometimes claimed to be less likely to cause irreversible damage. However, it is not entirely clear whether a lower concentration of the more potent thiocyanate ion would be equivalent.

A third type of eluant includes urea and guanidine-hydrochloride, which are potent denaturants when used at high concentration. Typically, one might elute with 5–8 M urea or 4–6 M guanidine-hydrochloride. Elution with guanidine is extremely effective, but may result in severe damage to the antibodies or antigen. It should probably be reserved for situations in which all other eluants fail.

Hydrophobic interactions are often important in the binding of antigen to antibody. As mentioned earlier, the slight detergent action of 1M propionic acid may result in more effective elution than glycine-hydrochloride at the same pH. Elution with propionic acid is relatively harsh and may result in aggregation and irreversible damage to the antigen. Hydrophobic interactions may also be weakened by the use of polarity-lowering agents such as dioxane (up to 10%) or ethylene glycol (up to 50%) in conjunction with extremes of pH (Table 11.2).

Tsang and Wilkins (1991) examined 13 different buffers for the elution of goat antibodies to human IgG that had been coupled to Affi-prep 10 beads (Bio-Rad). The best recovery occurred with a buffer consisting of 3.0 M $MgCl_2.6H_2O$, 0.075 M Hepes/NaOH, with 25% ethylene glycol pH 7.2.

Table 11.2 Elution of antibodies from immunoadsorbents

Elution conditions	References	Comments
Diethylamine, 0.05 M, pH 11.5	3	Compatible with deoxycholate; often the method of choice for membrane antigens
Glycine-hydrochloride, pH 2.2–2.8	1	Incompatible with deoxycholate. Poor recoveries and aggregation of some membrane antigens.
Sodium thiocyanate, 3.5 M	8	Traces will inhibit iodination and coupling with fluorescein; remove by extensive dialysis.
Guanidine-hydrochloride, 6 M	7	Very strongly denaturing; efficient eluant but may cause severe aggregation.
Urea, 2–8 M, pH 7	5	Heating in urea causes carbamylation of proteins.[6]
Magnesium chloride 2–5 M	9	Incompatible with deoxycholate.
3.0 M $MgCl_2.6H_2O$, 0.075 M Hepes/NaOH, with 25% ethylene glycol pH 7.2	13	
Propionic acid, 1 M	2	More effective than HCl at same pH, possibly owing to slight detergent action. May cause denaturation and irreversible aggregation.
Potassium or sodium iodide, 2.5–5M, pH 7.5–9.0	10	
1 M Ammonia, pH 11.5	4, 13	
Ethylene glycol (50%, v/v), pH 11.5	11	Polarity-reducing agent; disrupts hydrophobic interactions.
Dioxane (10%, v/v), at acid pH	11	Polarity-reducing agent; disrupts hydrophobic interactions.
Electrophoresis, isoelectric focusing	12	Avoids denaturing conditions; somewhat slow and cumbersome.

1. Kleinschmidt and Boyer (1952); numerous other papers.
2. Joniau *et al.* (1970); Johnson and Garvey (1977).
3. Read *et al.* (1974); Letarte-Muirhead *et al.* (1975); Cresswell (1977); Brodsky *et al.* (1979); McMaster and Williams (1979); Parham (1979); Sunderland *et al.* (1979); Schneider *et al.* (1982).
4. Chidlow *et al.* (1974).
5. Melchers and Messer (1970); Pikho *et al.* (1973); Stenman *et al.* (1981).
6. Stark *et al.* (1960); Tollaksen *et al.* (1981).
7. Dandliker *et al.* (1968); Weintraub (1970); O'Sullivan *et al.* (1979).
8. Dankliker *et al.* (1968); Zoller and Matzku (1976); George and Schenck (1983).
9. Avrameas and Ternynck (1969); Mains and Eipper (1976).
10. Avrameas and Ternynck, (1967); Lecomte and Tyrrell (1976).
11. Hill (1972); Andersson *et al.* (1978), (1979).
12. Brown *et al.* (1977); Morgan *et al.* (1978); Haff *et al.* (1979); Haff (1981).
13. Tsang and Wilkins (1991).

When starting a new experiment, a trial of several elution buffers is advisable. I suggest starting at the top of the Table 11.2 and working downwards.

It is important to remember that the combination of reducing agents with denaturing buffers is likely to cause dissociation of antibody light and heavy chains, and thus destruction of the affinity column. It should also be noted that IgA from BALB/c mice does not possess a disulfide bond between light and heavy chains, and that affinity matrices based on BALB/c IgA may not survive the first elution.

In a few cases, binding may be so tight that the necessary elution conditions may destroy the antigen or antibody. The use of monovalent Fab fragments of antibody (Section 9.5.1) may lower the effective affinity ('avidity') of binding (Fig. 2.4), and facilitate elution in these cases, particularly if the antigen possesses more than one identical antigenic determinant per molecule. The use of a lower degree of substitution of the matrix by antibody may also help.

11.5.1 Gradient Elution

Elution of antigen or antibodies by continuous gradients of denaturing agents has rarely been attempted. The wide range of affinities of polyclonal antibodies would result in an extremely broad elution profile. However, the homogeneity of monoclonal antibodies might suggest a reconsideration of gradient elution (Stenman *et al.*, 1981). The use of continuous gradients of pH or chaotropic ions might allow even better separations than can be achieved by step gradients. Gradient elution is widely used in other areas of affinity chromatography (Lowe and Dean, 1974; Lowe, 1979). Tsang and Wilkins (1991) used gradients to elute polyclonal goat antihuman IgG antibodies from a human IgG-Affi-prep 10 column. They found that broad elution peaks were obtained with linear gradients. Sharp peaks with good recoveries were only found using 'pulse-pause' step gradients in which the full-strength denaturing conditions were applied.

11.5.2 Affinity Chromatography of Membrane Antigens

The special problems of solubilization and handling membrane antigens have been discussed in Section 10.7. It is essential that all components in the antigen-containing mixture be adequately solubilized, because precipitation on the column will seriously degrade the separation and may ruin the column.

The extremely low abundance of membrane proteins puts large demands on the separation system, and purification to homogeneity is not always possible. As discussed in Section 11.2.1, preliminary enrichment of membrane antigens using Triton X-114 may be very helpful, and may substantially improve the purity of the final product (see Section 10.7.4 and also Symons *et al.*, 1994).

The need to maintain solubility means that it is obligatory to use adequate concentrations of a suitable detergent at all stages, including loading, washing and elution. Nonionic detergents may be used under a wide range of salt concentrations and pH, but deoxycholate will gel or precipitate at a pH below approximately 7.4, and in the presence of divalent cations. The properties of deoxycholate micelles are also subject to large changes depending on the ionic environment (Helenius and Simons, 1975; Helenius *et al.*, 1979). Triton X-100 and Nonidet P-40 absorb light very strongly at 280 nm, while deoxycholate and octyl glucoside do not. It is possible to exchange the detergent by washing the column with buffer containing the new detergent prior to elution, but care must be taken that the antigen remains bound, and is soluble in the new detergent.

Elution at alkaline pH has been used successfully in a number of instances, particularly for cell membrane antigens (Sunderland *et al.*, 1979; Parham, 1979; Parham *et al.*, 1979; Schneider *et al.*, 1982; Stearne *et al.*, 1985). The choice of alkaline elution conditions for membrane proteins stems partly from compatibility with deoxycholate, and partly from the empirical observation that membrane proteins appear to be less susceptible to aggregation at alkaline pH (Letarte-Muirhead *et al.*, 1975; Cresswell, 1977; Brodsky *et al.*, 1979; McMaster and Williams, 1979; Parham, 1979; Parham *et al.*, 1979; Sunderland *et al.*, 1979).

A practical example of the purification of a membrane protein by affinity chromatography is given by Stearne *et al.* (1985). Batches of cells were solubilized in 2% Triton X-100 in PBS (50 ml per 10 ml packed cells) for 10 min at 4°C, and debris removed by centrifugation at 800 ***g*** for 10 min. The supernatant was then centrifuged at 100 000 ***g*** for 60 min in a Beckman SW27 rotor, and passed over a column of transferrin-Sepharose, followed by a column of monoclonal anti-PC-1-Sepharose. The bed volume of the columns was 1.0 ml, and the flow rate 0.7 ml/min. The column was then washed with 20 ml 50 mM Tris-HCl, pH 8.0, containing 0.5 M NaCl with 0.5% Triton X-100. The PC-1 antigen was eluted with 50 mM diethylamine, pH 11.5, containing 0.05% Triton X-100. Preliminary experiments showed that nearly all the antigen was eluted in the second and third 1-ml fractions after application of the elution buffer. The sample was immediately neutralized by adding 200 μl of 1M Tris-HCl, pH 8.0, and lyophilized. A purification of *c.* 20 000-fold was obtained, and the antigen was found to be about 50% pure. Purification to homogeneity was achieved by preparative SDS–PAGE (see Chapter 15 for details).

11.5.3 Removal of the Elution Agent

There is evidence that some proteins subjected to denaturing environments may undergo conformational changes that continue for several hours

(Tanford, 1969, 1970; Shimizu *et al.*, 1974; Porath and Kristiansen, 1975). The longer that a protein is subjected to these environments, the greater the risk of difficulties in renaturation. The risk of irreversible damage is probably minimized by keeping the sample cold until the denaturing environment is completely removed.

If the protein is eluted by extremes of pH, it is best to neutralize as soon as possible after elution. Rapid removal of the denaturing agent may also be used for other eluants, including thiocyanate, urea and guanidine, although in these cases it is sometimes found that renaturation is more complete if the denaturing agent is removed very slowly over hours to days by dialysis against a series of solutions of decreasing concentration. In general, rapid restoration of physiological conditions is best (Hager and Burgess, 1980).

It is therefore common practice to remove the denaturing agent as soon as possible after the antigen is eluted. If elution has been by pH change, the antigen may be neutralized rapidly by collection into a tube containing 1 M Tris-HCl, pH 7.5–8.0. *Neutralization should never be carried out by adding Tris base, NaOH or strong acid directly to the antigen.* Chaotropic ions or urea may be diluted out. The eluted antigen may then be transferred into a more physiological environment by dialysis.

Alternatively, the eluted protein may be rapidly transferred into a more physiological buffer by gel filtration in Sephadex G-25. The pre-packed and precalibrated disposable PD-10 columns (Pharmacia) are ideal. The column is washed with 20–30 ml of the new buffer, and allowed to run dry. The protein (maximum volume 2.5 ml) is added to the top of the column, and allowed to run in. If the volume of the sample is less than 2.5 ml, buffer is then added to a final volume of 2.5 ml, and the effluent discarded. Then, 3.5 ml of buffer are added to the top of the column, allowed to run in, and the effluent collected until the top of the column is dry (3.5 ml effluent). The sample will be diluted by a factor of about 1.3, but this is usually unimportant.

It is not uncommon for some aggregation and precipitation to occur during the removal of the denaturing agent. This is more common when potent denaturants such as urea or guanidine-hydrochloride are used, and may sometimes be minimized by renaturing the protein very slowly by stepwise dialysis in decreasing denaturant concentrations over many hours. The only other solution is to try a different or milder elution procedure, or to centrifuge and discard the aggregates.

11.6 Storage of Affinity Columns

Affinity columns will last virtually indefinitely, provided they are protected from drying, microbial contamination, proteolysis, chemical attack and accumulation of extraneous insoluble material. These requirements are not difficult to meet. Columns should be stored in a neutral or slightly acidic buffer. PBS, pH 7.2–7.4, containing 0.005% merthiolate is suitable. Some authors prefer borate buffers because they are less likely to support microbial growth. Columns must be tightly sealed to prevent them from drying out. They should be stored at 4°C, but must not be frozen, because freezing causes irreversible disruption to the agarose.

11.7 Preparation of Protein for Amino Acid Sequencing

There are now many ways that antibodies can be used in gene cloning (Section 10.13). A common approach for low-concentration proteins involves the screening of libraries with oligonucleotides based on amino acid sequence information (see Section 10.13.4). Affinity chromatography is a very common method for purification of low-concentration proteins, and the eluted protein is well suited for amino acid sequencing. If the purity of the eluted protein is insufficient for sequencing, preparative SDS–PAGE is commonly used for the final purification (see Section 15.2).

The purification strategy requires a good deal of care to ensure that the protein is sequenceable. The first requirement is purity of the protein. Suffice it to say that 10% contamination of an M_r 100 000 protein with an M_r 10 000 peptide represents an equal molar concentration of the impurity, and sequencing would be impossible. In general, the purer the protein, the less difficulty will be found in obtaining an unambiguous sequence.

The second requirement concerns chemical purity. There are many potential contaminants that may interfere with sequencing. Especially important are aldehydes, which are common contaminants in many organic solvents, and nonvolatile amines. The sample should be in a buffer which is completely volatile and contains no nonvolatile salts. Suitable solutions include water, aqueous acetic acid, trifluoroacetic acid, ammonia, ammonium bicarbonate and ammonium acetate. The presence of small amounts of SDS are not detrimental, and may help maintain solubility and prevent losses from adsorption.

If the protein has been purified by preparative SDS–PAGE in the final step (see Section 15.2), contamination by high molecular weight acrylamide polymers is virtually certain. These cause massive artefacts in sequencing but may be removed by precipitating the protein with 90% methanol (Stearne *et al.*, 1985).

The amino-terminus of many proteins is blocked to Edman degradation. If it is desired to obtain protein sequence as a first step towards gene cloning, the most popular strategy is to digest the protein with trypsin, either in the gel or on a PVDF membrane, and to fractionate the tryptic peptides by reverse-phase HPLC (see Section 10.13.4). This strategy gives excellent recoveries and provides multiple chances of finding a region that is minimally degenerate and suitable for making oligonucleotides for probes or PCR primers.

The availability of multiple peptide sequences is also a great help in checking the validity of DNA sequences, because it allows checking for frameshift errors occurring during the reading of the DNA sequence. Such errors can be difficult to detect in the absence of a protein sequence.

References

Andersson, K., Benyamin, Y., Douzou, P. and Balny, C. (1978) Organic solvents and temperature effects on desorption from immunoadsorbents. DNP-BSA anti-DNP as a model. *J. Immunol. Methods* **23**, 17–21.

Andersson, K. K., Benyamin, Y., Douzou, P. and Balny, C. (1979) The effects of organic solvents and temperature on the desorption of yeast 3-phosphoglycerate kinase from immunoadsorbent. *J. Immunol. Methods* **25**, 375–381.

Avrameas, S. and Ternynck, T. (1967) Use of iodide salts in the isolation of antibodies and the dissolution of specific immune precipitates. *Biochem. J.* **102**, 37c–39c.

Avrameas, S. and Ternynck, T. (1969) The cross-linking of proteins with glutaraldehyde and its use for the preparation of immunoadsorbents. *Immunochemistry* **6**, 53–66.

Axèn, R., Porath, J. and Ernback. S. (1967) Chemical coupling of peptides and proteins to polysaccharides by means of cyanogen halides. *Nature* **214**, 1302–1304.

Bethell, G. S., Ayers, J. S., Hancock, W. S. and Hearn, M. T. W. (1979) A novel method of activation of cross-linked agaroses with 1,1′-carbonyldiimidazole which gives a matrix for affinity chromatography devoid of additional charged groups. *J. Biol. Chem.* **254**, 2572–2574.

Brodsky, F. M., Parham, P., Barnstable, C. J., Crumpton, M. J. and Bodmer, W. F. (1979) Monoclonal antibodies for analysis of the HLA system. *Immunol. Rev.* **47**, 3–61.

Brown, P. J., Leyland, M. J., Keenan, J. P. and Dean, P. D. G. (1977) Preparative electrophoretic desorption in the purification of human serum ferritin by immunoadsorption. *FEBS Lett.* **83**, 256–259.

Campbell, D. H., Leuscher, E. and Lerman, L. S. (1951) Immunologic adsorbents. I. Isolation of antibody by means of a cellulose-protein antigen. *Proc. Natl Acad. Sci. USA* **37**, 575–578.

Chidlow, J. W., Bourne, A. J. and Bailey, A. J. (1974) Production of hyperimmune serum against collagen and its use for the isolation of specific collagen peptides on immunosorbent columns. *FEBS Lett.* **41**, 248–252.

Cresswell, P. (1977) Human B cell alloantigens: separation from other membrane molecules by affinity chromatography. *Eur. J. Immunol.* **7**, 636–639.

Cuatrecasas, P. and Anfinsen, C. B. (1971) Affinity Chromatography. *Methods Enzymol.* **22**, 345–378.

Cuatrecasas, P. and Parikh, I. (1972) Adsorbents for affinity chromatography. Use of *N*-hydroxysuccinimide esters of agarose. *Biochemistry* **11**, 2291–2299.

Dalchau, R. and Fabre, J. W. (1982) The purification of antigens and other studies with monoclonal antibody affinity columns: the complementary new dimension of monoclonal antibodies. *In* 'Monoclonal Antibodies in Clinical Medicine' (A. J. McMichael and J. W. Fabre, eds), pp. 519–556. Academic Press, London.

Dandliker, W. B., de Saussure, V. A. and Levandoski, N. (1968) Antibody purification at neutral pH utilizing immunospecific adsorbents. *Immunochemistry* **5**, 357–365.

Dean, P. D. G., Johnson, W. S. and Middle, F. A. (1985) 'Affinity Chromatography: A Practical Approach'. IRL Press, Oxford.

Er-El, Z., Zaidenzaig, Y. and Shaltiel, S. (1972) Hydrocarbon-coated Sepharoses. Use in the purification of glycogen phosphorylase. *Biochem. Biophys. Res. Commun.* **49**, 383–390.

Ernst-Cabrera, K. and Wilchek, M. (1986) Silica containing primary hydroxyl groups for high-performance affinity chromatography. *Anal. Biochem.* **159**, 267–272.

George, S. and Schenck, J. R. (1983) Thiocyanate inhibition of protein iodination by the chloramine-T method and a rapid method for measurement of low levels of thiocyanate. *Anal. Biochem.* **130**, 416–419.

Gersten, D. M. and Marchalonis, J. J. (1978) A rapid, novel method for the solid phase derivatization of IgG antibodies for immune-affinity chromatography. *J. Immunol. Methods* **24**, 305–309.

Goding, J. W. and Burns, G. F. (1981) Monoclonal antibody OKT-9 recognizes the receptor for transferrin on human acute lymphocytic leukemia cells. *J. Immunol.* **127**, 1256–1258.

Goding, J. W. and Herzenberg, L. A. (1980) Biosynthesis of lymphocyte surface IgD in the mouse. *J. Immunol.* **124**, 2540–2547.

Haff, L. A. (1981) An investigation into the mechanism of electrophoretic desorption of immunoglobulin G from protein A-Sepharose. *Electrophoresis* 2, 287–290.

Haff, L. A., Lasky, M. and Manrique, A. (1979) A new technique for desorbing substances tightly bound to affinity gels: flat bed electrophoretic desorption in Sephadex via isoelectric focusing (FEDS-IEF). *J. Biochem. Biophys. Methods* **1**, 275–286.

Hager, D. A. and Burgess, R. R. (1980) Elution of proteins from sodium dodecyl sulfate-polyacrylamide gels, removal of sodium dodecyl sulfate, and renaturation of enzymatic activity: results with sigma subunit of Escherichia coli RNA polymerase, wheat germ topoisomerase, and other enzymes. *Anal. Biochem.* **109**, 76–86.

Harford, J. (1984) An artefact explains the apparent association of the transferrin receptor with a *ras* gene product. *Nature* **311**, 673–675.

Harris, E. L. V. and Angal, S. (1989) 'Protein Purification Methods: A Practical Approach'. IRL Press, Oxford.

Heinzel, W., Rahimi-Laridjani, I. and Grimminger, H. (1976) Immunoadsorbents: nonspecific binding of proteins to albumin-Sepharose. *J. Immunol. Methods* **9**, 337–344.

Helenius, A. and Simons, K. (1975) Solubilization of membranes by detergents. *Biochem. Biophys. Acta* **415**, 29–79.

Helenius, A., McCaslin, D. R., Fries, E. and Tanford, C. (1979) Properties of detergents. *Methods Enzymol.* **56**, 734–749.

Herrmann, S. H. and Mescher, M. F. (1979) Purification of the H-2K^k molecule of the murine major histocompatibility complex. *J. Biol. Chem.* **254**, 8713–8716.

Herrmann, S. H., Chow, C. M. and Mescher, M. F. (1982) Proteolytic modifications of the carboxy-terminal region of H-2K^k. *J. Biol. Chem.* **257**, 14181–14186.

Hermanson, G. T., Mallia, A. K. and Smith, P. K. (1992) 'Immobilized Affinity Ligand Techniques'. Academic Press, London.

Hill, R. J. (1972) Elution of antibodies from immunoadsorbents: effect of dioxane in promoting release of antibody. *J. Immunol. Methods* **1**, 231–245.

Houwen, B., Goudeau, A. and Dankert, J. (1975) Isolation of hepatitis B surface antigen (HBsAg) by affinity chromatography on antibody-coated immunoadsorbents. *J. Immunol. Methods* **8**, 185–194.

Jakoby, W. (ed.) (1984) 'Enzyme Purification and Related Techniques. Methods in Enzymology'. Vol. 104C, Academic Press, New York.

Jennissen, H. P. (1995) Cyanogen bromide and tresyl chloride chemistry revisited: The special reactivity of agarose as a chromatographic and biomaterial support for immobilizing novel chemical groups. *J. Mol. Recog.* **8**, 116–124.

Johnson, G. and Garvey, J. S. (1977) Improved methods for separation and purification by affinity chromatography. *J. Immunol. Methods* **15**, 29–37.

Joniau, M., Grossberg, A. L. and Pressman, D. (1970) Arginyl residues in the active sites of antibody against the 3-nitro-4-hydroxy-5-iodophenyl acetyl (NIP) group. *Immunochemistry* **7**, 755–769.

Kessler, S. W. (1981) Use of protein A-bearing staphylococci for immunoprecipitation and isolation of antigens from cells. *Methods Enzymol.* **73**, 442–459.

Kleinschmidt, W. J. and Boyer, P. D. (1952) Interaction of protein antigens and antibodies. I. Inhibition studies with the egg albumin-anti-egg albumin-system. *J. Immunol.* **69**, 247–255.

Kohn, J. and Wilchek, M. (1981) Procedures for the analysis of cyanogen bromide-activated Sepharose or Sephadex by quantitative determination of cyanate esters and imidocarbonates. *Anal. Biochem.* **115**, 375–382.

Kowal, R. and Parsons, R. G. (1980) Stabilization of proteins immobilized on Sepharose from leakage by glutaraldehyde cross linking. *Anal. Biochem.* **102**, 72–76.

Lecomte, J. and Tyrrell, D. A. J. (1976) Isolation of antihaemagglutinin antibodies with an influenza A virus immunoadsorbent. *J. Immunol. Methods* **13**, 355–365.

Letarte-Muirhead, M., Barclay, A. N. and Williams, A. F. (1975) Purification of the Thy-1 molecule, a major cell-surface glycoprotein of rat thymocytes. *Biochem. J.* **151**, 685–697.

Lowe, C. R. (1979) 'An Introduction to Affinity Chromatography'. North-Holland, Amsterdam and New York.

Lowe, C. R. and Dean, P. D. G. (1974) 'Affinity Chromatography'. John Wiley, London.

March, S. C., Parikh, I. and Cuatrecasas, P. (1974) A simplified method for cyanogen bromide activation of agarose for affinity chromatography. *Anal. Biochem.* **60**, 149–152.

McMaster, W. R. and Williams, A. F. (1979) Identification of Ia glycoproteins in rat thymus and purification from rat spleen. *Eur. J. Immunol.* **9**, 426–433.

Melchers, F. and Messer, W. (1970) The activation of mutant β-galactosidase by specific antibodies. Purification of eleven antibody activatable mutant proteins and their subunits on Sepharose immunosorbents. Determination of the molecular weights by sedimentation analysis and acrylamide gel electrophoresis. *Eur. J. Biochem.* **17**, 267–272.

Mescher, M. F., Stallcup, K. C., Sullivan, C. P., Turkewitz, A. P. and Herrmann, S. H. (1983) Purification of murine MHC antigens by monoclonal antibody affinity chromatography. *Methods Enzymol.* **92**, 86–109.

Morgan, M. R. A., Johnson, P. M. and Dean, P. D. G. (1978) Electrophoretic desorption of immunoglobulins from immobilised protein A and other ligands. *J. Immunol. Methods* **23**, 381–387.

Mosbach, K. and Nilsson, K. (1981) Immobilization of enzymes and affinity ligands to various hydroxyl group carrying supports using highly reactive sulfonyl chlorides. *Biochem. Biophys. Res. Commun.* **102**, 449–457.

Nilsson, K. and Mosbach, K. (1980) *p*-Toluenesulfonyl chloride as an activating agent of agarose for the preparation of immobilized affinity ligands and proteins. *Eur. J. Biochem.* **112**, 397–402.

Nilsson, K. and Mosbach, K. (1981) Immobilisation of enzymes and affinity ligands to various hydroxyl group carrying supports using highly reactive sulfonyl chlorides. *Biochem. Biophys. Res. Commun.* **102**, 449–457.

Nishikawa, A. H. and Bailon, P. (1975) Affinity purification methods. Improved procedures for cyanogen bromide reaction on agarose. *Anal. Biochem.* **64**, 268–275.

O'Carra, P., Barry, S. and Griffin, T. (1973) Spacer arms in affinity chromatography: the need for a more rigorous approach. *Biochem. Soc. Trans.* **1**, 289–290.

O'Sullivan, M. J., Gnemmi, E., Chieregatti, G., Morris, D., Simmonds, A. D., Simmons, S., Bridges, J. W. and Marks, V. (1979) The influence of antigen properties on the conditions required to elute antibodies from immunoadsorbents. *J. Immunol. Methods* **30**, 127–137.

Oi, V. T., Jones, P. P., Goding, J. W., Herzenberg, L. A. and Herzenberg, L. A. (1978) Properties of monoclonal antibodies to mouse Ig allotypes, H-2 and Ia antigens. *Curr. Top. Microbiol. Immunol.* **81**, 115–129.

Osborn, A. H., Kelleher, M., Handman, E. and Goding, J. W. (1994) Anti-galactosyl antibodies that react with unmodified agarose: a potential source of artifacts in immunoaffinity chromatography. *Anal. Biochem.* **217**, 181–184.

Parham, P. (1979) Purification of immunologically active HLA-A and -B antigens by a series of monoclonal antibody columns. *J. Biol. Chem.* **254**, 8709–8712.

Parham, P. (1983) Monoclonal antibodies against HLA products and their use in immunoaffinity purification. *Methods Enzymol.* **92**, 110–138.

Parham, P., Barnstable, C. J. and Bodmer, W. F. (1979) Use of a monoclonal antibody (W6/32) in structural studies of HLA-A,B,C antigens. *J. Immunol.* **123**, 342–349.

Parikh, I. and Cuatrecasas, P. (1975) Affinity chromatography in immunology. *Methods Prot. Sep.* **1**, 1–44.

Pihko, H., Lindgren, J. and Ruoslahti, E. (1973) Rabbit α-fetoprotein: Immunochemical purification and partial characterization. *Immunochemistry* **10**, 381–385.

Porath, J. and Kristiansen, T. (1975) Biospecific affinity chromatography and related methods. *In* 'The Proteins' (H. Neurath and R. L. Hill, eds), Vol. 1, pp. 95–178. Academic Press, London.

Read, R. J. D., Cox, J. C., Ward, H. A. and Nairn, R. C. (1974) Conditions for purification of anti-*Brucella* antibodies by immunoadsorption and elution. *Immunochemistry* **11**, 819–822.

Ruoslahti, E. (1976) 'Immunoadsorbents in Protein Purification'. *Scand. J. Immunol.* Suppl. 3.

Schneider, C., Newman, R. A., Sutherland, D. R., Asser, U. and Greaves, M. F. (1982) A one-step purification of membrane proteins using a high-efficiency immunomatrix. *J. Biol. Chem.* **257**, 10766–10769.

Scopes, R. L. (1993) 'Protein Purification. Principles and Practice'. 3rd edn. Springer Verlag, Berlin.

Secher, D. S. and Burke, D. C. (1980) A monoclonal antibody for large-scale purification of human leucocyte interferon. *Nature* **285**, 446–450.

Shimizu, A., Watanabe, S., Yamamura, Y. and Putnam, F. W. (1974) Tryptic digestion of immunoglobulin M in urea: conformational lability of the middle part of the molecule. *Immunochemistry* **11**, 719–727.

Smith, J. A., Hurrell, J. G. R. and Leach, S. J. (1978) Elimination of nonspecific adsorption of serum proteins by Sepharose-bound antigens. *Anal. Biochem.* **87**, 299–305.

Stallcup, K. C., Springer, T. A. and Mescher, M. G. (1981) Characterization of an anti-H-2 monoclonal antibody and its use in large-scale antigen purification. *J. Immunol.* **127**, 923–930.

Stark, G. R., Stein, W. H. and Moore, S. (1960) Reactions of the cyanate present in aqueous urea with amino acids and proteins. *J. Biol. Chem.* **236**, 3177–3181.

Stearne, P. A., van Driel, I. R., Grego, B., Simpson, R. J. and Goding, J. W., (1985) The murine plasma cell antigen PC-1: Purification and partial amino acid sequence. *J. Immunol.* **134**, 443–448.

Stenman, U.-H., Sutinen, M.-L., Selander, R.-K., Tontti, K. and Schröder, J. (1981) Characterisation of a monoclonal antibody to human alpha-fetoprotein and its use in affinity chromatography. *J. Immunol. Methods* **46**, 337–345.

Sunderland, C. A., McMaster, W. R. and Williams, A. F. (1979) Purification with monoclonal antibody of a predominant leukocyte-common antigen and glycoprotein from rat thymocytes. *Eur. J. Immunol.* **9**, 155–159.

Symons, F. M., Murray, P. J., Hong, J., Simpson, R. J., Osborn, A. H., Cappai, R. and Handman, E. (1994) Characterization of a family of integral membrane proteins in promastigotes of different *Leishmania* species. *Mol. Biochem. Parasitol.* **67**, 103–113.

Tanford, C. (1969) Protein denaturation. *Adv. Protein Chem.* **23**, 121–282.

Tanford, C. (1970) Protein denaturation. Part C. Theoretical models for the mechanism of denaturation. *Adv. Protein Chem.* **24**, 1–95.

Tesser, G. I., Fisch, H. U. and Schwyzer, R. (1974) Limitations of affinity chromatography: solvolytic detachment of ligands from polymeric supports. *Helv. Chim. Acta* **57**, 1718–1730.

Tollaksen, S. L., Edwards, J. J. and Anderson, N. G. (1981) The use of carbamylated charge standards for testing batches of ampholytes used in two-dimensional electrophoresis. *Electrophoresis* **2**, 155–160.

Tsang, V. C. W. and Wilkins, P. P. (1991) Optimum dissociating conditions for immunoaffinity and preferential isolation of antibodies with high specific activity. *J. Immunol. Methods* **138**, 291–299.

Turkewitz, A. P., Sullivan, C. P. and Mescher, M. F. (1983) Large-scale purification of murine I-A^k and I-E^k antigens and characterization of the purified proteins. *Mol. Immunol.* **20**, 1139–1147.

Weintraub, B. D. (1970) Concentration and purification of human chorionic somatomammotropin (HCS) by affinity chromatography. Application to radioimmunoassay. *Biochem. Biophys. Res. Commun.* **36**, 83–89.

Wilchek, M., Oka, T. and Topper, Y. J. (1975) Structure of a soluble super-active insulin is revealed by the nature of the complex between cyanogen-bromide-activated Sepharose and amines. *Proc. Natl Acad. Sci. USA* **72**, 1055–1058.

Wofsy, L. and Burr, B. (1969) The use of affinity chromatography for the specific purification of antibodies and antigens. *J. Immunol.* **103**, 380–382.

Zoller, M. and Matzku, S. (1976) Antigen and antibody purification by immunoadsorption: elimination of non-biospecifically bound proteins. *J. Immunol. Methods* **11**, 287–295.

12 Immunofluorescence

The use of fluorescent derivatives of antibodies to trace antigen was pioneered by Coons (Coons *et al.*, 1941, 1942; Coons and Kaplan, 1950; Coons, 1961). It was shown that antibodies could be coupled with β-anthracene or fluorescein isocyanate with retention of antigen-binding properties, and that the fluorescent antibodies could be used as very sensitive probes to detect and localize antigen.

Subsequently, Riggs *et al.* (1958) introduced the more stable and convenient fluorescein isothiocyanate (FITC), which has remained the most popular fluorochrome until the present time. One of the main disadvantages of FITC was its very rapid fading under intense illumination, but this problem is now solved (Johnson and Araujo, 1981; Johnson *et al.*, 1982; Giloh and Sedat, 1982).

In the last 10 years, there has been something of a revolution in the choice of fluorochromes. The introduction of extremely highly fluorescent proteins as fluorescent tracers (Oi *et al.*, 1982; Glazer and Stryer, 1984) has made an enormous improvement in sensitivity and flexibility, particularly for work involving two or more colours. The trend towards the use of fluorescent proteins has continued with the recent introduction of peridinin chlorophyll protein (PerCP; Rechtenwald *et al.*, 1990; Stewart and Stewart, 1993). The green fluorescent proteins (GFPs) from *Renilla reniformis* and from *Aequorea victoria* (Heim *et al.*, 1994, 1995) may also be exploited for fluorescence microscopy, and hold great promise because their absorption and emission spectra can be modified by site-directed mutagenesis.

In addition, there have been numerous new synthetic fluorochromes, notably Texas red (Titus *et al.*, 1982), Cy3 and Cy5 (Southwick *et al.*, 1990), AMCA (Khalfan *et al.*, 1986) and Cascade blue (Whitaker *et al.*, 1991). Three-colour fluorescence is now routine (Stewart and Stewart, 1993), and five-

Table 12.1 Spectral properties of fluorochromes

Fluorochrome	Excitation maximum (nm)	Emission maximum (nm)	Colour of fluorescence
Cy5	652	667	Deep red
Texas red	558,594	623	Deep red
Lissamine rhodamine B	539,574	602	Red
TRITC	520,554	582	Red
R-phycoerythrin	495	576	Orange
Cy3	552	570	Red
FITC, DTAF	494	528	Green
FluorX*	494	520	Green
Green fluorescent protein (GFP) from *Renilla reniformis*[†]	498	508	Green
Green fluorescent protein (GFP) from *Aequorea victoria*[‡]	395,475	508	Green
AMCA	347	456	Blue
Cascade blue	<400 nm	410–430	Blue
ELF (fluorescent substrate for alkaline phosphatase)	365	>515	Yellow-green

* FluoX is a succinimide ester derivative of carboxyfluorescein, with an aliphatic spacer between the fluorochrome and the succinimide group. It has a high quantum yield and forms a stable adduct with antibodies. It is available from Research Organics.

[†] From the sea pea *Renilla reniformis*. Only a single excitation peak is seen. The molar extinction coefficient is about 10 times greater than that of *Aequorea*.

[‡] From the jellyfish *Aequorea victoria*. Excitation at 475 nm is well suited to standard filters for fluorescein, and excitation at this wavelength is relatively resistant to photobleaching. A number of mutants with differing spectral properties and increased molar extinction coefficients have been described by Heim *et al.* (1994, 1995). Available from Clontech.

Cy3 and Cy5 are available from Biological Detection Systems or Research Organics. Cy3 is well excited by the 546 nm line of the mercury lamp, or at about 50% efficiency by the 514 nm line of the argon laser.

AMCA is available from Jackson ImmunoResearch.

Cascade Blue is available from Molecular Probes.

colour fluorescence is technically feasible, if not in widespread use (Beavis and Pennline, 1994). The properties of the commonly used fluorochromes are listed in Tables 12.1 and 12.2.

Immunofluorescence has now made the transition from a tricky and somewhat erratic technique to one of the highest precision (reviewed by Ploem and Tanke, 1987; Rost, 1992). The purity of the fluorochromes and the quality of the optics of fluorescence microscopes have improved enormously. The development of the fluorescence-activated cell sorter (FACS) has allowed the rapid, sensitive, quantitative and objective analysis of single cells, and also the possibility of cell separation on the basis of membrane antigens and/or other parameters such as DNA content, light scatter, calcium flux and intracellular pH, or indeed any other property that can be measured using optical techniques. Multiparameter analysis and sorting are now well-established procedures.

Table 12.2 'Third colour' fluorochromes that emit in the red region

Fluorochrome	Excitation wavelength (nm)	Emission wavelength (nm)	Overlap of emission with *R*-phycoerythrin
PerCP	488	peak at 677	±
Red 670	488	670	+
Cy-chrome,			+
Tri-color	488	>650; peak at 660–670	+
Quantum red,	488	670	+
Spectral red			
Red 613	488	613	++

PerCP (Peridinin Chlorophyll Protein) is available from Becton Dickinson. It is highly suitable for three-colour fluorescence with FITC and *R*-phycoerythrin, because of its very narrow emission spectrum. It requires virtually no compensation when used with phycoerythrin. PerCP may be rather susceptible to photobleaching by high-power lasers.

Tri-Color is available from Caltag Laboratories. It is suitable for use with FITC and *R*-phycoerythrin in three-colour experiments.

Red 613 is a conjugate of *R*-phycoerythrin and Texas red. Red 670 is a conjugate of *R*-phycoerythrin and Cy5. Both are available from Life Technologies, Inc. Not all the phycoerythrin emission energy is transferred to the long-wavelength fluorochrome, and as a result, a small peak of emission at about 575–600 nm occurs, which must be compensated for if *R*-phycoerythrin is used at the same time in multicolour experiments.

Quantum red and Spectral red are a tandem coupling of phycoerythrin and Cy5, such that they combine efficient excitation at 488 nm and deep red emission. They are available from Sigma and Southern Biotechnology Associates respectively.

As is the case for all procedures in which antibodies are used, the best results will only be obtained if there is an understanding of the basic principles and the important experimental variables. It is the purpose of this chapter to provide a conceptual and practical framework for the optimal use of monoclonal antibodies in immunofluorescence. The use of monoclonal antibodies with histological sections will be discussed in Chapter 13.

12.1 Principles of Immunofluorescence

When light is absorbed by certain molecules known as fluorochromes, the energy of the photons may be transferred to electrons, which assume a higher energy level. Some of the energy is liberated within 10^{-15} s as heat when the electron returns to the lowest vibrational energy of the excited state; the remainder is released after a few nanoseconds as a photon of lower energy (and therefore longer wavelength) than the initial photon. This phenomenon is called fluorescence and its efficiency is described by the term *quantum yield* (Crooks, 1978).

The wavelengths which are capable of causing a molecule to fluoresce are known as the *excitation spectrum*, and the wavelengths of emitted fluorescent light the *emission spectrum*. The emission spectrum is shifted to a slightly longer wavelength than the excitation spectrum (Stokes' Shift), although the two spectra often overlap. For practical purposes, the excitation spectrum is very similar to the absorption spectrum in the visible wavelength range. The most comprehensive listing of the spectra and other properties of fluorochromes is given by Haugland (1992). Additional information concerning fluorochromes may be found in Berlman (1971), Haugland (1995), Tsien and Waggoner (1990), and Waggonner (1990).

The intensity of fluorescence often varies depending on the physical environment (Crooks, 1978). Some fluorochromes are most highly fluorescent in an aqueous environment (e.g. FITC), while other fluoresce much more intensely in a nonpolar environment (e.g. dimethylamino naphthalene). The spectral characteristics of fluorescein are also dependent on pH, and this property has been exploited in the measurement of intracellular hydrogen ion concentration (Ohkuma and Poole, 1978).

Individual fluorochrome molecules which are in close proximity may interact and transfer energy from one to the other. This is often exploited to couple the emission of one fluorochrome to the absorption of another, allowing much larger Stokes' shifts and correspondingly better spectral separations in multicolour fluorescence (see Table 12.2). In a similar vein, the absorption and emission spectra of many fluorochromes often overlap, and it is also not uncommon for flurochromes to transfer energy among themselves, particularly if they are in very close proximity. When this occurs, there may be 'self quenching', leading to greatly reduced fluorescence. This can have drastic effects on the efficiency of fluorescent antibodies, and is discussed in more detail in Section 12.8.5

12.2 The Fluorescence Microscope

Virtually all immunofluorescence microscopy is now carried out using the vertical (incident or epi-) illumination system of Brumberg (1959) and Ploem (1967), in which the excitation light reaches the specimen via the objective (Fig. 12.1; reviewed by Ploem and Tanke, 1987; Rost, 1992). The advantages of epi-illumination include much higher intensity of illumination and brighter image, better image quality, ease of usage, and the ability to combine fluorescence simultaneously or in rapid alternation with visible transmission. The combination of immunofluorescence with phase-contrast transmission microscopy is particularly useful for examination of intact living cells (Nossal and Layton, 1976; Goding and Layton, 1976).

Figure 12.1 shows a typical fluorescence microscope. The high-pressure mercury lamp (B) is focused via a curved mirror (A) and a lens (C). The light passes through a heat filter (D) and an excitation filter (E) to dichroic mirror (H), where it is reflected downwards to the objective lens (I), which acts as condenser. Light emitted from the specimen (J) passes back upwards through

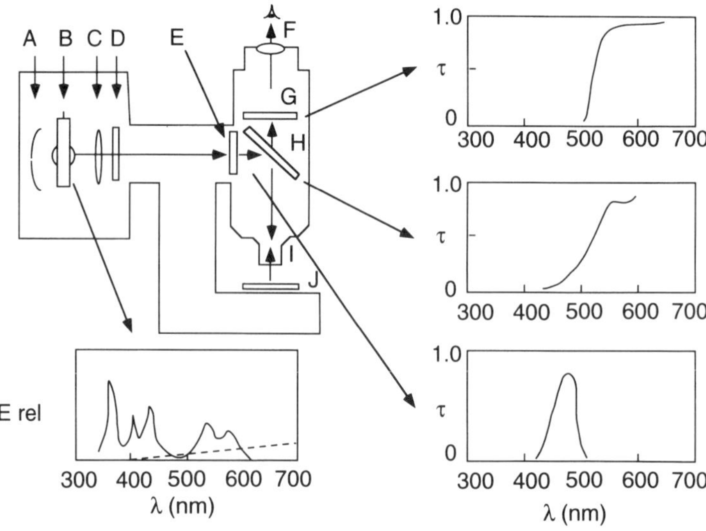

Fig. 12.1. The fluorescence microscope (epi-illumination). Light from the high-pressure mercury lamp (B) is concentrated by a mirror (A) and a lens (C), and passes through a heat filter (D), and a barrier filter (E). The barrier filter is usually a bandpass filter designed for optimal excitation but removing the shorter wavelengths that may cause autofluorescence. The light is then reflected by a dichroic mirror (beam splitter, H), and passes through the objective (I) to the specimen (J). The emitted light, which is of longer wavelength, passes through the dichroic mirror and the barrier filter (G), which removes any stray light of wavelengths for excitation. Light then passes through the ocular (F) to the eye. The schematized characteristics of the filters are for FITC.

the objective, but because of its longer wavelength it passes straight through the dichroic mirror. A barrier filter (G) prevents any residual stray excitation light from reaching the eyepiece (F), but freely transmits the emitted fluorescence.

Microscopes with epi-illumination generally employ a 50 W mercury lamp (HBO 50). The substantial improvement in filters and the much shorter light path of current microscopes renders the image as bright or brighter than that obtained with older 200 W lamps.

The mercury lamp has intense peaks of emission at 313, 334, 365, 405, 435, 546 and 578 nm; its output intensity in the range 450–525 nm is much less than that of the peaks. Unfortunately, it is the last range that is most suitable for excitation of FITC. However, the quantum yield for FITC is very high, and illumination with the mercury lamp is usually quite adequate. Quartz-halogen or high-pressure xenon lamps (60–100 W) provide a continuous spectrum of intensity comparable to the mercury lamp in this region (Fig. 12.1, dotted line) and are a viable alternative. They have the advantage that they are less expensive and may be turned on and off whenever desired. Quartz–halogen lamps are not nearly as efficient as mercury lamps in exciting rhodamine derivatives, as the mercury lamp has strong emission peaks at 546 and 578 nm (Fig. 12.1).

The introduction of the newer fluorochromes based on *R*-phycoerythrin and other protein molecules has had a major impact because they are often very efficiently excited at the same wavelength as fluorescein (490 nm), and have very high molar absorbance coefficients, very high quantum yields and large Stokes' shifts. Emission is often well into the red, and well separated from the emission of fluorescein (Tables 12.1 and 12.2).

The HBO 50 mercury lamp has a maximum life of about 200 h, and towards the end of its life its intensity diminishes. The pressure inside the lamp during operation is 40–70 atmospheres (4040–7070 kPa), and if used beyond the recommended life there is a chance of its exploding. Explosion of the lamp is not likely to escape the lamp housing, but will result in considerable damage to the microscope. The lifetime of the lamp is diminished by frequent starting and stopping and it is recommended that a log book of use be kept, and that once turned on, it be left on for several hours rather than turning it on and off more than once a day (Gardner and McQuillin, 1980). In particular, the lamp should not be restarted for at least 30 min after it has been turned off.

The *numerical aperture* (NA) of the objective lens is of prime importance in achievement of maximal fluorescence intensity. The numerical aperture is the sine of the half-aperture angle multiplied by the refractive index of the medium filling the space between the specimen and the front lens. Air has a refractive index of 1, so the theoretical maximum NA of an air objective is 1.0. Immersion oil has a refractive index of 1.515, making it possible to achieve a considerable improvement in NA.

In an epi-illumination system, both the intensity of excitation and efficiency of collection of fluorescent light increase as the square of the NA. It would therefore be expected that the observed fluorescence would increase as the fourth power of the NA, and this is the result obtained experimentally (Haaijman and Slingerland-Teunissen, 1978). Objective lenses for immunofluorescence should be chosen for the highest possible NA.

Gardner and McQuillin (1980) point out that with epi-illumination the most expensive oil immersion lens is not necessarily the most suitable for fluorescence microscopy. Complex apochromatic × 50 objectives with large numbers of lenses have greater internal light scattering than simple fluorite oil immersion lenses and may produce a worse background and less contrast.

It is preferable to use eyepieces of low magnification (e.g. × 8 rather than × 12.5), because fluorescence intensity decreases exponentially with increasing total magnification.

The recent introduction of laser confocal microscopy (White *et al.*, 1987; Wilson, 1990; Pawley, 1990; Lichtman, 1994) is having a major impact on immunofluorescence because one can generate extremely thin 'optical sections' and hence acquire much greater resolution, particularly in intact cells. In addition, the use of laser illumination creates many new opportunities for

Table 12.3 Suppliers of fluorochromes and fluorescent antibodies*

Supplier	Address	Reagent
Biological Detection Systems	955 William Pitt Way, Pittsburgh, PA 15238, USA	Cy3 and Cy5
Molecular Probes	PO Box 22010, Eugene, OR 97402-0414, USA	Cascade blue, FITC, TRITC, microspheres, phycoerythrin, anti-fade, ELF (enzyme labelled fluorescence) substrates
Jackson ImmunoResearch Laboratories	872 West Baltimore Pike PO Box 9, West Grove, PA 19390, USA	AMCA
Research Organics	4353 East 49th Street, Cleveland OH 44125–1083, USA	DTAF, FITC, FluorX, Cy3, Cy5, and many others
Vector Laboratories	30 Ingold Road, Burlingame, CA 94010, USA	FITC–avidin, TRITC–avidin, antifade
Becton Dickinson Immunocytometry Systems	2350 Qume Drive, San Jose, CA 95131–1807, USA	Phycoerythrin, PerCP
Life Technologies, Inc	PO Box 6009, Gaithersburg, MD 20884–9980, USA	Red 613, Red 670
Southern Biotechnology Associates, Inc.	PO Box 26221, Birmingham, AL 35226, USA	Spectral red
Clontech	4030 Fabian Way, Palo Alto, CA 94303, USA	Green fluorescent protein

* This listing is not exhaustive, and the catalogues of individual suppliers should be consulted. Some reagents are proprietary and only available from certain manufacturers.

efficient excitation of the newer fluorochromes, some of which are not well suited to the mercury lamp of the older technology.

12.3 Choice of Fluorochromes

Most investigators will probably purchase antibodies that are already conjugated with fluorochromes. However, it is important to have some understanding of the properties, advantages and limitations of individual fluorochromes, particularly if the sensitivity needs to be pushed to the limit, or if multicolour staining is to be used. There are now a large number of commercial suppliers of fluorochromes and fluorescent antibodies. These are listed in Table 12.3.

12.3.1 Fluorescein Isothiocyanate, Dichlorotriazinylamino Fluorescein (DTAF) and Related Fluorochromes

The choice of fluorochrome will be governed by a number of factors. If a single colour is needed, FITC will usually be the fluorochrome of choice (Fig. 12.2). FITC is inexpensive, has a high quantum efficiency and is relatively hydrophilic. The rapid fading under the intense ultraviolet illumination of fluorescence microscopes can be overcome by the use of chemicals such as DABCO, phenylenediamine or *n*-propyl gallate which greatly retard fading, although these reagents are rather toxic for living cells. A number of proprietary antifade reagents are also available (see Section 12.11.5).

As mentioned earlier (see Section 12.2), FITC is adequately excited by the mercury lamp, despite the fact that there is no discrete spectral line at its excitation maximum of 490 nm (Figs. 12.1 and 12.3). The 488 nm line of the argon laser used in the FACS is ideally suited to excitation of FITC (Table 12.1). The triazine derivative of fluorescein, dichlorotriazinylamino fluorescein (DTAF; Blakeslee and Baines, 1976; Blakeslee, 1977) has very similar optical properties to FITC (Table 12.1), but is more stable. DTAF conjugates

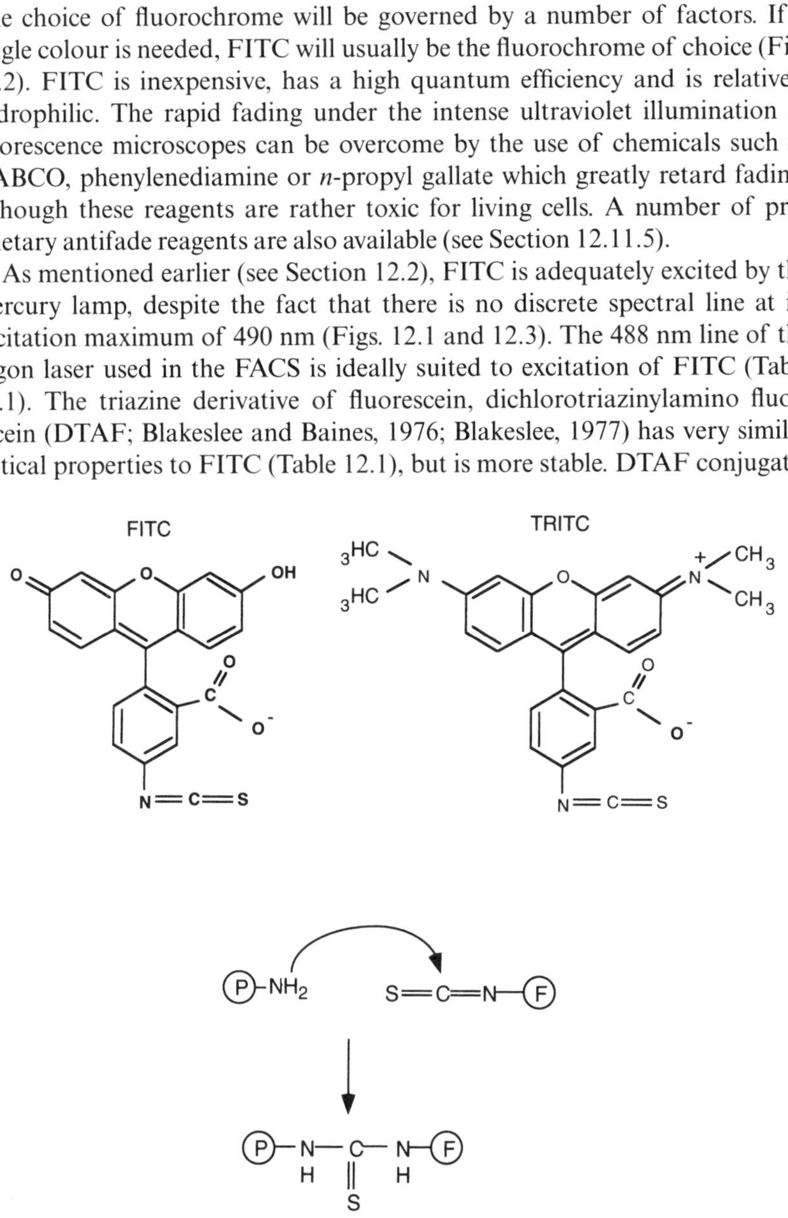

FITC

TRITC

$_3$HC N $_3$HC O OH O C O$^-$ N=C=S N$^+$ CH$_3$ CH$_3$

P-NH$_2$ S=C=N-F

P-N-C-N-F H S H

Fig. 12.2. Structures of commonly used fluorochromes, and mechanism of coupling to protein. The nucleophilic unprotonated ε-amino groups of lysine residues attack the isothiocyanate group, resulting in a thiourea bond.

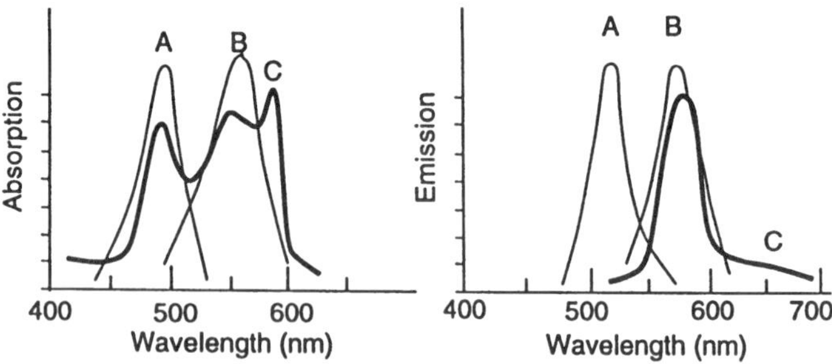

Fig. 12.3. Schematic illustration of the absorption and emission spectra of FITC (A), TRITC (B) and R*-phycoerythrin (C).*

may be fractionated by ammonium sulfate precipitation to remove over-conjugated molecules.

Recently, succinimide ester derivatives of fluorescein and tetramethyl rhodamine have become available from Molecular Probes and Research Organics. They are said to have very high quantum yields and similar spectral properties to the isothiocyanate forms. While fluorescein succinimide ester is more expensive than FITC, the tetramethyl rhodamine succinimide ester is comparable in price to TRITC. The succinimide ester of carboxyfluorescein is available with a spacer arm as 'FluorX' from Research Organics and is claimed to have less steric hindrance than more direct coupling chemistry. It has an absorption maximum at 494 nm and an emission maximum at 520 nm. The succinimide ester derivatives of fluorochromes are very easy to use and coupling proceeds smoothly at a pH of 8–8.5. The bond to protein is very stable, and these compounds may become popular.

12.3.2 *R*-Phycoerythrin

Oi *et al.* (1982) demonstrated that phycobiliproteins from red algae can be used as excellent fluorescent probes (see also Kronick and Grossman, 1983; Glazer and Stryer, 1984; Hardy, 1986; Kronick, 1986). In the last few years they have become very popular. Their molar absorbance coefficients are extremely high because of their multiple bilin chromophores, and they have high quantum yields. *R*-phycoerythrin is ideal for use with argon lasers, because it is efficiently excited at 488 nm. The emission spectrum of *R*-phycoerythrin begins at 550 nm and peaks at about 580 nm, and is ideally suited to two-colour fluorescence experiments in conjunction with FITC (Fig. 12.3). The phycoerythrins are highly soluble and stable and are easily purified. They are available commercially from many suppliers (e.g. Molecular Probes, Becton–Dickinson).

In view of its extremely favourable spectral properties, *R*-phycoerythrin is probably now the the fluorochrome of choice for two-colour fluorescence microscopy.

12.3.3 Tetramethyl Rhodamine Isothiocyanate

Another popular fluorochrome is tetramethyl rhodamine isothiocyanate (TRITC; Fig. 12.2). TRITC is much more highly fluorescent than rhodamine isothiocyanate. TRITC has an absorption and excitation maximum at 550 nm (Figs 12.3 and 12.4), and an emission maximum at 580 nm. The intense red fluorescence of TRITC is easily distinguished from the green fluorescence of FITC, and for many years these two fluorochromes were the combination of choice for two-colour fluorescence experiments (Nossal and Layton, 1976; Goding and Layton, 1976). Fading of TRITC is much slower than FITC (Giloh and Sedat, 1982). In recent years, however, TRITC has been superseded by *R*-phycoerythrin, particularly for flow cytometry (see Section 12.7.2).

The use of TRITC has a number of problems. It is more hydrophobic than FITC, and therefore has a tendency to bind nonspecifically to proteins and cells. The limited solubility of TRITC in water may be overcome by dissolving it in dimethyl sulfoxide (DMSO) prior to conjugation to antibodies (Bergquist and Nilsson, 1974; Goding, 1976), and it is strongly recommended that this procedure be followed. TRITC dissolves rapidly in DMSO, allowing it to be added to antibodies before significant hydrolysis occurs. The nonspecific adsorption of TRITC to proteins via hydrophobic interactions and subsequent slow release or exchange may cause significant background fluorescence; the use of DMSO as a solvent appears to diminish this problem (Goding, 1976). The hydrophobic nature of TRITC also causes denaturation and precipitation of antibodies if they are too heavily conjugated (see Section 12.3). Overconjugation of antibodies with TRITC and other rhodamine

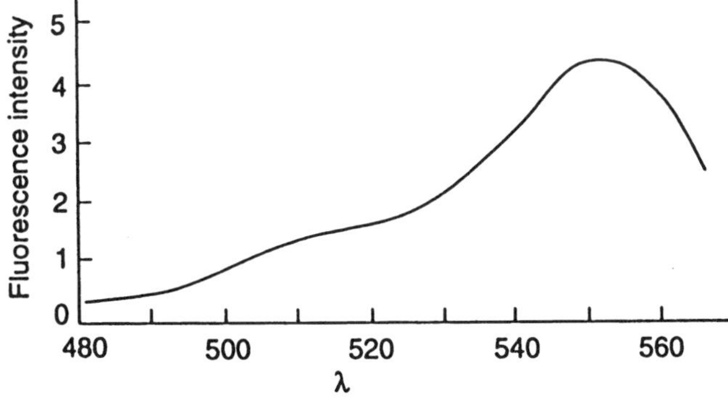

Fig. 12.4. Excitation spectrum of TRITC. Emission was measured at 575 nm.

derivatives may also lead to a severe reduction in fluorescence intensity owing to self-quenching (see Section 12.8.5).

Despite all the above, TRITC has proven to be a very useful fluorochrome. If attention is paid to these details, excellent results may be obtained. The succinimide ester of tetramethyl rhodamine is now available from Molecular Probes and may be a useful addition.

TRITC is now rarely used in two-colour immunofluorescence experiments with the FACS because its sensitivity is limited by its relatively poor excitation at the only available lines of the argon laser (488 nm and 514 nm).

12.3.4 Texas Red

'Texas Red' is a sulfonyl chloride derivative of sulforhodamine 101 (Haugland, 1992; Titus *et al.*, 1982; Table 12.1). Maximal excitation of protein-bound Texas Red is at 596 nm, and maximal emission at 615 nm (Titus *et al.*, 1982). There is very little overlap between the emission spectra of Texas red and that of FITC, so this combination should be well suited for two-colour fluorescence.

The main difficulty with Texas red is that its absorption maximum is not well suited to the mercury lamp or argon lasers, because it is too far into the red. Texas red is not adequately excited by the 488 nm or 514 nm lines available from argon lasers (Titus *et al.*, 1982), but is very efficiently excited by the 568 nm line of the krypton laser used in some dual-laser flow cytometers (Titus *et al.*, 1982) and the argon–krypton mixed-gas laser used on Bio-Rad confocal microscopes (Haugland, 1992).

Texas red has been used very effectively in combination with *R*-phycoerythrin to generate 'tandem' fluorochromes with good excitation at 490 nm but with emission well into the red and with minimal overlap with FITC (Table 12.2).

The microscope filter combination used for TRITC (Table 12.4) allows moderately efficient excitation of Texas Red, although the use of the 546 nm mercury line is probably not optimal. A set of filters exploiting the 578 nm line might be preferable.

The hydrolysis product of Texas red is a very water-soluble sulfonic acid derivative; it is thus easily removed with minimal risk of adsorption to proteins. Antibodies that are heavily conjugated with Texas red tend to precipitate. Because of its extreme susceptibility to hydrolysis, Texas red is sold in small sealed tubes, and it is recommended that each tube should be discarded after opening and not resealed.

12.3.5 Cy3 and Cy5

A new class of fluorochromes, the carboxymethylindocyanine succinimide esters, has recently been introduced (Southwick *et al.*, 1990). Of these, Cy3

Table 12.4 Typical filter combinations for epi-illumination

Fluorochrome	Excitation filter	Dichroic mirror	Barrier filter	Zeiss Filter Set
FITC, DTAF	BP 450–490	FT 510	LP 520	487909
TRITC, Texas red, phycoerythrin, Cy3	BP 546	FT 580	LP 590	487915
AMCA	BP 365	FT 395	LP 397	487901

FITC and DTAF give green fluorescence; phycoerythrin gives orange, and AMCA gives blue. Mercury lamps provide strong lines at 313, 334, 365, 405, 435, 546 and 578 nm. There is no discrete line in the optimal range for FITC excitation. The filters for TRITC may also be used for Texas red, XRITC and phycoerythrin.

The filter set for AMCA would be suitable for ELF (Enzyme Labeled Fluorescence) substrate for alkaline phosphatase (Molecular Probes).

Filter sets are also available from Omega Optical, distributed by Molecular Probes. The filter set O-5857 is a high-efficiency red, green and blue triple-band filter set, which works well for multicolour experiments using Texas red, FITC or DTAF and AMCA or Cascade blue. More details on filter sets are given by Haugland (1992).

and Cy5 are the most promising. They have high quantum yields, although they are prone to self-quenching if antibodies are over-conjugated (Southwick *et al.*, 1990, and see Section 12.8.5). Cy3 has an absorption maximum at 550 nm and an emission maximum at 565 nm; it can be excited at about 50% efficiency with the 514 nm line of the argon laser, or to about 75% by the 546 nm line of the mercury lamp. Cy3 has very similar spectral properties to TRITC, and so the filter combinations for TRITC are suitable (Table 12.4).

Cy5 has an absorption maximum at 652 nm and an emission maximum at 667 nm. It is suitable for specialized imaging systems, such as those using CCD cameras, which work in the long emission wavelength range. It can be excited efficiently by the krypton/argon laser at 647 nm or the He/Ne laser at 633 nm, or by mercury or tungsten–halogen lamps. Cy5 is sometimes used in conjunction with other fluorochromes in 'energy transfer' systems, where the emission of a shorter wavelength is converted to a longer wavelength by Cy5 (see Table 12.2).

12.3.6 Aminomethyl Coumarin Acetic Acid

The blue fluorochrome aminomethyl coumarin acetic acid (AMCA) was introduced by Khalfan *et al.* (1986). It has an absorption maximum at 347 nm and an emission maximum at 456 nm, and is a useful addition for multicolour work. It is said to be very susceptible to photobleaching, even by standard room lighting, and slides should be wrapped in foil when they are not being examined. A suitable filter set is given in Table 12.4.

12.3.7 Use of Fluorescent Microspheres

An interesting alternative to the use of fluorochrome molecules for antibody labelling is the use of fluorescent latex microspheres. They have a number of attractions, including the fact that a single microsphere can contain many thousands or even millions of fluorochrome molecules. The binding of a single particle may be detected with ease (Rembaum and Dreyer, 1980).

By avoiding the direct conjugation of antibodies with fluorochromes, one avoids the problem of antibody inactivation or destabilization by derivatization with aromatic compounds (see Wadsley and Watt, 1987). Fluorescent microspheres also avoid the problem of self-quenching of fluorochromes due to their close proximity to each other (see Sections 12.1 and 12.8.5). The fluorochromes inside microspheres are much more photostable and less influenced by buffer conditions than fluorochromes in solution.

Molecular Probes has recently released a range of 'TransFluoSpheres' that can be excited at 488 nm but have differing and essentially nonoverlapping emission spectra. A yellow particle that is 0.1 μm in diameter can have about 6500 unquenched fluorescein fluorophores, while a single 1 μm particle can have up to 10^7 fluorophores. The main problem with microspheres is their tendency to bind nonspecifically to cells and blots. It will also be necessary to refine the methods for coupling proteins to them. At present, microspheres are available uncoated or coated with avidin (see Sections 10.3 and 12.9). The surface of microspheres contains carboxyl groups, which are not particularly well suited for protein coupling, although water-soluble carbodiimides could be used. If these problems can be overcome, microspheres may become very useful in the future.

Parks *et al.* (1979) have shown that hybridoma cells may be analysed and fractionated using antigen-coated fluorescent microspheres. Higgins *et al.* (1981) have used fluorescent microspheres in a sensitive assay for cell-surface antigens.

12.4 Choice of Filters for Fluorescence Microscopy

The choice of filter combinations for immunofluorescence is a large subject, and the range of filters that are commercially available is constantly expanding. The magnitude of the problem can be gauged from the fact that Zeiss has offered no less than five different filter combinations for FITC alone. An extensive range of filters to fit standard Zeiss and Nikon microscopes is also available from Omega Optics and Molecular Probes. Custom-made filter sets with customer-specified optical properties are also available from Omega Optics. It is therefore possible to give only the most general advice, and the microscope manufacturers should be consulted for further details.

Virtually all current high-performance filters are of the interference type.

Interference filters have much higher transmission, steeper cutoff slopes and greater rejection of unwanted wavelengths than the older filters. The barrier filter (Fig. 12.1) usually consists of a long-pass (LP) filter which transmits all wavelengths longer than a certain value (the wavelength at which transmittance is 50%). The excitation filter may consist of a short-pass (KP, German *kurzpass*) filter in which all wavelengths shorter than a given wavelength are passed. More commonly, the excitation filter is a band-pass (BP) filter which passes a narrow band corresponding to a particular mercury line or portion of the spectrum. Band-pass filters are described by the centre wavelength and the bandwidth (e.g. BP546/10). Band-pass filters may be constructed by combining a long-pass and a short-pass filter. For example, the BP450–490 filter consists of a combination of LP 450 and KP 490.

The other major optical filter is the dichroic mirror or beam-splitter, which is an interference mirror set at 45° to the light path. The dichroic mirror is designed to reflect light of wavelengths shorter than a certain value (i.e. the excitation light), and to freely transmit light of wavelengths longer than this value (i.e. the emission). For example, the FT510 mirror reflects light of wavelength less than 510 nm.

Typical filters for use with the mercury lamp are given in Table 12.4. The excitation filter for FITC passes exciting light in the range 450–490 nm, all of which is reflected to the objective via the FT510 dichroic mirror. Emitted fluorescence (mostly longer than 510 nm) passes through the dichroic mirror. The LP 520 barrier filter provides additional exclusion of the excitation wavelengths.

The corresponding filters for TRITC are BP 546/10 (which selects the 546 nm mercury line), FT580 dichroic mirror and LP590 barrier filter. As mentioned earlier, the filter set for TRITC is reasonably well suited for Texas red, although better combinations could probably be obtained.

R-phycoerythrin is excited efficiently at 450–490 nm, and emits above 550 nm (Fig. 12.3; Oi *et al.*, 1982; Glazer and Stryer, 1984). The filter combination suggested for FITC would be suitable, because the barrier filter passes all wavelengths above 520 nm.

12.5 Direct and Indirect Immunofluorescence

Direct immunofluorescence involves the exclusive use of antibodies which have been covalently coupled with fluorochromes. The specimen is incubated with the labelled antibody, unbound antibody is removed by washing, and the specimen is examined.

In many cases, however, it may be preferable to use an indirect technique, in which the specimen is incubated with an unconjugated antibody, washed, and incubated with a fluorochrome-conjugated anti-immunoglobulin antibody. The second or 'sandwich' antibody thus reveals the presence of the first.

A major advantage of indirect immunofluorescence is that one fluorescent anti-immunoglobulin antibody will suffice for many first antibodies; it is not necessary to conjugate each new antibody individually. Indirect immunofluorescence is essential if the monoclonal antibody is only available in the form of a culture supernatant because direct conjugation of supernatants with fluorochromes is unsatisfactory. The indirect technique usually gives brighter fluorescence than the direct because many second antibodies may bind to the first. Amplifications of six- to eight-fold are possible. This may be a decided advantage where monoclonal antibodies are used, as their monospecificity results in smaller numbers of binding sites.

The main disadvantage of the indirect technique is that the anti-immunoglobulin sandwich reagent is unable to distinguish between exogenous and endogenous immunoglobulin. For example, a rabbit anti-mouse immunoglobulin sandwich reagent would detect membrane immunoglobulin on mouse B cells, regardless of whether other antibodies are present or not. The problem is not as severe when mouse monoclonal antibodies are used to stain human cells because rabbit or rat antimouse immunoglobulin will cross-react weakly or not at all with human immunoglobulin. Many commercial antisera to mouse immunoglobulins have been depleted of antihuman antibodies by passage over human immunoglobulin-coupled beads. None the less, the possibility of such cross-reactions is a real one, and controls must always be performed in which the first antibody is omitted.

The omission of the first antibody is also a useful control to check for non-specific binding of the second antibody via Fc receptors. Obviously, such a control does not address the possibility of Fc receptor binding by the first antibody.

12.6 The Fluorescence-activated Cell Sorter (FACS)

The development of the fluorescence-activated cell sorter (FACS) has been a major contribution to the field of immunofluorescence, and that of cell biology in general. Analysis by the FACS is often known as *flow cytometry*. The principles of operation have been reviewed in detail (Bonner *et al.*, 1972; Herzenberg *et al.*, 1976; Miller *et al.*, 1981; Kruth, 1982; Loken and Stall, 1982; Parks and Herzenberg, 1984; Darzynkiewicz and Crissman, 1990; Ormerod, 1990; Watson, 1991; Shapiro, 1994).

A single-cell suspension, labelled with fluorescent antibody or DNA stain, is passed in single file through a narrow laser beam, and the emitted fluorescence is collected by an optical system at right angles to the illumination. The emitted fluorescence of each individual cell is measured and stored. Other parameters may be measured and recorded on a cell-by-cell basis, such as low-angle forward light scatter (which generally reflects cell size and is also partly a measure of cell viability), side scatter (reflecting cell granularity), or Coulter volume. Many machines are also capable of separating the cells into individual fluid droplets which may be electrostatically deflected into tubes, depending on fluorescence, light scatter or a combination of parameters.

The power of the FACS may be exploited in many different ways. Although it was initially envisaged as a preparative machine, experience has shown that the major use in most laboratories is analytical. The FACS can analyse up to 5000 cells/s (some machines are even faster), and plot any desired combination of parameters. Unlike visual fluorescence microscopy, the readout is rapid, objective and quantitative.

The use of multiparameter analysis adds enormously to the power of flow cytometry. Simultaneous analysis of surface antigens detected by two or more different monoclonal antibodies, together with light scatter or Coulter volume, is now routine. Cells may also be separated on the basis of DNA content, or any other parameter which can be measured by optical methods.

It is possible to detect and isolate extremely rare cells, such as fetal cells in the human maternal circulation (Bianchi *et al.*, 1990, 1994). Very rare 'class-switch' mutants or variants of myeloma or hybridoma cells may be detected and isolated (Radbruch *et al.*, 1980; Holtkamp *et al.*, 1981; Neuberger and Rajewsky, 1981; Oi *et al.*, 1984). It is possible to use the FACS to sort and clone hybridomas on the basis of their binding of antigen (Parks *et al.*, 1979). Kavathas and Herzenberg (1983) have shown that the FACS may be used in conjunction with monoclonal antibodies to select DNA-transfected mouse L cells which express human surface antigens.

12.6.1 Suitable Fluorochromes for Flow Cytometry

The main constraint on the choice of fluorochromes for flow cytometry is the wavelength of the laser emission lines. The commonly used argon laser has a strong line at 488 nm, which is ideally suited to excitation of FITC and *R*-phycoerythrin (see Fig. 12.3). Because this pair of fluorochromes has excellent quantum yields and emission spectra that are well separated from each other, the use of FITC and *R*-phycoerythrin is now the standard combination for two-colour fluorescence in flow cytometry. Cross-channel spill is fairly small, and can be corrected electronically (Loken *et al.*, 1977; Ledbetter *et al.*, 1980a; Parks and Herzenberg, 1984; Ormerod, 1990). The use of multicolour fluorescence in flow cytometry is discussed in more detail in Section 12.7.3.

TRITC and Texas red are not very suitable for use with the argon laser, as their absorption/excitation spectra are too far into the red (see Figs 12.3 and 12.4), even when the 514 nm line is used. The use of FITC and TRITC for two-colour work in flow cytometry has now been entirely superseded by the *R*-phycoerythrin/FITC combination. Titus *et al.* (1982) have shown that the combination of FITC and Texas red can be used effectively in flow cytometers using an argon laser together with a krypton laser.

12.6.2 'Gating Out' Dead Cells

Dead cells may bind large amounts of fluorescent antibody in a nonspecific manner. The presence of dead cells will seriously degrade the precision with which cells may be analysed, particularly when identification and sorting of very rare cells is desired.

Low-angle light scatter correlates well with cell viability (reviewed by Loken and Stall, 1982). Living cells tend to have a larger degree of scatter than dead cells. It is possible to choose a 'window' of low-angle light scatter which includes the majority of living cells and excludes dead cells. However, the discrimination achieved by light scatter gating is not absolute.

The DNA stain propidium iodide, which only enters the nucleus of dead cells, can be used as an adjunct to low-angle light scatter for excluding dead cells. Propidium iodide bound to DNA is strongly excited at 488 nm and emits intense fluorescence at 570 nm (Krishnan, 1975). Addition of 2 μg/ml propidium iodide to the buffer during the final incubation period will label all dead cells. The intensity of fluorescence of dead cells is so great that they may be gated out without affecting the profiles of living cells (Layton, 1980). A stock solution of 100 μg/ml propidium iodide in PBS plus 0.1% azide will last for many months when stored at 4°C. It should be noted that residual propidium iodide in the medium may stick in the tubing, and give rise to varying degrees of staining of dead cells in later samples. Propidium iodide-treated cells cannot be fixed because transfer then occurs between labelled and unlabelled cells. Like all DNA stains, propidium iodide may be mutagenic, and should be handled with care.

12.6.3 The Logarithmic Amplifier

The range of fluorescence between the brightest positive and dullest negative cells may span three or even four orders of magnitude, making it impossible to display this range on linear coordinates. All flow cytometers now use amplifiers which perform logarithmic transformation of the fluorescence signals. The use of logarithmic presentation allows clear-cut distinction of populations of cells which appear to merge into each other on linear displays (Ledbetter *et al.*, 1980a, b; Hardy *et al.*, 1982a, b).

12.7 Multicolour Fluorescence

12.7.1 General Principles

It is frequently desirable to examine biological specimens simultaneously for two or more independent antigens. Multicolour immunofluorescence is not a great deal more difficult than single-colour, but a few simple principles should be observed.

The use of two or more colours imposes a number of constraints on the experimental system. The chosen fluorochromes must be able to be excited efficiently by the available light source, and should have emission spectra that overlap as little as possible. These criteria can usually be met by the judicious choice of fluorochromes and filter combinations. Often, compromises will be needed. 'Spill' between fluorochromes owing to overlapping emission spectra can be reduced by the use of narrow-band filters, but at the expense of reduced light intensity. In the case of flow cytometry, cross-channel spill can be corrected electronically (see below), but it is generally best to choose fluorochromes and filters to minimize the extent to which this is needed as the correction is seldom perfect.

The other strategic point that must be considered in multicolour immunofluorescence is that of sandwich reagents. If both antibodies are directly conjugated with fluorochromes, there is no problem. However, if a sandwich procedure is used, the second-step reagents must be chosen such that they will only react with the appropriate first step. Anti-immunoglobulin sandwich reagents would bind to both first antibodies. A common strategy is to couple one antibody directly to its fluorochrome, and couple the other with biotin. The biotinylated antibody may then be detected with avidin coupled to a second fluorochrome (see Section 12.9). If the first antibodies are of different classes, one can use class-specific fluorescent antibodies as sandwich reagents (Walliker *et al.*, 1987).

The opportunities for unexpected artefacts and nonspecificity are greatly increased in two-colour systems and it is mandatory to include a wide range of controls, such as the omission of each first-stage antibody.

If one finds that there is a perfect correlation between the intensity of staining of two colours on individual cells, this is a sign that the antibodies are detecting the same molecule, and may often be a warning sign of artefactual staining, as may occur if the first or second antibodies are interacting in an unplanned way.

12.7.2 Multicolour Fluorescence Microscopy

Until the introduction of *R*-phycoerythrin, the most common combination of fluorochromes for two-colour immunofluorescence was FITC and TRITC (Goding and Layton, 1976). This combination is capable of excellent results with epifluorescence microscopy using a mercury lamp. However, the excitation and emission spectra of FITC and TRITC overlap somewhat (Fig. 12.3), and care is needed to select filters that provide maximal selectivity (see also Kearney and Lawton, 1975). It may be necessary to choose a somewhat more selective filter combination at the expense of a certain degree of loss of fluorescence intensity.

The combination of FITC and *R*-phycoerythrin is now widely used (see Section 12.7.3). *R*-phycoerythrin is efficiently excited by the same wavelength spectrum as FITC (i.e. around 490 nm), and their emission spectra are well separated (Fig. 12.3). Various other fluorochromes that are excited efficiently at 490 nm but emit even further into the red are listed in Table 12.2; however, these are more commonly used in flow cytometry.

If a third colour is required, the use of blue fluorochromes such as AMCA (Khalfan *et al.*, 1986) or Cascade blue (Whitaker *et al.*, 1991) may be considered (Table 12.1). These dyes require excitation by ultraviolet light, but suitable filter combinations are now available (Table 12.3). AMCA, in particular, is extremely susceptible to photobleaching, even from ambient room light, and the slides should be kept shielded from light by wrapping in foil.

12.7.3 Multicolour Fluorescence with the FACS

As discussed previously (see Section 12.6.1), the most common combination of flurochromes for two-colour flow cytometry is FITC and *R*-phycoerythrin, using the 488 nm argon laser line for excitation. If a third colour is to be added, there is now a range of possibilities (Table 12.2; reviewed by Stewart and Stewart, 1993). Most of the current 'third colour' reagents use phycoerythrin that has been coupled with a synthetic fluorochrome such as Cy5 or Texas red, such that the emission of phycoerythrin is captured and rereleased at a wavelength that is further into the red. The main problem with such 'tandem' conjugates is that not all the emission from phycoerythrin is captured, and there is some 'leakage' resulting in cross-channel spill (Stewart and Stewart, 1993). According to the manufacturer, compensation of about 20% is required to eliminate the FITC overlap into the *R*-phycoerythrin signal, while only 0–3% compensation is required to eliminate the FITC overlap into the Quantum Red signal.

The introduction of peridinin chlorophyll protein (Rechtenwald *et al.*, 1990; Stewart and Stewart, 1993; marketed as PerCP by Becton-Dickinson) has been a significant improvement. It is efficiently excited at wavelengths from 380–560 nm, and its emission spectrum is particularly narrow and well into the red (650–700 nm). This fluorochrome is now widely used.

12.8 Conjugation of Antibodies with Fluorochromes

Previously, it was common for individual investigators to prepare their own fluorescent antibody conjugates. This time has now almost passed. The quality of commercially available fluorescent antibodies has improved greatly, and in most cases it is best to purchase ready-made conjugates. None the less, procedures for conjugation will be given here, because there is still sometimes the need for preparation of conjugates in special circumstances.

The conjugation of antibodies with synthetic fluorochromes is technically very simple to perform, but the best results will only be obtained if care is taken to follow certain principles. Bright, specific fluorescence is possible only if the antibodies are optimally conjugated.

12.8.1 General Principles

The conjugation of antibodies with the isothiocyanate derivatives of fluorescein or rhodamine proceeds by nucleophilic attack of the unprotonated ε-amino group of lysine on the fluorochrome, resulting in a thiourea bond (Fig. 12.2). Maximal efficiency of conjugation is achieved at pH 9.5, where a large fraction of lysines are unprotonated. Extraneous nucleophiles, such as Tris, amino acids, ammonium ions or azide will inhibit conjugation (Lachman, 1964).

Thiourea bonds are not completely stable, and the recent availability of succinimide esters of fluorescein and tetramethyl rhodamine from Molecular Probes may be a significant advance, as they form a more stable amide bond. In addition, the quantum yield of fluorescence from these compounds is significantly higher than from the isothiocyanate derivatives, perhaps because of less self-quenching (Haugland, 1992).

12.8.2 Susceptibility of Fluorochromes to Hydrolysis

The isothiocyanate and succinimide esters of fluorochromes are very susceptible to hydrolysis, even by moisture in the air. They must be stored in a dessicator, protected from light. If stored in the cold, they must be allowed to warm to room temperature before the bottle is opened to avoid condensation. It is more important that they be kept dry than cold. Almost all preparations of fluorochromes contain a certain proportion of hydrolysed material, which necessitates adding somewhat more total fluorochrome to achieve the desired conjugation ratio.

During conjugation of fluorochromes to antibodies, there is a competing hydrolysis reaction. At high protein concentrations (i.e. >10 mg/ml), conjugation to protein is strongly favoured, and efficiencies of up to 70% coupling may be achieved. At lower protein concentrations, the competing hydrolysis becomes very significant (The and Feltkamp, 1970 a, b). It is perfectly feasible to conjugate antibodies at a protein concentration of 1–2 mg/ml, but it is necessary to add much more fluorochrome, and greater care will be needed to make sure that all unreacted fluorochrome is removed. This is easily achieved by gel filtration, typically using Sephadex G-25 or equivalent. Any residual unconjugated fluorochrome will cause background nonspecific fluorescence, degrading the specificity that is achieved.

12.8.3 Solubility of Fluorochromes

Some fluorochromes, notably the rhodamine derivatives, are rather insoluble in water. During attempts to dissolve them, hydrolysis takes place, leading to erratic results. Addition of the fluorochrome to the protein as a solid is not recommended because it will lead to uneven labelling, and in any case does not solve the solubility problem. In addition, it is much easier to deliver a measured volume of a solution than to weigh accurately a very small mass of fluorochrome.

Conjugations will be found to be much easier and predictable if the fluorochrome is made up as a stock solution (10 or 1 mg/ml) in dimethyl sulfoxide (DMSO), and the desired volume added to the protein dropwise with stirring (Bergquist and Nilsson, 1974; Goding, 1976). Even FITC is more easily handled in this way. The solution should be made up immediately before use because the DMSO will usually contain some water. Unlike most organic solvents, DMSO does not denature proteins when used in low concentrations (1–10% v/v).

Sulfonyl chlorides react with DMSO, and this solvent must not be used with Texas red. Good results are obtained by adding Texas red as a powder, perhaps because its hydrolysis product is very hydrophilic. If a solvent is needed, dimethyl formamide or acetonitrile are suitable.

12.8.4 Achievement of Optimal Fluorochrome:Protein Ratio

The molar ratio of fluorochrome:protein (F/P) is one of the most important parameters in immunofluorescence. If this ratio is too low, the intensity of fluorescence will be poor. The ratio should certainly not be less than 1.0, because unlabelled antibodies would compete with labelled antibodies for binding sites, reducing the sensitivity. The upper limit is determined by a number of factors.

In the case of fluorescein, conjugation results in a marked lowering of the isoelectric point of the antibody, and overconjugation may produce highly charged acidic species which tend to stick nonspecifically to cells, especially if they are fixed. Optimally FITC-conjugated antibodies have a molar F/P ratio of *c*.2–3 for fixed cells, and *c*.4–6 for intact living cells. Self-quenching due to overconjugation (Sections 12.1 and 12.8.5) is not a serious problem with FITC conjugates.

The molar F/P ratio for FITC may be calculated from a formula proposed by The and Feltkamp (1970a, b):

$$\text{F/P ratio} = \frac{2.87 \times OD_{495}}{OD_{280} - 0.35 \times OD_{495}}$$

Similarly, the concentration of IgG may be calculated:

$$[IgG] = \frac{OD_{280} - 0.35 \times OD_{495}}{1.4}$$

These formulae correct for the contribution of FITC to the total absorbance at 280 nm and produce a figure that is sufficiently accurate for practical purposes.

Determination of the optimal conjugation ratios for other fluorochromes is not so easy. For reasons that will become apparent in the next section, it is not possible to calculate exact molar F/P ratios from the absorption spectra of rhodamine conjugates. The best basis for comparison of conjugation ratios is the ratio of absorbance at the peak in the visible region to the absorbance at 280 nm.

The various rhodamine derivatives are more hydrophobic than FITC, and even moderately conjugated antibodies tend to precipitate. Precipitation of some of the antibody after conjugation with rhodamine derivatives is not disastrous. The precipitated material probably represents the more heavily conjugated molecules, and the supernatant after centrifugation may contain highly active fluorescent antibodies. An additional problem with rhodamine derivatives is that they are very susceptible to self-quenching if antibodies are overconjugated (see next section). Self-quenching is easily mistaken for inactivation of antigen-binding.

12.8.5 Self-quenching of Rhodamine Derivatives and Other Fluorochromes

When two or more fluorochromes are in very close physical proximity, their complex light-absorbing and light-emitting properties interact, resulting in altered fluorescence. A frequent outcome of this interaction is self-quenching, in which the quantum yield is drastically reduced (see Southwick *et al.*, 1990). The most important practical implication of this effect is that the molar ratio of fluorochrome:antibody must be carefully optimized because excessive conjugation will result in greatly diminished fluorescence. The following discussion illustrates the problem in the case of TRITC, but the same principles apply for many other fluorochromes, notably Texas red, XRITC and Cy5. Self-quenching is not a problem with protein fluorochromes such as phycoerythrin.

The absorption spectrum of unbound TRITC in the visible range shows a major peak at 550 nm with a small shoulder at 515–520 nm. However, when TRITC is conjugated to IgG, the shape of the visible absorption peak changes (Fig. 12.5). As the molar F/P ratio increases, the shoulder at 515–520 nm becomes progressively more prominent until it almost reaches the same height

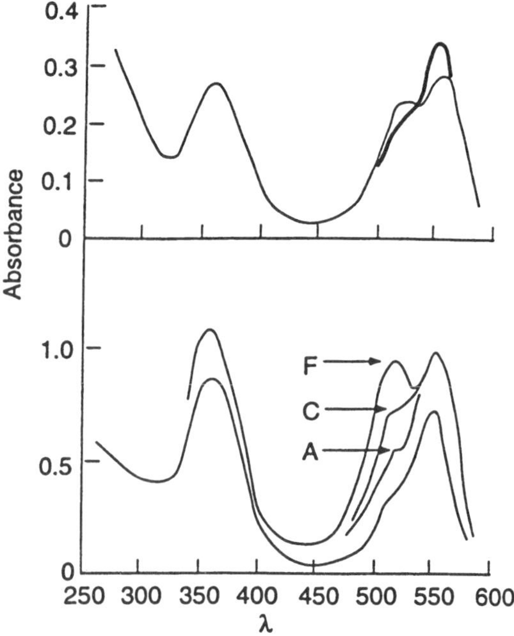

Fig. 12.5. Effect of close proximity of TRITC molecules to each other. Upper panel: absorption spectrum of moderately heavily conjugated TRITC conjugated protein before pronase digestion (thin line) and after pronase digestion (thick line). Lower panel: absorption spectrum of free TRITC (unlabelled line), lightly conjugated protein (A), moderately conjugated protein (C), and heavily conjugated protein (F). Conjugates were diluted such that their absorbances at 550 nm were equal.

as the 550 nm peak. A similar absorption spectrum has been observed for stacked dimers of fluorescein (Forster and Konig, 1957), for fluorescein bound to monoclonal antifluorescein antibody 20–30–3 (Kranz and Voss, 1981) and also for fluorescein in its monomeric protonated form at pH 5.5 (Alexandra *et al.*, 1980). When heavily conjugated TRITC–IgG is digested with pronase, the 515 nm peak decreases and the 550 nm peak increases (Fig. 12.5), suggesting that the spectral changes are due to rhodamine–rhodamine interactions. It is obvious from these results that simple calculations based on spectral data cannot allow accurate assessment of the molar F/P ratios of TRITC conjugates.

Rhodamine–rhodamine interactions cause severe self-quenching. As the molar F/P ratio increases, the quantum yield of fluorescence declines drastically (Fig. 12.6). Digestion of heavily conjugated TRITC-IgG with pronase results in a very large increase in fluorescence (Fig. 12.6), indicating that the decline is due to self-quenching by rhodamine–rhodamine interactions.

The foregoing makes it clear that heavily rhodamine-conjugated antibodies

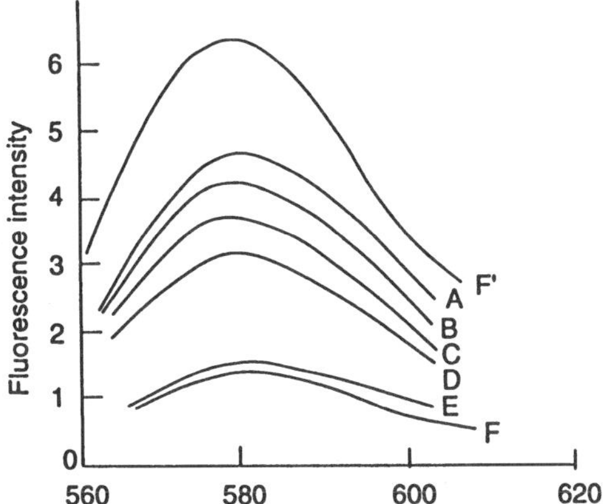

Fig. 12.6. Demonstration of self-quenching of TRITC. A series of conjugates of TRITC to IgG were prepared, with increasing ratios of fluorochrome (A–F), and their concentration adjusted to give equal absorbance at 550 nm. They were then excited at 550 nm, and the fluorescence emission spectrum measured. It is apparent that the quantum efficiency of emission decreases markedly with increasing conjugation ratio. When the most heavily conjugated preparation (F) was treated with pronase, there was a marked increase in quantum yield (F'), showing that self-quenching occurs due to proximity of TRITC molecules to each other. The OD_{550}:OD_{280} ratios of the various conjugates were as follows: A, 0.27; B, 0.30; C, 0.40; D, 0.49; E, 0.58; F, 0.60.

may have much weaker fluorescence than more lightly-conjugated antibodies, and practical experience bears out this point. The phenomenon of self-quenching has probably been seen repeatedly by immunologists, but erroneously ascribed to antibody inactivation. Similar phenomena are observed with Texas red conjugates (Titus *et al.*, 1982).

A quantitative study of self-quenching of fluorochromes was carried out by Southwick *et al.* (1990) as part of a wider study on the cyanine dyes. It was found that self-quenching was very severe with the rhodamine derivative XRITC, moderately severe with the red-emitting dye Cy5.12, but minimal for FITC. In the case of FITC, increasing the number of fluorophores per IgG leads to increasing fluorescence per IgG—up to about 12 FITC molecules/IgG. In the ascending part of the curve, the self-quenching was compensated by the increased number of fluorochromes per IgG, although the quantum yield per FITC molecule steadily decreased. Beyond about 12 FITC molecules/IgG, the fluorescence per IgG molecule decreased and self-quenching predominated. For XRITC, peak fluorescence is at about 2–3 molecules/IgG, and for Cy5.12, about 4 molecules/IgG (Southwick *et al.*, 1990).

The chemical properties of commonly used fluorochromes are listed in Table 12.5, and recommendations for optimal conjugation ratios are given in Table 12.6.

Table 12.5 Chemical properties of fluorochromes

Fluorochrome	Supplier*	Comments
FITC	RO, MP	pKa ≈ 6.5. Absorbance and emission maximal at pH > 8
DTAF	RO	Similar spectral properties to FITC. Active form more stable
TRITC	RO, MP	Rather hydrophobic; dissolves easily in dimethyl sulfoxide
Texas red	MP, RO	Hydrolyses rapidly. Hydrolysis product very soluble in water and not prone to adsorption to protein. Incompatible with dimethyl sulfoxide
Phycoerythrin	BD, MP	Protein from red algae. Virtually no emission below 550 nm. Very high absorbance and quantum yield. Requires special and somewhat complex procedures for coupling to proteins
Succinimide esters of fluorescein, tetramethyl rhodamine	MP, RO	Very stable bond to protein; competing hydrolysis reaction means that conjugation is inefficient at low protein concentrations
Cy3	BDS, RO	Supplied as succinimide ester (see above)
Cy5	BDS, RO	Supplied as succinimide ester (see above). Absorption and emission is at longer wavelengths, and not very useful for visual assessment. Not efficiently excited by mercury lamp or argon laser, but well excited by krypton laser. Often used in tandem with *R*-phycoerythrin to give long wavelength emission for three-colour staining
Cascade blue	MP	Supplied as succinimide ester (see above). Must be excited in ultraviolet region. More stable than AMCA to photobleaching. Intense blue emission
AMCA	JI	Supplied as succinimide ester. Blue emission. Must be excited in ultraviolet region. Extremely sensitive to photobleaching. Slides must be protected from ambient room light

* RO, Research Organics; MP, Molecular Probes; BD, Becton-Dickinson; BDS, Biological Detection Systems; JI, Jackson ImmunoResearch.

Table 12.6 Suggested conditions for labelling antibodies with fluorochromes

Fluorochrome	Protein concentration (mg/ml)	Mass of fluorochrome* (μg/mg protein)	Ideal absorbance ratio	Conditions
FITC	>10	10–20	495:280 = 0.5–1.0	pH 9.5, 20°C, 2h
	3	≈50	495:280 = 0.5–1.0	pH 9.5, 20°C, 2h
	1	≈100	495:280 = 0.5–1.0	pH 9.5, 20°C, 2h
DTAF	>10	<15	495:280 = 0.5–1.0	pH 9.0, 20°C, 2h
TRITC	>10	5–15	550:280 = 0.2-0.3	pH 9.5, 20°C, 2h
XRITC	>10	5–15	Unknown	pH 9.5, 20°C, 2h
Texas red	1–2	75–150	596:280 = 0.8–1.0	pH 9.0, 4°C, 1–2h
R-PE			565:280 = 5.3	See Text

* The figures given in this table are guidelines only; the mass of fluorochromes may need to be adjusted to achieve the recommended absorbance ratios.

12.8.6 Practical Procedure for Conjugation of Antibodies with FITC or TRITC

Antibodies for conjugation should be at least partially purified. Ammonium sulfate fractionation is adequate, provided that it is performed carefully (see Section 9.2.1). However, non-immunoglobulin proteins remaining after crude purification may result in higher background staining. In particular, the presence of transferrin might cause artefactual labelling of cells with transferrin receptors (Goding and Burns, 1981). The best results are obtained with highly purified antibodies (Goding, 1976).

Antibodies to be conjugated with FITC, TRITC or XRITC should be dialysed overnight against 2–3 changes of pH 9.5 carbonate/bicarbonate buffer (8.6 g Na_2CO_3 and 17.2 g N_aHCO_3 to 1 l H_2O), while antibodies to be conjugated with DTAF or Texas red should be dialysed against 0.025 M borate buffer, pH 9.0 (4.76 g $Na_2B_4O_7$ H_2O in 1 l H_2O, titrated to pH 9 with *c*.46 ml 0.1 M HCl). In addition to obtaining the optimal pH for conjugation, dialysis removes any inhibitors of conjugation such as amines or azide.

The optical density of the protein at 280 nm should be measured, and the concentration calculated. The concentration of IgG in mg/ml is obtained with sufficient accuracy by dividing the OD_{280} (1 cm cell) by 1.4. The total protein mass should be calculated, and the desired amount of fluorochrome calculated using the guidelines set out in Table 12.6.

The fluorochrome must be allowed to reach room temperature before the bottle is opened, to prevent condensation. The fluorochrome is weighed out and made up to 1.0 or 10.0 mg/ml in DMSO (except Texas red; see Section 12.3.4), and the desired volume added to the protein dropwise, with stirring.

The reaction is allowed to proceed at room temperature, shielded from light. Thermal movement is quite sufficient and stirring serves no purpose once the initial mixing has taken place.

After 1–2 h at room temperature the reaction will be essentially complete. The reaction proceeds somewhat more rapidly at high protein concentrations, and an incubation time of 2–3 h may be desirable if the protein concentration is less than 1–2 mg/ml. If the protein concentration is high (more than about 10 mg/ml), the initial fluorochrome:protein ratio should be held at the lower end of the recommended range, and the main risk is one of over-conjugation. At low protein concentrations (1–2 mg/ml), over-conjugation is not very likely, and much higher initial ratios are needed to achieve adequate conjugation (Table 12.6). Older batches of fluorochrome may require higher initial ratios.

After conjugation, unreacted or hydrolysed dye must be separated from the protein. Gel filtration on Sephadex G-25 in PBS containing 0.1% NaN_3 is convenient and effective. The disposable PD-10 columns (Pharmacia) are ideal. The first coloured band to emerge is the conjugated protein. Removal of free dye by dialysis is very slow and inefficient, and is not recommended. Even small traces of the free dye may cause significant nonspecific fluorescence. If a clear separation between free and protein-bound dye cannot be achieved using PD-10 columns, longer columns will be necessary.

In the older literature, it is frequently recommended that the conjugate be further purified by ion exchange chromatography to remove over-conjugated molecules (Cebra and Goldstein, 1965; Goding, 1976). This procedure is seldom used today. Provided that care has been taken in all the aforementioned aspects, ion exchange chromatography is seldom necessary. However, ion exchange chromatography will sometimes be required when fixed and permeabilized cells are used.

Principles and practical details of ion exchange chromatography are discussed in Chapter 9. The conjugate should be passed over a Sephadex G-25 or Biogel P-6 column equilibrated with 10 mM Tris-HCl, pH 8.0. (The same procedure can be used to simultaneously remove unbound fluorochrome.) A column of DEAE-Sepharose or DEAE-Sephacel equilibrated with the same buffer will bind virtually all conjugates. As the salt concentration is raised, progressively more heavily conjugated antibodies will be eluted (Cebra and Goldstein, 1965; Goding, 1976).

The absorbance of each conjugate should be measured at 280 nm and at its peak in the visible region. Conjugates are more susceptible to aggregation than unconjugated antibodies, and are best stored at 4°C, protected from light.

12.8.7 Coupling of *R*-phycoerythrin to Antibodies—General Aspects

The coupling of fluorescent proteins such as phycoerythrins to antibodies is more complex than that of synthetic dyes (see Hardy, 1986). The most common way to couple phycoerythrin to antibodies is via heterobifunctional agents such as succinimidyl *trans*-4-(*N*-maleimidylmethyl) cyclohexane-1-carboxylate (SMCC; see Kitagawa and Aikawa, 1976; Liu *et al.*, 1979; O'Sullivan *et al.*, 1979; Yoshitake *et al.*, 1979; Lee *et al.*, 1980; Youle and Neville, 1980; Vallera *et al.*, 1982; Volkman *et al.*, 1982; Ji, 1983; Wong, 1991). The use of heterobifunctional reagents minimizes the formation of like–like coupling, but it does not prevent the formation of higher molecular weight polymers. These polymers may not be deleterious; in fact they may cause a very useful increase in sensitivity.

Earlier procedures for coupling phycoerythrin to antibodies involved the formation of disulfide bonds (Oi *et al.*, 1982). Thiolated phycoerythrin was prepared by treatment with iminothiolane hydrochloride, and antibodies were coupled with protected thiol groups using *N*-succinimidyl 3–2(*p*-pyridylthio)-propionate (SPDP), and the two proteins were mixed to allow formation of disulfide bonds. Disulfide bonds prepared in this way are not very stable, and this was borne out by experience. For this reason procedures that generate stable nonreducible thioether bonds (see next Section) are to be preferred.

12.8.8 Procedure for Coupling of *R*-phycoerythrin to Antibodies

This procedure is slightly modified from that of R. Haugland, as described in a brochure from Molecular Probes (see also Haugland, 1995). It proceeds in three basic steps. First, the antibody is derivatized with a succinimidyl ester maleimide derivative such as SMCC (Molecular Probes, cat. no. S-1534). This reacts with lysine residues by the same mechanism as biotin succinimide ester (see Section 10.3.1), and adds thiol-reactive maleimide groups to the antibody. The second step produces thiolated phycoerythrin by derivatizing it with *N*-succinimidyl 3-(2-pyridyldithio-) propionate (SPDP; Molecular Probes or Pharmacia), which also couples to lysine residues in the same way, but leaves a protected disulfide bond. This disulphide bond is then reduced with dithiothreitol (DTT) to remove the protecting group and create a free thiol group. After removal of the reducing agent and reaction byproducts, the third step is performed by mixing the thiolated phycoerythrin with the maleimide-derivatized antibodies. The free thiol group on the phycoerythrin attacks the maleimide leaving group on the antibody and forms a stable thioether bond. Finally, the conjugate is chromatographically separated from unreacted proteins.

It is important to perform the first two steps separately but simultaneously to avoid the loss of maleimide reactivity due to hydrolysis and to avoid the oxidation of the thiol group. The succinimide esters must be stored dry, and made up in dimethyl sulfoxide immediately before use, as they are very susceptible to hydrolysis. All procedures should be carried out with no delay between steps. Antibodies and phycoerythrin must be dialysed before use, as any residual ammonium, Tris or azide ions will inhibit the coupling of succinimide esters.

(1) Dissolve the antibody at about 5 mg/ml in PBS, pH 7.5. Dissolve SMCC (Ji, 1983) to a concentration of about 5 mM in dimethyl sulfoxide. Add an amount equal to 10–20 mol equivalents to the antibody solution and incubate for 2 h at room temperature.

(2) Couple the phycobiliprotein with SPDP by the same protocol as for step 1, but aim to achieve an average of about 2 moles of protected disulfides per mole of phycobiliprotein. The guidelines in Section 10.3.1 should be helpful. If the phycobiliprotein is at 5 mg/ml, the reaction will be reasonably efficient (Fig. 10.2). Remove excess SPDP and its hydrolysis products by filtration on a PD-10 column (Sephadex G-25; Pharmacia), or by dialysis (less than 24 h). Thiolated phycoerythrin is available from Molecular Probes.

(3) Treat 5 mg of the phycoerythrin that has been coupled with the protected thiol groups (step 2) with about 35 μl of freshly prepared aqueous 0.5 M DTT per mg of SPDP modified protein. After 15 min, remove the thiolated protein from the DTT by dialysis (less than 24 h) or gel filtration on a Sephadex G-25 column (PD-10; Pharmacia).

(4) Mix the antibody and the thiolated phycoerythrin prepared as above,

and leave in the cold for 16–20 h to form thioether bonds. Terminate the reaction by adding a 20-fold excess of *N*-ethyl maleimide to 'cap' any remaining thiol groups.

(5) Chromatograph the conjugate on Sephacryl S-300 or Bio-Gel A. 1.5m (10 × 750 mm), as described by Kronick and Grossman (1983) and Haugland, (1995; section 9.2.4). The coloured conjugate will elute before the individual proteins because it has a higher molecular weight. The second coloured peak will be unconjugated phycoerythrin, with a third (noncoloured) peak being unconjugated antibody.

(6) Collect small fractions, and measure the optical density at 280 nm and at 565 nm. For *R*-phycoerythrin, optimally conjugated antibodies will have an OD_{565}:OD_{280} ratio of about 5.3.

(7) Store the conjugates at 4°C in PBS with 0.05% azide to inhibit microbial growth.

Alternatively, the conjugates can be chromatographed on hydroxyapatite (Hardy, 1986). Dialyse the crude conjugate against 1 mM sodium phosphate in 0.1 M NaCl, pH 7.0, and then pass over a small hydroxyapatite column in the same buffer. Elute with a linear gradient of 0.001 M phosphate to 0.05 M phosphate in 0.1 M NaCl. The unconjugated phycoerythrin and antibody elute first, followed by the higher molecular weight conjugates in order of increasing molecular weight. The highest molecular weight conjugates will elute at 0.05 M phosphate or higher. Analyse conjugates by optical density as above.

Finally, all conjugates must be tested functionally at several different concentrations for activity and specificity.

12.8.9 Storage of Phycobiliproteins and Phycoerythrin-Antibody Conjugates

Phycobiliproteins are very stable in 60% saturated ammonium sulfate at 2–5°C, and can be recovered by centrifugation and resuspension in PBS, then dialysed against the same buffer. They must not be frozen. Phycoerythrin-conjugated antibodies should be stored at 4°C, and also must not be frozen.

12.8.10 Bispecific Monoclonal Antibodies: 'Quadromas'

When two antibody-secreting cells are fused, the random assembly of light and heavy chains results in mixed molecules (Cotton and Milstein, 1973; see Section 7.2.3). This observation can be used to generate 'bispecific' antibodies which can have many uses, particularly in immunocytochemistry (Milstein and Cuello, 1984, Suresh *et al.*, 1986a, b; reviewed by Fanger, 1995).

Milstein and Cuello (1984) described a system in which HAT sensitivity was reintroduced into an anti-peroxidase hybridoma. Upon fusion with spleen cells, hybrid antibodies containing one anti-peroxidase arm and one anti-X arm were produced. This procedure allows very simple and efficient coupling of the enzyme to antibodies and could easily be adapted for use with phycoerythrin (see Milstein and Cuello, 1984).

When two hybridomas are fused to produce a bispecific cell line, the resulting cell line has been termed a 'quadroma' (see also Staerz and Bevan, 1986; Lanzavecchia and Scheidegger, 1987). There is a need for improved selection procedures for such lines, but the approach is powerful.

12.9 The Biotin–Avidin System

The biotin-avidin system has been discussed in detail in Section 10.3. It may be regarded as a 'lock and key' system that can be exploited in many different ways, and is particularly useful in multicolour immunofluorescence. The structures of biotin and its *N*-hydroxysuccinimide ester are shown in Fig. 10.2.

The biotin–avidin system is particularly attractive as a sandwich system used in conjunction with antibodies (Heitzmann and Richards, 1974; Heggeness and Ash, 1977; Wilchek and Bayer, 1988, 1990; Savage *et al.*, 1992). Biotin is relatively polar, and can be coupled to antibodies under very mild conditions with little disruption to their structure. Avidin may then be used as a stable high-affinity second-step reagent which may be coupled with fluorochromes, enzymes, gold or other molecules. The use of avidin as a sandwich reagent also avoids the use of anti-immunoglobulin antibodies. Finally, the biotin–avidin system is appealing because only one conjugate needs to be prepared and characterized for all affinity systems, and both components are commercially available and inexpensive.

12.9.1 Exploitation of the Tetrameric Structure of Avidin and Streptavidin in Sandwich Systems

Avidin and streptavidin are both tetrameric proteins made up of four identical subunits, each of which binds one biotin molecule. It is therefore possible to devise 'sandwich' systems in which an antigen is labelled with a biotin-conjugated antibody, washed, incubated with avidin or streptavidin, then incubated with a fluorescent or enzyme-labelled biotinylated protein, which will bind to the 'free' biotin-binding sites of the avidin or streptavidin. This capability gives the system much flexibility.

12.9.2 Enhancement of Fluorescence of Avidin and Streptavidin by Addition of Free Biotin

Fluorophores coupled to avidin or streptavidin can apparently interact with amino acids in the biotin-binding pockets, causing quenching of fluorescence.

The incubation of the avidin-biotinylated antibody complexes with 10 μM free biotin after all washes may cause a substantial increase in fluorescence (Haugland, 1992). This is presumably possible because avidin binds biotinylated proteins multivalently, and therefore free biotin does not cause uncoupling of the complexes.

12.9.3 Conjugation of Antibodies with Biotin Succinimide Ester

The procedures for coupling of biotin succinimide ester to proteins are described in Section 10.3. When biotinylating antibodies, the most important parameter is the ratio of biotin to IgG. If too little biotin is used, the sensitivity will be poor. If too much is used, there comes a point where the antibody may become inactivated and/or aggregated, with reduction in both sensitivity and specificity (Wadsley and Watt, 1987; Muzykantov *et al.*, 1995). Because of the competing hydrolysis reaction, the amount of biotin succinimide ester to be added to the antibody must be matched to the antibody concentration as well as its mass. Unlike fluorochromes, biotin does not absorb light in the visible region of the spectrum, and there is no easy way to measure the biotin:protein ratio of conjugated antibodies. It is therefore best to adhere closely to the conditions described in the following section.

12.9.4 Practical Procedure for Coupling Biotin to Antibodies

The antibody to be conjugated should be at least partly purified by careful ammonium sulfate precipitation followed by two washes in 40–50% saturated ammonium sulfate (see Section 9.2.1). More highly purified antibodies will give better results.

The antibody should then be dialysed overnight against 0.1 M $NaHCO_3$, pH 8.0–8.3. If precipitation with ammonium sulfate has been used, it is advisable to change the dialysis fluid once or twice. Following dialysis, the IgG concentration should be measured from the absorbance at 280 nm (OD_{280} of 1.4=1.0 mg/ml), and the protein concentration adjusted to 1.0 mg/ml by dilution with 0.1 M $NaHCO_3$.

The biotin succinimide ester should be warmed to room temperature before the bottle is opened. The ester is then weighed out and dissolved in DMSO to 1.0 mg/ml immediately before use. It is freely soluble in DMSO, which is important because any delays in dissolving it in water will result in hydrolysis. The half-life of succinimide esters in 0.1 M $NaHCO_3$ is of the order of minutes.

The ester solution is then added to the antibody, and mixed immediately. If the conditions of pH and protein concentration given above are followed strictly, it will be found that 120 μl of the ester solution per ml of antibody will usually result in optimal conjugation. Sometimes it may help to vary the ratio of biotin: protein in the range 60 μl/ml to 240 μl/ml. Under-conjugation results in weak staining; over-conjugation results in damage to the antibody and loss of specificity (Wadsley and Watt, 1987; Muzykantov *et al.*, 1995).

The mixture should be kept at room temperature for 1–2 h, by which time the reaction will have gone to completion. The biotin-conjugated protein should then be dialysed overnight against PBS with 0.1% sodium azide, and stored at 4°C. It is best not to freeze conjugated proteins, as they are susceptible to aggregation.

Avidin conjugated with FITC or TRITC is available from many suppliers. Optimally conjugated FITC-avidin has an OD_{496}:OD_{280} ratio of 1.4–1.6, while optimally conjugated TRITC-avidin has an OD_{550}:OD_{280} ratio of about 0.5.

12.9.5 Problems with the Avidin–Biotin System

When used to label intact living cells, the biotin–avidin system is capable of excellent results. The background nonspecific binding is very low, and the fluorescence is bright. However, the extremely basic nature of avidin (pI = 10.5) may cause it to bind electrostatically to acidic structures. When FITC–avidin is used to stain permeabilized and fixed cells, binding to condensed chromatin is observed (Heggeness, 1977). The chromosomes are beautifully outlined, and banding of *Drosophila* salivary gland chromosomes is vividly illustrated.

The nonspecific binding of avidin to DNA may be diminished by raising the salt concentration (0.3 M KCl; Heggeness, 1977), and by competition with a large excess of the basic protein cytochrome C. A combination of KCl (0.3–0.5 M) and cytochrome *c* (1 mg/ml) lowers the background considerably. Finn *et al.* (1980) have shown that the nonspecific binding of avidin may also be diminished by exhaustive succinoylation of lysine groups. Alternatively, it is possible to use anti-biotin antibodies (Barger, 1979). Monoclonal antibodies to biotin would be ideal.

Avidin is a glycoprotein, and the binding of its glycan moieties to tissue lectins can also sometimes cause nonspecific staining. Various forms of chemically modified avidin, with its carbohydrate removed and/or chemically treated to raise its isoelectric point, are commercially available (e.g. Neutravidin, Pierce cat. no. 3100).

An alternative to avidin is streptavidin, a biotin-binding protein from *Streptomyces*. Streptavidin has four identical chains of similar molecular weight to those of avidin, but a much less basic isoelectric point. It is not glycosylated. However, it has been found that streptavidin can also exhibit 'nonspecific' binding to cells. This has been shown to be caused by the presence of an amino acid sequence TG<u>RYD</u>SAPAT, which closely resembles the sequence TG<u>RGD</u>SPAS that is found in fibronectin, which binds to many cell surface receptors (Alon *et al.*, 1990).

Another potential problem with the avidin–biotin system is the fact that most tissue culture media contain biotin. It is therefore important to wash the cells in a biotin free buffer prior to use of avidin, and to avoid the use of tissue culture medium at any stage during staining. Failure to observe this point may result in very erratic results.

12.10 Fluorescent Substrates for Alkaline Phosphatase

Molecular Probes has recently introduced a new substrate for alkaline phosphatase, which is nonfluorescent but forms an intensely fluorescent yellow-green micro-precipitate upon removal of a phosphate group (Enzyme-Labeled Fluorescence; ELF™). Its properties are described by Larison *et al.* (1995). The ELF precipitate has an unusually large Stokes' shift, with optimal excitation at 365 nm and emission at >515 nm. The precipitate is extremely fine and the resolution is comparable to that of fluorescent antibodies. However, the intensity of fluorescence is much greater than that of fluorescent antibodies because each enzyme can produce hundreds or possibly thousands of fluorescent products. The resistance of the product to photobleaching is said to be much greater than for fluorescein (Larison *et al.*, 1995).

ELF is well suited to two-colour fluorescence with Cascade blue, and can also be used for three-colour fluorescence with Texas red or TRITC, using appropriate filters and double-exposure photography. Standard blue filter systems for DAPI/Hoechst long pass are suitable (Zeiss 487901; Omega Optics O-5702, O5703; see Table 12.4). It seems likely that this system will become very popular for immunofluorescence on fixed cells and histological sections.

12.11 Staining Cells with Fluorescent Antibodies

The staining of cells for immunofluorescence is very simple. Cells are held with the antibody for 30–60 min, washed and examined. Provided the antibodies have the correct specificity and are not under- or over-conjugated, and provided the microscope is suitable, little difficulty will be experienced at this stage.

If cells bearing Fc receptors are present, it may be necessary to deaggregate antibodies just prior to staining. Deaggregation should be at 100 000 ***g*** for 5–10 min, and the top half of the liquid used. The bottom half may be saved and recycled. The Beckman Airfuge or TL-100 benchtop ultracentrifuge are ideal, because of their rapid acceleration and deceleration, and because their rotors are well suited to tubes with small volumes.

Alternatively, binding to Fc receptors may be abolished by use of Fab or $F(ab)_2$ fragments (see Chapters 5 and 9). The use of such fragments might need to be considered for both first and second steps. The binding of the second-step reagent via Fc receptors may be suspected if controls omitting the first antibody are not totally negative. Protocols for the generation of $F(ab')_2$ and Fab fragments of sheep and goat antibodies are given in Chapter 15.

It is strongly advisable to titrate all reagents, and to use concentrations that are slightly higher than those needed to cause maximal fluorescence. Use of higher concentrations is wasteful, and results in poorer signal:noise ratio. The logic of Fig. 11.1 is applicable to any affinity system, including immunofluorescence.

The kinetics of binding of individual monoclonal antibodies to cells varies considerably. In some cases, particularly if a large excess of antibody is used, as little as 15 min incubation may be sufficient to achieve saturation. However, the occasional monoclonal antibody may take much longer (see Chapter 7). *If quantification of the intensity of fluorescence is needed, it is strongly recommended that the incubation times be held constant from one experiment to the next.*

Antibodies should be diluted in buffers containing irrelevant protein, such as 5% fetal calf serum or 1% bovine serum albumin (BSA). As a general rule, it is advisable to include 0.1% sodium azide in all buffers. This will discourage microbial growth and inhibit metabolism of the cells to be stained.

If indirect immunofluorescence is to be used, some thought should be given to the quality of the sandwich reagent. It makes no sense to use an extremely specific monoclonal antibody for the first step, and have a poorly defined second step. The amplification given by polyclonal second-step antibodies may be very useful in increasing the brightness, but care should be taken to use the optimal dilution. Whenever possible, affinity-purified second antibodies are preferred because they will have the least background.

12.11.1 Membrane Immunofluorescence

The membranes of intact, living cells are impermeable to antibodies. If the living cells are kept cold, labelling will be restricted to the membrane. The use of azide to inhibit cellular metabolism is not as effective as keeping the cells cold. If the cell is metabolically active, it is possible and indeed common for serum components to be taken up by fluid-phase endocytosis. In addition, antibodies attached to membrane structures can be endocytosed if the cell is metabolically active (Taylor *et al.*, 1971). These factors can be the source of experimental artefacts if they are not recognized. Whether or not antibodies can cross the membrane of endosomal structures and into the cytoplasm or nucleus remains unclear and controversial.

Dead cells often take up large amounts of fluorescent antibody in a non-specific manner. Care should be taken to ensure that the cells are highly viable, and dead cells should be excluded from any count. Dead cells are usually distinguishable from living cells under phase-contrast microscopy. They are a dull grey colour, while living cells are usually bright, refractile and smooth. The use of phase-contrast visible illumination in conjunction with immunofluorescence is strongly recommended.

12.11.2 Practical Procedure

(1) Prepare cell suspension, and wash 2–3 times in PBS containing 0.1–1.0% bovine serum albumin and 0.1% azide (medium). All steps should be carried out at 4°C.
(2) Count cells: adjust to 10^7/ml.
(3) To 100 μl cells, add desired volume of first antibody. Leave on ice for 30–90 minutes.
(4) Resuspend cells in 1–5 ml medium; centrifuge at 400 ***g*** for 5 min.
(5) Repeat step 4.
(6) Resuspend in 100 μl medium; add desired volume of second antibody. Leave on ice 30–90 minutes.
(7) Wash as before.
(8) Resuspend in 20–30 μl medium. The absorption and emission of FITC is markedly pH-dependent, and the mounting fluid should be neutral or slightly alkaline for maximal sensitivity. Place on a glass slide, and add coverslip. Seal edges with nail polish. Antifade reagents are useful, but are toxic for living cells (see Section 12.11.5)
(9) Examine immediately.

Cells may be fixed with 1% paraformaldehyde in 0.85% NaCl after staining, and examined immediately or kept at 4°C for at least a week (Lanier and Warner, 1981).

An alternative protocol, using adherent cells in 96-well trays, is well suited to the rapid screening of large numbers of monoclonal antibodies (Stitz *et al.*, 1988; see Sections 8.10.11 and 8.10.12).

12.11.3 Cytoplasmic Fluorescence

Reaction of antibodies with the cytoplasmic components of cells requires that they be fixed and made permeable to proteins. There are a large number of different ways of fixing cells, and it may be necessary to adapt the fixation procedure for each individual problem. In many cases, fixation and permeabilization are performed simultaneously by the use of organic reagents such as ethanol, methanol, acetone, or acetic acid. A variety of more sophisticated procedures are available. If the antigen to be detected by a monoclonal antibody is destroyed by one method of fixation, a method based on a different principle should be tried. Fixation of cells is discussed in more detail in Chapter 13.

In general, cytoplasmic staining is more prone to problems of nonspecific binding than is the case with living cells. It may be necessary to wash for prolonged periods (30–120 min) between steps. The addition of extra salt to the wash buffer (say 0.5 M final concentration) may help lower the background. It is often recommended that antibodies with lower F/P ratios should be used, and removal of over-conjugated molecules by ion exchange chromatography may sometimes be necessary. Absorption of conjugates with 'acetone liver powder' is a time-honoured but poorly understood procedure which may lower the background. It is one of the last of the 'black magic' methods of the 1950s to survive into the 1990s, and may still be found useful as a last resort.

Cells may be grown on coverslips or in 96-well flat-bottom tissue culture trays and processed *in situ* (see Stitz *et al.*, 1988, and Sections 8.10.11 and 8.10.12). Single-cell suspensions may be deposited onto slides as smears or using a cytocentrifuge (25 000 cells per slide in 100 μl of 50% fetal calf serum). Plasma cells may be air-dried and stored for long periods dessicated at –20°C. Other cells may require different handling.

Fixation by organic solvents

The simplest fixation procedure is a brief dip of the slide into acetone at –20°C (Gutman and Weissman, 1972; Rouse *et al.*, 1979). Alternatively, the slides may be fixed in 95% ethanol, 5% acetic acid at –12°C for 15 min (Kearney and Lawton, 1975). The exact temperature is not critical. A third procedure involves fixation in absolute methanol for 15 min (Goldstein *et al.*, 1982).

Whichever method is chosen, it is advisable to wash the slide thoroughly in PBS after fixation, and to keep it moist until staining. If the slide is allowed to dry out after fixation, the subsequent nonspecific staining may be severe.

Fixation by formaldehyde with subsequent permeabilization

Lazarides (1976) fixed cells in 3.5% formaldehyde in PBS for 30 min at room temperature, followed by three washes in PBS. The cells were then permeabilized by treatment with acetone (3 min in 50% acetone/water at 4°C; 5 min in pure acetone; 3 min in 50% acetone/water; 3 min in PBS). After brief drying, cells were exposed to antiserum.

Heggeness *et al.* (1977) fixed in 3% formaldehyde in PBS for 20–45 min at 20°C, rinsed in PBS and quenched any remaining aldehyde groups by washing in PBS containing 0.1M glycine. After a further rinse with PBS, cells were permeabilized by treatment with 0.1% Triton X-100 in PBS, followed by washing in PBS.

Treatment of cells with glutaraldehyde usually results in severe autofluorescence, but the reduction of aldehyde groups by sodium borohydride may abolish the problem.

12.11.4 Staining Procedure for Cytoplasmic Fluorescence

(1) Without allowing them to dry, transfer the rack of fixed slides into a dish containing PBS + 1% BSA and wash for about 1 h, stirring gently with a magnetic 'flea'. Change the medium a few times during washing.
(2) Place the slides horizontally in a humidified plastic box, and add 10–20 μl of appropriately diluted first antibody. Leave at room temperature for at least 30 min.
(3) Wash off excess antibody with PBS from a squirt bottle, then wash in PBS/BSA as in step 1.
(4) Repeat step 2, using second antibody.
(5) Repeat step 3.
(6) Add 1 drop of mounting medium, followed by a coverslip. Mounting medium may consist of 50% glycerol in PBS. The use of antifade reagents is highly recommended (see Section 12.11.5). As mentioned previously, the pH of the mounting medium should be neutral or slightly alkaline.
(7) Seal edges of coverslip with nail polish.

12.11.5 Photobleaching and its Prevention

Intense illumination of fluorochromes results in rapid fading of the image. The problem is particularly severe for FITC, but also occurs at a slower rate when TRITC is used. It is claimed that Texas Red is more resistant to photobleaching than TRITC (Titus *et al.*, 1982).

Many chemical reagents retard fading. The first of these to be discovered was phenylenediamine (Johnson and Araujo, 1981; Johnson *et al.*, 1982). The *o*- and *p*-isomers appear to be equally effective, and the effective concentration may be as little as 1–10 μg/ml (J. W. Goding, unpublished). Giloh and Sedat (1982) subsequently made a detailed and quantitative study of photobleaching, and found that n-propyl gallate (0.1–0.25 M, in glycerol) was effective for both FITC and TRITC. Another reagent that is very effective is DABCO, used at a concentration of 2.5% (Johnson *et al.*, 1982).

A number of newer proprietary anti-fade reagents are also available, such as SlowFade, SlowFade-Light and ProLong (Molecular Probes), or Vectashield (Vector Laboratories). Some of them have fairly specific advantages or disadvantages, such as almost totally preventing fading but at the cost of significant quenching of fluorescence (SlowFade), or slowing fading while preserving the initial intensity of fluorescence (SlowFade-Light; see Haugland, 1992). As a general rule, antifade reagents are toxic for living cells.

The mechanism of photobleaching probably involves molecular oxygen or oxygen-induced free radicals. Compounds that are effective antifade reagents probably work by 'mopping up' free radicals. Even in the absence of phenylenediamine or *n*-propyl gallate, photobleaching is a relatively slow process, with half-times measured in seconds. The duration of exposure of cells to laser light in the FAC is measured in microseconds, and so fading is unlikely to be a problem in flow cytometry.

12.12 Nonspecific Fluorescence

Specificity in biological systems is seldom absolute, and nowhere is this more true than in immunofluorescence. The various factors that tend to degrade specificity usually conspire to produce artefactual positive results (Winchester *et al.*, 1975; Goding, 1978), and it is absolutely crucial that rigorous negative controls be included in every experiment.

The causes of nonspecific fluorescence are manifold, and include inadequate removal of unconjugated fluorochrome, excessively high F/P ratios, binding of aggregated IgG to Fc receptors, binding of fluorescent transferrin to transferrin receptors, binding of streptavidin to receptors for RGD, electrostatic binding of basic proteins such as avidin to acidic cellular structures such as DNA, presence of fluorochrome-conjugated nonimmunoglobulin contaminants such as albumin, dead cells in membrane-staining experiments, contaminating antibodies (including antibodies against calf serum or albumin), inadequate washing, autofluorescence, and light-scattering by lenses.

Prevention of nonspecific fluorescence depends on the recognition that it is occurring (not always a trivial point), and the diagnosis of its cause. Some causes of nonspecific fluorescence with avidin or streptavidin have been discussed in Section 12.9.3.

12.12.1 Problems Due to the Antibody Preparation

If the antibody contains aggregates, binding to Fc receptors may occur (see Section 5.3.4). This problem may be made worse by the use of harsh purification procedures (e.g. elution from affinity columns with strongly denaturing buffers), or by multiple freeze–thaw cycles. Some antibodies (e.g. mouse IgG2b and IgG3) are intrinsically more susceptible to aggregation than others. The solution is to prevent aggregation if possible, to deaggregate before use (see Section 12.10), or to make Fab or $F(ab')_2$ fragments (see Chapters 9 and 15).

Contaminating antibodies are not a problem when hybridoma supernatants are used, but the serum of hybridoma-bearing mice will always contain small amounts of 'natural antibodies' against almost any antigen. If the serum from hybridoma-bearing mice is used at the optimal dilution (typically $1{:}10^{4-5}$), these contaminating antibodies are unlikely to be a problem. Purification by ion-exchange chromatography with careful gradient elution will also reduce the proportion of irrelevant antibodies.

A more serious problem exists for the sandwich anti-immunoglobulin reagents used for indirect immunofluorescence. If these are not first-rate, they will degrade the specificity of staining. It should be remembered that antibodies against BSA (Sidman, 1981) or fetal calf serum components (Johnsson *et al.*, 1976) are common in anti-immunoglobulin sera and may cause severe nonspecific fluorescence.

A third problem with the antibody preparation is the presence of non-immunoglobulin contaminant proteins. These are almost always more acidic than immunoglobulin. Conjugation with fluorochromes makes them even more acidic, and therefore susceptible to nonspecific electrostatic interactions.

Transferrin is a frequent contaminant in immunoglobulin preparations, and fluorochrome-labelled transferrin may cause artefacts by binding to transferrin receptors on proliferating cells (Goding and Burns, 1981). Similarly, antibodies to transferrin are present in many anti-immunoglobulin antisera, and may bind to cell-bound transferrin. Low-density lipoprotein also binds to cellular receptors, and could probably cause similar problems.

Some commercial antisera have soluble antigens added to them to 'absorb out' unwanted antibodies. This practice virtually guarantees that the antisera will contain antigen–antibody complexes that will bind to Fc receptors. The consumer should make certain that any absorption was done using immobilized antigen. Even if the free antigen does no harm on its own, unexpected reactions may occur when mixtures are used (e.g. in two-colour fluorescence). Liquid-phase absorption is no longer an acceptable practice.

Most of these problems are avoided by use of good quality antibodies, preferably highly purified before conjugation, and use of antibodies at optimal dilution. If necessary, anti-immunoglobulin sandwich reagents may be affinity-purified on immobilized immunoglobulin.

12.12.2 Problems Due to Conjugation

The presence of unconjugated fluorochrome in the antibody preparation is a potent source of nonspecific fluorescence. Removal of free fluorochrome by dialysis is very slow and inefficient, and gel filtration is preferable. There must be a clear area of gel lacking colour between the protein and the free dye. If there is not, the column is not big enough.

A second problem results from loose adsorption of free dye to protein. This problem is more severe with the more hydrophobic dyes such as TRITC, which tend to bind to protein and gradually leach off during staining. The result is a very uniform nonspecific fluorescence. The problem is not usually encountered with FITC or Texas red, because their hydrolysis products are very hydrophilic. Solving this problem is not easy; it is minimized by using highly purified IgG for conjugation (albumin binds many hydrophobic small molecules noncovalently) and by ensuring that the dye is dissolved in DMSO before use. Dialysis of conjugates against PBS containing 10% dimethyl sulfoxide may help (Goding, 1976).

Aggregation of IgG may occur during conjugation for a variety of reasons. The older practice of dissolving fluorochromes in acetone before addition to the antibody is a potent cause of aggregation, as is the adjustment of pH of antibody solutions by direct addition of NaOH. Aggregation may also occur if an excessive F/P ratio is used, and is particularly likely to occur with the more hydrophobic rhodamine derivatives.

12.12.3 Problems Due to the Antigen

If fluorescent anti-immunoglobulin is used, any immunoglobulin present will stain, regardless of its source. For example, roughly half of spleen cells and 10–20% of peripheral blood lymphocytes have membrane immunoglobulin (mostly IgM and IgD). Some cells may possess cytophilic immunoglobulin adsorbed from serum. Staining of immunoglobulins is avoided by using directly conjugated antibodies, or by use of the avidin–biotin system (see Sections 10.3 and 12.9). B cells expressing membrane IgG are rare, and anti-γ antibodies may often be used in situations where more broadly reactive anti-immunoglobulin antibodies would give excessively high background (Goding and Layton, 1976).

Some tissues or cells may autofluoresce under the near-ultraviolet light used in fluorescence microscopy. The problem may be lessened by use of good quality band-pass filters in the excitation path. Autofluorescence often gives a more yellow colour than the fluorescence of fluorescein. In some cases, counter-staining with other dyes may be used to disguise or eliminate autofluorescence (Gardner and McQuillin, 1980).

Cells grown in tissue culture may show autofluorescence, especially when used in the FACS. It is suspected that this is partly due to components of the tissue culture medium, notably riboflavin and phenol red.

12.12.4 Problems Due to Staining Technique

The problem of nonspecific staining of dead cells in membrane immunofluorescence has already been discussed (see Sections 12.6.2 and 12.11.1). In general, it is wise to include 'carrier' protein in all buffers, to saturate nonspecific protein-binding sites. Suitable proteins include BSA (0.1–1.0%) or fetal calf serum (1–10%), but the possibility that one or other antibody might cross-react with the carrier protein (see Section 12.12.1) should not be forgotten.

Inadequate washing between steps may also result in nonspecific fluorescence owing to free conjugate or resultant antigen–antibody complexes. The amount of washing that is necessary will be kept to a minimum if all reagents are used at their optimal dilutions. If a single-cell suspension is used, two washes are usually sufficient between steps. Conversely, cytoplasmic fluorescence of fixed cells or histological sections may require washes over a period of 30–120 min to remove nonspecifically bound antibody.

In addition to nonspecific staining of the specimen, inadequate washing may give rise to a very uniform fluorescent background over the entire field. A similar problem may occur owing to light scattering in the microscope, especially if objective lenses with large number of elements (apochromatic) are used (Gardner and McQuillin, 1980). This problem is easily revealed by examining the background with a blank slide on the stage.

Very little experimentation has been carried out on the composition of the wash buffers (see Section 10.8.3). In some cases, nonspecific binding of antibody may be due to electrostatic effects. This is particularly likely for antibodies heavily conjugated with FITC, and a trial of wash buffers with added NaCl (up to 0.5 M) or of different pH (5–9) may be rewarding.

12.13 Lack of Staining

In some cases, immunofluorescence with monoclonal antibodies may be weak or totally negative. Assuming that the antibodies can be proved to be active prior to conjugation, it is worth considering whether the conjugation has destroyed activity.

Inactivation of polyclonal antisera by conjugation with fluorochromes is extremely unlikely because it would require the simultaneous destruction of hundreds of different binding sites. In contrast, inactivation of monoclonal antibodies by conjugation is likely to be an 'all-or-none' process. In practice, most monoclonal antibodies may be coupled with fluorochromes or biotin without damage, but an occasional one is totally destroyed.

Virtually all the commercially available fluorochromes use lysine residues for coupling. It is fortunate that lysine does not often appear to be a crucial residue in antigen-combining sites. It would probably be possible to label monoclonal antibodies by partial reduction followed by reaction with iodoactamidofluorescein, but this approach has seldom been used.

If it is proved that the antibody is inactivated by coupling with fluorochromes, a trial of coupling with biotin is worthwhile. Although lysine residues are involved in both cases, it may be found that the biotin–conjugated antibodies retain activity. If this approach also fails, the main alternative is to use unconjugated monoclonal antibodies followed by fluorescent anti-immunoglobulin.

Other problems that may be traced to the antibody include a fluorochrome: protein ratio that is too low or too high (Table 12.6).

Yet another cause of failure to detect fluorescence is destruction of the antigen by fixation. In many cases, cells fixed with acetone (Rouse *et al.*, 1979) or methanol (Goldstein *et al.*, 1982) still react with monoclonal antibodies. High concentrations of aldehyde fixatives will often result in destruction of individual antigenic determinants (Gatti *et al.*, 1974), particularly those recognized by monoclonal antibodies. If it is suspected that fixation has destroyed the antigenic determinant recognized by a particular monoclonal antibody, the empirical trial of a few alternative methods of fixation is the only possible approach, except for using a different antibody (Brandtzaeg, 1982; Hancock *et al.*, 1982; Hancock and Atkins, 1986; see also Chapter 12 for a detailed discussion of fixation methods).

Another cause of unexpected absence of staining is the presence of biotin in tissue culture medium, when the biotin–avidin system is used.

Finally, it should be remembered that many monoclonal antibodies are of low affinity, and may be removed by excessive washing. A trial of fewer washes, longer incubation periods with antibody and shorter delays during processing is indicated if this problem is suspected.

References

Alexandra, I., Kells, D. I. C., Dorrington, K. J. and Klein, M. (1980) Noncovalent association of heavy and light chains of human immunoglobulin G: studies using light chain labelled with a fluorescent probe. *Molec. Immunol.* **17**, 1351–1363.

Alon, R., Bayer, E. A. and Wilchek, M. (1990) Streptavidin contains an RYD sequence which mimics the RGD receptor domain of fibronectin. *Biochem. Biophys. Res. Comm.* **170**, 1236–1241.

Barger, M. (1979) Antibodies that bind biotin and inhibit biotin-containing enzymes *Methods Enzymol.* **62**, 319–326.

Beavis, A. J. and Pennline, K. J. (1994) Simultaneous measurement of five-cell surface antigens by five-colour immunofluorescence. *Cytometry* **15**, 371–376.

Bergquist, N. R. and Nilsson, P. (1974) The conjugation of immunoglobulins with tetramethyl rhodamine isothiocyanate by utilization of dimethylsulfoxide (DMSO) as a solvent. *J. Immunol. Methods* **5**, 189–198.

Berlman, I. B. (1971) 'Handbook of Fluorescence Spectra of Aromatic Compounds' 2nd edn. Academic Press, New York.

Bianchi, D. W., Flint, A. F., Pizzimenti, M. F., Knoll, J. H. M. and Latt, S. A. (1990) Isolation of fetal DNA from nucleated erythrocytes in maternal blood. *Proc. Natl Acad. Sci. USA* **87**, 3279–3283.

Bianchi, D. W., Shuber, A. P., DeMaria, M. A., Fougner, A. C. and Klinger, K. W. (1994) Fetal cells in maternal blood: Determination of purity and yield by quantitative PCR. *Am. J. Obstet. Gynaecol.* **171**, 922–926.

Blakeslee, D. (1977) Immunofluorescence using dichlorotriazinylamino-fluorescein (DTAF). II. Preparation and fractionation of labelled IgG. *J. Immunol. Methods* **13**, 305–320.

Blakeslee, D. and Baines, M. G. (1976) Immunofluorescence using dichlorotriazinyl aminofluorescein (DTAF). I. Preparation and fractionation of labelled IgG. *J. Immunol. Methods* **13**, 305–320.

Bonner, W. A., Hulett, H. R., Sweet, R. G. and Herzenberg, L. A. (1972) Fluorescence activated cell sorting. *Rev. Sci. Instrum.* **43**, 404–409.

Brandtzaeg, P. (1982) Tissue preparation methods for immunochemistry. *In* 'Techniques in Immunocytochemistry' (G. R. Bullock and P. Petrusz, eds), Vol. 1, pp. 1–75, Academic Press, London.

Brumberg, Ye. M. (1959) Fluorescence microscopy of biological objects using light from above. *Biophysics* **4**, 97–104.

Cebra, J. J. and Goldstein, G. (1965) Chromatographic purification of tetramethyl-rhodamine-immune globulin conjugates and their use in the cellular localization of rabbit γ-globulin polypeptide chains. *J. Immunol.* **95**, 230–245.

Coons, A. H. (1961) The beginnings of immunofluorescence. *J. Immunol.* **87**, 499–503.

Coons, A. H. and Kaplan, M. H. (1950) Localization of antigen in tissue cells. II. Improvements in a method for the detection of antigen by means of fluorescent antibody. *J. Exp. Med.* **91**, 1–13.

Coons, A. H., Creech, H. J. and Jones, R. N. (1941) Immunological properties of an antibody containing a fluorescent group. *Proc. Soc. Exp. Biol.* (NY) **47**, 200–202.

Coons, A. H., Creech, H. J. and Jones, R. N. and Berliner, E. (1942) The demonstration of pneumococcal antigen in tissues by the use of fluorescent antibody. *J. Immunol.* **45**, 159–170.

Cotton, R. G. H. and Milstein, C. (1973) Fusion of two immunoglobulin-producing myeloma cells. *Nature* **244**, 42–43.

Crooks, J. E. (1978) 'The Spectrum in Chemistry'. Academic Press, London.

Darzynkiewicz, Z. and Crissman, H. (1990) 'Flow Cytometry. Methods in Cell Biology', Vol. **33**. Academic Press.

Fanger, M. W. (1995) 'Bispecific Antibodies'. Springer Verlag and R. G. Landes Biomedical Publishers, Austin, USA.

Finn, F. M., Titus, G., Montibeller, A. and Hofman, K. (1980) Hormone–receptor studies with avidin and biotinyl insulin-avidin complexes *J. Biol. Chem.* **255**, 5742–5746.

Forster, Th. and Konig, E. (1957) Absorption Spektren und Fluoreszenzeigen-schaften Konzentrieter Lonsungen organischer Farbstaffe. Z. *Electrochem.* **61**, 344–348.

Gardner, P. S. and McQuillin, J. (1980) 'Rapid Virus Diagnosis. Application of Immunofluorescence', 2nd edn. Butterworth, London.

Gatti, R. A., Ostborn, A. and Fagraeus, A. (1974) Selective impairment of cell antigenicity by fixation. *J. Immunol.* **113**, 1361–1368.

Giloh, H. and Sedat, J. W. (1982) Fluorescence microscopy: reduced photobleaching of rhodamine and fluorescein protein conjugates by n-propyl gallate. *Science* **217**, 1252–1255.

Glazer, A. N. and Stryer, L. (1984) Phycofluor probes. *Trends Biochem. Sci.* **9**, 423–427.

Goding, J. W. (1976) Conjugation of antibodies with fluorochromes: modifications to the standard methods. *J. Immunol. Methods* **13**, 215–226.

Goding, J. W. (1978) Allotypes of IgM and IgD receptors in the mouse: a probe for lymphocyte differentiation. *Contemp. Top. Immunobiol.* **8**, 203–243.

Goding, J. W. and Burns, G. F. (1981) Monoclonal antibody OKT-9 recognizes the receptor for transferrin on human acute lymphocytic leukemia cells. *J. Immunol.* **127**, 1256–1258.

Goding, J. W. and Layton, J. E. (1976) Antigen-induced co-capping of IgM and IgD-like receptors on murine B cells. *J. Exp. Med.* **144**, 852–857.

Goldstein, L. C., McDougall, J., Hackman, R., Meyers, J. D., Thomas, D. and Nowinski, R. C. (1982) Monoclonal antibodies to cytomegalovirus; rapid identification of clinical isolates and preliminary use in diagnosis of cytomegalovirus pneumonia. *Infec. Immun.* **38**, 273–281.

Gutman, G. and Weissman, I. L. (1972) Lymphoid cell architecture. Experimental analysis of the origin and distribution of T-cells and B-cells. *Immunology* **23**, 465–479.

Haaijman, J. J. and Slingerland-Teunissen, J. (1978) Equipment and preparative procedures in immunofluorescence microscopy: Quantitative studies. *In* 'Immunofluorescence and Related Staining Techniques' (W. Knapp, K. Holubar and G. Wick, eds), pp. 3–10. Elsevier/North Holland, Amsterdam.

Hancock, W. W. and Atkins, R. C. (1986) Immunohistological studies with monoclonal antibodies. *Methods Enzymol.* **121**: 828–848.

Hancock, W. W., Becker, G. J. and Atkins, R. C. (1982) A comparison of fixatives and immunohistochemical technics for use with monoclonal antibodies to cell surface antigens. *Am. J. Clin. Pathol.* **78**, 825–831.

Hansen, P. A. (1967) Spectral data of fluorescent tracers. *Acta Histochem.* **7** *(Suppl.)*, 167–180.

Hardy, R. R. (1986) Purification and coupling of fluorescent proteins for use in flow cytometry. *In* 'Handbook of Experimental Immunology'. (D. M. Weir, L. A. Herzenberg, C. Blackwell, and L. A. Herzenberg, eds), Vol. 1, pp. 31.1–31.12. Blackwell Scientific Publications, Edinburgh.

Hardy, R. R., Hayakawa, K., Haaijman, J. and Herzenberg, L. A. (1982a) B-cell subpopulations identified by two-colour fluorescence analysis. *Nature* **297**, 589–591.

Hardy, R. R., Hayakawa, K. and Herzenberg, L. A. (1982b) B-cell subpopulations identifiable by two-color fluorescence analysis using a dual-laser FACS. *Ann. NY Acad. Sci.* **399**, 112–121.

Haugland, R. P. (1995) Coupling of monoclonal antibodies with fluorophores. *In* 'Monoclonal Antibody Protocols' (W. C. Davis, ed.), pp. 205–221. Humana Press, New Jersey.

Haugland, R. (1992) Handbook of Fluorescent Probes and Research Chemicals. Molecular Probes Inc., Eugene, OR.

Heggeness, M. H. (1977a) Avidin binds to condensed chromatin. *Stain Technology* **52**, 165–169.

Heggeness, M. H. (1977b) Use of avidin-biotin complex for localization of actin and myosin with fluorescence microscopy. J. *Cell.* Biol. **73**, 783–789.

Heggeness, M. H. and Ash, J. F. (1977) Use of avidin–biotin complex for the localization of actin and myosin with fluorescence microscopy. *J. Cell Biology* **73**, 783–789.

Heggeness, M. H., Wang, K. and Singer, S. J. (1977) Intracellular distributions of mechanochemical proteins in cultured fibroblasts. *Proc. Natl Acad. Sci. USA* **74**, 3883–3887.

Heim, R., Cubitt, A. B. and Tsien, R. Y. (1995) Improved green fluorescence. *Nature* **373**, 663–664.

Heim, R., Prasher, D. C. and Tsien, R. Y. (1994) Wavelength mutations and post-translational autoxidation of green fluorescent protein. *Proc. Natl Acad. Sci. USA* **91**, 12501–12504.

Heitzmann, H. and Richards, F. M. (1974) Use of the avidin-biotin complex for specific staining of biological membranes in electron microscopy. *Proc. Natl Acad. Sci. USA* **71**, 3537–3541.

Herzenberg, L. A., Sweet, R. G. and Herzenberg, L. A. (1976) Fluorescence-activated cell sorting. *Scientific American* **234**, 108–117.

Higgins, T. J., O'Neill, H. C. and Parish, C. R. (1981) A sensitive and quantitative fluorescence assay for cell surface antigens. *J. Immunol. Methods* **47**, 275–287.

Holtkamp. B., Cramer, M., Lemke, H. and Rajewsky, K. (1981) Isolation of a cloned cell line expressing variant H-2K^k using fluorescence-activated cell sorting. *Nature* **289**, 66–68.

Ji, T. H. (1983) Bifunctional reagents. *Methods Enzymol.* **91**, 580–609.

Johnson, G. D. and Araujo, G. M. (1981) A simple method of reducing the fading of immunofluorescence during microscopy. *J. Immunol. Methods* **43**, 349–350.

Johnson, G. D., Davidson, R. S., McNamee, K. C., Russell, G., Goodwin, D. and Holborow, E. J. (1982) Fading of immunofluorescence during microscopy: a study of the phenomenon and its remedy. *J. Immunol. Methods* **55**, 231–242.

Johnsson, M. E., Bergquist, N. R. and Grandien, M. (1976) Antibodies to calf serum as a cause of unwanted reaction in immunofluorescence. *J. Immunol. Methods* **11**, 265–272.

Kavathas, P. and Herzenberg, L. A. (1983) Stable transformation of mouse L cells for human membrane T cell differentiation antigens, HLA and β2-microglobulin: selection by fluorescence-activated cell sorting. *Proc. Natl Acad. Sci. USA* **80**, 524–528.

Kearney, J. F. and Lawton, A. R. (1975) B lymphocyte differentiation induced by lipopolysaccharide. I. Generation of cells synthesizing four major immunoglobulin classes. *J. Immunol.* **115**, 671–676.

Khalfan, H., Abuknesha, R., Rand-Weaver, M., Price, R. G. and Robinson, D. (1986) Aminomethyl coumarin acetic acid: a new fluorescent labelling agent for proteins. *Histochem. J.* **18**, 497–499.

Kitagawa, T. and Aikawa, T. (1976) Enzyme coupled immunoassay of insulin using a novel coupling reagent. *J. Biochem* (Tokyo) **79**, 233–236.

Kranz, D. M. and Voss, E. W. (1981) Partial elucidation of an anti-hapten repertoire in BALB/c mice: comparative characterization of several monoclonal anti-fluorescyl antibodies. *Mol. Immunol.* **18**, 889–898.

Krishnan, A. (1975) Rapid flow cytofluorometric analysis of mammalian cell cycle by propidium iodide staining. *J. Cell Biol.* **66**, 188–193.

Kronick, M. N. (1986) The use of phycobiliproteins as fluorescent labels in immunoassay. *J. Immunol. Methods* **92**, 1–13.

Kronick, M. N. and Grossman, P. D. (1983) Immunoassay techniques with fluorescent phycobiliprotein conjugates. *Clin. Chem.* **29**, 1582–1586.

Kruth, H. S. (1982) Flow cytometry: rapid biochemical analysis of single cells. *Anal. Biochem.* **125**, 225–242.

Lachman, P. J. (1964) The reaction of sodium azide with fluorochromes. *Immunology* **7**, 507–510.

Lanier, L. L. and Warner, N. L. (1981) Paraformaldehyde fixation of hematopoietic cells for quantitative flow cytometry (FACS) analysis. *J. Immunol. Methods* **47**, 25–30.

Lanzavecchia, A. and Scheidegger, D. (1987) The use of hybrid hybridomas to target human cytotoxic T lymphocytes. *Eur. J. Immunol.* **17**, 105–111.

Larison, K. D., Bremiller, R., Wells, K. S., Clements, I. and Haugland, R. P. (1995) Use of a new fluorescent phosphatase substrate in immunohistochemical applications. *J. Histochem. Cytochem.* **43**, 77–83.

Layton, J. E. (1980) Anti-carbohydrate activity of T cell-reactive chicken anti-mouse immunoglobulin antibodies. *J. Immunol.* **125**, 1993–1997.

Lazarides, E. (1976) Actin, α-actinin, and tropomyosin interaction in the structural organization of actin filaments in nonmuscle cells. *J. Cell Biol.* **68**, 202–219.

Ledbetter, J. A., Rouse, R. V., Micklem, H. S. and Herzenberg, L. A. (1980a) T cell subsets defined by expression of Lyt-1,2,3 and Thy-1 antigens. Two-parameter immunofluorescence and cytotoxicity analysis with monoclonal antibodies modifies current views. *J. Exp. Med.* **152**, 280–295.

Ledbetter, J. A., Goding, J. W., Tokuhisa, T. and Herzenberg, L. A. (1980b) Murine T-cell differentiation antigens detected by monoclonal antibodies. *In* 'Monoclonal Antibodies. Hybridomas: A New Dimension in Biological Analyses' (R. H. Kennett, T. J. McKearn and K. B. Bechtol, eds), pp. 235–249. Plenum Press, New York.

Lichtman, J. W. (1994) Confocal microscopy. *Scientific American* **271**, 30–35.

Liu, F.-T., Zinnecker, M., Hamaoka, T. and Katz, D. H. (1979) New procedures for preparation and isolation of conjugates of proteins and a synthetic copolymer of D-amino acids and immunochemical characterization of such conjugates. *Biochemistry* **18**, 690–697.

Loken, M. R. and Stall, A. M. (1982) Flow cytometry as an analytical and preparative tool in immunology. *J. Immunol. Methods* **50**, R85-R112.

Loken, M. R., Parks, D. R. and Herzenberg, L. A. (1977) Two-color immunofluorescence using a fluorescence-activated cell sorter (FACS). *J. Histochem. Cytochem.* **25**, 899–907.

Miller, R. G., Lelande, M. E., McCutcheon, M. J., Stewart, S. S. and Price G. B. (1981) Usage of the flow cytometer-cell sorter. *J. Immunol. Methods* **47**, 12–24.

Milstein, C. and Cuello, A. C. (1984) Hybrid hybridomas and the production of bi-specific monoclonal antibodies. *Immunol. Today* **5**, 299–304.

Muzykantov, V. R., Gavriluk. V. D., Reinecke, A., Atochina, E. N., Kuo, A., Barnathan, E. S. and Fisher, A. B. (1995) The functional effects of biotinylation of anti-angiotensin converting enzyme monoclonal antibody in terms of targeting *in vivo*. *Anal. Biochem.* **226**, 279–287.

Neuberger, M. S. and Rajewsky, K. (1981) Switch from hapten-specific immunoglobulin M to immunoglobulin D secretion in a hybrid mouse cell line. *Proc. Natl Acad. Sci. USA* **78**, 1138–1142.

Nossal, G. J. V. and Layton, J. E. (1976) Antigen-induced aggregation and modulation of receptors on hapten-specific B lymphocytes. *J. Exp. Med.* **143**, 511–528.

O'Sullivan, M. J., Gnemmi, E., Morris, D., Chieregatti, G., Simmonds, A. D., Simmons, M., Bridges, J. W. and Marks, V. (1979) Comparison of two methods of preparing enzyme-antibody conjugates: application of these conjugates for enzyme immunoassay. *Anal. Biochem.* **100**, 100–108.

Ohkuma, S. and Poole, B. (1978) Fluorescence probe measurement of the intra-lysosomal pH in living cells and the perturbation of pH by various agents. *Proc. Natl Acad. Sci. USA* **75**, 3327–3331.

Oi, V. T., Glazer, A. N. and Stryer, L. (1982) Fluorescent phycobiliprotein conjugates for analyses of cells and molecules. *J. Cell. Biol.* **93**, 981–986.

Oi, V. T., Vuong, M., Hardy, R., Reidler, J., Dangl, J., Herzenberg, L. A. and Stryer, L. (1984) Correlation between segmental flexibility and effector function of antibodies. *Nature* **307**, 136–140.

Ormerod, M. G. (1990) 'Flow Cytometry. A Practical Approach'. IRL Press, Oxford.

Parks, D. R., Bryan, V. M., Oi, V. T. and Herzenberg, L. A. (1979) Antigen-specific identification and cloning of hybridomas with a fluorescence-activated cell sorter. *Proc. Natl Acad. Sci. USA* **76**, 1962–1966.

Parks, D. R. and Herzenberg, L. A. (1984) Fluorescence-activated cell sorting: theory, experimental optimization, and applications in lymphoid cell biology. *Meth. Enzymol.* **108**, 197–241.

Pawley, J. B. (ed) (1990) 'Handbook of Biological Confocal Microscopy'. Plenum Press, New York.

Ploem, J. S. (1967) The use of a vertical illuminator with interchangeable dichroic mirrors for fluorescence microscopy with incident light. Z. Wiss. *Mikroscopie* **68**, 129–142.

Ploem, J. S. and Tanke, H. J. (1987) 'Introduction to Fluorescence Microscopy'. Oxford University Press, Oxford.

Radbruch, A., Liesegang, B. and Rajewsky, K. (1980) Isolation of variants of mouse myeloma X63 that express changed immunoglobulin class. *Proc. Natl Acad. Sci. USA* **77**, 2909–2913.

Rechtenwald, D. B., Prezelin, B., Chen, C. H. and Kimura, J. (1990) Biological pigments as fluorescent labels for cytometry. *Proc. Int. Soc. Opt. Eng.* **1206**, 106–111.

Rembaum, A. and Dreyer, W. J. (1980) Immuno-microspheres: reagents for cell labeling and separation. *Science* **208**, 364–368.

Riggs, J. L., Seiwald, R. J., Burckhalter, J. H., Downes, C. M. and Metcalf, T. G. (1958) Isothiocyanate compounds as fluorescent labeling agents for immune serum. *Am. J. Pathol.* **34**, 1081–1098.

Rost, F. W. D. (1992) 'Fluorescence Microscopy'. Vol. 1. Cambridge University Press, Cambridge.

Rouse, R. V., van Ewijk, W., Jones, P. P. and Weissman, I. L. (1979) Expression of MHC antigens by mouse thymic dendritic cells. *J. Immunol.* **122**, 2508–2515.

Savage, M. D., Mattson, G., Desai, S., Nielander, G. W., Morgensen, S. and Conklin, E. J. (1992) 'Avidin–Biotin Chemistry: A Handbook'. Pierce Chemical Company, Rockford.

Shapiro, H. M. (1994) 'Practical Flow Cytometry'. 3rd edn. Wiley-Liss, New York.

Sidman, C. L. (1981) Lymphocyte surface receptors and albumin. *J. Immunol.* **127**, 1454–1458.

Smit, J. W., Meijer, C. J. L. M., Décary, F. and Feltkamp-Vroom, T. M. (1975) Paraformaldehyde fixation in immunofluorescence and immunoelectron microscopy. *J. Immunol. Methods* **6**, 93–98.

Southwick, P. L., Ernst, L. A., Tauriello, E., Parker, S. R., Mujumdar, R. B., Mujumdar, S. R., Clever, H. A. and Waggoner, A. S. (1990) Cyanine dye labeling reagents—carboxymethylindocyanine succinimidyl esters. *Cytometry* **11**, 418–430.

Staerz, U. D. and Bevan, M. J. (1986) Use of anti-receptor antibodies to focus T-cell activity. *Immunol. Today* **7**, 241–245.

Sternberger, L. A. (1979) 'Immunocytochemistry'. John Wiley, New York.

Stewart, C. C. and Stewart, S. J. (1993) Immunological monitoring utilizing novel probes. *Ann. NY Acad. Sci.* **677**, 94–112.

Stitz, L., Hengartener, H., Althage, A. and Zinkernagel, R. M. (1988) An easy and rapid method to screen large numbers of antibodies against internal cellular determinants. *J. Immunol. Methods* **106**, 211–216.

Suresh, M. R., Cuello, A. C. and Milstein, C. (1986a) Advantage of bispecific hybridomas in one step immunocytochemistry and in immunoassays. *Proc. Natl Acad. Sci. USA* **83**, 7989–7993.

Suresh, M. R., Cuello, A. C. and Milstein, C. (1986b) Bispecific monoclonal antibodies from hybrid hybridomas. *Methods Enzymol.* **121**, 210–228.

Taylor, R. B., Duffus, W. P. H., Raff, M. C. and DePetris, S. (1971) Redistribution and pinocytosis of lymphocyte surface immunoglobulin induced by anti-immunoglobulin antibody. *Nature* (New Biol.) **223**, 225–229.

The, T. H. and Feltkamp, T. E. W. (1970a) Conjugation of fluorescein isothiocyanate to antibodies. I. Experiments on the conditions of conjugation. *Immunology* **18**, 865–873.

The, T. H. and Feltkamp, T. E. W. (1970b) Conjugation of fluorescein isothiocyanate to antibodies. II. A reproducible method. *Immunology* **18**, 875–881.

Titus, J. A., Haugland, R., Sharrow, S. O. and Segal, D. M. (1982) Texas Red, a hydrophilic, red-emitting fluorophore for use with fluorescein in dual parameter flow microfluorometric and fluorescence microscopic studies. *J. Immunol. Methods* **50**, 193–204.

Tsien, R. Y. and Waggoner, A. (1990) Fluorophores for confocal microscopy: photophysics and photochemistry. *In* Handbook of Biological Confocal Microscopy (J. B. Pawley, ed), pp. 169–178, Plenum Press, New York.

Vallera, D. A., Youle, R. J., Neville, D. M. and Kersey, J. H. (1982) Bone marrow transplantation across major histocompatibility barriers. V. Protection of mice from lethal graft-vs.-host disease by pretreatment of donor cells with monoclonal anti-Thy-1.2 coupled to the toxin ricin. *J. Exp. Med.* **155**, 949–954.

Volkman, D. J., Ahmad, A., Fauci, A. S. and Neville, D. M. (1982) Selective abrogation of antigen-specific human B cell response by antigen-rich conjugates. *J. Exp. Med.* **156**, 634–639.

Wadsley, J. J. and Watt, R. M. (1987) The effect of pH on the aggregation of biotinylated antibodies and on the signal-to-noise observed in immunoassays utilizing biotinylated antibodies. *J. Immunol. Methods*. **103**, 1–7.

Waggoner, A. S. (1990) Fluorescent probes for cytometry. *In* 'Flow Cytometry and Sorting'. (M. R. Melamed, T. Lindmo, and M. L. Mendelsohn, eds), pp. 209–225, Wiley-Liss, New York.

Walliker, D., Quakyi, I. A., Wellems, T. E. McCutchan, T. F., Szarfman, A., London, W. T., Corcoran, L. M., Burkot, T. R. and Carter, R. (1987) Genetic analysis of the human malaria parasite *Plasmodium falciparum*. *Science* **236**, 1661–1666.

Watson, J. V. (1991) 'Introduction to Flow Cytometry'. Cambridge University Press, Cambridge.

Weber, K., Bibring, T. and Osborne, M. (1975) Specific visualization of tubulin containing structures in tissue culture cells by immunofluorescence. *Exp. Cell. Res.* **95**, 111–120.

Whitaker, J. E., Haugland, R. P., Moore, P. L., Hewitt, P. C., Reese, M. and Haugland, R. P. (1991) Cascade blue derivatives: Water soluble, reactive, blue emission dyes evaluated as fluorescent labels and tracers. *Anal. Biochem*. **198**, 119–130.

White, J. G., Amos, W. B. and Fordham, M. (1987) An evaluation of confocal versus conventional imaging of biological structures by fluorescence light microscopy. *J. Cell. Biol.* **105**, 41–48.

Wilchek, M. and Bayer, E. A. (1988) The avidin-biotin complex in bioanalytical applications. *Anal. Biochem*. **171**, 1–32.

Wilchek, M. and Bayer, E. A. (eds) (1990) Avidin–biotin technology. *Methods Enzymol.* **184**.

Wilson, T. (1990) 'Confocal Microscopy'. Academic Press.

Winchester, R. J., Fu, S. M., Hoffman, T. and Kunkel, H. G. (1975) IgG on lymphocyte surfaces: technical problems and the significance of third cell population. *J. Immunol.* **114**, 1210–1212.

Wong, S. S. (1991) 'Chemistry of Protein Conjugation and Cross-Linking'. CRC Press, Boca Raton.

Yoshitake, S., Yamada, Y., Ishikawa, E. and Masseyeff, R. (1979) Conjugation of glucose oxidase from *Aspergillus niger* and rabbit antibodies using *N*-hydroxy-succinimide ester of *N*-(4-carboxycycloehexyl-methyl)-maleimide. *Eur. J. Biochem.* **101**, 395–399.

Youle, R. J. and Neville, D. M. (1980) Anti-Thy-1.2 monoclonal antibody liked to ricin is a potent cell-type-specific toxin. *Proc. Natl Acad. Sci. USA* **77**, 5483–5486.

13 Immunohistology

Monoclonal antibodies have become standard reagents for immunohistochemistry, where they have had revolutionary impact. However, the production of high-quality results requires that attention be paid to numerous details, particularly tissue preparation, the cutting of sections, and the choice of individual antibodies. For further detail, the reviews by Hancock *et al.* (1982) and Hancock and Atkins (1986) are highly recommended.

13.1 Effects of Fixation on Antigen Recognition by Antibodies—The Balance Between Morphology and Antigenicity

In immunohistology, fixation and protein denaturation are closely related. One of the central issues is the balance between preserving antigenicity and the need to maintain good tissue structure, keeping the relevant structures stable and insoluble so that they are not washed away during cutting and staining.

At one extreme, one can cut frozen sections and stain them unfixed; however, water-soluble antigens will simply wash out of the section and therefore be unavailable for binding to antibody. At the other extreme, one may fix in extremely harsh reagents that are potent protein denaturants and/or cross-linkers, which will render the tissue stable and insoluble and give good morphology, but at the cost of the destruction of many antigenic determinants (Brandtzaeg, 1982; Hancock *et al.*, 1982; Hancock and Atkins, 1986).

As discussed in detail in Section 4.5, antibodies detect the shapes (conformation) of proteins. While a polyclonal antiserum against a protein antigen will almost certainly contain individual antibody molecules that react with the native protein and others that react with the denatured protein, the situation with monoclonal antibodies is different. Some monoclonal antibodies may only react with the native protein, and not react at all with the denatured protein. Other monoclonal antibodies may react only with the denatured protein. Antibodies to short peptides or recombinant proteins expressed in bacteria are particularly likely to detect the denatured form of the antigen but not the native form (see Section 4.5).

Antibodies that react with the denatured antigen are particularly useful in immunohistology. However, not all mechanisms of denaturation produce the same final structure. Chemical modification of the antigen by aldehydes or picric acid (Bouin's solution) will modify antigenic determinants in ways that are quite different to denaturation produced by heat, organic solvents or paraffin wax. The usefulness of a given monoclonal antibody with a particular fixation procedure can only be determined empirically.

Success in immunohistology will therefore require careful choice of fixation as well as choice of individual monoclonal antibodies. Perhaps the most rational approach to the problem might be to prepare a set of monoclonal antibodies that are made by immunizing mice with the antigens that have been fixed in the same way as the tissues are treated for histology. There is good evidence that such an approach can be very successful (Harrach and Robenek, 1991).

A more common strategy is to try a variety of monoclonal antibodies in the hope of finding ones that work on fixed tissues, or to experiment with different fixation procedures in the hope that an acceptable compromise can be achieved between morphology and antigenicity (see Brandtzaeg, 1982; Leenen *et al.*, 1985; Hancock and Atkins, 1986). Many monoclonal antibodies that work on paraffin wax sections are now commercially available.

In this Chapter, three methods for the production of histological sections will be described. Paraffin wax-embedded sections, usually fixed in formalin, are the 'workhorse' of histopathology laboratories and give excellent morphology, but at the expense of major loss of antigenicity. Frozen sections give rapid results and have minimal loss of antigenicity, but require a high degree of technical skill, and the morphology is not as good as paraffin wax sections. The use of tissue embedding in polyester wax is intermediate between paraffin wax sections and frozen sections in terms of morphology and antigenicity. A variety of other embedding procedures using proprietary acrylic plastic resins based on glycol methacrylate are now available, and these allow the cutting of sections as thin as 1 μm. The use of such resins is beyond the scope of this book.

13.2 Tissue Preparation and Fixation

13.2.1 General Factors Affecting Choice of Strategy

As mentioned above, fixatives such as formaldehyde cause extensive covalent modification of proteins. In addition, the exposure to organic solvents and hydrophobic waxes during embedding causes protein denaturation, which may be only partly reversible. The melting point of standard embedding wax is 56°C, and many proteins are denatured at this temperature. The combination of heat, organic solvents and additional, poorly understood 'masking' of antigenic determinants by paraffin wax embedding are responsible for the great majority of available monoclonal antibodies not achieving satisfactory staining when applied to paraffin wax sections. Very few randomly chosen monoclonal antibodies work on the resin-embedded sections used for electron microscopy.

The use of high temperatures can be avoided by embedding the tissue in polyester wax (melting point of about 39°C; see Section 13.4). Morphology with polyester wax is similar to that of paraffin wax, and many monoclonal antibodies which give satisfactory labelling on frozen sections will also work on polyester wax-embedded tissues. Unfortunately, there is no way of knowing which antibody will or will not work.

If paraffin wax-embedded tissues must be used, it is often found that antigenic determinants are 'masked' in some way, and can sometimes be revealed by treating the sections with carefully graded amounts of proteases such as trypsin (Finley and Petrusz, 1982), or by heating the slides in a microwave oven. These techniques are entirely empirical and require careful optimization (Hancock and Atkins, 1986). They may often fail to work with monoclonal antibodies.

Good fixation almost inevitably means protein denaturation and cross-linking, because this is what causes the insolubilization (fixing). There is a need for a new generation of monoclonal antibodies that have been selected to work on paraffin wax sections, and these are gradually becoming available.

Any buffers used must not interact with the fixative, or both the buffering and the fixing will be compromised. For example, aldehyde fixatives react with amino groups, so amino-containing buffers such as Tris must be avoided if formaldehyde or glutaraldehyde are used as fixative.

13.2.2 Fixation of Tissue Blocks Prior to Cutting Sections

If tissues are to be processed for paraffin wax embedding, they are usually fixed with 10% buffered formalin, although Bouin's fluid is occasionally used (Table 13.1). Penetration of fixative is a slow process, and the blocks must be small to allow adequate fixation. The depth penetrated is time dependent, and the fixed tissue may act as a barrier to further penetration of the fixative. Thus, the rate of penetration may slow down with time. Inadequate fixation can lead to heterogeneity of areas of the tissue and result in artefacts. In general, blocks should be no more than 5–6 mm thick, and preferably less. It is common for tissues to shrink during fixation. Subsequent dehydration and embedding may often cause further changes in volume. Tissue fixed in formalin and embedded in paraffin wax shrinks by about 33%. Excessively prolonged fixation in

Table 13.1 Fixatives for histology

Buffer	Recipe
Stock phosphate buffer (10x)	650 g anhydrous Na_2HPO_4 400 g $NaH_2PO_4.2H_2O$. Make up to 20 l.
10% Buffered formalin (4% formaldehyde in PBS)	2 l 10x stock phosphate buffer 2 l concentrated formalin (i.e. 40% formaldehyde). Make up to 20 l.
Bouin's fixative	75 ml saturated picric acid (2,4,6 trinitrophenol) in water 25 ml formalin (40% formaldehyde) 5 ml glacial acetic acid.

Table 13.2 Tissue processing for paraffin wax sections (standard cycle)

Cycle	Agent	Time (h)
1	10% Buffered formalin	2
2	50% Ethanol	1
3	70% Ethanol	1
4	95% Ethanol	1
5	Absolute ethanol	1
6	Absolute ethanol	1
7	Absolute ethanol	1
8	Histoclear	1
9	Histoclear	1
10	Histoclear	1.5
11	Paraffin wax	1.5
12	Paraffin wax	3

This 16-h schedule is suitable for medium-sized tissue blocks (up to 5–6 mm thick).

formaldehyde may lead to excessive shrinkage and hardening of the tissue. Overnight fixation is usually sufficient.

The embedding protocol for paraffin wax sections involves fixation, followed by dehydration and replacement of the organic solvent with paraffin wax. Tables 13.2 and 13.3 give details of a standard overnight cycle (16 hour), and a more rapid cycle that can be used for smaller blocks (not more than 2–3 mm thick).

Fixation of tissue blocks is usually carried out at room temperature. Although fixation at 4°C will slow autolysis and diffusion of cellular components, it will also slow down diffusion and reaction of the fixative.

Fixation is improved if samples are gently agitated on a rocking platform. Some protocols, such as periodate–lysine–paraformaldehyde (PLP) fixation require adequate washing out of the fixative by multiple changes of buffer; this step is also facilitated by the use of a rocker (Hancock and Atkins, 1986).

Table 13.3 Rapid processing cycle for paraffin wax embedding of small tissue blocks

Cycle	Agent	Time (min)
1	70% Ethanol	20
2	90% Ethanol	20
3	Absolute ethanol	30
4	Absolute ethanol	30
5	Absolute ethanol	30
6	Histoclear	15
7	Histoclear	30
8	Histoclear	30
9	Paraffin wax	30
10	Paraffin wax	20

This rapid 4-h schedule is for blocks that are no more than 2–3 mm thick.

13.2.3 Bouin's Fixation

Bouin's fixative consists of a mixture of picric acid (2,4,6 trinitrophenol), formaldehyde, acetic acid and water (Table 13.1). It is very rarely used for immunohistology, because the great majority of antigenic determinants are destroyed.

13.2.4 Fixation After Cutting Sections

An important choice is whether to fix the tissues before or after cutting sections. As discussed above, sections that are to be embedded in paraffin wax or plastic are routinely fixed before embedding. Conversely, frozen sections are not usually fixed before cutting. It must be borne in mind that some soluble antigens may be washed out of the tissues if no fixation is used. Frozen sections are therefore often fixed after cutting but prior to staining with antibodies. Even if frozen sections are not fixed prior to staining with antibodies, it is common practice to fix them after staining, as this will help preserve the tissues for longer-term storage and, at this stage, preservation of antigenic determinants is irrelevant.

13.2.5 A Single Specimen May be Fixed in Several Different Ways to Maximise the Chances of Success

Because of the somewhat unpredictable effects of fixation on antigenicity, several different protocols may be used on a specimen. For example, a piece of tissue can be sliced into three or four portions for: (1) quick freezing: (2) fixation in PLP for 3–4 h at 4°C, followed by progressive changes in PBS/70% sucrose followed by snap freezing (Hancock and Atkins, 1986), (3) ethanol fixation prior to polyester wax embedding; and (4) formalin fixation prior to paraffin wax embedding. The PLP-fixation protocol (i.e. fixed frozen sections) gives much better morphology than samples simply quick frozen, in part because of the effects of the fixation and in part because of the cryopreservative effect of the sucrose, which helps minimize deterioration in tissue section morphology by minimizing ice-crystal formation during freezing. The same samples are also adequate for immuno-electron microscopy.

13.3 Automated Processing for Paraffin Wax Embedding

The aim of tissue processing is to embed the tissue in a solid medium that is firm enough to give support to the tissue and enable thin sections to be cut with little damage to the tissue or to the knife. The most satisfactory material for routine histology is paraffin wax.

Before tissue can be processed to paraffin wax, it needs to be adequately fixed. Although dehydrating alcohols may complete the fixation of small fragments of tissue, larger pieces may not be fixed because of the slow rate of penetration of alcohol. Also, alcohols are less effective in dehydrating if called upon to act the dual role of dehydrant and fixative.

It is essential that the embedding material thoroughly permeates in fluid form, and that it solidifies without damaging the tissue. Paraffin wax is not miscible with water, and the fixed tissue must be gently but completely dehydrated to remove aqueous fixative and water, and cleared with a substance which is totally miscible with the dehydrant and embedding agent, such as xylene. 'Histoclear' is a nontoxic xylene substitute made from citrus fruit, and smells accordingly. The tissue is then embedded in paraffin wax in the correct orientation in metal moulds, and the wax cooled rapidly to minimize crystal size.

Tissue processing is carried out automatically in an Autotechnicon or similar instrument. This machine holds the 12 reagents required for processing and transports the tissue to each according to a programme controlled by a timing clock. Controlled heat, vacuum and vertical oscillation speed the exchange of fluids between tissues and reagents (Tables 13.2 and 13.3).

13.4 Polyester Wax Sections

The morphology observed in paraffin wax and polyester wax sections is better than that of frozen sections. However, for immunochemistry, polyester wax is preferable as it is available with a melting point of 37°C or 43°C, while the higher melting point of paraffin wax (56°C) is more likely to result in protein denaturation and thus loss of antigenic determinants.

Polyester wax for histology is available from BDH. It consists of 400 polyethylene glycol distearate, stabilized by the addition of 10% cetyl alcohol, and is miscible with 95% ethanol, absolute ethanol, methanol and acetone. Polyester wax is hygroscopic and should be stored in a dry location.

The protocols for polyester wax sections have been provided to me by Dr Helen Ramm, Department of Pathology and Immunology, Monash University.

13.4.1 Fixation for Polyester Wax Sections

Fixation may be in two changes of 19:1 v/v absolute ethanol/glacial acetic acid for 12–24 h at 4°C, depending on the size of the tissue. The inclusion of glacial acetic acid, although improving nuclear staining, can diminish antigenicity still further, and can be replaced with water if desired. The tissue is then transferred to absolute ethanol for 3 h (two changes, 4°C).

13.4.2 Embedding of Blocks in Polyester Wax

Infiltrate the fixed tissue with polyester wax for 3 h (three changes, 37°C). Then, embed in fresh polyester wax as follows:

(1) Coat the tissue mould with Tissue Tek releasing spray.
(2) Fill the mould with polyester wax.
(3) Arrange tissue within the mould.
(4) Place mould on ice tray and rapidly cover with prewarmed cassette; then add more polyester wax.
(5) As soon as wax has set, release cassette from mould.

Storage of freshly prepared blocks overnight at 4°C may facilitate sectioning.

13.4.3 Cutting Polyester Sections

Sections of 3 μm thickness are cut, floated on cold water and rapidly collected on gelatinized slides. Polyester wax sections are best treated as if they were quick-frozen blocks and sectioned at about –25°C using a standard cryostat. Handling of sections should be avoided, as the melting point of wax is 37°C. If flattened sections are not rapidly removed from the water bath, the wax will continue to expand and sections will pull apart. Resealing of blocks with molten polyester wax following sectioning has been found to be desirable for immunological staining procedures. Sections are then dried for 2–3 h at 37°C, and stored at 4°C. Loss of antigenicity has sometimes been observed on storage.

13.5 Frozen Sections

These take rather more practice and technical skill to achieve good morphology than do paraffin wax sections. However, they have the advantages of speed and preservation of antigenicity.

13.5.1 Freezing the Tissues for Frozen Sections

Label the bottles or other containers and then precool them in a –70°C freezer. It is vital to prevent the tissues warming during handling.

Cut the tissues into small blocks, no more than 3–5 mm per side, to ensure that the whole of the block freezes as rapidly as possible. Provided that care is taken to ensure snap freezing, larger blocks (up to 1.5 cm^3) can sometimes be used with good results.

Freezing the tissues in Tissue Tek OCT compound (4 fluid ounce bottles; cat. no. 4583, Miles Diagnostics, Elkart, Indiana 46515, USA) will help avoid thawing while handling or mounting the block on the cryostat chuck. OCT compound consists of 10% w/w polyvinyl alcohol, 4% polyethylene glycol and 85% w/w 'nonreactive ingredients'; its function is to bind the tissue to the specimen block and to cover the tissue specimen.

Snap freeze the block by placing it in a 'boat' made of aluminium foil on dry ice or in a bath of isopentane–liquid nitrogen. Avoid contact of the tissue with the isopentane–liquid nitrogen slurry, as this could cause damage to antigenic determinants by denaturation of proteins. Snap-freezing by direct immersion in liquid nitrogen has been performed successfully in some hands. Blocks for frozen sections may be stored at –70°C.

Ice crystals cause shearing of the tissue, and air spaces may cause folding of the tissue during cryosectioning. Care must therefore be taken to avoid the presence of excessive amounts of water around the tissue during freezing. Drying the tissue on filter paper is unsatisfactory, because it adheres strongly. Drying can be achieved by placing the tissue in OCT compound or by rolling it in talcum powder.

Avoid thawing the tissue once it is frozen. Do not leave the frozen blocks on the cryostat for longer than necessary. During the cutting process, the temperature of the cryostat may rise to –20°C or even higher.

13.5.2 Cutting Frozen Sections

It is essential to be familiar with the operation of the cryostat. It is highly recommended to read the manual, which contains a great deal of information of practical value, and will save many hours of frustration. Thin cryosections (about 4 microns) give the best morphology. Choice of section thickness depends on the sensitivity required. Thicker sections have more antigen, so they may give stronger signals but worse resolution.

(1) The knife must be sharp and with no nicks or other damage. Nicks will cause tears in the sections. A blunt knife will cause multiple small tears parallel to the knife edge and/or alternate thick and thin sections.

(2) The knife and tissue block holder must be firmly screwed in place. This seemingly trivial point is not an uncommon source of problems.

(3) The knife must be at the correct angle (consult the Cryostat manual). If the angle is too steep, multiple small tears parallel to the knife edge may develop. If the angle is too shallow, it may result in the blade leaving the block surface before the whole of the block is sectioned.

(4) Operation should be smooth—a consistent cutting speed will provide consistent thickness of serial sections.

(5) It is essential that the cryostat mechanism be functioning well. Lubrication is important and takes only seconds to perform. The cryostat mechanism should be kept ice-free because the presence of ice may cause irregular advancement of the block.

(6) Keep the glass door of the cryostat closed when not in use to minimize condensation and dust.

(7) Do not allow the cryostat and tissue to warm up excessively. Individual tissues cut best at different temperatures; consult the cryostat manual. Warming of the block surface can cause curling of sections.

(8) Frost and/or tissue debris on the knife and Perspex anti-roll shield may cause sticking and folding of sections. Wipe the knife clean with a thick wad of paper tissues, away from blade edge. The use of a brush will almost inevitably result in damage to the blade edge. The blade edge will also be damaged by trimming the tissue block while it is mounted adjacent to the knife. It is a good idea to wear a face-mask while cutting sections, to avoid breath causing frosting of the knife and Perspex shield.

(9) Curling of sections is usually due to warming of either the tissue block surface or the Perspex shield. Close the door of the cryostat and wait for the temperature to equilibrate.

(10) Transfer the sections to a microscope slide (preferably gelatinized to facilitate adhesion) kept in a slide rack at room temperature.

(11) If the block of tissue is uneven, a sloping side can cause twisting or tilting of one side of the section. This can be avoided by embedding in OCT and trimming the excess OCT to give square sides.

(12) Trimming the upper edge of the tissue block to have a pointed shape (vertex upwards) may make it easier to detach the sections for manoeuvring into position onto the glass slide.

(13) Grit and other hard debris in blocks will result in tearing of sections and may cause nicks in the knife.

(14) Check the quality every 10–12 sections using the rapid 'Paragon' multistain. The macroscopic appearance of the cut, unstained section on the

slide can give much information as to the quality of the section. Comparing it with a satisfactory Paragon-stained section is useful, looking at the outline as well as for easily recognized shapes within the section.

(15) The Perspex antiroll shield must be parallel to and slightly ahead of knife edge. Begin cutting with the Perspex just behind knife edge, then advance it forward until perfect sections are cut. Care must be taken to adjust the Perspex shield accurately over the knife, otherwise the section may roll over it or fold up on itself under it.

(16) Electrostatic sticking of sections to the Perspex shield may occur if there is low humidity in the air. The remedy may be to discharge the static by 'earthing'.

Cutting cryostat sections of fixed material such as PLP-fixed blocks requires more skill and perseverance than for straight unfixed, snap-frozen tissues.

13.5.3 Transfer of Frozen Sections to Glass Slides

It is a good idea to coat the glass slides with gelatin prior to cutting the sections. Gelatin helps the section to stick firmly on the slide, and stops them lifting off or rolling back on themselves. The coating is done by dipping the slides in a solution consisting of 2% gelatin in water for 30 seconds at room temperature. Gelatin solution is kept at 4°C and is solid at this temperature. Liquify by heating at 37°C for about 30 min. After dipping, the gelatin slides may be fixed in 10% formalin in water for 30 s (this is optional; residual formalin may decrease antigenicity of the specimen). Coated slides are dried overnight in a staining rack at room temperature to allow adequate ventilation, and stored in a box at room temperature.

If the sections detach from the gelatin-coated slides during washing, trials of other slide adhesives such as poly-L-lysine, starch, or proprietary glue preparations may be required. Gelatin coating should not be used if sections are to be protease-digested, the sections will lift off almost immediately; adherence of sections to the slide in this case may be achieved by drying at 37°C or use of a glue solution.

13.5.4 Drying the Sections

Opinions vary about how to dry the sections. If the sections are to be dried, this should probably be done before fixing and staining. They may be dried at room temperature for at least 1 hour or overnight, but it is probably best to fix and stain as soon as possible after cutting, as time-dependent loss of antigenicity has sometimes been observed. Alternatively, the slides may be air-dried in the cold room for 30 min to 2 h. If slides are dried in the cold room, they should be placed in a position where the air circulates freely, to avoid the accumulation of condensation.

After sections are prepared, the key factors that determine the extent and rate of deterioration in antigenicity are moisture and ambient temperature. Once dried, sections that are placed in slide boxes with dessicant and stored at –70°C are stable for years.

13.5.5 Should Frozen Sections be Fixed Before Antibody Staining?

This is a vexed question. In some cases, fixation is essential to prevent the relevant antigen from washing out during staining. However, as discussed at the beginning of this chapter, all fixation procedures, by their very nature, will cause some damage to antigenicity. In most cases, the fixation procedure will have to be determined empirically for each antigen and monoclonal antibody.

If the sections are to be fixed prior to staining, this should be done as soon as possible after the section is dry. Store the fixed section sealed in a plastic bag containing dessicant (silica gel) at 4°C or –20°C.

The most popular fixation procedure is probably a brief dip in acetone, usually pre-cooled in the –20°C freezer in a Coplin jar. Time may be 1–30 s, followed by aqueous wash in isotonic saline or drying in air for 5–30 m at room temperature. Fixation with organic solvents such as acetone will remove most of the membrane lipids and precipitate membrane proteins *in situ*. Fixation with organic solvents such as acetone also permeabilises cells, allowing antibodies access to the cytoplasm.

A brief dip in acetone is compatible with retention of many antigenic determinants. There is a greater chance of loss of antigenicity using ethanol, and even greater chance if the sections are fixed in methanol. PLP-fixed cryostat sections often give better morphology than fixation in organic solvents (see Hancock and Atkins, 1986).

13.6 Antibody Staining Strategy

There are many choices of strategy for immunological staining of tissue sections. In recent years, there has been a marked trend towards the use of enzyme-labelled antibodies rather than immunofluorescence, because antigens can be localized at the same time as viewing the architecture of the tissues, and a permanent record is obtained. However, the recent development of highly fluorescent substrates for alkaline phosphatase (see Sections 12.10 and 13.8.6) has provided a powerful new option.

Multilayer systems give greatly increased sensitivity due to the fact that each first antibody can bind several second antibodies. If a third antibody is added, the sensitivity increases still further, but at the risk of more 'background' staining and experimental artefacts (see Hancock and Atkins, 1986 for discussion).

13.6.1 Simultaneous Detection of Two or More Antigens

If two or more antigens are to be detected simultaneously, careful consideration will need to be given to the choice of enzymes and substrates, or to the use of fluorochromes that have easily separable emission spectra. The possibility of a combination of conventional substrates and the new fluorescent substrates may be particularly useful. *Great care must be taken to ensure that the antibody systems used in multicolour work do not interact in unplanned ways.* Many of these considerations have been discussed in Section 12.7.1 and other parts of Chapter 12 (see also Mason and Woolston, 1982).

13.6.2 The Avidin–Biotin System in Immunohistology

The biotin–avidin system is particularly helpful in the staining for several different antigens because it allows one antigen to be detected with anti-immunoglobulin and a second to be detected with labelled avidin or streptavidin (see Sections 10.3 and 12.9). The tetrameric structure of avidin and streptavidin allows the use of a 'bridging' technique in which the tissue is stained with biotinylated antibody, washed, incubated with avidin or streptavidin, washed again, and incubated with a biotinylated enzyme or other biotinylated reagent which can bind to unoccupied biotin-binding sites.

As discussed elsewhere (see Section 12.9.3), avidin is very basic and may bind nonspecifically to tissues. It binds particularly strongly to DNA and other acidic structures. This problem may be particularly severe in immunohistology, but may sometimes be overcome by raising the salt concentration to 0.3–0.5 M and/or adding cytochrome *c* as a competitor. Alternatively, one may use a modified form of avidin that has been chemically treated to raise its isoelectric point (e.g. Neutravidin from Pierce Chemicals). The use of streptavidin overcomes many of these problems, but can create new problems because it can bind to certain receptors which have specificity for the tripeptide Arg–Gly–ASP RGD and related motifs (Section 12.9.3).

If the tissue sections are to be incubated with hybridoma supernatants, it must be remembered that most tissue culture medium contains biotin, which could inhibit binding to avidin if it is not removed by washing. It should also be noted that some proteins in cells may contain attached biotin, and this could cause nonspecific staining with avidin or streptavidin.

13.7 Processing of Slides for Antibody Staining

13.7.1 De-waxing

Before paraffin wax sections can be stained with antibodies, they must be rapidly but adequately 'dewaxed' in xylene or Histoclear followed by 95% ethanol, then 70%, then water before adding the antibodies.

Polyester wax sections must also be dewaxed prior to staining with antibody. This is done by immersing them in absolute ethanol for 5–10 min, then 90% ethanol for 5 min, then 70% ethanol for 10–30 min, then water. If desired, the 70% ethanol step can include 3% H_2O_2 to destroy endogenous peroxidases, but this harsh oxidizing agent will destroy many antigenic determinants recognized by monoclonal antibodies.

13.7.2 Preparation of Slides for Antibody Staining

Using a diamond-tipped marking pencil, label slides clearly and circle around tissue sections.

One can use 6–8 sections per slide, with each section receiving a different antibody. Mixing of primary antibodies is prevented by applying boundaries with a wax pen or Vaseline. These bridges dissolve during the dehydration and clearing steps if a permanent mountant is used. Alternatively, the bridges can be scraped off using a blade.

Place the slides horizontally in a humidified chamber, such as a Perspex box with wet foam rubber on the bottom. Slides can be placed directly on the foam rubber, so that the tray can be tilted a little without the slides slipping. Cover the sections with PBS or other medium for 5–10 min to dissolve the OCT Tissue Tek compound, and then wipe around the sections using paper tissues.

13.7.3 'Preblocking' Nonspecific Binding of Antibodies to Tissue Sections

Given the enormous variation in immunohistological methods and order of addition of antibodies, a great deal of individualization of blocking is needed. Only the general principles can be dealt with here. Tissue sections are likely to bind proteins nonspecifically, and these nonspecific protein binding sites must be saturated with irrelevant protein to prevent high background staining when antibodies are applied.

If the second antibody is of sheep, goat or rabbit origin, nonspecific background tissue staining may often be reduced by incubating the section for 15 min with undiluted normal serum from the same species. Such blocking steps may need to be tested first to make sure that they do not cause problems. For example, some primary antibodies may cross-react with antigens in the normal serum.

Alternatively, one may block with PBS/0.2% gelatin and use the same fluid for dilutions and for washes. Although normal serum of the species from which the tissue was obtained can be added to subsequent layers to decrease background, this is best avoided, as many undesirable cross-reactions may occur. As far as I am aware, the use of nonfat powdered milk (BLOTTO, see Section 10.10.6) has not been tried as a blocker for immunohistology, but in view of the fact that it is extremely effective in other situations, it may be worth trying.

13.7.4 Deaggregation of Antibody

If binding to Fc receptors is likely to be a problem, the antibodies may be deaggregated by centrifugation at 100 000 ***g*** for 10 min, or $(Fab')_2$ fragments may be used (see Section 5.3.4 and 9.5.2).

13.7.5 Controls for Specificity of Antibody Staining

There are many possible causes of nonspecific staining of histological sections. The first and most important thing is to be able to recognize that it is happening. This is not always a trivial point, but nonspecific staining will usually be revealed if each experiment has adequate controls.

A common cause of nonspecific staining is the presence of endogenous immunoglobulin, either from blood that permeates all tissues or membrane immunoglobulin on B lymphocytes. Endogenous immunoglobulin from blood may be removed by extensive washing prior to antibody staining, but it is difficult to remove completely. Other causes of nonspecific staining include binding to Fc receptors (see above) and the presence of 'natural' antibodies to tissue components. Natural antibodies to carbohydrate antigens are commonly found in sera from rabbits when tested on mouse or human tissues, and can be particularly troublesome. They may sometimes be removed by absorption with red blood cells, or depleted by the use of affinity-purified antibodies.

The ideal control for specificity is obtained when one uses anti-allotype sera, such that one can keep all the antibodies constant but vary the strain of mouse from which the tissue was obtained. Unfortunately, this control is not always possible, and is generally impossible when the target tissues are of human origin. Although the option is not commonly available, the use of anti-immunoglobulin allotype antibodies as sandwich reagents may also be very helpful in eliminating background staining caused by endogenous immunoglobulin.

Other controls include leaving out the first antibody, which should completely abolish all staining, using an irrelevant first antibody of the same class as the experimental antibody, and testing on known negative and positive tissues. It is generally wise to include all possible controls. It is better to be safe than to be misled.

13.7.6 Controls for Endogenous Enzymes

When starting a new project, it is vital to test for endogenous enzyme activity by incubating the tissue sections with the substrate without antibody (see Section 8.10.6). When faced with a 'positive' result in one of the 'control' slides, review of this endogenous control slide allows quick distinction between endogenous enzymes and background due to other causes.

Endogenous peroxidase is present mainly in cells of the myeloid/macrophage lineage. Pseudoperoxidase activity occurs in red blood cells, but this is seldom a cause for confusion. Unfortunately, endogenous peroxidase is a tough enzyme and difficult to inactivate. The measures that inactivate endogenous peroxidase generally involve the use of strong oxidizing agents such as hydrogen peroxide, which often cause destruction of desired antigenic determinants (see Farr and Nakane, 1981, and Section 13.7.1).

A number of tissues express endogenous alkaline phosphatase, including B lymphocytes and gut epithelium (Culvenor *et al.*, 1981; Garcia-Rozas *et al.*, 1982). The alkaline phosphatase used to label antibodies is usually derived from gut epithelium, and is resistant to the drug levamisole, while alkaline phosphatase in other organs including lymphocytes is strongly inhibited by this drug (Ponder and Wilkinson, 1981; Mason *et al.*, 1986; see Sections 8.10.6 and 13.8.2).

Endogenous alkaline phosphatase can also be completely abolished by treatment with 20% acetic acid in water or ethanol (15 s at 4°C), but this very harsh treatment may destroy many antigenic determinants.

13.7.7 Incubation with Antibody

After incubation with the preblocking solution, let it run down the slide by holding it vertically, then gently wipe the slide dry without touching the section itself. Dilute antibody in Tris-buffered saline (TBS; 150 mM NaCl, 10 mM Tris-HCl, pH 7.4) or PBS containing 1% normal sheep serum or other irrelevant protein to help block nonspecific protein binding. The same buffer is also used for washes.

The appropriate concentration of antibody must be determined empirically. The optimal concentration is that which causes maximal intensity of staining, but no more. Higher concentrations of antibody will increase nonspecific staining, and is also wasteful. Lower concentrations could result in errors owing to insufficient intensity of staining of sparse antigens and lack of reproducibility.

Hybridoma supernatants generally contain 10–50 μg/ml antibody, and are often used undiluted. No agitation is required during incubation with antibody; diffusion is enough. For a typical 5 × 5 mm section, one might use 30–50 μl undiluted culture supernatant or antibody diluted in PBS containing 1% normal sheep serum. Transfer the antibody solution onto the horizontal slide, while holding the pipette vertically.

Incubation time is typically 30–60 min, usually at room temperature, and is chosen to give saturation staining with the concentration of antibody used. The kinetics of binding may vary considerably between monoclonal antibodies, and it is unwise to shorten the incubation time without first checking results. Incubation times may be shortened if one uses a larger excess of antibody, but it is generally best to keep conditions constant from one experiment to another to give consistent results.

For maximal sensitivity (e.g. to detect many of the extremely low concentration cytokines in tissue sections) it is sometimes worthwhile incubating overnight with the primary antibody. This facilitates reaching equilibrium even for low affinity antibodies, and allows conservation of precious samples, since primary antibodies can usually be diluted at least a further five-to tenfold compared with the dilution for 30–60 min incubations.

Ensure that the tissues do not dry out during staining, as this could cause gross nonspecific staining. Similarly, do not spend too much time with the slides uncovered with liquid between steps to avoid drying. The box must be adequately humidified, and the lid should be on during staining. The slides should be horizontal.

13.7.8 Washing

Wash sections on a horizontal shaker in PBS or TBS containing 1% normal sheep serum, 0.2% gelatin or other blocking agent. The whole slide should be covered in liquid. Each wash should be of about 5 min duration. The first wash liquid should be tipped off and fresh buffer added from a 'squirter' bottle and the process repeated. Do not squirt directly onto the section, or the tissue may be damaged. Aim for the centre of slide and let medium gently cover entire slide.

The best results are obtained by washing in a large volume (e.g. 400 ml) in a slide rack placed on a magnetic stirrer. If washing damages the sections or causes them to come off the slide, the washing time may be shortened and/or shaking avoided, as long as it is verified that this does not increase the background. Some laboratories do not use a shaker, but simply wash in Coplin jars.

13.8. Enzyme Staining

13.8.1 Horseradish Peroxidase

A typical protocol might involve using 1 mg/ml diamino benzidine tetrahydrochloride dihydrate (DAB; Sigma cat. no. D 5637) plus 0.003–0.009% H_2O_2 in PBS. This solution is unstable and should be made up just prior to use.

Development of colour may be done in a Coplin jar containing 50 ml of 1 mg/ml DAB in PBS and 5 μl of 30% H_2O_2. Typical incubation time is 8–10 min, but should be optimized for each experiment. The reaction forms a brown insoluble precipitate in the tissues, and allows good localization of antigens (Harahap and Goding, 1988).

The sensitivity might be increased by adding 0.5% $CuSO_4$ or $NiCl_2$ to the DAB solution, although commercial kits using nickel to enhance the intensity of staining with peroxidase have been disappointing. A silver enhancement step after the DAB will greatly increase the sensitivity (Hancock and Atkins, 1986). Adding imidazole (final concentration 0.01 M) to the DAB solution has also been claimed to increase sensitivity.

Azide inhibits peroxidases, and should not be present in buffers used with this enzyme. It is possible that diamino benzidine may be a carcinogen, and it should be handled with care. For convenience and safety, it is kept as a stock solution at 50 mg/ml in water or PBS.

After staining for peroxidase activity, the sections may be lightly counterstained with haematoxylin (see Section 13.10) to allow visualization of the tissues. Finally, a drop of Tissue Tek coverslipping resin is added, followed by a cover slip.

A simple and highly effective method for conjugating horseradish peroxidase to antibodies and to staphylococcal protein A is given in Section 8.10.7.

13.8.2 Alkaline Phosphatase—Direct Technique

In some cases, particularly blood, it may be necessary to avoid peroxidase-conjugated antibodies because of the high endogenous peroxidase activity of myeloid cells. The use of alkaline phosphatase as a histochemical label for antibodies has become very popular.

Some haemopoietic cells possess endogenous alkaline phosphatase (Culvenor *et al.*, 1981; Garcia-Rozas *et al.*, 1982), but this enzyme can be selectively inhibited with levamisole, which inhibits all endogenous alkaline phosphatases except those in the gut. Because the antibodies are conjugated with the gut form of the enzyme, they remain active in levamisole (Ponder and Wilkinson, 1981; Mason *et al.*, 1986).

Incubation with monoclonal antibody, washing, and incubation with alkaline phosphatase-conjugated anti-mouse immunoglobulin are done exactly as in Sections 13.7.7 and 13.7.8. Then, the alkaline phosphatase substrate (see Section 13.8.4) is added and incubated for enough time to develop strong colour, but not long enough to result in excessive background staining. Typical incubation time is 10 min (see Section 13.8.4).

13.8.3 The Alkaline Phosphatase–Anti-Alkaline Phosphatase (APAAP) Technique

The availability of monoclonal antibodies to calf intestine alkaline phosphatase makes an elegant sandwich method of labelling possible, avoiding the need for complex chemical coupling methods for conjugating alkaline phosphatase to antibodies. Tissues are incubated with monoclonal antibody against the desired antigen, washed, incubated with antimouse immunoglobulin, washed, and then incubated with complexes of AP-monoclonal anti-AP. The 'free arms' of the anti-mouse immunoglobulin allow the AP–anti-AP complexes to bind. Incubation with substrate is done in exactly the same manner as for the direct method. This technique is known as the alkaline phosphatase–anti-alkaline phosphatase technique (APAAP; see Mason, 1985; Mason *et al.*, 1986).

Alkaline phosphatase is a dimeric enzyme, and as the monoclonal antibodies to it are also dimeric, the complexes of alkaline phosphatase and anti-alkaline phosphatase could form large linear aggregates. This may contribute to the sensitivity of the method, but it also raises the possibility of Fc receptor binding because of the multivalent presentation of Fc regions of the antibody. Evidently, this has not been a major problem.

13.8.4 Substrates for Alkaline Phosphatase

Tables 13.4 and 13.5 give recipes for two common substrates for alkaline phosphatase. Each gives a strong red insoluble precipitate in sections. The procedure given in Table 13.4 gives stronger staining in our hands. The protocol using new fuchsin (Table 13.5) generates a product that is insoluble in organic solvents, and allows the sections to be dehydrated, cleared and covered with permanent mountant.

Incubation time may vary from one occasion to the next, depending on the rate of colour change. Rapid and intense red colour can sometimes develop as quickly as 5 min, but slower weaker colour (for low-density antigens) can sometimes take up to 30 min to develop. Beyond 30 min, background staining (often yellow-brown) becomes a limiting factor. Timing can be judged by the macroscopic appearance of the slide and also by using an inverted microscope, which allows one to watch the colour develop while the substrate is still

Table 13.4 Fast red substrate for alkaline phosphatase

Naphthol AS-MX Phosphate (Sigma cat. no. N4875)	2 mg
Dimethyl formamide	200 μl
0.1 M Tris buffer pH 8.2	9.8 ml
1 M Levamisole	10 μl
Fast-red TR salt (Sigma cat. no. F1500)	10 mg

Dissolve the naphthol AS-MX phosphate in dimethyl formamide in a glass tube. Dilute to 10 ml with Tris buffer. Add levamisole to block endogenous alkaline phosphatase activity. Solution may be stored for several weeks at 4°C, but for longer storage freeze at –20°C. The product is insoluble in aqueous solution but soluble in organic solvents, so an aqueous mountant must be used.

Table 13.5 New fuchsin/$NaNO_2$ substrate for alkaline phosphatase

Naphthol AS phosphate (Sigma cat no N2250)	10 mg
Dimethyl formamide	600 μl
0.2 M Tris buffer pH 9.0	100 μl
1 M levamisole (only necessary for tissues that have endogenous alkaline phosphatase)	100 μl
4% $NaNO_2$ (freshly prepared)	500 μl
4% New fuchsin (Merck cat no 4041) in 2 M HCl	200 μl

To the 4% solution of new fuchsin add the 4% solution of sodium nitrite. Mix. Add the Tris buffer. Add levamisole to block endogenous phosphatase. Dissolve the naphthol AS phosphate in the dimethyl formamide and add to the above solution. Mix well and filter directly onto the slides. (Provided that one works fairly quickly, filtering may not be essential.) Glycerol/PBS solution should not be used with this substrate. Instead, use Coverslipping Resin from Miles Laboratories as mountant. Note that the order of constituents is crucial. Add the sodium nitrite to the new fuchsin, mix, then add Tris, and finally add naphthol AS phosphate, already dissolved in dimethyl formamide. Finally, filter the mixture through Whatman no. 1 filter paper.

This procedure is known as a 'simultaneous capture' method. In the initial reaction, new fuchsin is converted to possess a diazonium group. As phosphate is cleaved from naphthol AS phosphate by the enzyme, the diazonium group on new fuchsin is 'captured' by Naphthol AS.

Naphthol AS Phosphate should be weighed out immediately before use. The substrate solution has been observed to deteriorate within 1 h.

New fuchsin produces a product that is insoluble in ethanol, xylene and Histoclear, and is suitable for permanent mounting media such as Tissue Tek resin.

present. If the slides are incubated with substrate in a Coplin jar, one can 'fish' them out every 10 min or so to monitor the development of colour. The reaction may be terminated by tipping off the substrate and briefly washing in TBS or PBS.

13.8.5 Simultaneous Immunoenzymic Staining with Alkaline Phosphatase and Horseradish Peroxidase

It is possible to perform simultaneous immunoenzymic staining of histological sections for two different antigens (Mason and Woolston, 1982). The main limitation is that it is difficult to evaluate regions where both antigens are present, because of the interaction between the colours of the two substrates.

Table 13.6 Preparation of naphthol AS-MX/fast blue BB substrate for alkaline phosphatase

Naphthol AS-MX	Dissolve 2 mg in 200 µl dimethyl formamide, in a glass vial. Add 9.8 ml 0.1 M Tris-HCl, pH 8.2.
Fast blue BB	Dissolve 10 mg in the naphthol AS-MX solution (above) immediately prior to use.

This substrate produces a blue colour. The protocol is from Mason and Woolston (1982). If the solution is cloudy, it should be filtered directly onto the slide. Otherwise, pipette the mixture directly onto the slide. Incubation time is typically 10–15 min, and may be observed under the microscope, and terminated by washing in PBS, followed by extensive washing in water. The product is insoluble in water, and must be used with an aqueous mounting fluid, as it is soluble in organic solvents.

This is particularly problematical if one antigen is present in trace amounts while the other is present in large amounts, but could be overcome by the use of the newly introduced fluorescent substrate for alkaline phosphatase (see Section 13.8.6). As in any two-colour system where multiple antibodies are used, great care must be taken to include all possible controls to check for unwanted interactions between the two sets of antibodies.

Mason and Woolston (1982) described a dual colour system in which horseradish peroxidase is used with DAB, producing a brown reaction product, and alkaline phosphatase is used with Naphthol AS-MX and fast blue BB to produce a blue reaction. Details of the preparation of the latter substrate are given in Table 13.6.

The reactions with substrates are carried out sequentially, with the peroxidase substrate first. When adequate colour has developed with DAB/H_2O_2, the reaction is terminated by washing in TBS for a few minutes, and then the alkaline phosphatase substrate is added, and incubated for 10–15 min, until adequate colour develops. The reaction is terminated by washing with TBS, followed by water, and then an aqueous mountant (Mason amd Woolston, 1982).

It should be noted that this system is generally used with no counterstaining to reveal cell nuclei. Counterstaining with haematoxylin can be used if the red alkaline phosphatase substrate described in Table 13.5 is used, but at the expense of reduced ability to identify mixed staining reactions (Mason and Woolston, 1982).

13.8.6 A Fluorescent Substrate for Alkaline Phosphatase Immunohistochemistry

As discussed in Section 12.10, Molecular Probes has recently introduced a new fluorescent substrate for alkaline phosphatase, which forms an intensely fluorescent yellow-green submicroscopic precipitate upon cleavage of its phosphate group (Enzyme-Labeled Fluorescence; ELF™). The extremely high sensitivity of this substrate and its resistance to photobleaching suggest that it may become very popular (Larison *et al.*, 1995). It would be particularly well suited to the simultaneous detection of two or more antigens when used together with another enzyme.

13.9 Postfixation of Slides after Antibody Staining

'Postfixation' after staining is strongly recommended to maintain long-term morphology. At this stage, preservation of antigenicity is no longer an issue, and the use of harsher fixatives such as aldehydes can be used safely.

Postfixation with formaldehyde or glutaraldehyde is strongly recommended if the slides are to be stored more than 1 or 2 days. Postfixation should stop autolysis and preserve morphology virtually indefinitely. Acetone is a poor fixative, and acetone 'fixation' prior to staining with antibody is not a substitute for postfixation with aldehyde fixatives. The latter cause covalent cross-linking of proteins, and are much more effective. A suitable postfixing solution is 10% formalin in PBS. Glutaraldehyde (1–10% in PBS) would probably be at least as good or better. Postfixation should be carried out before counterstaining with haematoxylin.

13.10 Counterstaining

It is often desirable to counterstain the nuclei with haematoxylin to allow better visualization of tissue structure. It is necessary to choose conditions that give the right degree of intensity of staining so that the enzyme reaction is not obscured. The intensity of staining can be varied by altering either the time or the concentration of haematoxylin. Varying the time is most satisfactory. Using the standard recipe of Harris' haematoxylin (Table 13.7), a time of about 75 s will usually be satisfactory. Harris' haematoxylin gives particularly clear nuclear staining. After counterstaining, the slides are washed briefly in water in a slide rack.

13.11 Mounting with Cover Slips

13.11.1 Aqueous Mounting

Dry the slides briefly, and place a drop of aqueous mounting solution (90% glycerol–10% PBS) on the cover slip, then put cover slip on the slide, taking care to avoid bubbles. Let the weight of the cover slip spread the mounting solution between the layers of glass. It helps to have the mounting solution at room temperature. Cold mounting solutions will be more viscous and have a greater probability of producing bubbles. Seal the edges of the cover slip with nail polish to prevent drying.

Table 13.7 Harris' haematoxylin

Haematoxylin	2.5 g
Absolute alcohol	25 ml
Potassium alum	50 g
Distilled water	500 ml
Mercuric oxide	1.25 g
Glacial acetic acid	20 ml

Dissolve the haematoxylin in absolute alcohol. Dissolve the alum in water. Add the two solutions to each other, bring rapidly to the boil, then add the mercuric oxide. Cool rapidly by plunging the flask into ice-cold water. When the solution is cold, add the acetic acid. The stain is then ready for immediate use. Glacial acetic acid is optional, but its inclusion gives a more precise and selective staining of nuclei.

The use of aqueous mountants has some disadvantages. They tend to dry out, may allow microbial growth, and do not have the optimal refractive index for good optical properties when used with high-power oil-immersion objectives.

13.11.2 Nonaqueous (Permanent) Mounting

Tissue Tek coverslipping resin is nonaqueous and does not need nail polish to provide a seal. Prior to mounting in Tissue Tek Resin, slides must be dehydrated by passaging through ethanol and Histoclear or xylene. This procedure is obviously only suitable for substrates whose reaction products are insoluble in these solvents. The substrate given in Table 13.5 is compatible with Tissue Tek mounting resin.

13.12 Immunofluorescence on Tissue Sections

The technique for immunofluorescence on tissue sections is virtually identical to the enzymic methods, with the substitution of fluorescent antibodies. The relevant techniques for production of fluorescent antibodies and immunofluorescence microscopy are given in detail in Chapter 12.

In recent years, the use of enzymic techniques has tended to become more popular than immunofluorescence, because one can obtain a better simultaneous view of tissue morphology and antigen distribution, and the mounts can be more permanent. The recent development of highly fluorescent substrates for alkaline phosphatase (see Section 13.8.6) may represent a very powerful combination of the two techniques.

After labelling with fluorescent antibodies, sections are mounted in a slightly alkaline medium such as 50% glycerol in PBS, because fluorescein gives stronger fluorescence when its carboxyl group is ionized (see Chapter 12). The fluorescence of FITC fades within seconds under the intense ultraviolet light used in typical modern fluorescence microscopes, but fading can be easily prevented by the inclusion of 10 μg/ml phenylenediamine or 2.5% DABCO (Johnson *et al.*, 1982) or 0.1–0.25 M *n*-propyl gallate (Giloh and Sedat, 1982) in the mounting medium. Of these, DABCO is one of the best. Many commercial antifade products are now available, and are discussed in Section 12.11.5. Glutaraldehyde should not be used for postfixation, as it causes autofluorescence.

References

Brandtzaeg, P. (1982) Tissue preparation methods for immunohistochemistry. *In* 'Techniques in Immunocytochemistry'. Vol. 1 (G. R. Bullock and P. Petrusz, eds) pp. 1–75. Academic Press, London.

Culvenor, J. G., Harris, A. W., Mandel, T. E., Whitelaw, A. and Ferber, E. (1981) Alkaline phosphatase in hematopoietic tumor cell lines of the mouse: high activity in cells of the B lymphoid lineage. *J. Immunol.* **126**, 1974–1977.

Farr, A. G. and Nakane, P. K. (1981) Immunohistochemistry with enzyme labeled antibodies: a brief review. *J. Immunol. Methods* **47**, 129–144.

Finley, J. C. W. and Petrusz, P. (1982) The use of proteolytic enzymes for improved localization of tissue antigens with immunocytochemistry. *In* 'Techniques in Immunocytochemistry'. Vol. 1 (G. R. Bullock and P. Petrusz, eds) pp. 239–250. Academic Press, London.

Garcia-Rozas, C., Plaze, A., Diaz-Espada, F., Kreisler, M. and Martinez-Alonso, C. (1982) Alkaline phosphatase activity as a membrane marker for activated B cells. *J. Immunol.* **129**, 52–55.

Giloh, H. and Sedat, J. W. (1982) Fluorescence microscopy: reduced photobleaching of rhodamine and fluorescein protein conjugates by *n*-propyl gallate. *Science* **217**, 1252–1255.

Hancock, W. W. and Atkins, R. C. (1986) Immunohistological studies with monoclonal antibodies. *Methods Enzymol.* **121**: 828–848.

Hancock, W. W., Becker, G. J. and Atkins, R. C. (1982) A comparison of fixatives and immunohistochemical technics for use with monoclonal antibodies to cell surface antigens. *Am. J. Clin. Pathol.* **78**, 825–831.

Harahap, A. R. and Goding, J. W. (1988) Distribution of the murine plasma cell antigen PC-1 in non-lymphoid cells. *J. Immunol.* **141**, 2317–2320.

Harrach, B. and Robenek, H. (1991) Immunocytochemical localization of apolipoprotein A-I using polyclonal antibodies raised against the formaldehyde-modified antigen. *J. Microscopy* **161**, 97–108.

Johnson, G. D., Davidson, R. S., McNamee, K. C., Russell, G., Goodwin, D. and Holborow, E. J. (1982) Fading of immunofluorescence during microscopy: a study of the phenomenon and its remedy. *J. Immunol. Methods*. **55**, 231–242.

Larison, K. D., Bremiller, R., Wells, K. S., Clements, I. and Haugland, R. P. (1995) Use of a new fluorogenic phosphatase substrate in immunochemical applications. *J. Histochem. Cytochem.* **43**, 77–83.

Leenen, P. J. M., Jansen, A. M. A. C. and van Ewijk, W. (1985) Fixation parameters for immunocytochemistry: the effect of glutaraldehyde or paraformaldehyde fixation on the preservation of mononuclear phagocyte differentiation antigens. *In* 'Techniques in Immunocytochemistry'. Vol. 3 (G. R. Bullock and P. Petrusz, eds) pp. 1–24. Academic Press,. London.

Mason, D. Y. (1985) Immunocytochemical labeling of monoclonal antibodies by the APAAP immunoalkaline phosphatase technique. *In* 'Techniques in

Immunocytochemistry'. Vol. 3 (G. R. Bullock and P. Petrusz, eds) pp. 25–42. Academic Press, London.

Mason, D. Y. and Woolston, R-W. (1982) Double immunoenzymatic labelling. *In* 'Techniques in Immunocytochemistry'. Vol. 1 (G. R. Bullock and P. Petrusz, eds) pp. 135–153. Academic Press, London.

Mason, D. Y., Erber, W. N., Falini, B., Stein, H. and Gatter, K. C. (1986) Immunoenzymatic labelling of haematological samples with monoclonal antibodies. *In* 'Methods in Haematology. Monoclonal Antibodies' (P. C. L. Beverley, ed.) pp. 145–181. Churchill Livingstone, Edinburgh.

Ponder, B. A. and Wilkinson, M. M. (1981) Inhibition of endogenous tissue alkaline phosphatase with the use of alkaline phosphatase conjugates in immunohistochemistry. *J. Histochem. Cytochem.* **29**, 981–984.

14 Construction, Screening and Expression of Recombinant Antibodies

R. A. Irving, P. J. Hudson and J. W. Goding

The 'classical' methods of production of monoclonal antibodies by cell fusion have not changed greatly since their original descriptions. However, a new generation of technologies is emerging, based upon recombinant DNA, and may eventually surpass or even replace the older approaches. This chapter focuses on the use of bacteria and bacteriophages to display a diverse library of antibodies equivalent to the mammalian immune repertoire. The new technology combines production, selection and affinity maturation in numerous imaginative ways. It is, however, rather more complex than cell fusion, and still has some technical and logistical problems. None the less, it holds great promise, especially for the production of human monoclonal antibodies (reviewed by Winter and Milstein, 1991; Burton and Barbas, 1994; Winter *et al.*, 1994).

14.1 The Bacteriophage Display Library System

It is now possible to engineer bacteriophage genomes such that the phage display on their surface, antibody-like proteins which are encoded by the DNA that is contained within the phage. The power of this approach derives from the fact that the selection of the highest-affinity antibody is coupled to the recovery of the gene encoding that antibody. The selection process is based on protein affinity, and can be applied to any high-affinity binding reagents such as antibodies, antigens, receptors and ligands (Winter and Milstein, 1991). Obtaining the gene also makes it easy to design systems for its expression using bacterial or mammalian host cells.

Bacterial expression systems also offer an ethical advantage over ascites fluid production in live mice and a significant cost advantage over hybridoma cell cultures (Boss *et al.*, 1984; Bird *et al.*, 1988; Skerra and Plückthun, 1988; Ward *et al.*, 1989; Skerra *et al.*, 1991). They can be exploited for the rapid synthesis and selection of reagents that were hitherto difficult or impossible to produce by hybridoma technology. These advantages, coupled to the development of an affinity maturation process that mimics the immune system, delineates a process that has the potential to supersede hybridoma technology and bypass the limitations inherent in producing mammalian antibodies.

It will be apparent, however, that this approach is far more complex than the production of monoclonal antibodies by cell fusion, and requires a great deal of specialist expertise. It is by no means a routine procedure.

The key components that are needed for the new library technology are:

(1) A repertoire of antibody genes, typically derived by polymerase chain reaction (PCR) technology.
(2) A method for producing a stable library, in which each member displays a unique antibody and also carries the gene encoding that antibody.
(3) A method for selecting the highest affinity antibody from the library.
(4) An affinity maturation strategy for improving and manipulating the selected antibody.
(5) An expression system for large-scale production of the antibody.

The vectors for phage display are extremely effective for the selection of the desired antibody specificity, but are generally not efficient for high-level expression of the antibody gene segment (often because of the toxicity of gene III fusions). Various bacterial secretion systems for high-level expression are described in more detail later in this chapter. Once an appropriate antibody gene has been selected by phage display, it is usually fairly straightforward to transfer the gene to a plasmid system in which high level expression is facilitated (see Section 14.14).

14.2 The Choice of Fv, scFv or Fab Fragments

Antibodies bind to their target antigen using the outer surface of the associated V_H and V_L domains (Fig. 14.1). Any antibody fragment, including Fab and Fv molecules, that retains the correct alignment of the associated V_H and V_L domains will possess the same monovalent binding specificity and affinity as the parent antibody. There has been considerable debate on the affinity properties of single V_H or V_L domains, but now it is generally accepted that Fv fragments are the smallest fragments showing equivalent binding affinity to the parent Fab fragment (Denzin *et al.*, 1991; Milenic *et al.*, 1991; Malby *et al.*, 1993).

Bacteria are well suited to the expression of small nonglycosylated proteins such as the Fab or Fv fragments of antibodies (Fig. 14.1). The evidence to date is that bacterially produced antibody fragments have the same affinity as the

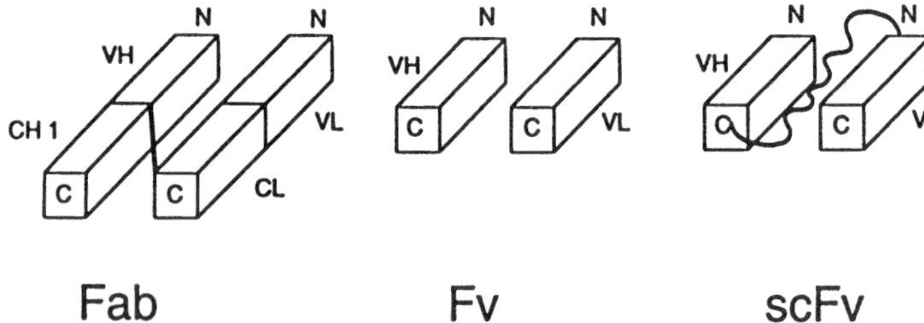

Fig. 14.1. A Fab antibody fragment comprises one light and a portion of one heavy chain. An Fv fragment consists of a pair noncovalently associated variable region domains, one from a light chain and one from a heavy chain. A single-chain Fv fragment (scFv) consists of one V_H and one V_L domain joined by a peptide linker.

corresponding antibodies produced in mammalian cells. A further advantage of the Fab and Fv antibody fragments is the ease of bacterial expression systems for large-scale production (Cabilly, 1984; Power *et al.*, 1992; Malby *et al.*, 1993).

Fv fragments produced by genetic engineering sometimes suffer from aggregation and/or instability, which may sometimes be improved by chemical crosslinking or engineering of disulfide bonds between chains (see Glockshuber *et al.*, 1990; Brinkmann *et al.*, 1993, 1995; Pack *et al.*, 1993; Jung *et al.*, 1994; Reiter *et al.*, 1994, 1995). However, most phage libraries to date have not used crosslinking for Fv fragments.

14.3 Single-chain Fv (scFv) Antibodies

F_v fragments from existing mouse monoclonal antibodies can be expressed as single-chain molecules (scFv) in which a peptide permanently links the V_H and V_L domains (Bird *et al.*, 1988; Huston *et al.*, 1988; Skerra *et al.*, 1991; Malby *et al.*, 1993, Lilley *et al.*, 1994; see Fig. 14.1). Generally, the levels of production of scFvs are higher than for Fvs, as the V_H and V_L domains are fused and do not have to find each other in the bacterial periplasm. ScFv molecules have been shown to have similar binding specificity and affinity to the parent antibody, and can be used to replace whole antibodies in many applications. The polypeptide linker must span at least 35Å (3.5 nm) between the *C* terminal of one variable (V) domain to the *N* terminal of the other V domain without compromising the fidelity of the V_H–V_L pairing and scFv binding site. The lengths of the linker used in different experiments have varied between 10 and 25 amino acid residues, with a preferred size of 15 residues.

The *N* termini of both V_H and V_L are near to, but not part of, the protein-binding interface, so there is little to choose in orientation between V_H–V_L or V_L–V_H. ScFvs have been assembled in either domain order as a V_H–linker–V_L fusion protein (Glockshuber *et al.*, 1990; Anand *et al.*, 1991; Cheadle *et al.*,

1992; Malby *et al.*, 1993) or as a V_L–linker–V_H fusion protein (Anand *et al.*, 1991; Denzin *et al.*, 1991; Gibbs *et al.*, 1991; Milenic *et al.*, 1991; Pantoliano *et al.*, 1991). The linker should preferably be hydrophilic in nature to prevent it from associating with the hydrophobic V domains (Bird *et al.*, 1988; Huston *et al.*, 1988; Batra *et al.*, 1990; Tai *et al.*, 1990; Glockshuber *et al.*, 1990). It can be designed with features such as incorporation of a diagnostic 'tag' or residues for precise proteolytic excision in subsequent purification steps (Huston *et al.*, 1991; Whitlow and Filpula, 1991). A commonly used linker sequence is $(Gly_4Ser)_3$.

It should be noted that the linker is an artificial addition to the antibody, and as such may be regarded as nonself and antigenic if the scFv antibody were injected into animals or humans. In principle, the linker could be substituted by a human sequence of similar size and properties, although the boundaries between the linker and the rest of the molecule would still be nonself.

The propensity of scFv molecules to form dimers and the potential for the design and production of stable bispecific scFvs (covalently joined by short 5-mer linkers) is not discussed in this chapter, but has recently been discussed by several authors (Holliger at al, 1993; Kortt *et al.*, 1994; Neri *et al.*, 1995).

14.3.1 The Pharmacia Recombinant Phage Antibody System

An outline of the use of the Fd phage expression system for Fab or scFv antibodies is shown in Fig. 14.2 (Hoogenboom *et al.*, 1991).

A kit for the production of scFv antibodies using the Fd phage system is available from Pharmacia Biotech (Recombinant Phage Antibody System). The kit is designed to make cDNA from mouse spleen or hybridomas, and to amplify immunoglobulin variable region genes by PCR, and clone them into the phagemid vector pCANTAB 5 E using an *Sfi* I site at the 5′ end and a *Not*I site at the 3′ end. These sites occur rarely in immunoglobulin genes. A short polypeptide linker consisting of $(Gly_4Ser)_3$ is used to join the heavy and light chains. The scFv antibody is produced as a fusion protein to the *N* terminus of the gene III coat protein, but separated from it by an amber stop codon. After transformation into an amber suppressor mutant strain of *Escherichia coli*, phage are rescued using KO7 helper phage, and selected by binding to antigen.

Once the phage with the desired specificity are isolated, they can be used to transfect wild-type *E. coli* that lack the amber suppressor, resulting in a stop codon at the end of the scFv, such that the gene III protein is not produced. In some cases, the resultant antibody is secreted into the medium, while in others, it will be retained in the periplasmic space. A 'tag' sequence (E Tag) that is recognizable by a special antibody has been incorporated at the *C* terminus of the scFv, and may be used to identify and purify the products of the clone (see Section 14.15.1). Many of the special features used in this system will be explored later in this chapter.

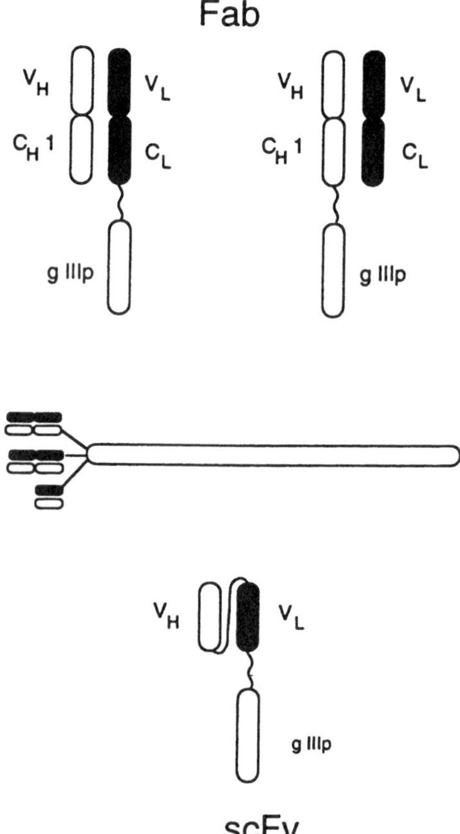

Fig. 14.2. Antibody fragments such as Fab and scFv are shown displayed on the surface of filamentous Fd bacteriophage by covalent fusion to the gene III protein.

14.4 Libraries of Antibody Genes from Naive and Immunized Sources

In mammals, the diversity of antibodies is brought about both by reassortment and splicing of DNA fragments encoding specific regions of the antibody genes as well as random sequences added without DNA templates and also somatic mutation (Chapter 6; see also Winter and Milstein, 1991). In order to mimic this repertoire, PCR technology may be used to extract and amplify the antibody genes. The coding sequences for constant regions flanking the variable domains have been well characterized. There are two possible strategies for primer design. First, one may design matched sets of 5′ and 3′ primers for each variable region subgroup that allow the amplification of antibody gene cDNAs from the donor tissue mRNA (Marks *et al.*, 1991a; see

Tables 14.1 and 14.2). Alternatively, degenerate PCR primers can be devised which cover most or all variable region subgroups (Larrick *et al.*, 1989a, b). The definitive sets of PCR primers that are needed to amplify mouse antibody genes are described by Orlandi *et al.* (1989) and Kettleborough *et al.* (1993), and for human genes by Marks *et al.* (1991a).

A complete library based on the natural naive immune repertoire requires the ability to extract all possible germline antibody genes (Tomlinson *et al.*, 1992 and Cox *et al.*, 1994). The ultimate goal is to construct very large libraries ($>10^{12}$ different specificities), which provide greater complexity than the primary immune repertoire of a mouse or human. Antibody genes can be derived from cells of unimmunized individuals for the construction of a naive library of human antibody fragments. Such an IgM-based library would be least biased, and would represent a naive source for human antibodies.

The construction of the first naive human library was based on IgM scFv fragments and was of limited size (about 10^7 specificities), but gave a clear demonstration that antibodies could be selected by this procedure (Marks *et al.*, 1991b). As might have been expected, the antibodies recovered from this library were of relatively low affinity. Griffiths *et al.* (1994) and Nissim *et al.* (1994) have described an improved system with potential diversity in the region of 10^{14} specificities after recombination.

Naive primary libraries from unimmunized individuals exhibit predominantly low affinity antibody fragments, but provide the realistic possibility of directly selecting a high affinity antibody (K_D of 10^{-8} M; Figini *et al.*, 1994). In contrast, the library size from pre-immunized animals depends on the extraction of rearranged V-genes from both heavy and light chain cDNA by PCR (Huse *et al.*, 1989; Larrick *et al.*, 1989a, b; Sastry *et al.*, 1989; Marks *et al.*, 1991a, b; Persson *et al.*, 1991). The heavy and light chain genes are assembled randomly to encode either scFv antibody fragments or Fab fragments in phage or phagemid vectors, resulting in individual libraries containing in excess of 10^8 different members.

An alternative library of antibody genes can be derived by a partially synthetic route, as the diversity is primarily determined through amino acid sequence variations encoded in the six CDR loops that form the antigen combining site (see Chapter 5). This diversity can be generated during DNA synthesis by specifying the random base incorporation at particular positions in the DNA (Barbas *et al.*, 1992; Huse *et al.*, 1993). Diversity can also be encoded into the framework regions, which are the protein scaffold that forms the central core of the antibody domains; this will often alter the affinity or specificity of the antibody.

14.5 Bacteriophage Display Libraries

Smith (1985) first demonstrated that peptide libraries could be displayed on the surface of filamentous Fd-bacteriophages. This work has subsequently

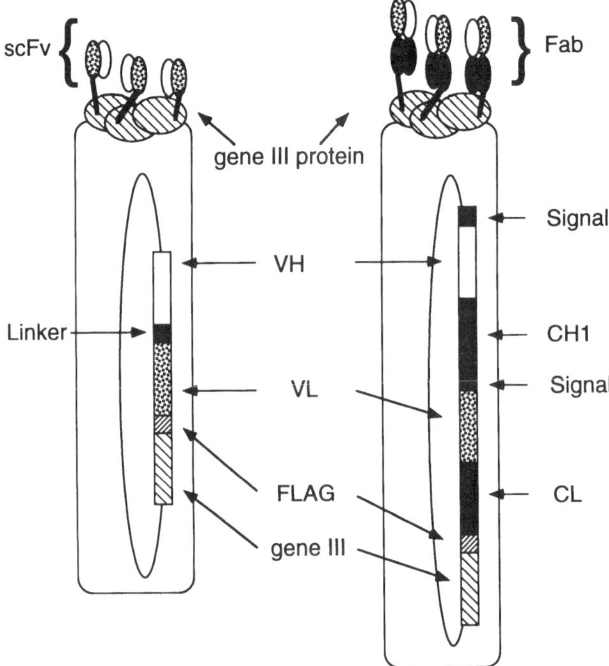

Fig. 14.3. Typical constructs for expression of scFv or Fab fragments as fusion proteins with gene III coat protein of Fd phage, using a single vector to encode both chains.

been extended to show that protein domains, such as scFvs, could also be displayed by Fd-bacteriophage, superseding the use of live bacterial cells as the display system (Huse *et al.*, 1989; McCafferty *et al.*, 1990; Hoogenboom *et al.*, 1991). Fd-bacteriophages have an enormous advantage over bacterial cells in that they can be easily grown to extremely high titres, in excess of 10^{13} plaque-forming units (pfu) per ml. As mentioned earlier, the fact that each phage particle displaying an antibody contains within it the gene for that antibody provides the means for selecting such genes from a complex library by direct binding to antigen.

Libraries of antibody gene fragments can be displayed on the surface of Fd bacteriophage, covalently coupled to phage coat proteins encoded by gene III, or less commonly gene VIII (Figs 14.2 and 14.3). Insertion of antibody genes into the gene III coat protein gives expression of only about 3–5 protein molecules per phage, situated at the ends. In contrast, insertion of antibody genes into gene VIII has the potential to display about 2000 copies of the recombinant protein per phage particle, but suffers from the potential drawback of being a multivalent system which could mask the affinity of a single displayed fragment.

Antibody-displaying phage may be selected by binding to antigen covalently coupled to beads (see Chapter 11) or adsorbed to plastic surfaces in a manner similar to ELISA or solid phase radioimmunoassays (see Sections 8.10.3 to 8.10.6). While almost any plastic surface will adsorb protein antigens, some commercial products are specially formulated for this purpose, such as Nunc Immunotubes. The choice of affinity matrix can often determine the kinetics of the selected antibody gene fragment (Griffiths *et al.*, 1993; Marks *et al.*, 1992a).

Fd phagemid vectors are now the vectors of choice because they can be easily switched from the display of functional antibody fragments on the surface of Fd-bacteriophage to secreting soluble antibody fragments in *E. coli*. Phage-displayed scFv or Fab antibody-gene fusions with the *N* terminus of the gene III coat protein are made possible by an amber codon strategically positioned between the two protein genes. In amber supressor strains of *E. coli*, the resulting antibody–gene III fusions become anchored in the phage coat. Expression of soluble forms of the antibody fragments is achieved by simply transferring the vector to non-*supE* strains of *E. coli* which recognize the amber translation stop codon. The *E. coli* strains which have been successfully used for soluble expression include *HB2151, MC1061, BL21* and *TOPP 1–6*.

14.6 Experimental Details

14.6.1 The Source of Antibody Genes: mRNA or DNA

The repertoire of antibody genes from pre-immunized animals may be amplified from either RNA or DNA extracted from hybridomas, peripheral blood lymphocytes (PBLs) or spleen. For the amplification of the genes from cells expressing immunoglobulin, RNA is the template of choice, using a cDNA copy as the primary template.

Isolation of RNA by the method of Chomczynski and Sacchi (1987) is easy and reliable and less expensive than commercial kits. It does not require expensive reagents such as caesium chloride or the use of an ultracentrifuge, and produces excellent quality RNA, even from tissues that contain large amounts of ribonuclease. Once the RNA is isolated, it is very helpful to resuspend it in formamide rather than aqueous solutions because ribonucleases are inactive in this solvent (Chomczynski, 1992). If the RNA is resuspended in formamide, it can be directly added to the reverse transcriptase reaction using Moloney virus Superscript II reverse transcriptase (Life Technologies), provided that the final concentration of formamide in the mix does not exceed 5–10% v/v (Gerard, 1994). It has been found that RNA can also be resuspended in dimethyl formamide with minimal inhibition of reverse transcriptase at these concentrations (J. Goding, unpublished).

Reverse transcription of the RNA template can be achieved by any of the commercial kits available or the published methods. It is generally preferable to use reverse transcriptase from Moloney virus, such as Superscript II (Life Technologies). Using oligo dT to prime the reaction rather than specific primers or random hexamers may result in better yields of cDNA. This is critical for the subsequent PCR amplifications with the specific primers.

DNA may be isolated from lymphocytes by any of the standard methods (Sambrook *et al.*, 1989).

14.6.2 Amplification of Ig Heavy and Light Chain Variable Region Genes from Lymphocyte DNA or cDNA by the Polymerase Chain Reaction

Variable region genes may be rapidly isolated from genomic DNA or cDNA by PCR using one primer at the 5′ end of the V region and a second primer at the 3′ end of the V region. If the sequence of a particular desired gene is known, then the primers can be designed based on that sequence. This may be determined by cloning and sequencing of the corresponding cDNA or by amino acid sequencing of the antibody protein. However, the *N* termini of many antibody heavy and light chains are blocked to Edman degradation.

If desired, the 3′ primer may be based on the sequence of the first part of the constant regions of the heavy and light chain V regions. This has the advantage that only a single 3′ primer needs to be synthesized for each isotype. Typical sets of primers are listed in Tables 14.1 and 14.2.

The primers often have restriction sites or other features added at their 5′ ends to facilitate cloning of the product or other manipulations. It has been found that the shortest possible oligonucleotide primers often provide the best results, especially when constructing libraries by PCR from naive sources. The incorporation of the restriction enzyme sites for subsequent cloning has sometimes led to poorer results, perhaps owing to an adverse effect of the extra length of noncomplementary sequence encoding the restriction site and extension. Tables 14.1 and 14.2 contain a selection of the oligonucleotides that have been successful in the amplification of murine V_H and V_L domains and includes the appropriate extensions for subsequent construction of scFvs. These extensions are lower case and italicized.

If it is desired to make more general libraries in which many specificities are represented, it will be necessary to use primers that will prime all the members of particular V gene ‘families’.

Table 14.1 Oligonucleotide sequences for the amplification of mouse V_H and V_L domains prior to scFv construction

	Type-specific sequences
5′ V_H primers	
5′-tta tta ctc gcg gcc cag ccg gcc atg gcc	GAT GTG AAG CTT CAG GAG TC-3′
5′-tta tta ctc gcg gcc cag ccg gcc atg gcc	CAG GTG CAG CTG AAG GAG TC-3′
5′-tta tta ctc gcg gcc cag ccg gcc atg gcc	CAG GTG CAG CTG AAG CAG TC-3′
5′-tta tta ctc gcg gcc cag ccg gcc atg gcc	CAG GTT ACT CTG AAA GAG TC-3′
5′-tta tta ctc gcg gcc cag ccg gcc atg gcc	GAG GTC CAG CTG CAA CAA TCT-3′
5′-tta tta ctc gcg gcc cag ccg gcc atg gcc	GAG GTC CAG CTG CAG CAG TC-3′
5′-tta tta ctc gcg gcc cag ccg gcc atg gcc	CAG GTC CAA CTG CAG CAG CCT-3′
5′-tta tta ctc gcg gcc cag ccg gcc atg gcc	GAG GTG AAG CTG GTG GAG TC-3′
5′-tta tta ctc gcg gcc cag ccg gcc atg gcc	GAG GTG AAG CTG GTG GAA TC-3′
5′-tta tta ctc gcg gcc cag ccg gcc atg gcc	GAT GTG AAC TTG GAA GTG TC-3′
3′ V_H primers	
5′-tga acc gcc tcc acc	TGC AGA GAC AGT GAC CAG AGT-3′
5′-tga acc gcc tcc acc	TGA GGA GAC TGT GAG AGT GGT-3′
5′-tga acc gcc tcc acc	TGA GGA GAC GGT GAC TGA GGT-3′
5′-tga acc gcc tcc acc	TGA GGA GAC GGT GAC CGT GGT-3′
5′ V_L (κ) primers	
5′-tct ggc ggt ggc gga tcg	GAT GTT TTG ATG ACC CAA ACT-3′
5′-tct ggc ggt ggc gga tcg	GAT ATT GTG ATG ACG CAG GCT-3′
5′-tct ggc ggt ggc gga tcg	GAT ATT GTG ATA ACC CAG-3′
5′-tct ggc ggt ggc gga tcg	GAC ATT GTG CTG ACC CAA T-3′CT
5′-tct ggc ggt ggc gga tcg	GAC ATT GTG ATG ACC CAG TCT-3′
5′-tct ggc ggt ggc gga tcg	GAT ATT GTG CTA ACT CAG TCT-3′
5′-tct ggc ggt ggc gga tcg	GAT ATC CAG ATG ACA CAG ACT-3′
5′-tct ggc ggt ggc gga tcg	GAC ATC CAG CTG ACT CAG TCT-3′
5′-tct ggc ggt ggc gga tcg	GAC ATC CAG CTG ACT CAG TCT-3′
5′-tct ggc ggt ggc gga tcg	CAA ATT GTT CTC ACC CAG TCT-3′
3′ V_L (κ) primers	
5′-atg agt ttt tgt tct gcg gcc gc	CCG TTT CAG CTC CAG CTT G-3′
5′-atg agt ttt tgt tct gcg gcc gc	CCG TTT TAT TTC CAG CTT GGT-3′
5′-atg agt ttt tgt tct gcg gcc gc	CCG TTT TAT TTC CAA CTT G-3′
5′-atg agt ttt tgt tct gcg gcc gc	GGA TAC AGT TGG TGC AGC ATC-3′

The regions in lower case italics at the 5′ ends of the listed oligonucleotides are extensions for cloning and other manipulations, and do not form part of the antibody genes.

Table 14.2 Oligonucleotide sequences for the amplification of human V_H and V_L domains prior to scFv or Fab construction

5′ V_H primers	5′-CAG GTG CAG CTG GTG CAG TCT GG-3′
	5′-CAG GTC AAC TTA AGG GAG TCT GG-3′
	5′-GAG GTG CAG CTG GTG GAG TCT GG-3′
	5′-CAG GTG CAG CTG CAG GAG TCG GG-3′
	5′-GAG GTG CAG CTG TTG CAG TCT GC-3′
	5′-CAG GTA CAG CTG CAG CAG TCA GG-3′
3′ V_H primers	5′-TGA GGA GAC GGT GAC CAG GGT GCC-3′
	5′-TGA AGA GAC GGT GAC CAT TGT CCC-3′
	5′-TGA GGA GAC GGT GAC CAG GGT TCC-3′
	5′-TGA GGA GAC GGT GAC CGT GGT CCC-3′
5′ V_κ primers	5′-GAC ATC CAG ATG ACC CAG TCT CC-3′
	5′-GAT GTT GTG ATG ACT CAG TCT CC-3′
	5′-GAA ATT GTG TTG ACG CAG TCT CC-3′
	5′-GAC ATC GTG ATG ACC CAG TCT CC-3′
	5′-GAA ACG ACA CTC ACG CAG TCT CC-3′
	5′-GAA ATT GTG CTG ACT CAG TCT CC-3′
3′ V_κ primers	5′-CAG TCT GTG TTG ACG CAG CCG CC-3′
	5′-CAG TCT GCC CTG ACT CAG CCT GC-3′
	5′-TCC TAT GTG CTG ACT CAG CCA CC-3′
	5′-CAC GTT ATA CTG ACT CAA CCG CC-3′
	5′-CAG GCT GTG CTC ACT CAG CCG TC-3′
	5′-AAT TTT ATG CTG ACT CAG CCC CA-3′
5′ V_λ primers	5′-ACG TTT GAT TTC CAC CTT GGT CCC-3′
	5′-ACG TTT GAT CTC CAG CTT GGT CCC-3′
	5′-ACG TTT GAT CTC CAC CTT GGT CCC-3′
	5′-ACG TTT GAT ATC CAC TTT GGT CCC-3′
	5′-ACG TTT AAT CTC CAG TCG TGT CCC-3′
3′ V_λ primers	5′-ACC TAG GAC GGT GAC CTT GGT CCC-3′
	5′-ACC TAG GAC GGT CAG CTT GGT CCC-3′
	5′-ACC TAA AAC GGT GAG CTG GGT CCC-3′

14.6.3 PCR Amplification of V_H and V_L Sequences

Standard volume of PCR is 50 µl.

oligonucleotide primers	5 µl (25 pmol each)
10 × dNTP mix (2.5 mM of each dNTP; 10 mM total)	5 µl
H_2O	24 µl
10 × PCR buffer*	5 µl
template DNA (50–200 ng)	10 µl

(10 × PCR buffer = 100 mM Tris pH 9.3 at 20°C, 500 mM KCl, 15 mM $MgCl_2$, 0.1% gelatin.) Suitable buffers are usually supplied with the thermostable polymerases.

Mix well and centrifuge briefly to ensure that no liquid remains on the tube walls or lid. Heat to 94°C, and then add 0.5 μl *Taq* polymerase (hot start). Mix again by flicking the tube, and then quickly add a drop of mineral oil without allowing the tube to cool. A typical PCR protocol uses the following programme: 1 min at 94°C, 2 min at 65°C, 1.5 min at 72°C, for 30 cycles, then holding for 5 min at 72°C to ensure complete elongation and 5 min at 60°C to ensure renaturation of the amplified fragment.

Occasionally, V_H primer-dimers form and the product does not amplify efficiently. This problem is much less likely to occur if one uses a 'hot start' as above. The problem may sometimes be circumvented using a primer that is shortened at its 3′ end.

The ideal annealing temperature of these PCRs varies with the base composition of the oligonucleotides. Although the above conditions give a general reaction that is effective, the appearance of extra PCR products or lack of product may be remedied by calculating the specific annealing temperature for the oligonucleotide and/or by varying the annealing temperature in 5°C steps. A 10 μl sample of the reaction product should be analysed by electrophoresis on a 2% agarose gel in the presence of ethidium bromide.

Amplification products may be easily isolated from normal agarose gels using Qiaex beads.

14.7 Methods for Producing Libraries

14.7.1 scFv-based Libraries

The gene pools encoding the V-domains may be used as template in the construction of scFv genes by a second PCR using two linker oligonucleotides (Fig. 14.4; Table 14.3) to join the V_H and V_L domains (Davis *et al.*, 1991; Clackson *et al.*, 1991). This results in the random joining of the V_H and V_L domains via the linker, and provides a source of diversity in the production of a scFv library. It may be advantageous to construct the scFv in a single step (Clackson *et al.*, 1991; Lake *et al.*, 1995).

PCR assembly of V_H and V_L PCR bands in scFv

This PCR protocol has been optimized for the construction of a specific scFv, with single V_H and V_L genes amplified from a hybridoma cell line rather than a library of scFv molecules. It is advisable to take a number of precautions to prevent amplification of contaminating sequences in the PCR reactions,

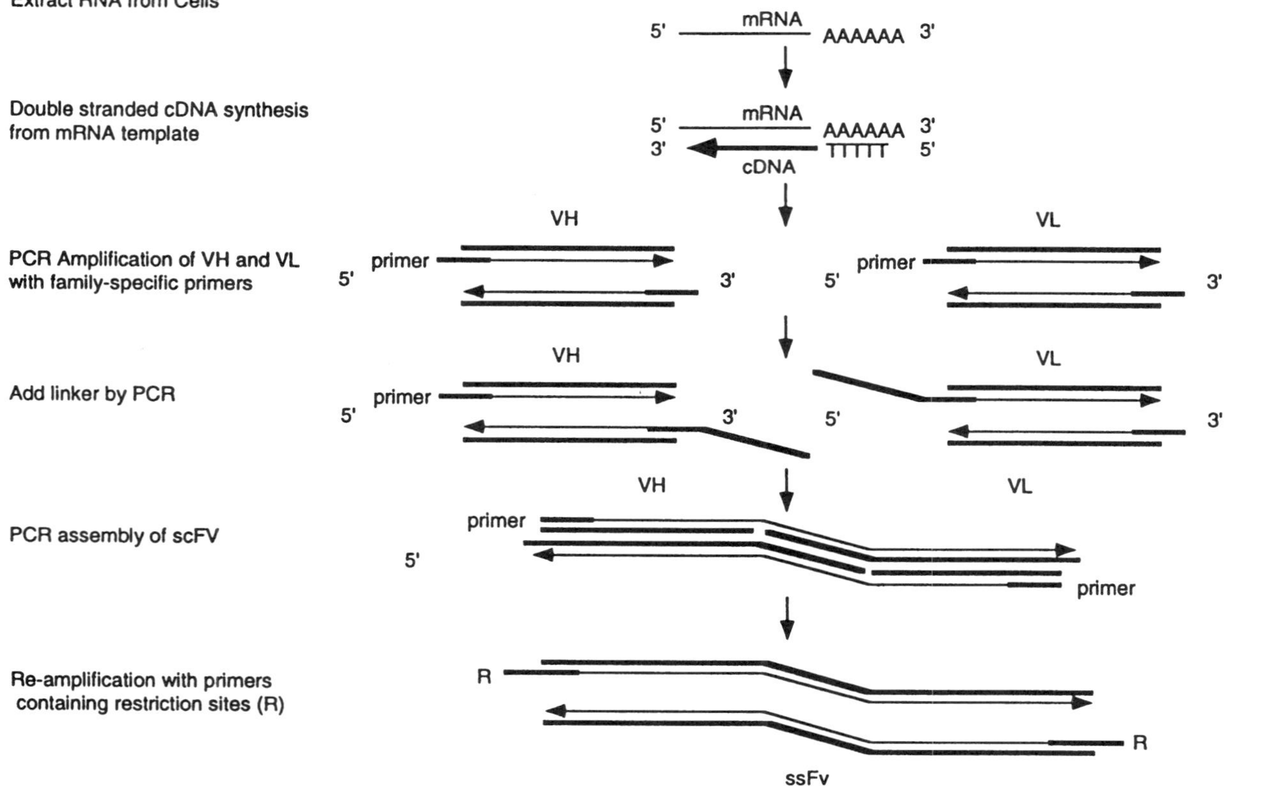

Fig. 14.4. Strategy for use of polymerase chain reaction to add a linker between V_H and V_L domains for the construction of scFv fragments.

Table 14.3 Typical linker oligonucleotides for mouse scFv construction

5′ linker	5′-GTC TCC TCA *GGT GGA GGC GGT TCA* GGC GGA GGT GGC TCT GGC GGT GGC GGA TCG-3′
3′ linker	5′-*CGA TCC GCC ACC GCC AGA* GCC ACC TCC GCC TGA ACC GCC TCC ACC TGA GGA GAC-3′

This pair of oligonucleotides produces a 15 amino acid linker with the sequence $(Gly_4Ser)_3$

especially where the templates are present at low concentration. Irradiation of buffers for 5 min on a short-wave transilluminator prior to the addition of oligonucleotides, templates and enzymes is recommended, as is the use of sterile aerosol-resistant pipette tips. The linkers given in Table 14.3 are used in a one-step joining reaction.

PCR conditions for scFv construction

Water	20 μl
10 × PCR buffer	5 μl
10 × dNTPs	5 μl
BSA (10 mg/ml)	0.5 μl

Irradiate tubes with short-wavelength u.v. light for 5 min then add:

V_H linker 3′ primer	20 ng in 2.5 μl
V_L linker 5′ primer	20 ng in 2.5 μl
V_H primary PCR band	100 ng in 5 μl (template)
V_L primary PCR band	100 ng in 5 μl (template)
V_H 5′ primer(s)	20 ng in 2 μl
V_L 3′ primer(s)	20 ng in 2 μl

Incubate at 94°C for 5 min, then add 0.5 μl *Taq* polymerase.
Mix, then overlay with oil without allowing to cool.
Cycle at 94°C, 1 min, 72°C, 2.5 min for 25 cycles.
Analyse on 2% agarose gels. The scFv should be about 800 bp long.

14.8 Vectors for Expression of Recombinant Fab Antibody

The phage display vectors pHFA (Fig. 14.5; Lah *et al.*, 1994) and $pHFA^2$ (Dolezal *et al.*, in press) are derived from a previously described phagemid vector, pHEN1 (Hoogenboom *et al.*, 1991). pHFA is designed to express scFv antibodies, while $pHFA^2$ can be used either for scFv or for production of Fab fragments. PCR is used to generate a fragment bounded by an *Sfi*I site at the *N* terminus and by a *Xho*I or *Not*I site at the *C* terminus. This PCR product is

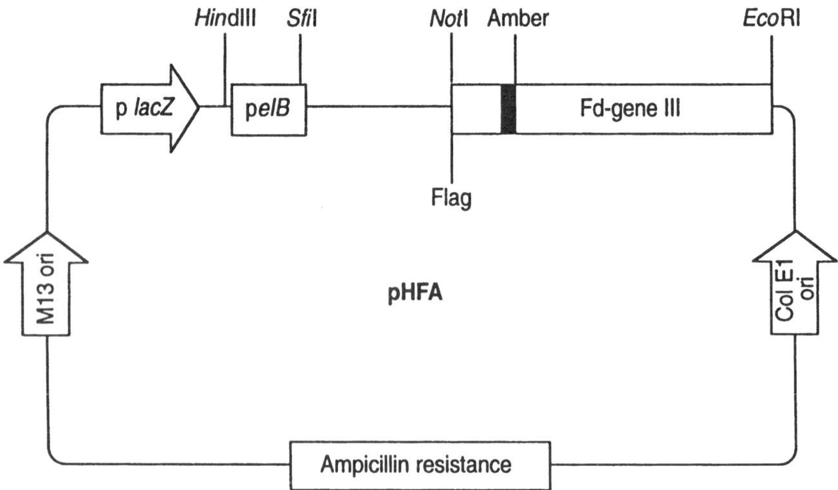

Fig. 14.5. The phagemid display vector pHFA.

inserted into pHFA[2] between *Sfi*I and *Xho*I restriction sites using standard cloning procedures.

The pHEN and pHFA systems express a polycistronic Fab coding region under control of the *lacZ* promoter to produce functional Fabs in the *E. coli* culture medium (Hoogenboom *et al.*, 1991; Dolezal *et al.*, 1995). A typical construct is shown in Fig. 14.6. The heavy chain is driven from the *lacZ* promoter, and is preceded by the *pelB* signal sequence, which directs the heavy chain into the periplasmic space. The heavy chain ends with a short tag region which can be recognized by antibodies. The tag is followed by a stop codon. The light chain is preceded by another signal sequence, which is typically the signal sequence from the *pelB* protein, and directs the light chain into the periplasmic space. Immediately following the light chain is a further tag region and the gene III coat protein sequence.

Between the *C* terminus of the light chain and the *N* terminus of the gene III protein is an amber stop codon. When the gene is expressed in wild-type *E. coli*, this stop codon is recognized, the light chain terminates at this point, and a Fab molecule is secreted into the periplasmic space. However, if the plasmid is expressed in an amber suppressor strain of *E. coli*, translation continues through the amber stop codon into the gene III, such that a fusion protein is produced, and the light chain is attached to the tip of the phage, fused to the gene III-encoded coat protein.

PCR protocols can be designed to generate the light and heavy chain gene fragments with restriction sites which give in-frame fusion of the scFv or Fab expression unit in pHFA and pHFA[2] respectively (Fig. 14.6). For Fab construction the κ light chain cDNA encoding the variable (V_L) and constant domain (C_L) can be amplified by PCR, gel purified and joined

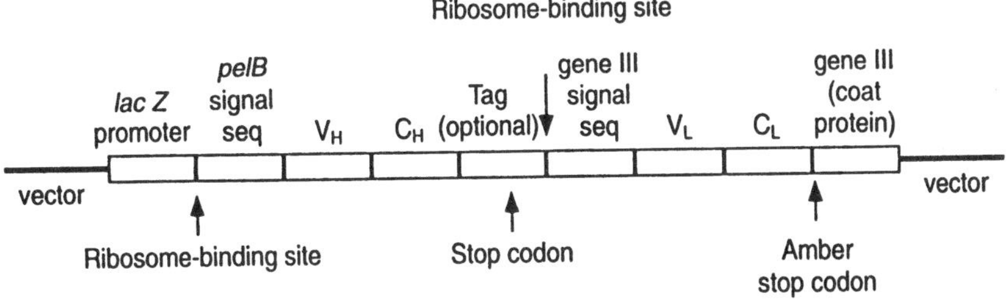

Fig. 14.6. A typical construct for the expression of Fab antibodies in Fd phage and plasmids. Note that the amber stop codon between C_L and gene III is ignored in amber suppressor strains of Escherichia coli, *so that the C_L is produced as a fusion protein with the coat protein product of gene III, which is expressed at the tips of the phage (see Figs 14.2 and 14.3). When expressed in wild-type* E. coli, *translation stops at the amber codon and the Fab protein is either secreted or retained in the periplasmic space. The tag sequence is incorporated to facilitate purification and also to allow the Fab to be probed with anti-tag antibodies.*

together by PCR (Horton *et al.*, 1989) to generate the full-length κ light chain gene fragment flanked with *Nco*I (5′ end) and *Not*I (3′ end) restriction sites. The cDNA is used as a template in a PCR designed to amplify heavy chain gene fragments and to add *Sfi*I (5′ end) and *Sal*I (3′ end) restriction sites. The κ and heavy chain PCR fragments are then inserted sequentially into pHFA[2].

14.8.1 Cloning of PCR-Generated Fragments into Expression Vectors

The amplified DNA may be purified from PCR mixture using Qiaex beads, 'Wizard PCR Preps' (Promega) or similar products. The DNA is then digested with the appropriate restriction enzymes, run on a 1.5% low-melting point agarose gel and purified from the gel again. The digested and purified PCR fragment is ligated into an expression vector such as pPOW (see Section 14.14) that has been prepared in a similar way. The ligation mixture is electroporated (Dower *et al.*, 1988) into *E. coli* TG1 cells, plated on YT medium containing 100 μg/ml ampicillin and 1% glucose and incubated at 37°C overnight. Transformed clones are screened by restriction analysis of minipreps or by colony PCR for the presence of DNA inserts of expected size using primers flanking the insert (Gussow and Clackson, 1989). The correct DNA sequence of selected clones must be confirmed, for example using the dideoxy chain termination method on double-strand DNA constructs in expression vectors after alkaline lysis (Sambrook *et al.*, 1989).

14.9 Isolation of Antibody-displaying Phage by Panning

14.9.1 General Principles

Phage displaying specific Fv regions on their surface may be selected by binding to antigen-coated tubes, dishes or columns. After washing off unbound phage, the specifically bound phage may be eluted by a change in pH, which disrupts the antigen–antibody bond. This process is known as panning.

Amazingly, the phage can withstand 5–10 min at a pH as high as 10 or as low as 3.8. Whether or not they drop in titre after this harsh treatment is not known, but evidently enough survive in infective form for the method to work.

14.9.2 Coating Plastic Tubes and Dishes with Antigen for Phage Selection

The selection of phage-displayed antibodies by panning can involve the use of a variety of support matrices for coating with antigen, such as ordinary tissue culture dishes (35 × 10 mm Falcon 3001 tissue culture dish), ELISA trays (for small-scale selections), or Nunc Immunotubes (Polysorb or Maxisorb) which vary in their hydrophobic properties and for which different antigens have different affinities.

Panning plates are coated overnight with 1–4 ml of antigen. Typical concentrations of protein antigen for coating of tubes is 5–10 μg/ml in PBS or $NaHCO_3$ at pH 9.5. For most antigens, the buffer makes little difference. It is vital, however, that there be no other protein present, as it will compete for the binding sites on the plastic. The coating of tubes with antigen is essentially the same as for plates. Nunc Immunotubes (75 × 12 mm) are specially designed for this purpose. Maxisorb and Polysorb tubes have somewhat different binding specificity. If one type of tube does not work, one can try the other type. They are coated by incubating overnight at 4°C with about 4 ml of antigen in PBS.

The tubes should then be washed three times with PBS, blocked with 2% skimmed milk in PBS (BLOTTO) for 2 h to saturate all nonspecific protein-binding sites, and washed a further three times in PBS.

14.9.3 Growth of Phage for Selection

Selection of phages displaying antibodies has been described by Marks *et al.* (1991b). Phages are rescued from *E. coli* amber suppressor cells with the replication-deficient helper phage VCS-M13 (Stratagene) in an overnight incubation at 37°C, using 2×10^{10} pfu of helper phage per ml of 2 × YT broth containing 100 μg/ml of ampicillin and 25 μg/ml kanamycin. Phage particles are purified by PEG precipitation (Sambrook *et al.*, 1989) and resuspended to 10^{12} cfu per ml.

14.9.4 Binding of Phage to Plates or Tubes

Phage particles (typically 10^{12} to 10^{13} cfu in 4 ml of 2% BLOTTO) are added to the coated tubes or dishes and incubated for 2 h at ambient temperature with constant agitation on a wheel or rocker.

14.9.5 Washing to Remove Unbound Phage

Each washing step is performed by pouring buffer in and out immediately and is repeated up to 20 times with PBS, 0.1% Tween 20 then up to 20 washes in PBS for higher stringency.

14.9.6 Elution of Phage from Antigen-coated Plates or Tubes

Elution of bound phage may be with about 1 ml of 100 mM triethylamine with gentle agitation for no longer than 5–10 min. The eluted material is immediately neutralized with an equal volume of 1.0 M Tris-HCl (pH 7.4). The phage may be stored at 4°C.

14.9.7 Affinity Purification/Selection of Fd Phage Expressing Antibodies on Antigen-Sepharose Columns

A typical column might have a bed volume of about 1.5 ml beads. The column should be washed extensively prior to each use, typically with 2% BLOTTO (2% nonfat powdered milk in PBS). To 10^{12} phage, suspended in 1 ml water, add 9 ml BLOTTO, and load all 10 ml onto the column. Wash with 10 ml BLOTTO, then 200 ml PBS. Elute phage by competition with 5 ml of 1 mM antigen in PBS, or by acid elution and neutralization as above. Competitive elution has the advantage that it is very gentle, and is unlikely to damage the phage. This system is particularly suited to subsequent transfection into *E. coli.* Store phage-containing eluate at 4°C.

14.9.8 Amplification of Phagemid Libraries and Rescue of Phage for Screening by Binding to Antigen—A Practical Example

Escherichia coli TG1 can be infected with libraries, either in the form of phage particles or by transformation with the plasmids, and cells grown on plates containing 100 μg/ml ampicillin and 1% glucose (AG plates). Cells are collected by scraping in 5 ml 2 × YT containing 100 μg/ml ampicillin and 1% glucose (2 × YT/AG). Scraping of two nearly-confluent 15-cm circular plates will yield about 4×10^{11} bacteria; in 5 ml, this is about 10^8 bacteria/μl or 100 OD_{600} units.

For an antibody library of 10^7 clones, the starting inoculum should contain 10^8 clones (1 μl). Inoculate 1 μl of this mix (OD_{600} = 100) into 10 ml 2 × YT/AG. Grow for 1.5 h at 37°C with shaking. The phagemids are rescued by adding replication-deficient helper phage. For rescue, the OD_{600} of culture should be around 0.05 (approximately 4×10^8 cells total). Superinfect with 8 μl VCSM13 helper phage (10^{11} pfu, giving a multiplicity of infection of about 20:1). Leave for 30 min in 37°C water bath without shaking, and then dilute into 90 ml of 2 × YT containing 100 μg/ml ampicillin and 25 μg/ml kanamycin, 1% glucose. After 1 h at 37°C with shaking, pellet the cells by centrifugation 15 min at 4000 r.p.m., in a 50 ml Falcon tube. Resuspend pellet in 100 ml × 2 YT-AMP without glucose.

Grow the phage overnight at 37°C with shaking. Centrifuge to remove bacteria, and then pellet the phage particles (containing the packaged phagemid) by two PEG-precipitations using standard procedures (Sambrook *et al.*, 1989) and resuspend the phage in 1 ml PBS. Yield is about 2×10^{13} cfu.

Phagemids at 10^{10}–10^{11} cfu in 2% Blotto PBS are applied to the antigen-coated plate and rocked for 2 h at room temperature. Plates are then washed five times in PBS, PBS-Tween (0.02% Tween-20); 50 mM Tris-HCl pH 7.5; 500 mM NaCl, 50 mM Tris-HCl pH 8.5; 500 mM NaCl, 50 mM Tris-HCl pH 9.5; 500 mM NaCl. Bound phage-antibody are eluted with 100 mM triethylamine for 5 min. Phagemid viability decreases with longer elution times. They should be immediately neutralized with one volume of 1 M Tris-HCl, pH 7.4.

14.10 Increasing Library Complexity through Recombination—Production of Fab Fragments of Antibodies in Bacteriophage

Based on the observation that heterodimeric Fab molecules can be assembled on phage surfaces in a form that retains binding affinity (Hoogenboom *et al.*, 1991), it has been possible to construct Dual-Combinatorial Libraries (DCLs) in which random pairing of the chains provides an additional source of variation, affinity and specificity (Zebedee *et al.*, 1992; Barbas *et al.*, 1991; Hoogenboom *et al.*, 1991; Kang *et al.*, 1991a, b).

Further complexity can be introduced by cross-transfection of two smaller libraries. For example, one gene library could provide light chains anchored to the Fd-phage surface (McCafferty *et al.*, 1990) via GeneIII fusions, and a second library constructed in nonrecombining phagemids could provide the heavy chains secreted into the periplasmic space of the host bacterium. The soluble heavy chains associate with the light chains to form a functional Fab fragment on the phage surface (Fig. 14.7). The DCL approach can produce

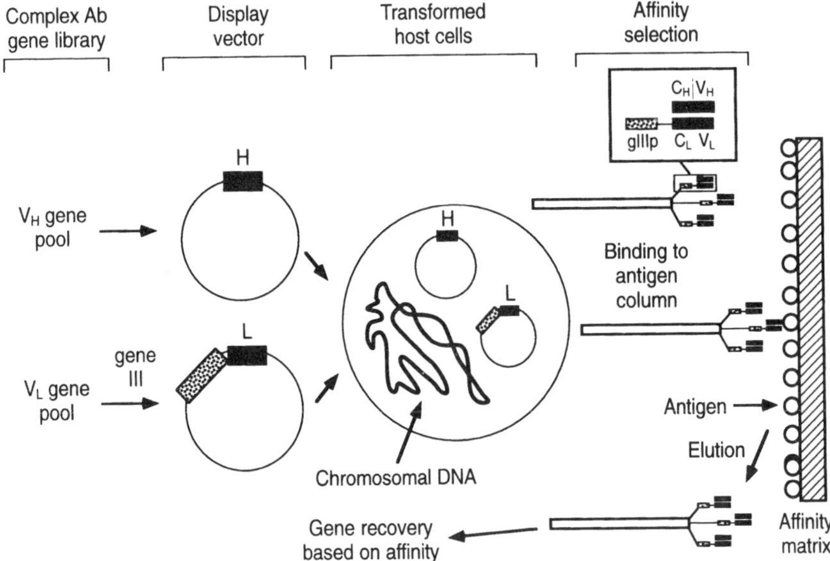

Fig. 14.7. Overall strategy for combinatorial expression of light chain and heavy chain variable regions as Fab fragments in association with the gene III coat protein of Fd phage. Pools (libraries) of heavy and light variable chains are constructed into a phage display vector. The display vector is transfected into host cells to generate a dual-combinatorial library. Each host cell produces viable Fd phage in which the antibody fragment is displayed on the phage surface and the gene encoding the antibody is packaged with the viral genome. Affinity purification of the phage allows simultaneous recovery of the gene encoding the antibody. Alternative strategies include the construction of hierarchical libraries in which one chain is held constant and displayed with a library of the second chain to select the highest-affinity paired chains.

highly complex libraries. A library of 10^7 V_Ls crossed with 10^7 V_Hs will produce a complexity of 10^{14} specificities.

Thus far all the systems discussed in this chapter have been based on non-recombining Fd phage and phagemid vectors. More recently the *Cre-lox* system has been utilized for the recombination of the heavy and the light chain genes (Waterhouse *et al.*, 1993). This system, which is still based on Fd vectors, offers the potential for the largest base library diversity that is currently available.

14.11 Affinity Maturation

Antibodies selected either by the primary immune response in a mouse or from bacterial expression libraries cloned and amplified from a naive source will most likely possess relatively low affinities. The greater the library size, the the greater the possibility that the selected antibodies will have relatively higher affinities. To date, however, there have been few reports of successful screening directly from a large library (Griffiths *et al.*, 1994; Nissim *et al.*,

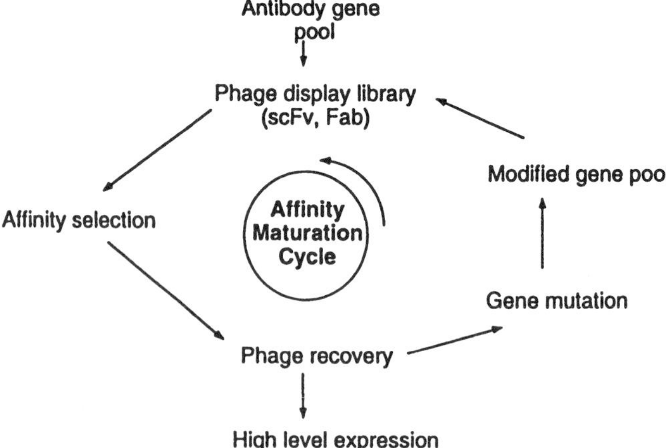

Fig. 14.8. Phage bearing the scFv genes may be cycled through multiple rounds of mutation to isolate variants with higher affinity scFv or Fab displayed on the surface of phage which bind to antigen are selected by panning and tested in an ELISA.

1994). There is thus an urgent need for the development of ways of increasing the affinity of engineered antibodies. Some very promising strategies for affinity maturation are currently being developed, and bear certain similarities to *in vivo* somatic hypermutation mechanisms during normal immune responses (Fig. 14.8).

14.11.1 Somatic Hypermutation in the Immune Response

During the development of the mammalian immune response, antibody affinity is progressively increased by somatic mutation effecting amino-acid changes throughout the variable region (see Chapters 3 and 6). Although the pairing of heavy and light chains in lymphocytes is usually considered to be a stochastic process (Kaushik *et al.*, 1990), biochemical analyses and phage-displayed recombinant antibody fragment library reconstructions have shown some preferential associations, almost as if there is imprinting of the partnered framework domains.

Considerable functional immunoglobulin diversity is created independently of antigen contact by gene rearrangements and junctional diversity, whereas after antigenic stimulation the immunoglobulin repertoire is even further increased as the variable regions undergo somatic hypermutation (Chapter 6). This is an important mechanism for generating diversity in immunoglobulin variable regions of all chain types, eventually leading to high-affinity antibodies. In the early stages of the immune response, any B-cells that express surface immunoglobulin with some affinity for the immunizing antigen can bind the antigen and participate in the response. Later, as the antigen becomes more limiting, only B-cells expressing immunoglobulin with the highest affinity for the antigen will be activated (see Section 3.3.9).

At the molecular level the site of mutation within the molecule is critical. Mutations within the 'structural' (framework) regions have different effects from mutations at the antigen-combining site. The ratio of replacement:silent mutations is higher in the complementarity determining regions than the framework regions, which can be explained by framework region mutations perhaps being more likely to be deleterious to affinity of binding by disrupting the overall structure of the variable region (Manser, 1989). Conversely, mutations in the CDRs may result in an altered antigen binding site, but the overall variable region conformation would stay the same (Chothia *et al.*, 1989). Accordingly, mutations in the CDRs might be more likely to be reflected in a change in affinity and would, if the affinity is high enough, be preferentially selected.

14.11.2 Affinity Maturation of Recombinant Molecules

Low affinity antibodies selected from phage libraries by the methods outlined in the previous sections can have their affinities increased by emulating the natural somatic hypermutation process that occurs in animals. In the laboratory there are numerous alternative approaches to introduce mutations into single Ab-genes or complex libraries of binding molecules such as scFvs and Fabs (Table 14.4). One general approach for phage-displayed antibodies is to select antigen-binding reagents from the libraries by affinity to antigen. The genes encoding these antibodies are first modified by a number of *in vivo* and *in vitro* mutation strategies, and constructed as a new gene pool which is converted into a new phage-display library (Fig. 14.8). A second round of affinity screening using more stringent conditions selects for the highest-binding reagents in the new gene pool. These cyclical rounds of mutation and selection in phage-display libraries mimic the process of somatic mutation.

14.11.3 *In vitro* Mutagenesis

There are currently two competing but complementary technologies for improving antibody affinity. In the first method, proteins of known structure can be modified at precise positions by site-directed or regional mutagenesis. An example of this is the process of 'humanization' of monoclonal antibodies by grafting CDR loops from rodent antibodies (the antigen-binding surface) onto the surface of a human antibody framework (Reichmann *et al.*, 1988). This is achieved by splicing the relevant coding regions into a DNA vector and transferring this to a host cell for protein expression.

Table 14.4 Mutation strategies for increasing affinity of engineered antibodies

Source	Mechanism	Reference
Reverse transcriptase random mutation	Increased replication errors by SIV and HIV reverse transcriptase	Takeuchi *et al.* (1988)
Site-directed mutagenesis	Oligonucleotide-coded mutations	Hawkins *et al.* (1993) Yamagishi et al. (1990)
Randomized oligonucleotide mutagenesis	Spiked (random sequence at specific sites)	Hermes *et al.* (1989)
Saturation mutagenesis	PCR to synthesize a randomized V_H CDR3	Barbas *et al.* (1992)
Chain shuffling	Antigen-selected pairing of phage displayed fragments	Marks *et al.* (1992b)
Random PCR mutagenesis	Choice of PCR conditions that increase mutation rate (e.g. addition of manganese)	Leung *et al.* (1989) Gram *et al.* (1992)
In vivo: mutator strains	Errors in proofreading in DNA replication	Irving and Hudson (1994)

A second technology involves random mutation, combined with powerful rational selection processes (RAMSEL) to isolate the individual products with desired properties. The power of the latter technique resides in the ability to modify a protein or family of proteins by simple gene manipulations, and to isolate the coding region for one molecule from a complex mix of over 10^{12} variants. Various mutation strategies have been applied to modifying antibody fragment affinities. A selection is shown in Table 14.4.

14.11.4 Chain Shuffling

One successful technique for maturation is the use of a hierarchical approach of 'chain-shuffling' (Clackson *et al.*, 1991; Marks *et al.*, 1992b), in which one of the molecules (e.g. V_H) is displayed as a gene III protein fusion on the phage surface in association with a library of V_L molecules. Screening on the basis of affinity will select the V_L molecule with highest affinity. Then, by holding the phage V_L-gene III protein complex constant and displaying a library of V_H-molecules, a high-affinity V_H domain can be selected. Repeated cycles of chain shuffling and affinity selection will produce the desired high-affinity antibodies.

An exciting application of this technology has been termed 'Epitope imprinting' (Figini *et al.*, 1994; Winter *et al.*, 1994) in which a mouse Fab can be replaced by a human Fab. One V-region of a mouse monoclonal antibody (e.g. V_H) is displayed on phage, together with a library of human V_L genes. Selection by antigen produces a high-affinity human V_L paired to the original mouse V_H. The human V_L is then displayed with a library of human V_H genes and antigen selection will now produce a totally human Fab fragment that replaces the original mouse antibody. In this way, a mouse antibody for a given specificity can be converted, one step at a time, into a human antibody.

14.11.5 Site-directed Mutagenesis

Site-directed mutagenesis facilitates changes at specific sites or domains within the antibody genes, such as in specific CDR loops or framework regions, allowing the design of new molecules in a 'rational' strategy. It involves the use of oligonucleotides to incorporate either specific or random sequence changes at defined sites. One example is the random (saturation) mutagenesis to the CDR loops followed by phage display and selection (Barbas *et al.*, 1992, 1994). In cases where the three-dimensional structures of the protein molecules have been determined by crystallography, an improved interface between molecules could be designed, or domains grafted from one molecule to another.

One model system for the evaluation of the potential applications of this technology has been the antibodies with specificity for influenza neuraminidase. The binding surfaces of two native antibodies (NC10 and NC41; Malby *et al.*, 1994) and the recombinant scFv NC10 (Malby *et al.*, 1993; Kortt *et al.*, 1994) have been resolved by X-ray crystallography, making this an ideal system. Mutants have been designed in order to fill the cavities at the planar interface of the antibody–antigen combining site, or to modify the salt-bridge interactions. The antibody fragment coding regions were then transferred to a high-level expression vector such as pPOW (Power *et al.*, 1992) in order to produce sufficient amounts of the proteins for further structural analysis and immunodiagnostic and therapeutic applications.

Sharon (1990) used site-directed mutation of three amino acids in the heavy chain to increase the affinity of a low-affinity monoclonal antibody to the hapten *p*-azophenylarsonate. The changes were based on a high-affinity monoclonal antibody to the hapten. Of the 19 amino-acid residues that differed between the two antibodies, only three were necessary and sufficient to bring about a 200-fold increase to produce equivalent affinities. Two mutations were in CDR2 and the other in CDR3. This illustrates how antigen-driven affinity maturation, in this case under *in vivo* conditions, provides the information to allow the design of specific mutations to produce a higher affinity antibody.

Barbas *et al.* (1994) have shown that directed sequential site-directed mutagenesis in the CDR3 region of a heavy chain resulted in 3 amino acid changes, and increased the affinity 1000-fold. However, single-chain molecules with high initial affinities (such as when taken from hybridomas) can be intransigent to affinity maturation in the laboratory (Hawkins *et al.*, 1992).

14.11.6 *In vitro* Random Mutagenesis

Alanine scanning, a method to replace systematically each residue in turn by alanine, has been successful in identifying amino acid residues involved in receptor–ligand binding. However, predictions of the roles of individual amino acid residues in antigen binding can often be fraught with difficulty. Mutation of the heavy chain aspartate 101 of the monoclonal antibody to hen egg lysozyme, HyHel to alanine resulted in a 9000-fold drop in affinity, despite it not being involved in H-bonding or salt bridges (Lavoie *et al.*, 1992). It was suggested that this residue was important for the conformation of CDR3.

Random mutagenesis of contact residues should be performed simultaneously so that they may co-vary. Additivity principles suggest that the sum of the free-energy effects for residues in contact may not equal that of the combined mutant because these side-chains can interact.

It would be expected that residues affecting the tightness of binding would be found predominantly at the interface surface of the antibody which contacts antigen. However, in growth hormone a few of the buried side chains were found, by alanine scanning, to enhance binding (Cunningham and Wells, 1989; Fuh *et al.*, 1992). The change in affinity was due to a slowing of the off-rate of the hormone (60-fold) and an increase in the on-rate (fourfold).

Gram *et al.* (1992) used error-prone PCR (Leung *et al.*, 1989) to introduce random mutations into naive Fab V-region genes to achieve up to a 30-fold increase in affinity of binding to antigen. This high level of error in PCR amplification can be brought about by high cycle numbers, by unusually high or low levels of one nucleotide, by adding manganese, or by changing the magnesium concentrations. PCR mutagenesis can produce about two mutations per V gene, but it is not truly random. It has equal numbers of transitions and transversions, but G/C or C/G changes are rare.

14.11.7 Mutation PCR—Practical Procedure

The following protocol for PCR with high mutation rate is designed for construction of very large libraries from which higher-affinity mutants can be selected. The volume of the reaction mix is 1 ml, but in view of the difficulties in rapidly heating and cooling such a large volume, it is recommended that it be processed in 100-μl aliquots. Of course, it is not essential to work on this scale, and for many purposes a single 100-μl reaction may suffice.

PCR buffer (10 ×)	100 μl
dNTPs (10 mM stock)	100 μl
$MgCl_2$ (100 mM stock)	45 μl
$MnCl_2$ (100 mM stock)	10 μl
5′ primer (10 pmol/μl)	50 μl
3′ primer (10 pmol/μl)	50 μl
Template	(1–20 ng ssDNA)
Water	to 1.0 ml total

Heat to 94°C. Add 10 μl *Taq* or other thermostable polymerase.
Cycle at 94°C for 1 min, annealing temperature and time as required, 72°C for 4 min, for 25–55 cycles.

Using similar techniques, Deng *et al.* (1994) showed that it is possible to mutate a single-chain Fv into a library with 10^5 members, from which one individual could be selected with an increase in affinity. This improved scFv had a faster K_{on} due to amino acid changes in CDR1 and CDR2, and bound to an epitope larger than that recognized by the wild-type scFv.

14.12 *In vivo* Mutation and Rational Selection

The preceding sections have described methods for the manipulation of isolated DNA molecules to increase antibody affinity. It is also possible to perform random mutagenesis *in vivo*, by replicating the DNA in cells which are prone to errors (mutator strains).

This approach can be applied either to the maturation of a single antibody molecule, or to the production of an entire library of antibody genes. The *mutD5* mutator strain of *E. coli* is a conditional mutant which has been shown to preferentially produce single base substitutions (Fowler *et al.*, 1986; Schaaper, 1988) with both transitions and transversions occuring at high frequency in rich media (Dengen and Cox, 1974).

Using *in vivo* techniques such as *E. coli* mutator strains, mutations are produced at random throughout the gene pool. To generate large libraries, recombinant phagemids are transfected into the mutator strains of *E. coli* which, during the exponential growth phase, introduce errors into the gene sequences of high-copy-number episomal DNA. Mutagenesis is particularly useful to improve the binding properties of a low-affinity antigen-binding molecule selected from a library. These mutation techniques, either singly or in combination, also provide greater diversity to complex libraries, and therefore may emulate or exceed the capacity of the B cell in affinity maturation. A significant advantage of *in vivo* mutation is that the number of mutants generated is not limited by the efficiency of transformation.

14.12.1 Practical Approach

Mutant strains of *E. coli* have been used to induce random mutations in plasmid-based genes (Yamagishi *et al.*, 1990). However, in order to produce an effective library mutation and selection scheme, it is necessary to handle extremely large numbers of mutants (in excess of 10^{10} members). This requirement can be met by using a phage-based system. *E. coli* strains are used that are not only defective in proof-reading during DNA replication but also competent to be transfected by and synthesize M13 phage for surface display of the encoded scFvs (R. A. Irving, P. Holliger, G. Winter, and P. Hudson, in preparation). *In vivo* mutation of plasmids and phagemids involves transformation or transfection of antibody-encoding phagemids into the mutator strains (*mutD5* or *mutT1*) followed by exponential growth at 37°C for 16 hours in a rich culture medium such as a YT based medium (Power *et al.*, 1992).

Recombinant phage are rescued and transfected into *E. coli* TG1 for amplification and phage surface display of antibody fragments without further modification. Care must be taken to ensure that the mutator cells are maintained on minimal media under conditions in which mutation of the *E. coli* chromosomal DNA is minimized. Affinity selection may be achieved by panning in Immunotubes (see Section 14.9.2) as described by Marks *et al.* (1991b). Alternating mutation and selection cycles are used, with passage through TG1 cells for amplification of the phage (Fig. 14.8), and the affinities of the selected phage-displayed antibodies are generally increased by one or two orders of magnitude. Generally, three rounds of mutation and selection generates a library in which >90% of the molecules carry at least one point mutation.

Of the point mutations that resulted in a significant increase in binding affinity, more than 80% were in the framework regions of the V_L domains. The most dramatic increase in affinity (10^2–10^3) was associated with two point mutations at the junction between CDR2 and framework 3. The progressive sequential amino acid changes arising from these mutations within this 'Vernier zone' (Foote and Winter, 1992) only marginally affect the charge and volume of the amino acids at this site, yet are capable of producing large increases in affinity.

14.12.2 Analysis of Generated Mutants

'Filter sandwich' assays

A convenient method for colony screens suitable for low complexity expression libraries, which was initially published by Skerra *et al.* (1991), involves the plating of newly transformed or freshly grown *E. coli* onto a nitrocellulose membrane. Colonies should not be too large or crowded on the nitrocellulose; about 10^4 colonies and 6–8 h growth is ideal. The colonies are then induced to express the recombinant antibodies by transferring the nitrocellulose disc onto agar containing an appropriate inducing agent such as IPTG if the gene is under the control of the *lacZ* promoter.

A second nitrocellulose disc is coated with antigen, and then all nonspecific protein-binding sites are saturated with sterile BLOTTO. This filter is then gently overlaid on top of the first filter upon which the bacteria are growing, and the expressed antibodies are thereby captured. After washing, the second filter is probed with labelled anti-immunoglobulin antibodies and processed in a manner that is essentially identical to that for Western blots (see Section 10.10).

As an alternative to the use of anti-immunoglobulin antibodies, it is possible to incorporate a tag sequence in the engineered antibodies, such as the FLAG octapeptide (Hopp *et al.*, 1988) which is DYKDDDDK (single-letter amino acid code), and to use antibodies to this tag (see Brizzard *et al.*, 1994).

Binding of phage-displayed antibodies to intact cells

One of the major advantages of the phage display technology is that library selection/affinity maturation can be applied not only to purified antigens, but also to cell-surface proteins (Marks *et al.*, 1993). For example, phage-displayed scFvs can be bound to red blood cells in PBS. After washing with PBS and eluting by a large change in pH such as Tris-glycine buffer pH 3.8 or ethanolamine pH 10, the specifically bound and eluted phage may be transfected into *E. coli* for amplification and rescue. Selected antibodies have specificity for red blood cell membrane proteins such as glycophorin. A similar strategy has been used to produce Fab fragments of human antibodies against the rhesus D (Rh) antigen (Dziegiel *et al.*, 1995).

Fab phage selection using human erythrocytes as solid phase

Glycophorin on human red cells is a useful model for phage selection, because of the ready access to both intact red cells, and the purified glycophorin. It has been possible to select with the phage displayed scFv on the red cells and confirm the binding activity on the purified glycophorin in ELISA wells. It could be adapted for other antigens and cell types with only minor modifications.

About 4×10^7 erythrocytes/ml in PBS bearing the desired antigen (in this case, glycophorin) are reacted with two volumes of a suspension of phage displaying Fab fragments of antibodies, and rocked for 30 min. About 10 volumes of ice-cold PBS are then added, and the erythrocytes pelleted by centrifugation and washed several times in PBS. The bound phage are eluted from the packed cells with 20 μl of elution buffer (pH 3.5), centrifuged at 6500 r.p.m. and the supernatant transferred to another Eppendorf tube and immediately neutralized.

It will sometimes be helpful to initially deplete a random library for phage that are reactive with erythrocyte membrane antigens with undesired specificities. This may sometimes be achieved if a source of cells can be found that lack the desired antigen, but are similar to the selecting cells in other respects. For example, erythrocytes may be selected from a donor who lacks glycophorin, and the phage library absorbed on these cells. This may be done by incubating about 10^{12} phage with the packed erythrocytes for 30 min at room temperature on a rocking platform, and then removing the cells by centrifugation. The supernatant, containing the depleted phage, may then be used for a round of positive selection on cells possessing the desired antigen. This procedure may help to reduce the amount of screening required to isolate rare clones.

Affinity selection of phage antibodies using magnetic beads

Isolation of antigen-binding phage may be achieved by binding to soluble biotinylated antigen followed by binding to streptavidin-coated beads. This is a faster process than either column or Immunotube selection, which effectively results in the selection of phage antibodies with desired on and off rates. This may be done by using the biotin–avidin system, as described in Section 10.3. Streptavidin-coated magnetic beads are available from Dynal (M280 streptavidin coated magnetic beads). The antigen is coupled with biotin as described in Section 10.3.1.

The phage library at suitable titre (typically about 10^{10}/ml, or 1:10 000 dilution of phage stock in 1 ml of PBS, 3% BSA, 0.5% Tween 20, 0.02% azide) is added to the biotinylated antigen which should be at a suitable final concentration. (A suitable concentration of antigen is approximately that concentration which is equivalent to the expected affinity of the finally selected antibody. If the best antibody is expected to have a K_D of 100 nM, then a final antigen concentration of 50–100 nM should be used.)

After allowing equilibrium to be reached (typically 1 h at room temperature), add 0.1 ml M280 streptavidin-coated magnetic beads (Dynal) for 5 min at room temperature. The magnetic particles are separated on a magnetic separator (MPC-E; Dynal) for 5 min at room temperature, and the supernatant carefully removed. The beads are washed 10 times in medium, and the phage-antibodies eluted with 100 mM Triethylamine. The beads are separated with a magnetic field, and the supernatant removed to a new tube and neutralized with 0.2 ml 1 M Tris pH 7.4. The eluted phage can be used to directly reinfect *E. coli*.

The use of two or more rounds of selection whereby the eluted phage are reselected before reinfecting is only recommended if the eluted titre is known to be high. Two rounds of selection significantly increases the proportion of higher-affinity antibodies, but runs the risk that the final phage titre will be too low to be workable. In practice, multiple rounds of selection are best reserved for selection of higher-affinity phage from a mutated library. If the antigen concentration is gradually lowered between rounds of screening, it may be possible to isolate higher-affinity phage. However, it is recommended that the titre of eluted phage be monitored, and the antigen concentration reduced gradually so that diversity is not lost. A reasonable eluted titre in such an experiment is 10^6 and it should increase to 10^{8-9} as the higher-affinity phage are selected.

14.13 Increasing the Complexity of Phage-displayed Recombinant Antibody Libraries

Although it is sometimes possible to construct scFvs with binding specificity equivalent to parent monoclonal antibodies (Chaudhary *et al.*, 1989; Denzin and Voss, 1992; Malby *et al.*, 1993; Skerra *et al.*, 1991), the affinity of scFvs selected from phage libraries is often lower than that of the parental antibody. This may sometimes be achieved by successive cycles of mutation and selection (Fig. 14.8). For example, the number of different antibodies binding to a specific antigen can be enhanced by affinity maturation of a phage-displayed antibody library by *in vivo* mutation with *E. coli mutD5* cells. Prior to mutation a naive scFv library contained only one individual scFv capable of binding to the antigen glycophorin. After three rounds of mutation and library amplification followed by selection, another four individual scFv anti-glycophorin molecules were identified; each had a different sequence and three κ gene families were represented (P. Hudson and R. Irving, unpublished data).

14.14 Expression Systems

We have discussed the advantages of antibody fragments, such as Fabs and scFvs, that may eventually supersede the use of whole antibodies in many applications. These small fragments can be produced cheaply, conveniently and consistently using bacterial expression systems, particularly *E. coli* (Boss *et al.*, 1984; Skerra 1993).

As both the V_H and V_L domains contain intrachain disulfide bonds, it is important to direct the nascent protein into an oxidizing environment where these bonds can form correctly. In an analogous strategy to eukaryotes, where the signal or presequences direct the protein through the endoplasmic reticulum, signal sequences on bacterial proteins direct the nascent polypeptide through the inner membrane. The protein then enters the oxidizing environment of the periplasm, where the presequence is removed by proteolytic enzymes, and the disulfide bonds form during folding. The protein can then either remain in the periplasm where it can build up to sufficiently high levels to aggregate as periplasmic inclusion bodies, or it may be secreted through the outer membrane into the cell culture media.

14.14.1 Inducible Promoters

Fab, Fv and scFv genes have been expressed in *E. coli* using numerous vectors. The choice of vector and host depends somewhat on the particular antibody fragment being expressed. The pPOW vector developed in our laboratory (Power *et al.*, 1992; Lilley *et al.*, 1994) has the advantage of having a strong promoter which is tightly regulated to prevent 'leakage' during the preliminary growth phase prior to induction, which could be important if the product is toxic for the host bacteria. The pPOW vector may be induced by heat shock (40°C) for up to 4 h, but we have found that 15 min of induction followed by 4–18 h of growth at 30°C gives best results.

We have successfully expressed antineuraminidase scFv fragments using pPOW vectors in *E. coli* TOPP or BL21, whereas for antiglycophorin scFv the best system was found to be the pHEN vectors in *E. coli* HB2151 (Lilley *et al.*, 1994). It is best to choose the host cell and optimize the growth conditions for each particular type of scFv.

Using optimized conditions, cells can be grown either in shake flasks or in fermenters. As a general rule, cells are grown to a moderately high density before induction. If induction is at 37°C, the induction time is usually not more than 4 h, but growth at lower temperatures for longer times (e.g. 20°C for 14 h) may generate a more soluble product in the culture supernatants (Ward *et al.*, 1989). It may sometimes be recovered from the periplasm by osmotic shock (Skerra *et al.*, 1991).

In other cases, particularly if expression is at a very high level, the antibody fragments will be found as insoluble intracellular or periplasmic inclusion bodies within the bacteria. Expression from pPOW results in exceptionally high levels of protein being deposited in the periplasm, and proteins must occasionally be recovered by solubilizing the aggregates (inclusion bodies) with denaturants, followed by a refolding step and column purification. This has yielded in excess of 10 mg of pure scFv fragment per litre of culture (Malby *et al.*, 1993).

Alternative vectors which incorporate chemical induction (such as *tac* or *lac* promoters as in pHEN) are often 'leaky', but this is only likely to be a problem if the product is toxic to the host cells and prevents growth. Chemically inducible promoters have an advantage over heat-inducible systems in permitting growth at lower temperature ranges. As mentioned above, induction at lower temperatures (e.g. 20°C) often results in a soluble product when induction at 37°C gives insoluble inclusion bodies (J. W. Goding and E. Handman, unpublished data). For a new scFv construct, it is recommended that several vectors be tested under a range of growth and induction conditions to optimize expression levels and protein solubility.

14.14.2 Small-Scale Expression of Recombinant Fab Antibody in *E. coli*

Recombinant pHEN/pHEA plasmids encoding Fab fragments may be expressed in *E. coli* HB2151, grown overnight at 37°C with agitation in 2 × YT medium (Sambrook *et al.*, 1989) containing 100 μg/ml ampicillin and 1% w/v glucose (Hoogenboom *et al.*, 1991). The next day, cultures are diluted 1:100 into 10 ml of the same medium and grown at 37°C with shaking until the OD_{600} is between 0.5 and 1.0. Cells are pelleted by centrifugation (4000 ***g***, 10 min), washed once in prewarmed 2 × YT medium, and resuspended in 10 ml of 2 × YT medium containing 100 μg/ml ampicillin and 1 mM IPTG. The mixture is left to incubate for a further 16 h at 30°C with shaking. The cells are recovered by centrifugation, and the culture supernatant collected.

In some cases, the antibody will be recovered in good yield from the supernatant, while in others, it will be necessary to extract it from the cells. Antibody recovered from inclusion bodies by lysis of the *E. coli* under denaturing conditions and refolding sometimes does not have as high a biological activity per unit mass as antibody recovered from culture supernatants.

14.14.3 Scale-up Production in Fermenters

Whether the choice to express antibodies results in shake-flask or fermenter (bioreactor) cultures, a knowledge and application of microbial physiology and biochemical engineering principles will improve the growth of the cells as well as the expression levels of the foreign protein, although a large amount of empirical experimentation will also be necessary.

The process of scale-up is usually explored by preliminary experiments using shake flasks, where factors such as the growth rate, the genetic stability of the cells, the optimal time for the induction of product synthesis and nutrient requirements can be determined before proceeding to larger-scale cultures. Scale-up of expression can mean growing cultures in bioreactors from a few litres to tens of thousands of litres or more.

In a bioreactor, the process parameters can be defined and controlled far more accurately and reproducibly than in shake flasks, thus enabling optimal conditions for growth and expression to be attained. Parameters such as pH, temperature and dissolved oxygen can be monitored and controlled at set levels. Nutrient requirements can also be met more effectively than in shake flasks as the growth medium can be fed continuously, preventing nutrient limitation. Cell densities of over 60 OD_{600} units are possible. The growth rate can be controlled by the nutrient feed rate as well as the other parameters, so that optimal rates for genetic stability of the culture and protein expression levels can be achieved. Protein solubility may also be influenced by many of the factors which influence the growth rate. Control of these parameters may also improve the product quality.

14.14.4 Heat induction of Vectors Containing Lambda PL/PR Promoters.

An overnight 7 ml culture of the recombinant *E. coli* at 30°C is used to inoculate 1 l of 2 × YT containing 100 μg/ml ampicillin in a 2-l flask. The culture is then diluted, and after growth at 30°C until the OD_{600} is about 1.0 to 1.5 (about 8 h), expression of the recombinant gene is induced by heating to 42°C. (This may be achieved by swirling the flask in a 70°C water bath, monitoring the temperature with a thermometer immersed in the culture.) Growth is then continued in a 30°C incubator for a further 4 h.

Cells are removed by centrifugation, and the scFv recovered from the supernatant by precipitation with ammonium sulfate. The supernatant is made up to 35% saturation in ammonium sulfate, centrifuged, and the pellet discarded. The supernatant is then brought to 70% saturated ammonium sulfate, centrifuged again, and the pellet collected. The precipitate is resuspended in PBS and dialysed against 2 l of PBS, 0.2 mM EDTA at 4°C, and then centrifuged at 10 000 r.p.m. for 10 min to remove any insoluble material. The recovery of scFv at this stage depends on the individual scFv, the vector, the mechanism of induction, and the host cell line. Typically, one might expect between 1 and 20 mg of antibody per litre of culture (Maltby *et al.*, 1993), although yields of up to 0.5 g/l have been reported (Rodrigues *et al.*, 1993; Wulfing and Plückthun, 1994).

14.14.5 IPTG Induction of Vectors

Overnight cultures of *E. coli* HB2151 harbouring the expression vector carrying a recombinant Fab or scFv coding region is diluted 1:100 into fresh 2 × YT medium containing 100 μg/ml ampicillin and 0.1% (w/v) glucose, and grown with shaking at 30°C until A_{600} = 0.8–1.0. IPTG is then added to a final concentration of 0.2–2.0 mM, and cultures are held for a further 4–16 h at ambient temperature (*c*.20–22°C) with gentle agitation. Cells are pelleted by centrifugation. If the antibody is secreted, the clarified culture supernatants may be used for purification of recombinant Fabs or scFvs.

14.15 Affinity Purification of Antibody Fragments from *E. coli* Supernatant

Many laboratories now include hexahistidine tags in their scFv constructs to aid purification on chelation columns. This approach has the advantage that binding to the column can be performed under denaturing conditions, if necessary. This may be useful if the antibody must be denatured before it can be recovered in a water-soluble form. Alternatively, one can use immunoaffinity chromatography, which must be performed under native conditions.

14.15.1 Affinity Purification Recombinant Fab Fragments Bearing the Flag Epitope

It is often convenient to incorporate a tag sequence in the vector such that the recombinant antibody may be purified by affinity chromatography on a general anti-tag monoclonal antibody column. One popular tag system involves an epitope called 'Flag', which consists of the amino acid sequence DYKDDDDK (Hopp *et al.*, 1988; Brizzard *et al.*, 1994). Elution from an anti-Flag column can be by competition with the Flag peptide, which is very gentle and nondenaturing but somewhat expensive, or by the use of acidic pH, which is inexpensive but depends on the ability of the recovered antibody to refold upon restoration of neutral pH.

Recombinant *E. coli* culture supernatants are centrifuged (10 000 ***g***, 30 min) and passed through a 0.45 μm filter (Millipore) to remove all bacteria and debris, and applied to an M2 anti-FLAG Sepharose column (Eastman, Kodak, New Haven, CT, USA) equilibrated in PBS. Bound Fab–FLAG conjugates are eluted with 0.1M glycine buffer, pH 3.0, and eluted fractions immediately neutralized with 1M Tris-HCl, pH 8.0. Fractions containing activity are combined and dialysed against PBS.

If a higher degree of purity of the antibody is required, a preliminary fractionation on a Sephadex G-100 column may be carried out. Washing the anti-Flag column with 0.1 M glycine-hydrochloride, pH 5.0 after loading the column but before elution at pH 3.0 may also help improve the purity of the final product.

14.16 Discussion

In this chapter we have presented a representative overview of the phage-display technology and some of the processes by which phage-libraries are made and handled. It is easy to see that we are now at a crossroads in the field of antibody technologies. The new technologies of production of antibodies in *E. coli* and using phage display are approaching maturity.

It may be anticipated that in the future the use of recombinant antibody libraries and reagent selection processes will be more common. The creation of libraries of adequate complexity is now possible. However, the selection and affinity maturation processes continue to be optimized and are still far from routine.

There are many attractions to the production of antibodies in *E. coli*. The problems of production of human antibodies discussed in Section 7.2.4 are largely overcome by phage display and expression in bacteria, although many problems remain.

If sufficiently large 'naive' libraries can be made and high-affinity antibodies derived from them by mutation, the need for immunization of humans may be avoided. The time may not be too far off when a single library carrying the complexity of the mammalian immune repertoire will be a resource that will be widely available; a 'pot' that can be dipped into an unlimited number of times by screening with simple affinity procedures.

Manipulation of sequences for altering the affinity and specificity is possible, both by random or site-directed mutagenesis. It is relatively straightforward to add extra sequences as tags or for effector functions. The new technology is particularly suited to the immunodiagnostic industry because antibodies can be rapidly manipulated. It is also relatively easy to design and covalently attach reporter groups, such as chromophores or metal ligands to produce novel new reagents for imaging and sensing.

Challenges that remain include improvements to affinity maturation, the production of even larger libraries, and the secretion of large amounts of soluble antibody into the culture medium. Potential difficulties that will require innovative solutions include the antigenicity of linker peptides, unexpected modifications of proteins in bacteria, and contamination with endotoxin. In many ways, the use of recombinant DNA technology may minimize batch to batch variation of antibodies and facilitate quality control. The *in vivo* half-life of scFv and Fab antibodies is very short. This may be an advantage or a disadvantage, depending on the situation. In some situations, it may be useful for unbound antibody to be rapidly cleared from the body, and rapid clearance may also facilitate the use of antibodies for imaging of organs or tissues. Although antibodies produced in bacteria may lack effector functions that depend on the Fc portion, such as fixation of complement and interaction with cellular Fc receptors, their small size may be a distinct advantage in improved tissue penetration. It appears likely that bacterially produced monoclonal antibodies will provide an alternative production strategy to cell fusion in the near future, and it is certain that the newer technology will find many applications where its unique advantages can be exploited.

Acknowledgements

We wish to thank our colleagues in the Protein Engineering Program at CSIRO and collaborators at the Medical Research Council-IRC, Cambridge, U.K. for their techniques presented in this chapter. This work has been supported in part by a grant from the Industrial Research and Development Board of Australia and the British Council.

References

Anand, N. N., Mandal, S., MacKenzie, C. R., Sadowska, J., Sigurskold, B., Young, N. M., Bundle, D. R. and Narang, S. A. (1991) Bacterial expression and secretion of various single-chain Fv genes encoding proteins specific for a *Salmonella serotyp*e B O-antigen. *J. Biol. Chem.* **266**, 21874–21879.

Barbas, C. F., Kang, A. S., Lerner, R. A. and Benkovic, S. J. (1991) Assembly of combinatorial antibody libraries on phage surfaces: The gene III site. *Proc. Natl Acad. Sci. USA* **88**, 7978–7982.

Barbas C. F., Bain J. D., Hoekstra M. D. and Lerner, R. D. (1992) Semisynthetic combinatorial antibody libraries: a chemical solution to a diversity problem. *Proc. Natl Acad. Sci. USA* **89**, 4457–4461.

Barbas C. F., Hu D., Dunlop N., Sawyer L., Cababa D., Hendry R. M., Nara P. L. and Burton, D. R. (1994) *In vitro* utilisation of a neutralizing human antibody to human immunodeficiency virus type 1 to enhance affinity and broaden strain cross reactivity. *Proc. Natl Acad. Sci. USA* **91**, 3809–13.

Batra, J. K., FitzGerald, D., Gately, M., Chaudhary, V. K. and Pastan, I. (1990) Anti-Tac(Fv)-PE40, a single chain antibody *Pseudomonas* fusion protein directed at interleukin 2 receptor bearing cells. *J. Biol. Chem.* **265**, 15198–15202.

Bird R., Hardman K. D., Jacobson J. W., Johnson S., Kaufman B. M., Lee S-M., Lee T., Pope S. H., Riordan, G. S. and Whitlow, M. (1988) Single-chain antigen-binding proteins. *Science* **242**, 423–426.

Boss M. A., Kenten, J. H., Wood, C. R. and Emtage, J. S. (1984) Assembly of functional antibodies from immunoglobulin heavy and light chains synthesised in *E. coli*. *Nucl. Acids Res.* **12**, 3791–806.

Brinkmann, U., Reiter, Y., Jung, S. H., Lee, B. and Pastan, I. (1993) A recombinant immunotoxin containing a disulfide-stabilized Fv fragment. *Proc. Natl Acad. Sci. USA* **90**, 7538–7542.

Brinkmann, U., Chowdhury, P. S., Roscoe, D. M. and Pastan, I. (1995) Phage display of disulfide-stabilized Fv fragments. *J. Immunol. Methods* **182**, 41–50.

Brizzard, B. L., Chubet, R. G. and Vizard, D. L. (1994) Immunoaffinity purification of FLAG epitope-tagged bacterial alkaline phosphatase using a novel monoclonal antibody and peptide elution. *BioTechniques* **16**, 730–734.

Burton, D. R. and Barbas, C. F. (1994) Human antibodies from combinatorial libraries. *Adv. Immunol.* **57**, 191–280.

Cabilly, S., Riggs, A. D., Pande, H., Shively, J. E., Holmes, W. E., Rey, M., Perry, L. J., Wetzel, R. and Heyneker, H. L. (1984) Generation of antibody activity from immunoglobulin polypeptide chains produced in Escherichia coli. *Proc. Natl Acad. Sci. USA* **81**, 3273–3277.

Chaudhary V. K., Queen C., Junghans, R. P., Waldmann, T. A., Fitzgerald. D. J. and Pastan, I. (1989) A recombinant immunotoxin consisting of two antibody variable domains fused to *Pseudomonas* exotoxin. *Nature* **399**, 394–397.

Cheadle, C., Hook, L. E., Givol, D. and Ricca, G. A. (1992) Cloning and expression of the variable regions of mouse myeloma protein MOPC315 in *E. coli*: Recovery of active Fv fragments. *Mol. Immunol.* **29**, 21–30.

Chomczynski, P. (1992) Solubilization in formamide protects RNA from degradation. *Nucl. Acids Res.* **20**, 3791–3792.

Chomczynski, P. and Sacchi, N. (1987) Single-step method of RNA isolation by acid guanidinium thiocyanate-phenol-chloroform extraction. *Anal. Biochem.* **162**, 156–159.

Chothia, C., Lesk, A. M., Tramontano, A., Levitt, M., Smith, G. S., Air, G., Sheriff, S., Padlan, E. A., Davies, D. and Tulip, W. R. (1989) Conformations of immunoglobulin hypervariable regions. *Nature* **342**, 877–83.

Clackson, T., Hoogenboom, H. R., Griffiths, A. D. and Winter, G. (1991) Making antibody fragments using phage display libraries. *Nature* **352**, 624–628.

Cox, J. P., Tomlinson, I. M. and Winter, G. (1994) A directory of human germ-line V kappa segments reveals a strong bias in their usage. *Eur. J. Immunol.* **24**, 827–836.

Cunningham, B. C. and Wells, J. A. (1989) High resolution epitope mapping of hGH-receptor interactions by alanine-scanning mutagenesis. *Science* **244**, 1081–1085.

Davis, G. T., Bedzyk, W. D., Voss, E. W. and Jacobs, T. W. (1991) Single chain antibody (SCA) encoding genes: one step construction and expression in eukaryotic cells. *Bio/Technology* **9**, 165–169.

Dengen, G. E. and Cox, E,C. (1974) A conditional mutator in *Escherichia coli*: isolation, mapping, and effector studies. *J. Bacteriol.* **117**, 477–487.

Deng, S-j., MacKenzie, R., Sadowska, J., Michniewicz, J., Young, N. M., Bundle, D. R. and Narang, S. A. (1994) Selection of antibody single-chain variable fragments with improved carbohydrate binding by phage display. *J. Biol. Chem.* **269**. 9533–9538.

Denzin, L., Whitlow, M. and Voss, Jr. E. W. (1991) Single-chain site-specific mutations of Fluorescein-amino acid contact residues. *J. Biol. Chem.* **266**, 14095–14103.

Denzin, L. K. and Voss, E. W. Jr. (1992) Construction, characterisation and mutagenesis of an anti-fluorescein single-chain antibody idiotype family. *J. Biol. Chem.* **267**, 8925–8931.

Dolezal, O. Coia, G., Guthrie, R. Lilley, G. L. and Hudson, P. J. (1995) *Escherichia coli* expression of a bifunctional Fab-peptide epitope reagent for the rapid diagnosis of HIV-1 and HIV-2. *Immunotechnology* **3/4**, in press.

Dower, W. J., Miller, J. F. and Ragsdale, C. W. (1988) High efficiency transformation of *E. coli* by high voltage electroporation. *Nucl. Acids Res.* **16**, 6127–6145.

Dziegiel, M., Nielsen, L. K., Andersen, P. S., Blancher, A., Dickmeiss, E. and Engberg, J. (1995) Phage display used for gene cloning of human recombinant antibody against the erythrocyte surface antigen, rhesus D. *J. Immunol. Methods* **182**, 7–19.

Figini, M., Marks, J. D., Winter, G. and Griffiths, A. G. (1994) *In vitro* assembly of repertoires of antibody chains on the surface of phage by renaturation. *J. Mol. Biol.* **239**, 68–78.

Foote, J. and Winter, G. (1992) Antibody framework residues affecting the conformation of the hypervariable loops. *J. Mol. Biol.* **224**, 487–499.

Fowler, R. G., Schapper, R. M. and Glickman, B, W. (1986) Characterization of mutational specificity within the lacI gene for a MutD5 mutator strain of *Escherichia coli* defective in 3′–5′ exonuclease (proofreading) activity. *J. Bacteriol.* **167**, 130–137.

Fuh, G., Cunningham, B. C., Fukunaga, R., Nagata, S., Goeddel, D. V. and Wells, J. A. (1992) Rational design of potent antagonists to the human growth hormone receptor. *Science* **256**, 1677–1680.

Gerard, G. G. (1994) Inhibition of SuperScript II reverse transcriptase by common laboratory chemicals. *Focus (Life Technologies)* **16**, 102–104.

Gibbs, R. A., Posner, B. A., Filpula, D. R., Dodd, S. W., Finkelman, M. A., Lee, T. K., Wroble, M., Whitlow, M. and Benkovic, S. J. (1991) Construction and characterization of a single-chain catalytic antibody. *Proc. Natl Acad. Sci. USA* **88**, 4001–4004.

Glockshuber, V. K., Malia, M., Pfitzinger, I. and Plückthun, A. (1990) A comparison of strategies to stabilize immunoglobulin Fv-fragments. *Biochemistry* **29**, 1362–1367.

Gram, H., Marconi, L-A., Barbas, C. F., Collet, T. A., Lerner, R. A. and Kang, A. S. (1992) *In vitro* selection and affinity maturation of antibodies from a naive combinatorial immunoglobulin library. *Proc. Natl Acad. Sci. USA* **89**, 3576–3580.

Griffiths, A. D., Malmqvist, M., Marks, J. D., Bye, J. M., Embleton, M. J., McCafferty, J., Baier, M., Holliger, K. P., Gorick, B. D., Hughes-Jones, N. C., Hoogenboom HR and Winter G. (1993) Human anti-self antibodies with high specificity from phage display libraries. *EMBO J.* **12**, 725–734.

Griffiths, A. D., Williams, S. C., Hartley, O., Tomlinson, I. M., Waterhouse, P. W., Crosby, W. L., Kontermann, R. E., Jones, P. T., Low, N. M., Allison, T. J., Prospero, T. D., Hoogenboom, H. R., Nissim, A., Cox, J. P. L., Harrison, J. L., Zaccolo, M., Gherardi, E. and Winter, G. (1994) Isolation of high affinity human antibodies directly from large synthetic repertoires. *EMBO J.* **13**, 3245–3260.

Gussow, D. and Clackson, T. (1989) Direct clone characterization from plaques and colonies by the polymerase chain reaction. *Nucl. Acids Res.* **17**, 4000.

Hawkins, R. E., Russell, S. J. and Winter, G. (1992) Selection of phage antibodies by binding affinity:mimicking affinity maturation. *J. Mol. Biol.* **226**, 889–896.

Hawkins, R. E., Russel, S. J., Baier, M. and Winter, G. (1993) The contribution of contact and non-contact residues of antibody in the affinity of binding to antigen: the interaction of mutant D1.3 antibodies with lysozyme. *J. Mol. Biol.* **234**, 958–964.

Hermes, J. D., Parekh, S. M., Blacklow, S. C., Koster, H. and Knowles, J. R. (1989) A reliable method for random mutagenesis: the generation of mutant libraries using spiked oligodeoxyribonucleotide primers. *Gene* **84**, 143–151.

Holliger, P., Prospero, T. and Winter, G. (1993) 'Diabodies': Small bivalent and bispecific antibody fragments. *Proc. Natl Acad. Sci. USA* **90**, 6444–6448.

Hoogenboom, H. R., Griffiths, A. D., Johnson, K. S., Chiswell, D. J., Hudson, P. and Winter, G. (1991) Multi-subunit proteins on the surface of filamentous phage: methodologies for displaying antibody (Fab) heavy and light chains. *Nucl. Acids Res.* **19**, 4133–4137.

Hoogenboom, H. R. and Winter, G. (1992) Bypassing immunisation human antibodies from synthetic repertoires of germ line V_H-gene segments rearranged *in vitro*. *J. Mol. Biol.* **227**, 381–388.

Hopp, T. P., Prickett, K. S., Price, V. L., Libby, R. T., March, C. J., Ceretti, D. P., Urdal, D. L. and Conlon, P. J. (1988) A short polypeptide marker sequence useful for recombinant protein identification and purification. *Biotechnology* **6**, 1204–1210.

Horton, R. M., Hunt, H. D., Ho, S. N., Pullen, J. K. and Pease, L. R. (1989) Engineering hybrid genes without the use of restriction enzymes: gene splicing by overlap extension. *Gene* **77**, 61–68.

Huse, W. D., Sastry, L., Iverson, S. A., Kang, A. S., Alting, M. M., Burton, D. R., Benkovic, S. J. and Lerner, R. A. (1989) Generation of a large combinatorial library of the immunoglobulin repertoire in phage lambda. *Science* **246**, 1275–1281.

Huse, W. W., Yelton, D. E. and Glaser, S. M. (1993) Increased antibody affinity and specificity by codon-based mutagenesis. *Int. Rev. Immunol.* **10**, 129–137.

Irving, R. and Hudson, P. (1994) Superseding hybridoma technology with phage-display libraries. *In* 'Monoclonal Antibodies: The Second Generation' (H. Zola, ed.) pp. 119–139. BIOS Scientific Publications, Oxford.

Jung, S.-H., Lee, B. and Pastan, I. (1994) Design of interchain disulfide bonds in the framework region of the Fv fragment of the monoclonal antibody B3. *Proteins: Struct. Funct. Genet.* **19**, 35–47.

Kang, A. S., Barbas, C. F., Janda, K. D., Benkovic, S. J. and Lerner, R. A. (1991a) Linkage of recognition and replication functions by assembling combinatorial antibody Fab libraries along phage surfaces. *Proc. Natl Acad. Sci. USA* **88**, 4363–4366.

Kang, A. S., Jones, T. M. and Burton, D. R. (1991b) Antibody redesign by chain shuffling from random combinatorial immunoglobulin libraries. *Proc. Natl Acad. Sci. USA* **88**, 11120–11123.

Kaushik, A., Schulze, D. H., Bonilla, F. A., Bona, C. and Kelsoe, G.(1990) Stochastic pairing of heavy-chain and κ light-chain variable gene families occurs in polyclonally activated B cells. *Proc. Natl Acad. Sci. USA* **87**, 4932–4936.

Kettleborough, C. A., Saldanha, J., Ansell, K. H. and Bendig, M. M. (1993) Optimisation of primers for cloning libraries of mouse immunoglobulin genes using PCR. *Eur. J. Immunol.* **23**. 206–211.

Kortt, A. A., Malby, R. L., Caldwell, J. B., Gruen, L. C., Ivancic, N., Lawrence, M. C., Howlett, G. J., Webster, R. G., Hudson, P. J. and Colman, P. M. (1994) Recombinant antineuraminidase single chain Fv antibody: Characterization, formation of dimer and higher molecular mass multimers and the solution of the crystal structure of the scFv-neuraminidase complex. *Eur. J. Biochem.* **221**, 151–157.

Lah, M., Goldstraw, A., White, J. F., Dolezal, O., Malby, R. and Hudson, P. J. (1994) Phage surface presentation and secretion of antibody fragments using an adaptable phagemid vector. *Hum. Antibody Hybridomas* **5**, 48–56.

Lake, D. F., Bernstein, R. M., Schluter, S. F. and Marchalonis, J. J. (1995) Generation of diverse single-chain proteins using a universal $(Gly_4\text{–}Ser)_3$ encoding oligonucleotide. *Bio Techniques* **19**, 700–702.

Larrick, J. W., Danielsson, L., Brenner, C. A., Abrahamson, M., Fry, K. E. and Borrebaeck, C. A. K. (1989a) Rapid cloning of rearranged immunoglobulin genes from human hybridoma cells using mixed primers and polymerase chain reaction. *Biochem. Biophys. Res Commun.* **160**, 1250–1256.

Larrick, J. W., Danielsson, L., Brenner, C. A., Wallace, E. F., Abrahamson, M., Fry, K. E. and Borrebaeck, C. A. K. (1989b) Polymerase chain reaction using mixed primers: Cloning of human monoclonal antibody variable region genes from single hybridoma cells. *Bio/Technology* **7**, 934–938.

Lavoie, T. B., Drohan, W. N. and Smith-Gill, S. J., (1992) Experimental analysis by site-directed mutagenesis of somatic mutation effects on affinity and fine specificity in antibodies specific for lysozyme. *J. Immunol.* **148**, 503–508.

Leung, D. W., Chen, E. and Goeddel, D.V. (1989) *J. Methods Cell. Mol. Biol.* **1**. 11–15

Lilley, G. G., Dolezal, O., Hillyard, C., Bernard, C. and Hudson, P. (1994) Recombined single chain antibody polypeptide conjugates expressed in *Escherichia coli* for the rapid diagnosis of HIV. *J. Immunol. Methods* **211**, 211–226.

Malby, R. L., Caldwell, J. B., Gruen, L. C. *et al.* (1993) Recombinant antineuraminidase single chain antibody: expression, characterization and crystallisation in complex with antigen. *Proteins: Struct. Funct. and Gen.* **16**, 57–63.

Malby, R. L., Tulip, W. R., Harley, V. R., McKimm-Breschin, J. L., Laver, W. G., Webster, R. G. and Colman, P. M. (1994) The structure of a complex between the NC10 antibody and the influenza virus neuraminidase and comparison with the overlapping binding site of the NC41 antibody. *Structure* **2**, 733–746.

Manser, T. (1989) The efficiency of antibody affinity maturation: can the rate of B-cell division be limiting? *Immunol. Today* **11**, 305–308.

Marks, J. D., Tristrem, M., Karpas, A. and Winter, G. (1991a) Oligonucleotide primers for polymerase chain reaction amplification of human immunoglobulin variable genes and design of family-specific oligonucleotide probes. *Eur. J Immunol.* **21**, 985–991.

Marks, J. D., Hoogenboom, H. R., Bonnert, T. P., McCafferty, J., Griffiths, A. D. and Winter, G. (1991b) Bypassing immunization: Human antibodies from V-gene libraries displayed on phage. *J. Mol. Biol.* **222**, 581–597.

Marks, J. D., Hoogenboom, H. R., Griffiths, A. D. and Winter, G. (1992a) Molecular evolution of proteins on filamentous phage: mimicking the strategy of the immune system. *J. Biol. Chem.* **267**, 16007–16010.

Marks, J. D., Griffiths, A. D., Malmqvist, M., Clackson, T., Bye, J. M. and Winter, G. (1992b) Bypassing immunisation: building high affinity human antibodies by chain shuffling. *Bio/Technology* **10**, 779–783.

Marks, J. D., Tristrem, M., Karpas, A. and Winter, G. (1993) Human antibody fragments specific for human blood group antigens from a phage display library. *Bio/Technology* **11**, 1145–1149.

McCafferty, J., Griffiths, A. D., Winter, G. and Chiswell, D. J. (1990) Phage antibodies: filamentous phage displaying antibody variable domains. *Nature* **348**, 552–554.

Milenic, D. E., Yokota, T., Filpula, D. R., Finkelman, M. A. J., Dodd, S. W., Wood, J. F., Whitlow, M., Snoy, P. and Schlom, J. (1991) Construction, binding properties, metabolism, and tumor targeting of a single-chain Fv derived from the pancarcinoma monoclonal antibody CC49. *Cancer Res.* **51**, 6363–6371.

Neri, D., Momo, M., Prospero, T. and Winter, G. (1995) High-affinity antigen binding by chelating recombinant antibodies. *J. Mol. Biol.* **246**, 367–73.

Nissim, A., Hoogenboom, H. R., Tomlinson, I. M., Flynn, G., Midgley, C., Lane, D. and Winter, G. (1994) Antibody fragments from a 'single pot' phage display library as immunochemical reagents. *EMBO J.* **13**, 692–698.

Orlandi, R., Gussow, D., Jones, P. T. and Winter, G. (1989) Cloning immunoglobulin variable domains for expression by the polymerase chain reaction. *Proc. Natl Acad. Sci. USA* **86**, 3833–3837.

Pack, R., Kujau, M., Schroeckh, V., Knupfer, U., Riesenberg, W. and Plückthun, A. (1993) Improved bivalent miniantibodies with identical avidity as whole antibodies, produced by high cell density fermentation of *Escherichia coli. Bio/Technology* **11**, 1271–1277.

Pantoliano, M. W., Bird, R. E., Johnson, L. S., Asel, E. D., Dodd, S. W., Wood, J. F. and Hardman, K. D. (1991) Conformational stability, folding, and ligand-binding affinities of single chain Fv immunoglobulin fragments expressed in *E. coli. Biochemistry* **30**, 10117–10125.

Persson, M. A. A., Caothien, R. H. and Burton, D. R. (1991) Generation of diverse high-affinity human monoclonal antibodies by repertoire cloning. *Proc. Natl Acad. Sci. USA* **88**, 2432–2436.

Power, B. E., Ivancic, N., Harley, V. R., Webster, R. G., Kortt, A. A., Irving, R. A. and Hudson, P. J. (1992) High-level temperature-induced synthesis of an antibody VH-domain in *Escherichia coli* using the *PelB* secretion signal. *Gene* **113**, 95–99.

Reiter, Y., Brinkmann, U., Kreitman, R. J., Jung, S.-H., Lee, B. K. and Pastan, I. (1994) Stabilization of the Fv fragments in recombinant immunotoxins by disulfide bonds engineered into conserved framework regions. *Biochemistry* **33**, 5451–5459.

Reiter, Y., Kurucz, I., Brinkmann, U., Jung, S.-H., Segal, D. M. and Pastan, I. (1995) Construction of a functional disulfide-stabilized TCR Fv indicates that antibody and TCR Fv frameworks are very similar in structure. *Immunity* **2**, 281–287.

Reichmann, L. and Weiller, M. (1993) Phage display and selection of site-directed randomized single-chain antibody Fv fragment for its affinity improvement. *Biochemistry* **32**, 8848–8855.

Reichmann, L., Clark, M., Waldmann, H. and Winter, G. (1988) Reshaping human antibodies for therapy. *Nature* **332**, 323–327.

Rodriguez, M. L., Snedecor, B., Chen, C., Wong, W. L., Garg, S., Blank, G. S., Maneual, D. and Carter, P. (1993) Engineering Fab′ fragments for efficient $F(ab')_2$ formation in *Escherichia coli* and for improved *in vivo* stability. *J. Immunol.* **151**, 6954–6961.

Sambrook, J., Fritsch, E. F. and Maniatis, T. (1989) 'Molecular Cloning. A Laboratory Manual' 2nd edn. Cold Spring Harbor Laboratory Press, Cold Spring Harbor.

Sastry, L., Alting, M. M., Huse, W. D., Short, J. M., Sorge, J. A., Hay, B. N., Janda, K. D., Benkovic, S. J. and Lerner, R. A. (1989) Cloning of the immunological repertoire in *Escherichia coli* for generation of monoclonal catalytic antibodies: construction of a heavy chain variable region-specific cDNA library. *Proc. Natl Acad. Sci. USA* **86**, 5728–5732.

Schaaper, R. M. (1988) Mechanisms of mutagenesis in the *Escherichia coli* mutator mutD5: role of DNA mismatch repair. *Proc. Natl Acad. Sci. USA* **85**, 8126–8130.

Sharon, J. (1990) Structural correlates of high antibody affinity: Three engineered amino acid substitutions can increase the affinity of an anti-*p*-azophenylarsonate antibody 200-fold. *Proc. Natl Acad. Sci. USA* **87**, 4814–4817.

Skerra, A. (1993) Bacterial expression of immunoglobulin fragments. *Curr. Opin. Immunol.* **5**, 256

Skerra, A. and Plückthun, A. (1988) Assembly of a functional immunoglobulin Fv fragment in *Escherichia coli. Science* **240**, 1038–1041.

Skerra, A., Pfitzinger, I. and Plückthun, A. (1991) The functional expression of antibody Fv fragments in *Escherichia coli*: improved vectors and a generally applicable purification technique. *Bio/Technology* **9**, 273–278.

Smith, G. P. (1985) Filamentous fusion phage: novel expression vectors that display cloned antigens on the virion surface. *Science* **228**, 1315–1317.

Tai, M-S., Mudgett-Hunter, M., Levinson, D., Wu, G-M., Haber, E., Opperman, H. and Huston, J. S. (1990) A bifunctional fusion protein containing Fc-binding fragment B of Staphylococcal protein A amino terminal to antidigoxin single-chain Fv′ *Biochemistry* **29**, 8024–8029.

Takeuchi, Y., Nagumo, T. and Hoshino, H. (1988) Low fidelity of cell-free synthesis by reverse transcriptase of Human Immunodeficiency Virus. *J. Virol.* **62**, 3900–3902.

Tomlinson, I. M., Walter, G., LLewelyn, M. B. and Winter, G. (1992) The repertoire of human germline V_H sequences reveals about fifty groups of V_H segments with different hypervariable loops. *J. Mol. Biol.* **227**, 776–798.

Ward, E. S., Gussow, D., Griffiths, A. D., Jones, P. T. and Winter, G. (1989) Binding activities of a repertoire of single immunoglobulin variable domains secreted from *Escherichia coli. Nature* **341**, 544–546.

Waterhouse, P., Griffiths, A. D., Johnson, K. S. and Winter, G. (1993) Combinatorial infection and in vivo recombination: a strategy for making large phage antibody repertoires. *Nucl. Acids Res.* **21**, 2265–2266.

Whitlow, M. and Filpula, D. (1991) Single chain Fv proteins and their fusion proteins. *Methods: A Companion to Methods in Enzymology*, Vol. **2**, pp. 97–105.

Winter, G., Griffiths, A. D., Hawkins, R. E. and Hoogenboom, H. R. (1994) Making antibodies by phage display technology. *Annu. Rev. Immunol.* **12**, 433–55.

Winter, G. and Milstein, C. (1991) Man-made antibodies. *Nature* **349**, 293–299.

Wulfing, C. and Plückthun, A. (1994) Protein folding in the periplasm of *Escherichia coli. Mol. Microbiol.* **12**, 685–692.

Yamagishi, J., Kawashima, H., Matsuo, N., Ohue, M., Yamayoshi, M., Fukui, T., Kotani, H., Furuta, R., Nakano, K. and Yamada, M. (1990) Mutational analysis of structure-activity relationships in human tumor necrosis factor-alpha. *Protein Eng.* **3**, 713–719.

Zebedee, S. L., Barbas, III C. F., Hom, Y-L., Caithien, R., Graff DeGraw, J., Pyati, J., LaPolla, R., Burton, D. R., Lerner, R. A. and Thornton, G. B. (1992) Human combinatorial antibody libraries to hepatitis B surface antigen. *Proc. Natl Acad. Sci. USA* **89**, 3175–3179.

15 Generation of Conventional Antibodies

It may seem somewhat paradoxical to end a book on monoclonal antibodies by describing the production of conventional polyclonal antibodies. However, there are many reasons for doing so.

The production of monoclonal antibodies involves a great deal of work. A suitable screening assay must be developed before the fusion, and hundreds or thousands of tests will have to be performed before the prized clone is immortalized. The sheer work and time involved in 'cell farming' is considerable. In comparison, the preparation of antibodies with nothing more than antigen, a rabbit, and a syringe, might be seen as a technological breakthrough! For many purposes, conventional antibodies will do the job adequately, with much less work.

In addition, there are some cases where the extreme monospecificity of monoclonal antibodies may be a liability rather than an asset, and it would appear that these problems will be with us for some time to come. Most involve experiments in which the antigen is not in its native conformation.

A common problem occurs in the screening of bacterial expression libraries for gene cloning (see Section 10.13.2). Most of the current expression vectors synthesize the cloned gene product as part of a fusion polypeptide. Cloned proteins expressed in *Escherichia coli* are very likely to be less than full length, will not be glycosylated or have most other post-translational modifications that are common in eukaryotic cells, and are often malfolded. It would be expected that only a minority of the original antigenic determinants would be detectable, and experience has borne this out (see Fig. 10.11). In cell-free translation systems, unless microsomal membranes are added, membrane and secretory proteins will retain the signal sequence that is normally removed as the nascent chain enters the lumen of the endoplasmic reticulum, and will not be glycosylated. Glycosylation influences the conformation of proteins. Similarly, the cell-free translation or cloning of a single subunit of a multimeric protein may result in a product that is different in conformation to that of the native, fully assembled protein. In each of these cases, a particular monoclonal antibody may or may not recognize the antigen. It may sometimes be possible to choose or make a monoclonal antibody specially for the purpose, but this is not always feasible or desirable.

For all these reasons, polyclonal antibodies will continue to have a place for a long time to come. One possible future scenario might involve the purification of a protein by affinity chromatography with monoclonal antibodies, and the subsequent generation of highly specific polyclonal antibodies, which would in turn be used to identify the gene after cloning in bacteria.

Useful background on antibodies in general will be found in the volumes by Di Sabato *et al.* (1985), Harlow and Lane (1988) and Catty (1989).

15.1 Strategies for the Preparation of Highly Specific Polyclonal Antibodies

15.1.1 Dose of Antigen

In the older literature, it was common to use 1–10 mg antigen for each immunization, but much lower doses (10–100 μg) are often sufficient (Hurn and Chantler, 1980; Vaitukaitis, 1981). In some instances, even lower doses may be used. Johansson *et al.* (1979) were able to generate highly specific antisera to adenovirus with as little as 50 ng of antigen coupled to affinity beads. Specific antisera against membrane IgD could be generated by six injections of 2×10^7 spleen cells (Goding *et al.*, 1976). Assuming 10^5 IgD molecules per cell, the dose of antigen was only 0.25 μg per injection. In general, the lower doses of antigen lead to higher-affinity antibodies, because of the increased competition for antigen in the lymphoid follicles (see Section 3.3.9).

If the antigen is readily available and extremely pure, higher doses (1–2 mg) may result in brisker and stronger responses. However, the risk of production of antibodies to impurities increases with dose. An old adage among immunologists is 'Always remember that the rabbit is a better immunologist than you are'.

15.1.2 Form of Antigen

The form of antigen and use of adjuvants is discussed in detail in Chapter 4. A high degree of aggregation favours good responses, and the use of Freund's adjuvant is virtually universal. Molecules of very low molecular weight (<1000) are generally very poorly immunogenic, but may evoke strong responses when coupled to an immunogenic carrier. Drugs and hormones may be used as haptens. Sometimes, even quite large molecules are poorly immunogenic, but good responses may be achieved by treating them as haptens and coupling them to strongly immunogenic carriers. Suitable carriers include keyhole limpet haemocyanin, bovine serum albumin (BSA), ovalbumin or fowl immunoglobulin. (In many cases, one may wish to avoid BSA because it is a component of fetal calf serum and many buffers.) The chemistry of coupling depends on the availability of suitable groups on the hapten (see Erlanger, 1980; Bauminger and Wilchek, 1980; Reichlin, 1980; Doolittle, 1986). Heterobifunctional cross-linking agents are now available for this purpose from many commercial suppliers (e.g. Pierce). Their use for protein–protein coupling is discussed in Section 12.8.7 (see also Wong, 1991).

Peptides may be coupled to proteins by glutaraldehyde using the procedure of Doolittle (1986).

(1) Dissolve 20 mg of carrier protein (e.g. keyhole limpet haemocyanin) in 0.5 M sodium phosphate pH 7.5.
(2) Add 15 μmol peptide in 1.5 ml water.
(3) Add 1.0 ml 20 mM glutaraldehyde dropwise with stirring for 30 min. The solution will turn yellow.
(4) Add 0.25 ml 1 M glycine to block unreacted glutaraldehyde. Stir for a further 30 min.
(5) Dialyse against PBS to remove excess glutaraldehyde.

15.1.3 Antigen Purity

The purity of antigen is crucial for generation of specific antisera. As a general rule, there should be no more than 1–2% contaminants, and antibodies to contaminants may still occur at this level. Methods of antigen purification could include any of the procedures outlined in this book, but will have to be individualized for each case. For best results, at least two procedures which fractionate on the basis of different parameters should be used sequentially. The greatest purity will be obtained by combinations of high-resolution methods such as isoelectric focusing, chromatofocusing, high performance liquid chromatography, affinity chromatography or polyacrylamide gel electrophoresis (Scopes, 1993).

15.1.4 Immunization with Proteins Purified by Polyacrylamide Gel Electrophoresis

SDS-treated proteins are capable of eliciting strong antibody responses, and a subset of the resulting antibodies can react with the native protein (Stumph *et al.*, 1974). Accordingly, the extreme resolving power of SDS–PAGE can be used to purify proteins for immunization to produce highly specific antisera (Tjian *et al.*, 1974; Dahl and Bignami, 1977; Springer *et al.*, 1977; Carrol *et al.*, 1978; Lane and Robbins, 1978; Granger and Lazarides, 1980; Boulard and Lecroisey, 1982).

In most cases, the Coomassie blue-stained protein band was excised, and the gel crushed and emulsified in complete Freund's adjuvant (see Section 8.2.1). Acrylamide is said to act as an adjuvant in its own right (Weintraub and Raymond, 1963), but it should be noted that antibodies may also be formed against the acrylamide itself, and against Coomassie blue (Granger and Lazarides, 1980).

15.2 Preparative SDS–PAGE

Sometimes, the escape of antigen from the acrylamide is too slow to allow sufficient stimulation. If this is suspected, the protein may be eluted electrophoretically prior to injection. Stephens (1975) described a simple method for electrophoretic elution into a dialysis sac. Figure 15.1 shows a modified version (see also Stearne *et al.*, 1985).

The polyacrylamide gel may be either one- or two-dimensional, exactly as described in Chapter 10. Molecular weight standards run at the sides of the gel may help identify the desired band. Gels may be stained for 5 min in Coomassie blue, followed by destaining in ethanol–acetic acid until bands are visible (see Chapter 10), or they may be 'stained' by precipitating the SDS–protein complexes with ice-cold 0.25 M KCl until the bands are clearly visible by oblique lighting (Hager and Burgess, 1980). The desired band is then cut with a scalpel, and equilibrated with elution buffer (see below) plus 5 mM dithiothreitol for 30–60 min. Losses due to diffusion in this time are negligible. The function of the dithiothreitol is to ensure that the protein does not form a large oxidized disulfide-bonded polymer that is trapped in the gel.

The acrylamide gel slice is then loaded into the elution tube (Fig. 15.1), and the tank and tube filled with elution buffer. The elution buffer may be 0.1% SDS in 50 mM NH_4HCO_3 (volatile buffer suitable for protein sequencing) or 50 mM $NaHCO_3$ or even SDS running buffer (1X; Table 10.3).

Elution is typically at 50 V for 6–12 h. The buffer in the tank should be replaced every 12 h to prevent pH changes. If the gel was stained with Coomassie blue, the dye will be observed to elute from the gel slice and concentrate in a narrow band at the bottom of the tube (Fig. 15.1, I). The dye is not covalently linked to the protein and elutes from the gel at a much faster rate than the protein.

When elution is complete, there will be approximately 200 μl of 10% SDS in the bottom of the tube. The build-up of SDS is caused by the trapping of SDS

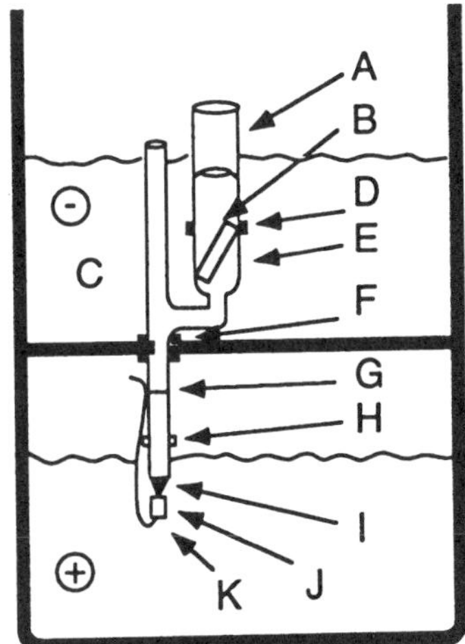

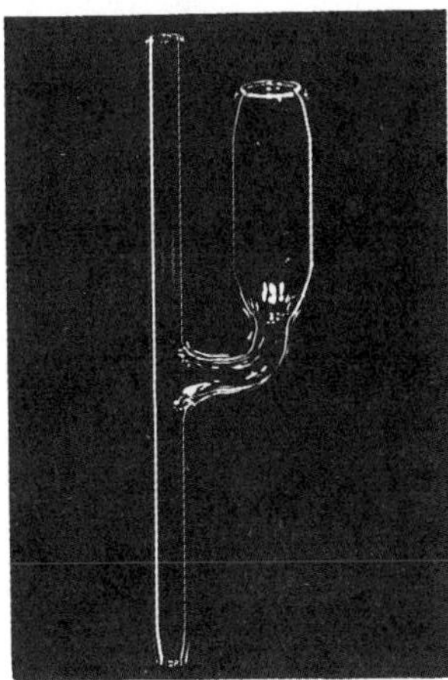

Fig. 15.1. Left Panel: *apparatus for elution of proteins from polyacrylamide gel slices. A standard tube gel tank is used, but with a modified glass tube (right). A, dialysis tubing, 25 mm flat width, to prevent entry and concentration of macromolecular contaminants from the upper tank buffer; B, gel slice; C, tank buffer; D, Silastic tubing, 12.7 mm inside diameter, cat. No. 601-625; E, elution tube; F, rubber grommet; G, rubber band; H, Silastic tubing, 6.4 mm inside diameter, cat. No. 601-441; I, eluted protein; J, Spectrapor clip, cat. No. 132734; K, dialysis tubing, 10 mm flat width. The straight portion of the tube is 110 mm in length and 7 mm in diameter. The wide section is 40 mm in length and 16 mm in diameter (see Fig. 15.2). The lower end of the dialysis tubing is held above the buffer by a rubber band (G), so even if clip J leaks, the sample will remain inside the dialysis tubing. Right panel: Glass tube for electro-elution See text for further details. Reproduced with permission from Stearne* et al. *(1985),* J. Immunol. ***134****, 443–448. Copyright 1985, The American Association of Immunologists.*

micelles ($M_r \approx 20\,000$) by the lower dialysis membrane, and can be seen by a sharp change in refractive index. If Coomassie blue is present, it will become concentrated in the region of concentrated SDS. Although the Coomassie blue is a small molecule, it does not pass through the dialysis membrane because it becomes inserted into the SDS micelles. The dye remains trapped even if no protein is present.

The build-up of such high SDS concentrations should not be seen as a problem. On the contrary, the increased density of the SDS-enriched layer stabilizes against convection during elution and removal of the sample. It is this dense layer that allows most of the buffer inside the tube to be removed at the end of elution without disturbing the protein.

Recovery of the eluted protein is achieved by gently removing the liquid above the SDS-enriched layer with a long Pasteur pipette. Virtually all this

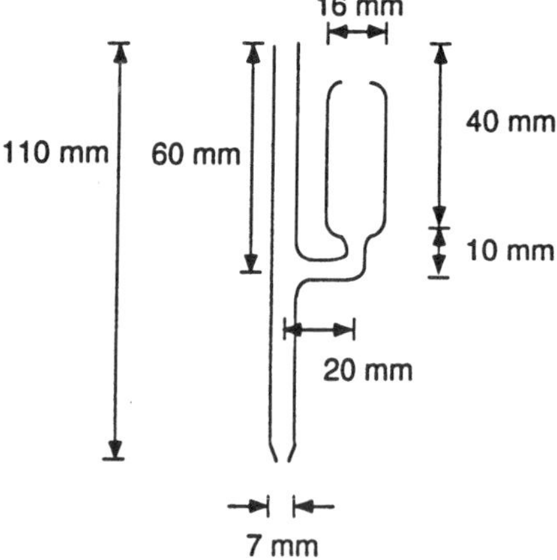

Fig. 15.2. Dimensions of glass electro-elution tube.

liquid can be removed without danger of disturbing the protein layer. Finally, a fresh Pasteur pipette is used to remove the protein.

Recovery of protein is usually better than 90%, even for submicrogram amounts of protein. The elution system is virtually leak-proof because the direction of the electric field is always away from the seals.

The protein may be recovered from the SDS and contaminating acrylamide polymers by precipitation with methanol (Stearne *et al.*, 1985). The protein-containing solution is transferred to a siliconized Corex tube, and nine volumes of methanol (precooled to –20°C) added and mixed. The tube is kept at –20°C overnight, and then centrifuged at 10 000 r.p.m. in a Sorvall HB-4 swing-out rotor at –5°C. The protein is precipitated as a compact transparent or slightly milky gelatinous pellet, leaving the SDS and acrylamide in solution. It may then be recovered for amino acid sequencing (see Sections 10.13.4 and 11.7) or emulsified in Freund's adjuvant for immunization (Section 8.2.1).

The above protocol was worked out after much trial and error. Of many organic solvents tried, only methanol was capable of keeping such large amounts of SDS in solution. Methanol also has the advantage that it is available commercially in high purity without contamination by aldehydes, which is particularly important if the protein is to be sequenced. It is essential to keep the sample cold at all times, because precipitation of microgram amounts of protein by organic solvents is more efficient at low temperatures. The temperature should not be lower than –20°C, however, because the SDS may precipitate. If the above protocol is followed exactly, recovery of microgram amounts of protein is usually more than 90%. A few very hydrophobic or low molecular weight proteins occasionally may fail to be precipitated.

15.3 Immunization

15.3.1 Choice of Species

Any species other than that of the antigen may be used, but in practice the commonly used recipients are goat, sheep and rabbit. It is widely believed that larger animals need more antigen, but there is little hard evidence (Hurn and Chantler, 1980). Goats, sheep and rabbits will all respond vigorously to 50–100 μg of protein emulsified in complete Freund's adjuvant, especially when given repeatedly. Sheep and goats may be preferred for their large size (Stewart-Tull and Rowe, 1975), but it has been shown that rabbits may be bled of 100 ml at weekly intervals with no mortality and only minimal haematological disturbances (Nerenberg *et al.*, 1978).

Rabbits are notoriously variable in their immune responsiveness between individuals. If rabbits are used as recipients, a minimum of three should be injected, and the sera tested individually. One has the impression that the responsiveness of goats and sheep is more uniform, and that it may often be possible to obtain a good antiserum from a single randomly chosen individual. Rabbit monoclonal antibodies have recently been produced (Spieker-Polet *et al.*, 1995).

15.3.2 Immunization Protocol

Many different immunization protocols have been used and there is little but anecdotal evidence to form a basis for comparison. In general, good results will be obtained by distributing 0.5–2.0 ml of the antigen, emulsified in complete Freund's adjuvant, over a number of sites. Freund's adjuvant is a potent inflammatory stimulus, but it is not at all clear that stimulation of inflammation and adjuvanticity can be separated (see Section 4.10). Subcutaneous or intramuscular injections probably cause the least distress. The use of foot-pad injections have not been justified for effectiveness by a controlled trial and is unnecessarily cruel to the animal. Similarly, the use of intradermal injections results in skin breakdown and ulcers which take weeks to heal. The dangers of Freund's adjuvant to laboratory personnel should also be remembered (see Section 8.2.1).

Over the years, there have been numerous claims for newer adjuvants to replace Freund's. These claims should be viewed with scepticism; most have not survived the test of time (see Stewart-Tull, 1995). In my opinion, complete Freund's adjuvant is still by far the most effective in terms of antibody responses. Provided that it is used in the way recommended above, there is little evidence that it causes distress to animals. Conversely, I have found that some of the newer adjuvants which are claimed to be less toxic are in fact more toxic than Freund's adjuvant.

In order to obtain a strong antibody response, it will usually be necessary to boost at least once, and possibly several times. It is customary to boost in incomplete Freund's adjuvant (i.e. lacking the mycobacteria) to avoid hypersensitivity reactions. However, many workers boost using complete Freund's adjuvant without encountering this problem. Boosting is also quite effective using water-soluble antigen without adjuvant, possibly because the existing antibody causes the formation of antigen–antibody complexes which bind avidly to Fc receptor-bearing antigen-presenting cells (see Section 3.3.9).

The dose of the boosting antigen is not critical, and 50–200 μg per injection are usually adequate. The frequency and timing of booster injections may also be varied over a wide range. Typically, one might boost at 3–4 weeks after the priming dose and perform a test bleed 5–7 days later.

It is the impression of many workers in the field that a prolonged time between priming and boosting leads to a much stronger secondary response, although I am not aware of any rigorous documentation of this point.

Subsequent boosts might be at intervals of 3–6 weeks, depending on the strength of the response. Blood may be collected 5–7 days after boosts, but the titre will usually remain high for several weeks. When a satisfactory response has been achieved, it is advisable to lethally bleed the animal under anaesthetic to obtain a large and uniform pool of serum.

15.3.3 Bleeding Rabbits

The rabbit should be wrapped up firmly in a tea-towel with its head protruding, and one ear *gently* warmed by placing it over a low-wattage (10–25 W) light bulb attached to a goose-neck reading lamp support without the shade. Alternatively, gentle warming with an infra-red lamp or spotlight at an appropriate distance may be used. Once the vessels are dilated, a small transverse cut in a marginal vein will usually allow 40–50 ml blood to be collected in 5–10 min.

Alternatively, blood may be collected with ease by vacuum aspiration from a nicked ear-vein (Nerenberg *et al.*, 1978). In this case, is not necessary to heat the ear or to use xylene or other irritants. A side-arm flask is connected to a water-pump, and the neck of the flask is placed over the ear and pressed gently against the head. Collection of blood into the flask is rapid and painless. Alternatively, the central ear artery may be punctured with a hubless 18-gauge needle (Gordon, 1981). In my experience, this method is less reliable and, in addition, arterial puncture is more painful than venous.

Anaesthetized rabbits may also be bled by cardiac puncture with an 18-gauge needle attached to a 50 ml syringe. In skilled hands, the mortality from cardiac puncture is very low.

15.3.4 Bleeding Sheep and Goats

It is easiest to bleed sheep and goats if two people are present. One person places the sheep on its rump, and holds it from behind. The other person feels for the jugular vein, using the anterior edge of the sternomastoid muscle as a landmark. The vein is usually made visible if its lower end is gently obstructed by finger pressure. A 12-gauge needle is inserted by pushing upwards through the skin over the vein, and will be felt to pop into the vein with a sudden release of resistance. Bleeds of 200–300 ml may be performed without risk (Stewart-Tull and Rowe, 1975).

The same general procedure is used for goats, but the goat is held against a fence and is bled standing up.

15.3.5 Collection of Serum

Blood collected into glass or plastic containers will clot rapidly. However, the maximum amount of serum is obtained by incubating at 37°C for an hour, and then leaving the blood at 4°C overnight to allow the clot to retract. The serum is recovered by centrifugation (1000 ***g*** for 10 min). It is important to be sure that there are no red cells present in the serum because freezing of red cells will cause haemolysis.

Serum may be stored frozen at –20°C or below. Alternatively, it may be kept at 4°C as a slurry in 50% saturated ammonium sulfate, or may be lyophilized and stored at 4°C (see Section 9.7).

15.4 Purification and Fragmentation of Rabbit, Sheep and Goat IgG

15.4.1 Binding of Rabbit, Sheep and Goat IgG to Staphylococcal Protein A and Streptococcal Protein G

The binding of IgG to protein A has been discussed in Sections 5.3.5 and 9.3.2. Rabbit antisera have the distinct advantage that all their IgG binds to protein A at pH 7.4, and all elutes at pH 4.0. This fact may be used to purify rabbit IgG, and to separate Fab and $F(ab')_2$ fragments of rabbit IgG from Fc or undigested IgG (Goding, 1976, 1978). Affinity chromatography on protein A is the method of choice for rabbit antisera.

Sheep and goats have two IgG subclasses. The major subclass (IgG1) binds very weakly to protein A at alkaline pH, and virtually not at all at neutral or acid pH (Delacroix and Vaerman, 1979; Duhamel *et al.*, 1980). The minor IgG2 subclass (10–20% of total IgG) binds firmly to protein A. Sheep, goat and bovine IgG will bind much more strongly to protein A in a buffer that combines high pH with high salt (e.g. 0.1 M Tris-HCl pH 8.5, with 1 M NaCl).

The subclass specificity of protein A for goat and sheep IgG2 has made this reagent somewhat unattractive for purification of IgG from these species. However, streptococcal protein G binds both IgG1 and IgG2 from goats and sheep, and gives excellent recoveries from these species (Boyle *et al.*, 1985; and Sections 5.3.5 and 9.3.3). Binding may be at pH 7.4 at physiological salt concentration, and elution at pH 2.7 in 0.1 M glycine-hydrochloride. The column should then be 'stripped' by washing with 0.1 M glycine-hydrochloride pH 2.0 to make sure that no trace of IgG remains, as it could contaminate subsequent purifications. Affinity chromatography on protein G may become the method of choice for purification of sheep and goat IgG.

15.4.2 Purification of Sheep and Goat IgG by Ion Exchange Chromatography

The purification of IgG from sheep and goat serum may also be performed by ion exchange chromatography (see Section 9.2.3). After precipitation in 40% saturated ammonium sulfate, the IgG is resuspended in PBS and dialysed overnight against 10 mM Tris-HCl, pH 8.0. It is then passed over a column of DEAE-Sephacel (or equivalent) equilibrated with the same buffer. Any dropthrough material will be pure IgG (those IgG molecules with the most basic isoelectric points). The remaining IgG (the majority) may be eluted with a linear salt gradient (0–250 mM NaCl in 10 mM Tris-HCl pH 8.0). The first peak to emerge will be IgG2, and it will be immediately followed by a peak containing IgG1. At this point, the IgG should be about 90% pure. If necessary, further purification may be achieved by gel filtration (see Section 9.2.4).

15.4.3 Production of Fab and $F(ab')_2$ Fragments of IgG from Rabbits, Goats and Sheep

The production of Fab fragments of IgG from rabbits, goats and sheep is quite straightforward. The basic principles and procedures described in Section 9.5 may be applied with little modification (see also Mage, 1980). The IgG may be purified by affinity chromatography on protein G (see Section 15.4.1), or by careful ammonium sulfate precipitation (see Section 9.2.1), and dialysed against 0.1 M Tris-HCl, 1 mM EDTA, pH 8.0. Further purification is usually unnecessary, but may be performed by ion exchange chromatography or gel filtration (see Chapter 9).

Digestion is carried out using papain in 0.1 M Tris-HCl, 1 mM EDTA pH 8.0 (or PBS plus 1 mM EDTA) plus a small amount of reducing agent (25 mM mercaptoethanol or 1–2 mM dithiothreitol). An enzyme:substrate ratio of

1:100 is customary, but 1:1000 may be sufficient. After incubation at 37°C for 1 h, the reaction is terminated by addition of an excess of iodoacetamide. The Fab fragment may be separated from the Fc by ion exchange chromatography (see Sections 9.2.3 and 9.5.1; Mage, 1980).

The production of $F(ab')_2$ fragments must be individualized depending on the species. Pepsin must be made up from the powder in 1 mM HCl or pH 4 acetate buffer, as it is ireversibly denatured at neutral or alkaline pH. Rabbit IgG is easily digested in pH 4–4.5 acetate buffer at 37°C overnight, using an enzyme:substrate ratio of 1:100. Sheep and goat IgGs are rather resistant to pepsin, but adequate digestion may be achieved at pH 4.0–4.5 using an enzyme:substrate ratio of 1:50 at 37°C for 48 h. Alternatively, cleavage of sheep IgG with trypsin may be used (Davies *et al.*, 1978).

Peptic digestion results in extensive degradation of the Fc, and simple dialysis against PBS at pH 7.4 will irreversibly terminate the reaction by denaturing the pepsin and remove most of the Fc products. A fragment of the Fc (pFc′) consisting of a noncovalent dimer of the C_H3 domains (total M_r 27 000) may be removed if necessary by gel filtration on Sephacryl S-200 or S-300.

Whichever method is chosen, it is strongly advisable to check the nature and completeness of digestion by SDS–PAGE.

15.4.4 Assessment of the Specificity of Polyclonal Antisera

The concentration of total IgG in serum ranges from 5–20 mg/ml. A very strong polyclonal antiserum might contain 1–3 mg/ml of specific antibody, or occasionally even more. More commonly, the antibody concentration may be 50–200 μg/ml, and an antiserum may be adequate for some purposes with even lower levels.

One cannot assume that because the antigen was 'pure', the antiserum will be specific. There are many reasons why this need not be the case (see Chapter 7). The principles of antigen analysis outlined in Chapter 10 may be applied with equal effectiveness to the analysis of antibodies.

If an antiserum is to be used in cell-binding assays, immunofluorescence or radioimmunoassay, the results of a crude test of specificity such as gel diffusion (Ouchterlony analysis) may be very misleading.

The specificity of an antiserum must be tested in a system that is at least as sensitive as the one in which it is to be used.

15.5 Specific Antibodies from Nonspecific Antisera

Even when the antigen has been purified with great care, some impurities are certain to remain. There is no such thing as absolute purity and antibodies against contaminants are an ever-present risk.

In addition, many occasions will arise in which immunological cross-reactions occur. Antisera against IgG are virtually certain to cross-react with other immunoglobulin classes, because the light chains are shared. Similarly, different glycoproteins often possess similar or identical carbohydrate moieties. Baird and Raschke (1982) have shown that chicken antimouse immunoglobulin cross-reacts with the envelope protein gp 70 of murine leukaemia virus, possibly through carbohydrate sharing (Layton, 1980).

These problems sometimes may be overcome by passing the antiserum over columns of immobilized antigen. For example, anti-IgG may be depleted of antibodies to light chains and thus rendered specific for γ heavy chains by passage over a series of columns containing other classes of immunoglobulin (see Chapter 11 for a detailed discussion of affinity chromatography). Conversely, the desired antibodies may be selectively enriched by affinity chromatography on immobilized antigen (Chapter 11).

All of the above presupposes that the antigen, contaminants, or cross-reacting substances are available in sufficient quantity and purity to allow the construction of suitable affinity columns. Specific antibodies may be affinity-purified on a micro-scale form nonspecific antisera by eluting antibodies from Western blots (see Section 10.10). The antigen-containing mixture is subjected to SDS–PAGE, and the resolved proteins transferred electrophoretically to a nitrocellulose or PVDF membrane. The paper is then cut into sections corresponding to the desired bands, and incubated with antiserum. Specific antibodies are eluted by 0.2 M glycine-hydrochloride pH 2.2, and rapidly neutralized (Olmsted, 1981; Coudrier, 1983; Talian *et al.*, 1983; Smith and Fisher, 1984). Any of the conditions for elution from affinity columns described in Chapter 11 (Table 11.2) may be considered; the use of 3.5 M sodium or potassium thiocyanate is particularly recommended. Recovery of antibody is facilitated by 'carrier' protein such as BSA or gelatin in the elution buffer. This technique is extremely useful for production of small quantities (a few hundred ng) of highly specific polyclonal antibodies, and might be ideal as a source of antibodies for screening of cloned genes. Elution of antibodies from nitrocellulose will certainly not be capable of producing enough antibodies to make a preparative affinity column.

Finally, it is sometimes possible to identify antigens by Western blotting using nonspecific antisera. The trick is to exploit the fact that antibodies are symmetrical and bivalent. The crude antigen-containing mixture is fractionated by polyacrylamide gel electrophoresis, transferred electrophoretically to nitrocellulose, and probed with nonspecific antibodies. The region of the membrane containing the desired antigen is identified by probing the membrane with labelled or enzymically active antigen, which binds to the 'free arms' of the relevant antibodies (Muilerman *et al.*, 1982; van der Meer *et al.*, 1983).

References

Baird, S. M. and Raschke, W. C. (1982) Murine T-lymphoma 'immunoglobulin' is identical to leukemia virus gp 70. *Mol. Immunol.* **19**, 1045–1050.

Bauminger, S. and Wilchek, M. (1980) The use of carbodiimides in the preparation of immunizing conjugates. *Methods Enzymol.* **70**, 151–159.

Boulard, Ch. and Lecroisey, A. (1982) Specific antisera produced by direct immunization with slices of polyacrylamide gel containing small amounts of protein. *J. Immunol. Methods* **50**, 221–226.

Boyle, M. D. P., Wallner, W. A., von Mering, G. O., Reis, K. J. and Lawman, M. J. P. (1985) Interaction of bacterial Fc receptors with goat immunoglobulins. *Mol. Immunol.* **22**, 1115–1121.

Carroll, R. B., Goldfine, S. M. and Mehero, J. A. (1978) Antiserum to polyacrylamide gel-purified simian virus 40 T antigen. *Virology* **87**, 194–198.

Catty, D. (1989) 'Antibodies. A Practical Approach' Vols 1 and 2. IRL Press, Oxford.

Coudrier, E., Reggio, H. and Louvard, D. (1983) Characterization of an integral membrane glycoprotein associated with the microfilaments of pig intestinal microvilli. *EMBO J.* **2**, 469–475.

Dahl, D. and Bignami, A. (1977) Effect of sodium dodecyl sulfate on the immunogenic properties of the glial fibrillary acidic protein. *J. Immunol. Methods* **17**, 201–209.

Davies, M. E., Barrett, A. J. and Hembry, R. M. (1978) Preparation of antibody fragments: conditions for proteolysis compared by SDS slab-gel electrophoresis and quantitation of antibody yield. *J. Immunol. Methods* **21**, 305–315.

Delacroix, D. and Vaerman, J. P. (1979) Simple purification of goat IgG1 and IgG2 subclasses by chromatography on protein A-Sepharose at various pH. *Mol. Immunol.* **16**, 837–840.

Di Sabato, G., Langone, J. J. and Van Vunakis, H. (1985) Immunochemical Techniques. Part H. Effectors and Mediators of Lymphoid Cell Functions. *Methods Enzymol.* **116**.

Doolittle, R. F. (1986) 'Of Urfs and Orfs'. University Science Books, Mill Valley, California 94941.

Duhamel, R. C., Meezan, E. and Brendel, K. (1980) The pH-dependent binding of goat IgG1 and IgG2 to protein A-Sepharose. *Mol. Immunol.* **17**, 29–36.

Erlanger, B. F. (1980) The preparation of antigenic hapten-carrier conjugates: A survey. *Methods Enzymol.* **70**, 85–104.

Goding, J. W. (1976) Conjugation of antibodies with fluorochromes: modifications to the standard methods. *J. Immunol. Methods* **13**, 215–226.

Goding, J. W. (1978) Use of staphylococcal protein A as an immunological reagent. *J. Immunol. Methods* **20**, 241–253.

Goding, J. W., Warr, G. W. and Warner, N. L. (1976) Genetic polymorphism of IgD-like cell surface immunoglobulin in the mouse. *Proc. Natl Acad. Sci. USA* **73**, 1305–1309.

Gordon, L. K. (1981) A reliable method for repetitively bleeding rabbits from the central artery of the ear. *J. Immunol. Methods* **44**, 241–245.

Granger, B. L. and Lazarides, E. (1980) Synemin: a new high molecular weight protein associated with desmin and vimentin filaments in muscle. *Cell* **22**, 727–738.

Hager, D. A. and Burgess, R. R. (1980) Elution of proteins from dodecyl sulfate polyacrylamide gels, removal of sodium dodecyl sulfate, and renaturation of enzymatic activity: results with sigma subunit of Excherichia coli RNA polymerase, wheat germ topoisomerase and other enzymes. *Anal. Biochem.* **109**, 76–86.

Harlow, E. and Lane, D. (1988) 'Antibodies. A Laboratory Manual'. Cold Spring Harbor Laboratory Press, New York.

Hurn, B. A. L. and Chantler, S. M. (1980) Production of reagent antibodies. *Methods Enzymol.* **70**, 104–142.

Johansson, M. E., Wadell, G., Jacobsson, P. A. and Svensson, L. (1979) Preparation of specific antisera against adenoviruses by affinity bead immunization (ABI). *J. Immunol. Methods* **26**, 141–149.

Lane, D. P. and Robbins, A. K. (1978) An immunochemical investigation of SV40 T antigens. 1. Production, properties and specificity of a rabbit antibody to purified simian virus 40 large-T antigen. *Virology* **87**, 182–193.

Layton, J. E. (1980) Anti-carbohydrate activity of T cell-reactive chicken anti-mouse immunoglobulin antibodies. *J. Immunol.* **125**, 1993–1997.

Mage, M. G. (1980) Preparation of Fab fragments from IgGs of different animal species. *Methods Enzymol.* **70**, 142–150.

Muilerman, H. G., ter Hart, H. G. J. and Van Dijk, W. V. (1982) Specific detection of inactive enzyme protein after polyacrylamide gel electrophoresis by a new enzyme-immunoassay method using unspecific antiserum and partially purified active enzyme: application to rat liver phosphodiesterase I. *Anal. Biochem.* **120**, 46–51.

Nerenberg, S. T., Zedler, P., Prasad, R., Biskup, N. S. and Pedersen, L. (1978) Hematological response of rabbits to chronic, repetitive, severe bleedings for the production of antisera. *J. Immunol. Methods* **24**, 19–24.

Olmsted, J. B. (1981) Affinity purification of antibodies from diazotized paper blots of heterogenous protein samples. *J. Biol. Chem.* **256**, 11955–11957.

Reichlin, M. (1980) Use of glutaraldehyde as a coupling agent for proteins and peptides. *Methods Enzymol.* **70**, 159–165.

Scopes, R. K. (1993) 'Protein Purification: Principles and Practice', 3rd edn. Springer-Verlag, New York.

Smith, D. E. and Fisher, P. A. (1984) Identification, developmental regulation, and response to heat shock of two antigenically related forms of a major nuclear envelope protein in *Drosophila* embryos: Application of an improved method of affinity purification of antibodies using polypeptides immobilized on nitrocellulose blots. *J. Cell Biol.* **99**, 20–28.

Spieker-Polet, H., Sethupathi, P., Yam, P.-C. and Knight, K. L. (1995) Rabbit monoclonal antibodies: Generating a fusion partner to produce rabbit-rabbit hybridomas. *Proc. Natl. Acad. Sci. USA,* **92**, 9348–9352.

Springer, T. A., Kaufman, J. F., Siddoway, L. A., Mann, D. L. and Strominger, J. L. (1977) Purification of HLA-linked B lymphocyte alloantigens in immunologically active form by preparative sodium dodecyl sulfate-gel electrophoresis and studies on their subunit association. *J. Biol. Chem.* **252**, 6201–6207.

Stearne, P. A., van Driel, I. R., Grego, B., Simpson, R. J. and Goding, J. W. (1985) The nurine plasma cell antigen PC–1: purification and partial amino acid sequence. *J. Immunol.* **134**, 443–448.

Stephens, R. E. (1975) High resolution preparative SDS-polyacrylamide gel electrophoresis: fluorescent visualization and electrophorectic elution-concentration of protein bands. *Anal. Biochem.* **65**, 369–379.

Stewart-Tull, D. E. S. and Rowe, R. E. C. (1975) Procedures for large-scale antiserum production in sheep. *J. Immunol. Methods* **8**, 37–45.

Stewart-Tull, D. E. S. (1995) 'The Theory and Practical Application of Adjuvants'. John Wiley and Sons, Chichester.

Stumph, W. E., Elgin, S. C. R. and Hood, L. E. (1974) Antibodies to proteins dissolved in sodium dodecyl sulfate. *J. Immunol.* **113**, 1752–1756.

Talian, J. C., Olmsted, J. B. and Goldman, R. D. (1983) A rapid procedure for preparing fluorescein-labelled specific antibodies from whole antiserum: its used in analyzing cytoskeletal architecture. *J. Cell Biol.* **97**, 1277–1282.

Tijan, R., Stinchcomb, D. and Losick, R. (1974) Antibody directed against *Bacillus subtilis* σ factor purified by sodium dodecyl sulfate slab gel electrophoresis. *J. Biol. Chem.* **250**, 8824–8828.

Vaitukaitis, J. L. (1981) Production of antisera with small doses of immunogen: Multiple intradermal injections. *Methods Enzymol.* **73**, 46–52.

van der Meer, J., Forssers, L. and Zabel, P. (1983) Antibody-linked polymerase assay on protein blots: a novel method for identifying polymerases following SDS-polyacrylamide gel electrophoresis. *EMBO J.* **2**, 233–237.

Weintraub, M. and Raymond, S. (1963) Antiserums prepared with acrylamide gel used as an adjuvant. *Science* **142**, 1677–1678.

Wong, S. S. (1991) 'Chemistry of Protein Conjugation and Cross-linking'. CRC Press, Boca Raton.

Glossary

Adjuvant substance that promotes immune responses in a nonantigen-specific manner.

Allelic exclusion expression of only one allele (maternal or paternal) in an individual cell; a characteristic feature of immunoglobulin and T cell receptor synthesis in lymphocytes.

Allotype allelic form.

Amphipathic consisting of hydrophilic and hydrophobic portions. Amphipathic molecules tend to form micelles and membranes in water.

Antibody antigen-specific molecule synthesized in response to stimulation; consists of light and heavy chains.

Antigen any substance which elicits a specific immune response. (Strictly speaking, this is the definition of an *immunogen*, because some substances such as haptens may sometimes only be revealed as antigenic by their reaction with the products of the immune system after a response, but may not elicit a response in their own right.)

Antigenicity degree to which an antigen stimulates an immune response (in this sense, synonymous with *immunogenicity)*; degree to which an antigen reacts with particular antibody or antibodies (in this sense, *not* synonymous with *immunogenicity*).

Antigen-combining site that portion of the antibody molecule which combines with antigen.

Antigenic competition concept that when an animal is immunized with two antigens at the same time, the response to each is diminished. The validity of the concept is questionable.

Antigenic determinant small site on antigen to which antibody binds; each antigen may have many antigenic determinants.

Avidity strength of binding of antibody to antigen; takes into account both affinity and valence.

Avidin protein from egg white that binds the vitamin biotin with very high affinity

B lymphocyte lymphocyte expressing immunoglobulin; produced in the bone marrow.

Class of antibody classification of antibody structure on the basis of the constant portion of the heavy chain; there are five classes, IgM, IgD, IgG, IgA, and IgE. See *Isotype*.

Clone group of cells with common ancestor; also used as verb to describe the growth of cells from a single ancestor; also used to describe the process of isolation and replication of individual genes.

Complement group of plasma proteins which interact with antibodies resulting in lysis of cells, removal of antigen and stimulation of inflammation.

Complementarity-determining region (CDR) portion of light and heavy chain polypeptide chains that forms part of antigen combining site; there are three CDRs for each chain, and they are encoded by hypervariable regions (CDR1–3) plus J regions (light chains) or hypervariable regions plus J plus D and N regions (antibody heavy chains and T cell receptor).

Critical micelle concentration concentration above which the monomer concentration of detergent remains constant.

Cross-reaction reaction of an antiserum with a molecule not present in the immunizing preparation; usually a manifestation of structural similarity.

Cryoglobulin immunoglobulin which precipitates in the cold (without antigen).

D gene diversity-generating gene element; a part of the immunoglobulin heavy chain and T cell receptor gene complexes; encodes portion of third hypervariable region and plays a major role in defining the specificity of the antigen-combining site.

Denaturation change in the conformation of a protein to a form that is different from the *native* conformation; may be partially or fully reversible; proteins may be denatured by many conditions including heat, extremes of pH, chemical modification, organic solvents, urea and certain ions and detergents.

Detergent amphipathic molecule which forms small micelles in aqueous solution; often used for solubilization of lipids and integral membrane proteins.

Domain structural unit of a protein, consisting of compact folded region, and often but not always encoded by a single exon. Some authors use the term to describe two closely juxtaposed *homology units* of immunoglobulin heavy and/or light chains.

Dynamic range ratio of the strongest possible signal to the weakest detectable.

Electrophoresis technique of separation of molecules on the basis of their mobility in an electric field.

Enhancer gene segment that causes increased transcription of adjacent gene; works regardless of orientation and is only minimally affected by its position in relation to the gene that it controls. Action is often tissue-specific.

Enhanced Chemiluminescence (ECL) method of increasing the intensity of light production in an enzyme-coupled system; increasingly used to replace radioactivity

Epitope chemically defined antigenic determinant.

Euglobulin protein which precipitates at low ionic strength.

Exon gene segment that is present in mature messenger RNA; often encodes discrete protein domain, but does not always code for protein (see *intron*).

Fab fragment proteolytic fragment of immunoglobulin containing antigen combining site; univalent; consists of intact light chain disulfide-bonded to *N*-terminal fragment (Fd) of heavy chain.

$F(ab')_2$ fragment proteolytic fragment of immunoglobulin containing two antigen-combining sites; contains intact light chains and *N*-terminal portion of heavy chains (Fd′).

Fc fragment homogeneous *C*-terminal portion of immunoglobulin heavy chain.

Fc receptor molecule which binds to Fc portion of immunoglobulins.

Fd region portion of immunoglobulin heavy chain consisting of variable region and first constant region domain

Fd phage filamentous bacteriophage that infects *Escherichia coli;* contains single stranded DNA genome surrounded by coat proteins. The M13 bacteriophage is derived from Fd phage.

Fluorography exposure of photographic film by light, as opposed to *autoradiography*, in which exposure is due to direct ionizing radiation.

Freund's adjuvant commonly used adjuvant consisting of an emulsion of the antigen in saline and a mixture of an emulsifying agent (e.g. Arlacel A) in mineral oil with killed mycobacteria (complete Freund's adjuvant) or without mycobacteria (incomplete Freund's adjuvant).

Fv fragment fragment of immunoglobulin consisting of noncovalently bound variable domains of heavy and light chains; the smallest antigen-binding fragment.

Gamma (γ) globulin obsolete term for serum proteins of low electrophoretic mobility; virtually synonymous with *immunoglobulin*.

Genetic polymorphism presence of multiple allelic forms.

Haplotype set of allelic forms of closely linked genes; usually inherited *en bloc.*

Hapten substance that can combine with antibody, but is not immunogenic unless coupled to an immunogenic 'carrier' molecule. Haptens are usually (but not always) low molecular weight (<1000), while carriers are usually proteins. Many drugs, hormones and other small molecules can act as haptens.

HAT selection technique in which mutant cells are unable to grow in a medium containing hypoxanthine, aminopterin and thymidine; hybridization to cells lacking the mutation allows growth.

Heavy chain component polypeptide of immunoglobulins; heavy chains are always glycosylated, and have molecular weights from 50 000–85 000.

Heterokaryon multinucleate cell containing at least two different types of nuclei.

Hinge portion of immunoglobulin molecule between Fab and Fc; generally open and flexible conformation, and often susceptible to proteolytic attack.

Homology unit segment of immunoglobulin polypeptide chain consisting of approximately 110 amino acids, and containing an intrachain disulfide bond; thought to reflect evolution of immunoglobulins by a series of gene duplications.

Hybridoma hybrid word combining elements of Greek and Latin; strictly speaking, the use of '-oma' should be confined to tumours in animals, but the word is generally used to describe any continuosly growing cell line which is a hybrid between a malignant cell and a normal cell.

Hydrophilic literally, water-loving; polar or ionic species freely soluble in water.

Hydrophobic literally water-fearing; substance which is poorly soluble in water due to its non-polar nature.

Hypervariable region portion of immunoglobulin light or heavy chain containing largest degree of variation between molecules; the intact immunoglobulin molecule is folded such that the hypervariable portions are part of the *antigen-combining site.*

Immunoblot see *Western Blot*

Immunogenicity degree to which a substance is capable of eliciting an immune response.

Immunoglobulin group of molecules consisting of κ or λ light chains and μ, δ, γ, ε or α heavy chains; all antibodies are immunoglobulins, but not all immunoglobulins have known antibody properties.

Integral membrane protein membrane protein which interacts strongly with the lipid bilayer, and can only be solubilized in an intact form by the use of detergents; contains hydrophobic portion embedded in lipid bilayer.

Intron gene segment which is transcribed into messenger RNA precursor but is absent from mature mRNA; introns are excised from messenger RNA during its maturation; their function is not known, but occasionally they contain *enhancer* segments. Introns are also known as *intervening sequences.*

Isoelectric focusing electrophoretic technique in which an electric field sets up a stable pH gradient; proteins move to the unique point where their net charge is zero; the technique is simple, yet capable of extremely high resolution. Separation is based on isoelectric point (pH at which net charge is zero).

Isotype group of heavy or light chains with similar or identical constant region sequence; sometimes used synonymously with *class* or *subclass*. The term is also sometimes used to describe families of variable region sequences.

J chain small, highly acidic glycosylated polypeptide disulfide-bonded to heavy chains of polymeric immunoglobulins (IgM and polymeric IgA); J chain appears to play a role in polymerization and possibly for its interaction with *secretory piece* in the transport of polymeric immunoglobulins across epithelia.

J gene short region of DNA in heavy and light chain gene clusters; encodes for amino acids at junction of variable constant regions; J genes are intimately involved in joining of variable constant regions.

Leader sequence (signal sequence) short and relatively hydrophobic *N*-terminal sequence of secretory and membrane proteins which guides them across the membrane of the endoplasmic reticulum; virtually always removed by proteolytic cleavage in the lumen of the endoplasmic reticulum.

Lectin carbohydrate-binding protein. Some antibodies and enzymes have lectin-like properties, although they are not usually defined as lectins.

Light chain polypeptide constituent of immunoglobulins, of molecular weight 22 000–25 000; usually not glycosylated; may be subdivided into κ and λ subsets.

Lymphocyte small, round cell with scanty cytoplasm and round nucleus, involved in immunity; may be subdivided into T (thymus derived) and B (bone-marrow derived). Lymphocytes can often enlarge and may develop cytoplasmic granules or pseudopodia under certain conditions.

Macrophage large vacuolated cell with irregular outline and specialized for phagocytic function.

Malignant property of tumours (and cells) characterized by invasion and metastasis; when malignant cells are transferred into animals that are not able to reject them, they often form tumours.

Micelle aggregate of amphipathic molecules in water; the hydrophobic portions face inwards, while the hydrophilic portions face outwards. The lipid bilayer of cell membranes is closely related to micelles in structure; the main difference being that micelles are small and curved, while membranes are large and planar.

Microcurie unit of radioactivity; 2.2×10^6 disintegrations per minute.

Mycoplasma diverse family of very small organisms capable of passing through a 0.45 μm filter, but capable of growth on artificial media in the absence of other cells.

Myeloma tumour consisting of malignant form of plasma cell.

Native conformation in which protein is synthesized or exists in nature or is biologically active; see *denaturation*.

Non-equilibrium pH gradient electrophoresis form of electrophoresis closely related to isoelectric focusing, but in which proteins do not stop moving; useful for analysis of basic proteins. The separation is based on net charge or isoelectric point.

Ouchterlony analysis form of analysis in which antigen and antibody are allowed to diffuse through agar gel; the presence of a reaction is indicated by the formation of visible precipitation lines. Synonymous with *double diffusion*. A test based on the same principle was described at the same time by Oudin.

Peripheral membrane protein membrane protein which may be released and solubilized by minor changes in pH or ionic strength, and does not require detergents to maintain solubility.

Phage shorthand expression for bacteriophage; a virus that grows in bacteria.

Plasma portion of blood remaining after removal of cells and platelets; in order to collect plasma, agents such as heparin or EDTA must be used to prevent clotting.
Plasma cell antibody-secreting cell with eccentric nucleus and large amount of basophilic cytoplasm; the term is now often applied to all antibody-secreting cells, even though many do not possess classical plasma cell morphology.
Plasmid relatively small circular DNA molecule capable of autonomous self-replication in bacteria.
Plasmacytoma malignant tumour of plasma cells; synonymous with *myeloma.*
Polymorphism literally, many shapes; see *genetic polymorphism.*
Pronase mixture of proteases from *Streptomyces griseus.*
Protein A protein of molecular weight 42000, made by most strains of *Staphylococcus aureus;* binds certain IgG subclasses in certain species.
Protein G protein made by certain strains of streptococci; binds IgG but its specificity regarding species and IgG subclasses is different to *Staphylococcal protein A.*
Receptor molecule which binds another molecule with high affinity and specificity; most receptors are proteins; in some cases, binding is followed by particular biological consequences.
Secretory piece glycosylated polypeptide of molecular weight 50 000–80 000; bound to secretory IgA; a proteolytic fragment of the receptor for IgA on secretory epithelial cells.
scFv single chain Fv fragment, usually made by genetic engineering; see *Fv fragment.*
Serum non-cellular components of blood which remain after clotting.
Signal sequence see *leader sequence.*
Signal:noise ratio ratio of desired information (signal) to undesired background (noise). The signal:noise ratio is thus dependent on the signal level, which is variable, and the noise level, which is more or less constant.
Streptavidin protein from *Streptomyces* that binds avidin with high affinity. Streptavidin has an isoelectric point closer to neutrality than avidin, and may show reduced nonspecific binding for this reason. However, streptavidin can bind to cellular receptors for RGD peptides.
Subclass group of antibodies with very similar characteristics (e.g. IgG2a, IgG2b). See also *class* and *isotype.*
Titre maximum dilution of antibody at which reaction with antigen can still be detected; an approximate measure of antibody concentration.
Tolerance acquired immunological nonresponsiveness caused by recognition of self.
T lymphocyte thymus-derived lymphocyte; the main role of T cells is to regulate the activities of other cells, notably B cells; a subset of T cells is also capable of killing other cells by direct contact (the ultimate in regulation).
Tunicamycin antibiotic which inhibits the glycosylation of asparagine residues.
Variable region *N*-terminal domain of antibody heavy and light chains; contains *antigen-combining site.*
Western blot (Immunoblot) electrophoretic technique in which proteins are separated in a polyacrylamide gel, and then transferred electrophoretically to a membrane (usually nitrocellulose). The membrane is then probed with antibodies.

For further details of these and other immunological terms, the reader is referred to Rosen, F. S., Steiner, L. and Unanue, E. (1989). Macmillan Dictionary of Immunology, (Macmillan Reference Books, London).

Index

Note: When a topic is listed several times, the major discussion is listed in **bold.**